Introduction to Multicopter Design and Control

Quan Quan

Introduction
to Multicopter Design
and Control

 Springer

Quan Quan
Department of Automatic Control
Beihang University
Beijing
China

ISBN 978-981-10-9859-8   ISBN 978-981-10-3382-7   (eBook)
DOI 10.1007/978-981-10-3382-7

Printed on acid-free paper

This Springer imprint is published by Springer Nature
The registered company is Springer Nature Singapore Pte Ltd.
The registered company address is: 152 Beach Road, #21-01/04 Gateway East, Singapore
189721, Singapore

To my parents

Preface

The flight is always the dream of mankind. The kite was invented in China, possibly as far back to the fifth century. Beginning with the last years of the 15th century, Leonardo da Vinci wrote about and sketched many designs for flying machines and mechanisms, including ornithopters, fixed-wing gliders, rotorcraft, and parachutes. On December 17, 1903, the Wright brothers invented and built the world's first successful airplane and made the first controlled, powered, and sustained heavier-than-air human flight. Since 1903, in the following one hundred years, many types of aircraft emerged. However, not too many people tried to pilot and enjoy their flying. Aircraft and pilots were still mysterious until small and micro multicopters were approaching consumers via the Radio Controlled (RC) toy market. This mainly due to the huge improvement on the user experience via their salient features: ease-of-use, high reliability, and easy maintainability. It hardly finds an aircraft as simple as multicopters. They do make more and more people really pilot and enjoy flying. Besides as an RC toy, multicopters demonstrate other commercial applications as drones, including surveillance, search, and rescue missions. So far, multicopters have taken and consolidated their dominance in the market of small aircraft. No matter defined as an RC toy or a drone, the multicopter is definitely an appropriate research target for students. This is because students have a chance to experience all phases from design to flight testing within a short time. One more reason is that students can read and interact with the open-source code of multicopters to understand the underline principle.

As an outcome of a course developed at Beihang University (Beijing University of Aeronautics and Astronautics, BUAA), this book is intended as a textbook or introductory guide for senior undergraduate and graduate students, and as a self-study book for practicing engineers involved in multicopters. With the intention of covering most design phases on hardware and algorithms of multicopters, it has fifteen chapters which is divided into five parts, including multicopter airframe and propulsion system design, modeling, perception, control, and decision. This book can also be used for a supplementary reading material for other unmanned flying systems. It aims to organize the design principles adopted in engineering practice into a discipline and to emphasize the power of fundamental concepts. This book is featured with four salient characteristics.

(1) *Basic and practical.* The most contents related to multicopters are self-contained, aiming at making this book understandable to readers with the background of Electronic Engineering (EE). For such a purpose, the components are introduced starting from their functions and key parameters. The introduction to the design process starts with the principle, while the modeling section starts with the theoretical mechanics. The state estimation section starts with the measurement principle of sensors. Before talking about the control, the notions of stability and controllability are introduced. In addition, most of basic and practical methods are presented. These methods are closely related to open-source autopilots which are often used now.

(2) *Comprehensive and systematic.* This book hopes to give a complete picture of multicopter systems rather than a single method or technique. Very often, the role of a single method or technique is not sufficient to meet the users requirements or to solve a practical complex problem. On the other hand, improving other related methods or techniques will reduce the difficulty in a single method. For example, the improvement of the state estimation performance or the mechanical structure can avoid dealing with delay or vibration as a control problem. Through this, some complex controller design can be avoided. For an undergraduate student, basic knowledge has been obtained, such as mathematics, aerodynamics, materials, structures, electronics, filtering, and control algorithms, which correspond to numerous courses. This book is hoped to combine them together to lay foundations for full stack developers.

The preparation and writing of this book have suffered me a lot. Fortunately, the Software and Control Lab, I stayed as a student (I have been staying here as a faculty since 2010), started to support the research on multicopters since about 2007. Ruifeng Zhang and I devoted ourselves to building quadcopters then. This made me almost witness the gold development period of small multicopters. Moreover, fortunately, public documents are shared selflessly by developer teams of open-source autopilots, such as APM and PX4, and numerous technique dissertations and papers are contributed by scholars all over the world. More importantly, this book could never have been written without the support and assistance from my students from BUAA Reliable Flight Control Group http://rfly.buaa.edu.cn/, which is a part of Software and Control Lab. Deep thanks go to graduate students Jinrui Ren, Zhiyao Zhao, Guang-Xun Du, Xunhua Dai, Zibo Wei, Heng Deng, Dongjie Shi, Yangguang Cai, Jiang Yan, Hongxin Dong, Jianing Fu, Zhenglong Guo, Jing Zhang, Yao Luo, Shuaiyong Zheng, Baihui Du and Xiaowei Zhang for material preparation, chapter revision, and simulations. I would like to thank my colleague Prof. Zhiyu Xi, graduate students Usman Arif and Hanna Tereshchenko, and undergraduate student Kun Xiao for their comments and hours of tireless proof reading. I would like to thank Prof. Wei Huo, Prof. Dawei Li at Beihang University, Mr. Yun Yu at DJI-Innovations,

Dr. Liang Zhu at AVIC Information Technology Co., Ltd for providing the valuable suggestions. This book has been used in manuscript form in several courses at Beihang University and National University of Defense Technology, and the feedback from students has been invaluable. Finally, I would like to thank Mr. Xiang Guo for the traditional Chinese-style illustrations at the beginning of each chapter.

Beijing, China   Quan Quan
December 2016

Contents

Acronyms

2D	Two-dimensional
3D	Three-dimensional
4D	Four-dimensional
ABCF	Aircraft-Body Coordinate Frame
AC	Alternating Current
AC	Automatic Control
ACAI	Available Control Authority Index
ANN	Artificial Neutral Network
AOD	Additive Output Decomposition
AOPA	Aircraft Owners and Pilots Association
API	Application Programming Interface
APM	ArduPilot Mega
ASD	Additive State Decomposition
ATE	Automatic Trigger Event
BMS	Battery Management System
C/A	Coarse/Acquisition
CCF	Camera Coordinate Frame
CES	Consumer Electronics Show
CF	Carbon Fiber
CFD	Computational Fluid Dynamics
CIFER	Comprehensive Identification from FrEquency Response
CoG	Center of Gravity
DC	Direct Current
DGPS	Differential Global Position System
DIC	Dynamic Inversion Control
DMS	Degree-Minute-Second
DoC	Degree of Controllability
DoF	Degree of Freedom
DSP	Digital Signal Processor
EE	Electronic Engineering
EFCF	Earth-Fixed Coordinate Frame
EFSM	Extended-Finite-State-Machine
EKF	Extended Kalman Filter
EMF	ElectroMotive Force
ESC	Electronic Speed Controller
ESI	External System Independence
FAC	Fully-Autonomous Control
FFT	Fast Fourier Transform

FM   Frequency Modulation
FOC   Field Oriented Control
FoE   Focus of Expansion
FPV   First-Person View
FSM   Finite-State-Machine
GCS   Ground Control System
GNSS   Global Navigation Satellite System
GNU   GNU's Not Unix
GPS   Global Position System
GWN   Gaussian White Noise
IARC   International Aerial Robotics Competition
ICF   Image Coordinate Frame
IEKF   Implicit Extended Kalman Filter
IMU   Inertial Measurement Unit
INCOSE   International Council On Systems Engineering
INS   Inertial Navigation System
IoD   Internet of Drones
IRF   Inertial Reference Frame
LGPL   Lesser General Public License
LIDAR   LIght Detection And Ranging
LiPo   Lithium Polymer
LQE   Linear Quadratic Estimate
MBSE   Model-Based Systems Engineering
MEMS   Micro-Electro-Mechanical System
MIE   Manual Input Event
MIMO   Multi-Input Multi-Output
NAVSOP   NAVigation via Signals of OPportunity
NED   North East Down
NiMH   Nickel Metal Hydride
OCV-SoC   Open Circuit Voltage and SoC
PCM   Pulse Code Modulation
PCS   PieCeS
PD   Proportional-Derivative
PEM   Prediction-Error Minimization
PI   Proportional-Integral
PID   Proportional-Integral-Derivative
PPM   Parts-Per-Million
PPM   Pulse Position Modulation
PRBS   Pseudo-Random Binary Signal
PTAM   Parallel Tracking And Mapping
PWM   Pulse Width Modulation
PWM   Pulse Width Modulated
RC   Remote Control
RC   Remote Controller
RC   Radio Controlled
RMAH   Reserved Maximum Ampere-Hour
RPM   Revolution Per Minute
RPV   Remote-Person View
RTK   Real-Time Kinematic

RTL	Return-To-Launch
SAA	Semi-Autonomous Autopilot
SAC	Semi-Autonomous Control
SCM	Single Chip Microcomputer
SDK	Software Development Kit
SF	Slow Flyer
SLAM	Simultaneous Localization And Mapping
SoC	State-of-Charge
SyNAPSE	Systems of Neuromorphic Adaptive Plastic Scalable Electronics
TDoA	Time Difference of Arrival
ToF	Time of Flight
TSE	Text-Based Systems Engineering
TTC	Time To Contact/Collision
UAS	Unmanned Aerial System
UAV	Unmanned Aerial Vehicle
UHF	Ultra High Frequency
UTM	Unmanned aerial system Traffic Management system
UWB	Ultra Wide Band
VLoS	Visual Line of Sight
VTOL	Vertical Take-Off and Landing

Symbols

$=$	Equality
$\approx$	Approximately equal
$\sim$	Same order of magnitude or probability distribution
$\triangleq$	Definition. $x \triangleq y$ means that x is defined to be another name for y, under certain assumption
$\equiv$	Congruence relation
$\in$	Belong to
$\perp$	Perpendicular
$\times$	Cross product
$\otimes$	Multiplication of quaternions
$\mathscr{B}$	$\mathscr{B}(\mathbf{o}, \delta) \triangleq \{\boldsymbol{\xi} \in \mathbb{R}^m \mid \|\boldsymbol{\xi} - \mathbf{o}\| \leq \delta\}$, and the notation $\mathbf{x}(t) \to \mathscr{B}(\mathbf{o}, \delta)$ means $\min\limits_{\mathbf{y} \in \mathscr{B}(\mathbf{o}, \delta)} \|\mathbf{x}(t) - \mathbf{y}\| \to 0$
$\mathbb{C}$	Set of complex numbers
$\mathscr{D}$	$\mathscr{D} = \{\mathbf{D} \mid \mathbf{D} = (d_{ij}), d_{ij} = 0 \; if \; i \neq j\}$, diagonal matrix set
$\mathscr{P}$	$\mathscr{P} = \{\mathbf{P} \mid \mathbf{P} > 0\}$, positive definite matrix set
$\mathbb{R}, \mathbb{R}^n, \mathbb{R}^{n \times m}$	Set of real numbers, Euclidean space of dimension n, Euclidean space of dimension $n \times m$
$\mathbb{R}_+$	Set of positive real numbers
$\mathscr{S}$	$\mathscr{S} = \{\mathbf{S} \mid \mathbf{S} = \mathbf{S}^\mathrm{T}\}$, symmetric matrix set
$\mathbb{Z}$	Integer
$\mathbb{Z}_+$	Positive integer
$SO(3)$	$SO(3) \triangleq \{\mathbf{A} \mid \mathbf{A}^\mathrm{T}\mathbf{A} = \mathbf{I}_3, \det(\mathbf{A}) = 1, \mathbf{A} \in \mathbb{R}^3\}$
$\mathbf{e}_1, \mathbf{e}_2, \mathbf{e}_3$	Unit vectors, $\mathbf{e}_1 = \begin{bmatrix} 1 & 0 & 0 \end{bmatrix}^\mathrm{T}, \mathbf{e}_2 = \begin{bmatrix} 0 & 1 & 0 \end{bmatrix}^\mathrm{T},$ $\mathbf{e}_3 = \begin{bmatrix} 0 & 0 & 1 \end{bmatrix}^\mathrm{T}$
$\mathbf{I}_n$	Identify matrix of dimension $n \times n$
$\mathbf{0}_{n \times m}$	Zero matrix of dimension $n \times m$
x	Scale
$\mathbf{x}$	Vector, x_i represents the ith element of vector $\mathbf{x}$
$\dot{\mathbf{x}}, \dfrac{\mathrm{d}\mathbf{x}}{\mathrm{d}t}$	The first derivative with respect to time t
$\hat{\mathbf{x}}$	An estimate of $\mathbf{x}$
$[\mathbf{x}]_\times$	$[\mathbf{x}]_\times \triangleq \begin{bmatrix} 0 & -x_3 & x_2 \\ x_3 & 0 & -x_1 \\ -x_2 & x_1 & 0 \end{bmatrix}_\times, \mathbf{x} \in \mathbb{R}^3$
$^i\mathbf{x}$	Vector $\mathbf{x}$ represented in frame i. For example, $^e\mathbf{x}, {}^b\mathbf{x}$ represent the vector $\mathbf{x}$ represented in frame $o_e x_e y_e z_e$ and frame $o_b x_b y_b z_b$, respectively

$\mathbf{A}$	Matrix, a_{ij} represents the element of matrix $\mathbf{A}$ at the ith row and the jth column
$\mathbf{A}^T$	Transpose of $\mathbf{A}$
$\mathbf{A}^{-T}$	Transpose of inverse $\mathbf{A}$
$\mathbf{A}^\dagger$	Moore-Penrose pseudoinverse of $\mathbf{A}$
$\det(\mathbf{A})$	Determinant of $\mathbf{A}$
$\operatorname{tr}(\mathbf{A})$	Trace of a square matrix $\mathbf{A}$, $\operatorname{tr}(\mathbf{A}) \triangleq \sum_{i=1}^{n} a_{ii}, \mathbf{A} \in \mathbb{R}^{n \times n}$
$\mathbf{q}$	Quaternion, $\mathbf{q}_b^e$ represents the rotation from the frame $o_b x_b y_b z_b$ to the frame $o_e x_e y_e z_e$
$\mathbf{q}^*$	Conjugate quaternion
$\mathbf{q}^{-1}$	Inverse quaternion
$(\cdot)^{(k)}$	The kth derivative with respect to time t
∇	gradient
$\operatorname{Cov}(\mathbf{x}, \mathbf{y})$	Covariance, $\operatorname{Cov}(\mathbf{x}, \mathbf{y}) \triangleq \operatorname{E}\left((\mathbf{x} - \operatorname{E}(\mathbf{x}))\left(\mathbf{y} - \operatorname{E}(\mathbf{y})^T\right)\right)$, $\operatorname{Cov}(\mathbf{x}) \triangleq \operatorname{E}\left((\mathbf{x} - \operatorname{E}(\mathbf{x}))\left(\mathbf{x} - \operatorname{E}(\mathbf{x})^T\right)\right)$
$\mathscr{C}(\mathbf{A}, \mathbf{B})$	Controllability matrix of pairs $(\mathbf{A}, \mathbf{B})$, $\mathscr{C}(\mathbf{A}, \mathbf{B}) = [\mathbf{B} \ \mathbf{A}\mathbf{B} \cdots \mathbf{A}^{n-1}\mathbf{B}]$
$\operatorname{E}(\mathbf{x})$	The expectation of the random variable $\mathbf{x}$
$L_f h$	The Lie derivative of the function h with respect to the vector field $\mathbf{f}$
$L_\mathbf{f}^i h$	The ith order derivative of $L_f h$, $L_\mathbf{f}^i h = \nabla\left(L_\mathbf{f}^{i-1} h\right)\mathbf{f}$
$\mathscr{O}(\mathbf{A}, \mathbf{C}^T)$	Observability matrix of pairs $(\mathbf{A}, \mathbf{C})$, $$\mathscr{O}(\mathbf{A}, \mathbf{C}^T) = \begin{bmatrix} \mathbf{C}^T \\ \mathbf{C}^T\mathbf{A} \\ \vdots \\ \mathbf{C}^T\mathbf{A}^{n-1} \end{bmatrix}$$
$\operatorname{vex}(\cdot)$	$\operatorname{vex}\left([\mathbf{x}]_\times\right) \triangleq \mathbf{x}$
$\operatorname{Var}(x)$	The variance of the random variable x, $\operatorname{Var}(x) \triangleq \operatorname{E}\left((x - \operatorname{E}(x))^2\right)$
$\|\cdot\|$	Absolute value
$\|\cdot\|$	Euclidean norm, $\|\mathbf{x}\| \triangleq \sqrt{\mathbf{x}^T\mathbf{x}}, \mathbf{x} \in \mathbb{R}^n$
$\|\cdot\|_\infty$	Infinity norm, $\|\mathbf{x}\|_\infty = \max\{|x_1|, \cdots, |x_n|\}, \mathbf{x} \in \mathbb{R}^n$
$\mathbf{a}$	Specific force of a multicopter, $\mathbf{a} \in \mathbb{R}^3$. For example, $^e\mathbf{a}, {^b\mathbf{a}}$ represent the vector $\mathbf{a}$ in frame $o_e x_e y_e z_e$ and frame $o_b x_b y_b z_b$, respectively. $a_{x_i}, a_{y_i}, a_{z_i}$ represent the specific force components along axes $o_i x_i, o_i y_i, o_i z_i$, respectively, $i = $ e,b. (m/s^2)
$^b\mathbf{a}_m$	Calibrated specific force measured accelerometer in frame $o_b x_b y_b z_b$, i.e., $^b\mathbf{a}_m = \begin{bmatrix} \mathbf{a}_{x_b m} & \mathbf{a}_{y_b m} & \mathbf{a}_{z_b m} \end{bmatrix}^T$ (m/s^2)
$b_{d_{baro}}$	Bias of the altitude estimated by a barometer (m)
$\mathbf{B}_f$	Control effectiveness matrix of a multicopter with efficiency parameters
c_M	Lumped parameter torque coefficient (N $\cdot$ ms^2/rad^2)
c_T	Lumped parameter thrust coefficient (N $\cdot$ s^2/rad^2)

C_R	Constant parameter. It can represent the slope of the linear relationship from the throttle command to the motor speed
d_{baro}	Height measured by a barometer (m)
d_{sonar}	Distance measured by an ultrasonic range finder (m)
f	Total thrust acting on a multicopter (N)
$\mathbf{f}$	Thrust vector acting on a multicopter, $\mathbf{f} \in \mathbb{R}^3$. For example, $^e\mathbf{f}, {}^b\mathbf{f}$ represent the vector $\mathbf{f}$ in frame $o_e x_e y_e z_e$ and frame $o_b x_b y_b z_b$, respectively. $f_{x_i}, f_{y_i}, f_{z_i}$ represent the component thrusts along axes $o_i x_i, o_i y_i, o_i z_i$, respectively, $i = $ e,b. Or, it represents the propeller thrust vector (N)
g	Acceleration of gravity (m/s^2)
$\mathbf{J}$	Multicopter moment of inertia, $\mathbf{J} \in \mathbb{R}^{3\times3}, J_{xx}, J_{yy}, J_{zz}$ are central principal moments of inertia, J_{xy}, J_{yz}, J_{xz} are products of inertia (kg$\cdot$m^2)
k_{drag}	Drag coefficient due to blade flapping, $k_{drag} \in \mathbb{R}_+$. It can be used to determine the drags applied to the rotating blades
m	Multicopter mass (kg)
M	Propeller torque (N$\cdot$m), or a positive integer
M_i	Reaction torque generated by the ith propeller (N$\cdot$m)
$^b\mathbf{m}_m$	Calibrated magnetic field measured by a magnetometer in frame $o_b x_b y_b z_b$, i.e., $^b\mathbf{m}_m = \begin{bmatrix} m_{x_b} & m_{y_b} & m_{z_b} \end{bmatrix}^T$
$^e\mathbf{m}$	Magnetic field in frame $o_e x_e y_e z_e$
$\mathbf{M}_{n_r}$	Control effectiveness matrix, $\mathbf{M}_{n_r} \in \mathbb{R}^{4\times n_r}$
n_r	Number of propulsors
N	Motor speed (RPM), or a positive integer
$o_b x_b y_b z_b$	Aircraft-body coordinate frame
$o_e x_e y_e z_e$	Earth-fixed coordinate frame
$\mathbf{p}$	Position of the center of a multicopter, $\mathbf{p} \in \mathbb{R}^3$. For example, $^e\mathbf{p}, {}^b\mathbf{p}$ represent the vector $\mathbf{p}$ in frame $o_e x_e y_e z_e$ and frame $o_b x_b y_b z_b$, respectively. $p_{x_i}, p_{y_i}, p_{z_i}$ represent the position components along axes $o_i x_i, o_i y_i, o_i z_i$, respectively, i = e,b. (m)
$\mathbf{p}_{GPS}$	Position measured by a GPS receiver, $\mathbf{p}_{GPS} \in \mathbb{R}^3$. For example, (p_{xGPS}, p_{yGPS}) represents the measured 2D location by a GPS receiver (m)
$\mathbf{p}_h$	Horizontal position, $\mathbf{p}_h = \begin{bmatrix} p_{x_e} & p_{y_e} \end{bmatrix}^T$ (m)
$\hat{\mathbf{R}}$	Rotation matrix estimate
$\tilde{\mathbf{R}}$	Error between $\mathbf{R}_m$ and $\hat{\mathbf{R}}$, i.e., $\tilde{\mathbf{R}} = \hat{\mathbf{R}}^T \mathbf{R}_m$
$\mathbf{R}_a^b$	$\mathbf{R}_a^b \in SO(3)$ represnets the rotation matrix rotating vectors from frame a to frame b
$\mathbf{R}_m$	Measured rotation matrix
T_m	Time constant. It can determine the dynamic response of motors
$\mathbf{v}$	velocity (also called linear velocity) of the center of a multicopter, $\mathbf{v} \in \mathbb{R}^3$. For example, $^e\mathbf{v}, {}^b\mathbf{v}$ represent the vector $\mathbf{v}$ in frame $o_e x_e y_e z_e$ and frame $o_b x_b y_b z_b$, respectively. $v_{x_i}, v_{y_i}, v_{z_i}$ represent the velocity components along axes $o_i x_i, o_i y_i, o_i z_i$, respectively, $i = $ e,b. (m/s)

$\dot{v}_{x_i}, \dot{v}_{y_i}, \dot{v}_{z_i}$	Acceleration components of a multicopter along axes $o_i x_i, o_i y_i, o_i z_i$, respectively, $i =$ e,b. (m/s^2)
$\mathbf{W}$	Matrix that represents the relationship between the attitude rate and the aircraft body's angular velocity
θ_m	Measured pitch angle (rad)
$\mathbf{\Theta}$	Eular angles, $\mathbf{\Theta} = [\phi \quad \theta \quad \psi]^{\mathrm{T}}$, where θ, ϕ, ψ are pitch angle, roll angle, and yaw angle, respectively
τ	Torque $\tau \in \mathbb{R}^3$, where τ_x, τ_y, τ_z represent the torque components along axes $o_b x_b, o_b y_b, o_b z_b$, respectively, (N · m)
σ	Throttle command
ψ_{GPS}	Yaw angle estimated by GPS (rad)
ψ_m	Measured yaw angle (rad)
ψ_{mag}	Yaw angle estimated by a magnetometer (rad)
ω	Angular velocity of a multicopter, $\omega \in \mathbb{R}^3$. For example, $^e\omega, {}^b\omega$ represent the vector ω in frame $o_e x_e y_e z_e$ and frame $o_b x_b y_b z_b$, respectively. $\omega_{x_i}, \omega_{y_i}, \omega_{z_i}$ represent the angular velocity components along axes $o_i x_i, o_i y_i, o_i z_i$, respectively, $i =$ e,b. (rad/s)
$^b\omega_m$	Calibrated angular velocity measured by a gyroscope in frame $o_b x_b y_b z_b$ (rad/s)
ϖ	Angular speed of a propeller (rad/s)
ϖ_b	Constant parameter. It is the constant term of the linear relationship from the throttle command to the motor speed
ϖ_k	Angular speed of the kth propeller, $\varpi_k \in \mathbb{R}_+$ (rad/s)
ϖ_{ss}	Steady-state speed of a motor (rad/s)

Introduction

1

Commonly used small aircraft (less than 20 kg [1, pp. 4–5] or 25 kg [2]) can be classified into fixed-wing aircraft, helicopters, and multicopters, among which multicopters are most popular so far. Before 2010, fixed-wing aircraft and helicopters took overwhelming dominance in the field of both aerial photography and model aircraft sports. However, in the following years, due to the ease-of-use, the multicopter became a new star. During this period, since the autopilot and the other components were sold separately, multicopters were often only assembled by some professional personnel. Moreover, the parameters had to be tuned in accordance with the payload. At the end of 2012, DJI (Da-Jiang Innovations Science and Technology Co., Ltd.) released an all-in-one solution, that is, the ready-to-fly[1] Phantom quadcopter. Users could pilot it in a short time. Moreover, DJI's Phantom only cost about one thousand dollars. It was a huge saving in comparison with some commercial multicopters, such as MD4-200 or MD4-1000 from Microdrones GmbH in Germany. Since the Phantom quadcopter reduced the difficulty and cost of aerial photography, its market share expanded rapidly and took dominance soon after. In the following 2 years, there are a lot of media released about multicopters in terms of technologies, products, applications, and investigations. Today, multicopters have consolidated its dominance in the market of small aircraft. The speed they occupy the market is similar to that of the mobile Internet defeating the traditional Internet, or the smart mobile phones defeating the traditional mobile phones. The development of multicopters is becoming increasingly popular because of several factors, i.e., stories of success, the advancement in related technologies, the promotion of open source autopilots, participation of talented professionals, popularization of multicopters, continual capital investment, and preferential policy support. This chapter aims to answer the question as below:

Why do people choose small multicopters eventually?

The answer to this question involves the introduction to multicopters, the performance evaluation of small aircraft, and the brief history of multicopters. Most contents in this chapter are the revision and extension of a magazine paper in Chinese published by the author [3].

---

[1]A ready-to-fly kit is a radio controlled aircraft that is fully built with no assembly required. It includes a transmitter and a receiver.

© Springer Nature Singapore Pte Ltd. 2017
Q. Quan, *Introduction to Multicopter Design and Control*, DOI 10.1007/978-981-10-3382-7_1

Bamboo-Copter

Long time ago, the ancient Chinese had known that rotation could produce thrust for flight. In the fourth century AD, Hong Ge, best known for his interest in Daoism, alchemy, and longevity techniques, had already talked about rotorcraft. At that time, there was a toy named bamboo-copter. This kind of toy consisted of bamboo-blades connected with one end of a stick by a thread, which was rapidly spun and released into flight by tugging the thread. The bamboo-copter had an important impact on European aeronautic pioneers. George Cayley, the inventor of modern aeronautics, made a Chinese-style bamboo-copter (which he later called "rotary wafts") in 1809, which could fly 7–8.33 m high. He then improved the rotor of the prototype, and the bamboo-copter could fly 30 m high. In 1853, George Cayley drew down his helicopter rotors. It was to be one of the key elements in the birth of modern aeronautics in the West.

1.1 Concepts

1.1.1 Classification of Commonly Used Small Aircraft

As shown in Fig. 1.1, commonly used small aircraft are classified into three types.

(1) Fixed-wing aircraft. As shown in Fig. 1.1a, wings are permanently attached to the airframe of the aircraft. Most civil aircraft and fighters are fixed-wing aircraft. Its propulsion system generates a forward airspeed, which produces the lift to balance the vehicle's weight. Based on this principle, fixed-wing aircraft must maintain a certain forward airspeed, and therefore cannot take off and land vertically. In comparison with traditional helicopters, a fixed-wing aircraft has a much simpler structure and is able to carry a heavier payload over a longer distance while consuming less power. The disadvantage of the fixed-wing solution is the requirement of a runway or launcher for takeoff and landing.

(2) Single rotor blade helicopter. As shown in Fig. 1.1b, a helicopter is a type of rotorcraft in which lift is supplied by rotors directly. A single rotor helicopter has four flight control inputs, which are the cyclic, the collective, the anti-torque pedals, and the throttle, respectively. The collective is used to control the rotor's angle of attack. Although the lift of helicopter is mainly controlled by the collective and the throttle, the fast dynamic response of the lift is adjusted by the collective control. From the introduction above, it is understood that the single rotor blade helicopter has the ability of Vertical Take-Off and Landing (VTOL) since the lift is not controlled by the velocity (also called linear velocity) of the airframe. Therefore, no runway or launcher is required for takeoff and landing. Compared with a fixed-wing aircraft, it does not have the advantage in terms of the time of endurance. Moreover, its complex structure incurs a high maintenance cost.

(3) Multicopter. Multicopter is also called *multirotor*[2] or *multirocopter*. It can be considered as a type of helicopter, which has three or more propellers. It also has the ability of VTOL. The most popular multicopter is the quadcopter, as shown in Fig. 1.1c. A quadcopter has four control inputs, which are the four propeller angular speeds. Unlike the single rotor helicopter, the rapid lift adjustment is realized by the control of propeller angular speeds. Due to multiple-rotor structure, their anti-torque moments can be canceled out by each other. Thanks to the simple structure, a multicopter is easy-to-use and features high reliability and low maintenance cost. But, its payload capacity and the time of endurance are both compromised. What are the differences between a multicopter and a quadcopter then? As shown in Fig. 1.2, a multicopter, such as a hexacopter, has multiple propellers to generate the

(a) Fixed-wing aircraft　　　　(b) Helicopter　　　　(c) Multicopter

Fig. 1.1 Commonly used small aircraft

---

[2]In fact, the definition of rotors are different from that of propellers. The airplane often uses propellers to produce thrust, whereas the helicopters use rotors to produce lift. Unlike the airplane propeller (often fixed blade pitch angle), the blade pitch angle of the rotor of the helicopter is controlled by swashplate. From the definition, it should be said that multicopters are often mounted with propellers rather than rotors [5, pp. 79–85], because the propellers of multicopters often have fixed blade pitch angles. However, the terms "quadrotor" and "multirotor" are widely used in the world. Therefore, the two terms "rotor" and "propeller" are not distinguished in this book strictly. But the term "propeller" is preferred.

thrust, pitching moment, rolling moment, and yawing moment. On the other hand, a quadcopter only has four propellers to generate the thrust and the three-axis moments. That means there are hardly any fundamental differences but allocation of the thrust and moments to these propellers.

Besides the three types of drones introduced above, there are some combinations of them. As shown in Fig. 1.3a, a type of compound helicopter combines a tricopter and a fixed-wing aircraft [4]. It has abilities of both high-speed forward flight and VTOL. A compound helicopter as shown in Fig. 1.3b combines a quadcopter (bottom), a fixed-wing aircraft, and a helicopter [6]. It also has abilities of both high-speed forward flight and VTOL. In Fig. 1.4, a compound multicopter is shown, which combines a coaxial rotor and a quadcopter [7]. The maximum allowable payload is increased as a result.

Fig. 1.2 Thrust and moments of quadcopter and hexacopter

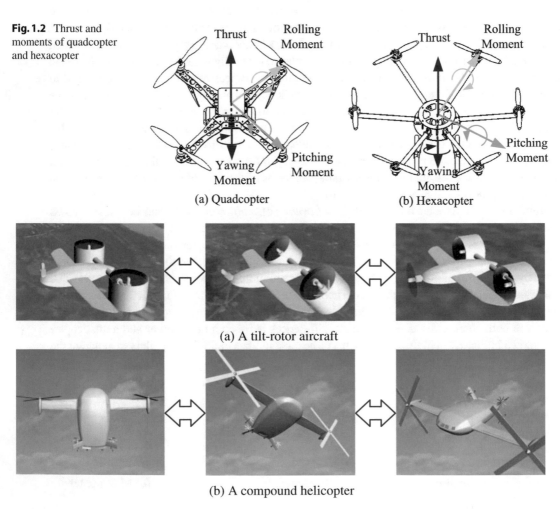

(a) A tilt-rotor aircraft

(b) A compound helicopter

Fig. 1.3 Transition of compound helicopters from the hover mode to the forward flight mode. **a** Rotors tilt to achieve mode transition; **b** rotors are all fixed to the airframe, while the airframe tilts to achieve mode transition

Fig. 1.4 A compound multicopter. The middle coaxial propellers with a slow dynamic response provides the main lift, while the surrounding quadcopter can change the propeller angular speeds rapidly to improve the dynamic response of attitude

1.1.2 Unmanned Aerial Vehicle and Model Aircraft

(1) Unmanned Aerial Vehicle or Uninhabited Aerial Vehicle (UAV), i.e., an aircraft without pilots on board. The flight of UAVs may be controlled either autonomously by onboard computers or by the remote control from a pilot on the ground or in another vehicle. UAVs are also called *drones*. In this book, small UAVs, namely small *drones*, are mainly considered.

(2) Model Aircraft. "An aircraft of limited dimensions, with or without a propulsion device, not able to carry a human being and to be used for aerial competition, sport or recreational purposes" is called a model aircraft [8]. It is also referred to as Radio Controlled (RC) model aircraft or RC aircraft. For the whole flight, an RC model aircraft must be within the Visual Line of Sight (VLoS) of the remote pilot. The statutory parameters of a model aircraft operation are outlined in [9, 10].

As shown in Table 1.1, differences between drones and model aircraft are summarized as follows:

(1) Composition
A small drone is more complex than a model aircraft in terms of its composition. A drone consists of an airframe, a propulsion system, an autopilot, a task system, a communication link system, and a Ground Control Station (GCS), etc. A model aircraft normally consists of an airframe, a propulsion system, a simple stabilizing control system, an RC transmitter and receiver system, etc.

(2) Operation
Drones are controlled either autonomously by onboard computers or by remote pilots on the ground or in another vehicle, whereas model aircraft are only controlled by remote pilots.

(3) Function
Drones are often used for the purpose of military or special civil applications. They are expected to carry out particular missions. Model aircraft are more like toys.

For most multicopters, they have two high-level control modes: Semi-Autonomous Control (SAC) and Fully-Autonomous Control (FAC). Many open source autopilots support both modes. The SAC mode implies that autopilots can be used to stabilize the attitude of multicopters, and also, they can help multicopters to hold the altitude and position. Taking the open source Autopilot Ardupilot Mega (APM)[3] for example, under the SAC mode, users are allowed to choose one of the following modes:

Table 1.1 Differences between drones and model aircraft

	Drones	Model aircraft
Composition	Complex	Simple
Operation	Autonomous control and remote control	Remote control
Function	Military or special civil applications	Entertainment

---

[3]Refer to http://ardupilot.org for details.

"stabilize mode," "altitude hold mode," or "loiter mode." Under such a high-level mode, a multicopter will be still under the control of remote pilots. Therefore, it is more like a model aircraft. On the other hand, the FAC mode implies that the multicopter can follow a pre-programmed mission script stored in the autopilot which is made up of navigation commands, and also can take off and land automatically. Under such a mode, remote pilots on the ground only need to schedule the tasks. The multicopter is then similar to a drone. Some multicopter autopilots support both modes which are able to switch by remote pilots, with each mode corresponding to some particular applications. In this book, multirotor drones and multirotor model aircraft are both called as multicopters for simplicity.

1.2   Remote Control and Performance Evaluation

Here comes a question then: Why do people choose multicopters? In order to answer this question, the remote control has to be understood first and then the performance evaluation of commonly used small aircraft.

1.2.1   Remote Control of a Multicopter

For simplicity, the introduction to the flight principle of a multicopter is based on a quadcopter, which is propelled by four propellers. A quadcopter can control four motors to change the propeller angular speeds which further change the thrust and moments. According to the classical mechanic theory, a force can be translated on the rigid body and turn into a force and an associated moment, while the moment can be translated directly into the center axis. As a result, the thrust and moments about its center are regulated. With the changed thrust and moments, the attitude and position of a quadcopter are controlled. As shown in Fig. 1.5a, at a hover position, all propellers are spinning at the same angular speed with propellers #1 and #3 rotating counterclockwise and propellers #2 and #4 clockwise. Taking the propeller #1 as an example, as shown in Fig. 1.5b, the reaction torque of this propeller can be translated directly into the center axis, while the translation of the thrust generates a pitching and a rolling moment simultaneously. As there are four thrusts to be regulated, the associated moments can be compensated each other. In the case of hovering flight, the sum of the four produced thrusts compensates for the weight of the quadcopter. The thrusts of four propellers are the same and the moments of four propellers sum to zero.[4]

Basically, the movement of a quadcopter falls into one of the following four basic types.[5]

(1) Upward-and-downward movement
As shown in Fig. 1.6, the propeller angular speeds are increased by the same amount. As a result, the thrust incurred will be increased, but moments of the four propellers still sum to zero. In this case, if

---

[4]If the blades are spinning counterclockwise, then the airframe will start to rotate clockwise due to the torque reaction. This is due to Newton's Third Law, which states that every action has an equal and opposite torque reaction. Therefore, most helicopters, namely the single rotor helicopter, have a small rotor placed vertically at the tail to compensate for the opposite reaction. Since a quadcopter has four propellers, the sum of the opposite torque reaction of four propellers can be zero, which is also used to control the yaw of a quadcopter.

[5]Altitude hold mode is considered. In fact, in most cases, the altitude will be held as same as possible.

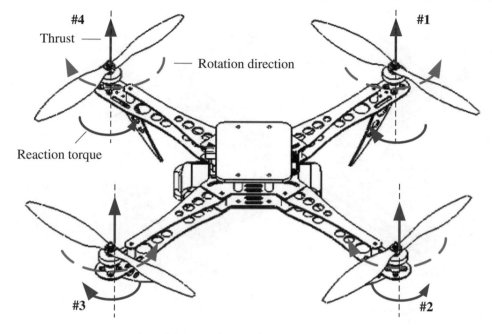

(a) A quadcopter in hovering flight

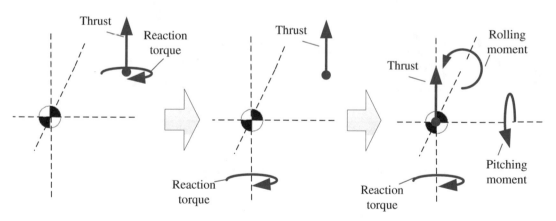

(b) The process of Propeller #1 thrust and moments acting on the center of the quadcopter

Fig. 1.5 A quadcopter in hovering flight

the quadcopter is placed on a leveled ground, it will move up once the thrust is greater than the weight of the quadcopter. Otherwise, if the propeller angular speeds are decreased by the same amount, the quadcopter will move down. Generally, the upward-and-downward movement is controlled by the RC transmitter as shown in Fig. 1.7.[6]

---

[6]Left stick operates upward-and-downward and yaw movements, and right stick operates forward-and-backward and leftward-and-rightward movements. This mode is commonly used in North America and other parts of the world including China. This is one RC transmitter mode, which can be changed to other modes.

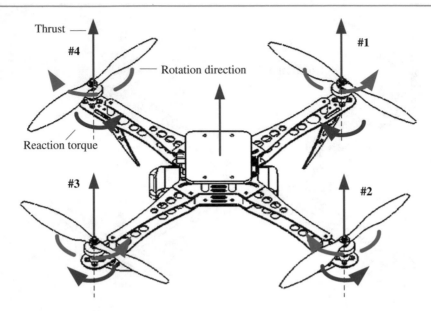

Fig. 1.6 Upward movement of a quadcopter

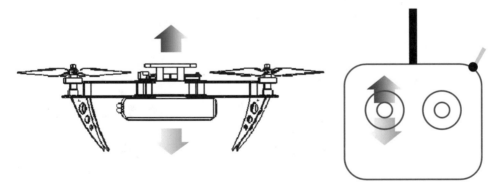

Fig. 1.7 Operation of an RC transmitter for the upward-and-downward movement

(2) Forward-and-backward movement
As shown in Fig. 1.8, the angular speeds of propellers #1, #4 are decreased by the same amount, while the angular speeds of propellers #2, #3 are increased by the same amount. This will lead to a moment which makes the quadcopter pitch forward. Then, the thrust has a forward component. However, at this moment, the vertical component of the thrust is decreased, which will not be equal to the weight of the quadcopter. So, based on the previous change, the four propeller angular speeds should be further increased by the same amount to compensate for the weight. Similarly, the backward movement can be also achieved. Generally, the forward-and-backward movement is controlled by the RC transmitter as shown in Fig. 1.9.

(3) Leftward-and-rightward movement
As shown in Fig. 1.10, the angular speeds of propellers #1, #2 are decreased by the same amount, while the angular speeds of propellers #3, #4 are increased by the same amount. This will lead to a moment

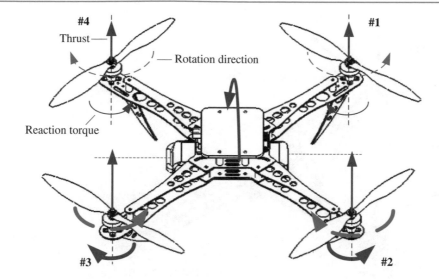

Fig. 1.8 Forward movement of a quadcopter

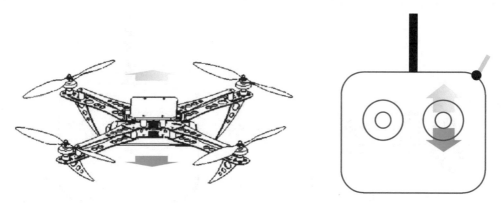

Fig. 1.9 Operation of an RC transmitter for the forward-and-backward movement

which makes the quadcopter roll to the right. Then, the thrust has a right component. However, the vertical component of the thrust is decreased, which will not be equal to the weight of the quadcopter. So, based on the previous change of the four propellers, the four propeller angular speeds should be further increased by the same amount to compensate for the weight. Similarly, the leftward movement can be also achieved. Generally speaking, the leftward-and-rightward movement corresponds to the operation of the RC transmitter as shown in Fig. 1.11.

(4) Yaw movement
As shown in Fig. 1.12, the angular speeds of propellers #2, #4 are decreased by the same amount, while the angular speeds of propellers #1, #3 are increased by the same amount. This will lead to zero moments both in the forward-and-backward and in the leftward-and-rightward directions. Owing to Newton's Third Law, every action has an equal and opposite torque reaction. Therefore, the clockwise yaw moment is increased because the angular speeds of propellers #1, #3 are increased in the counterclockwise direction. On the other hand, the counterclockwise yaw moment of the quadcopter is decreased because the angular speeds of propellers #2, #4 are decreased in the clockwise direction. Finally, this results in a clockwise yaw moment of the quadcopter. With this moment, the quadcopter

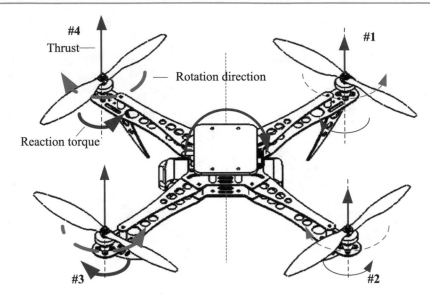

Fig. 1.10   Rightward movement of a quadcopter

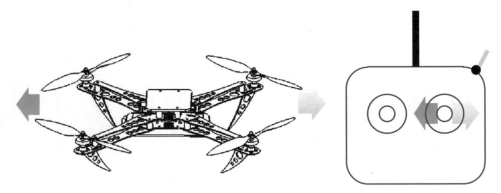

Fig. 1.11   Operation of an RC transmitter for the leftward-and-rightward movement

will turn clockwise and then change its orientation. Generally, the yaw movement corresponds to the operation of the RC transmitter as shown in Fig. 1.13.

1.2.2   Performance Evaluation

Performance evaluation of a small aircraft can be carried out from the following five factors.

(1) *Ease-of-use*: the ease of learning in order to operate the remote control of a small aircraft during hover and in maneuvers.

(2) *Reliability*: often quantified as the mean time between failures.

(3) *Maintainability*: the characteristic of design and assembly which determines the probability that a failed component can be restored to its normal operable state within a given time frame, following prescribed practices and procedures. Its two main components are serviceability (ease of conducting scheduled inspections and servicing) and reparability (ease of restoring service after a failure).

(4) *Time of endurance*: the maximum flight time of an aircraft with a prescribed payload and mission.

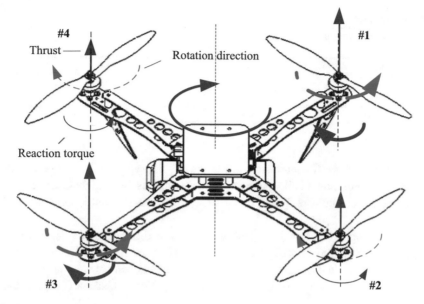

Fig. 1.12 Clockwise yaw movement of a quadcopter

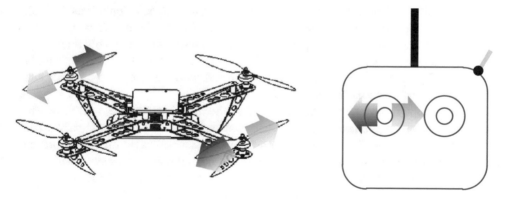

Fig. 1.13 Operation of an RC transmitter for the yaw movement

(5) *Payload capacity*: the maximum weight of payload which an aircraft can take under a prescribed throttle (80% of the total throttle for example) and mission.

Normally, the combination of the five performance factors in fact determines the *user experience*. By taking the three small aircraft mentioned in Sect. 1.1.1 for example, their user experiences are analyzed in the following.

(1) Ease-of-use. The remote control of a multicopter is the simplest because a multicopter has the ability of VTOL and can hover in the air. As shown in Sect. 1.2.1, the two sticks of an RC transmitter correspond to the forward-and-backward, leftward-and-rightward, upward-and-downward, and yaw movements, respectively. Since the four movements are decoupled from each other, the principle of the remote control of a multicopter is simple. Generally, an adult will understand and master the operation skills basically in several hours. Moreover, the controller design is easy and the controller parameters can also be tuned easily. The difficulty in controlling a helicopter lies in strongly coupled modes and highly nonlinear dynamics. These also make the autopilot design difficult. Moreover, it is hard to tune the controller parameters. The flying operation of a fixed-wing aircraft needs to be

Table 1.2 Comparisons of user experiences of three types of small aircraft (More "+" implies better)

	Fixed-wing aircraft	Single rotor blade helicopters	Multicopters
Ease-of-use	++	+	+++
Reliability	++	+	+++
Maintainability	++	+	+++
Time of endurance	+++	++	+
Payload capacity	++	+++	+

performed in a large air space. Since it cannot stay still in the air, the remote pilots have to perform the control actions frequently. For model helicopters and model airplanes, both of them will cost users a long time to learn to operate and control. Based on the analysis above, the multicopter has the best ease-of-use performance.

(2) Reliability. The multicopter has a high reliability in terms of the mechanical structure. Both fixed-wing aircraft and helicopters have turning joints in their airframes. As a result, there is always mechanical wear to some degree. On the contrary, there is hardly any mechanical wear in multicopters because no turning joints are involved in the airframe and brushless Direct Current (DC) motors are adopted.

(3) Maintainability. Multicopters are the easiest to maintain. They have simple structure and therefore can be assembled with little effort. For example, if motors, Electronic Speed Controllers (ESCs), batteries, propellers, or airframes fail, they can be replaced easily. Conversely, both fixed-wing aircraft and helicopters have more components and complex structures. As a result, their assembly is not easy.

(4) Others issues. The energy conversion efficiency of multicopters is the lowest. So their flight time and payload capacity do not have any advantages compared with the fixed-wing aircraft and helicopters. Their overall performances are shown in Table 1.2.

From Table 1.2, multicopters have remarkable advantages in terms of ease-of-use, reliability, and maintainability, while they have disadvantages in terms of time of endurance and payload capacity. The former three performance factors are fundamental and more important, whereas the time of endurance and payload capacity are more case sensitive and sometimes can be compromised or even sacrificed. For example, in the field of aerial photography, the flight time can be extended by replacing the battery. On the other hand, the payload capacity can be increased by increasing the number or the radius of propellers. In practice, remote pilots will choose different multicopters for particular tasks. Therefore, it can be concluded that multicopters beat other competitors in terms of *user experience*. Moreover, multicopters' user experience will continue to be updated. For the three types of small aircraft, the ease-of-use characteristic is related to the flight principle; meanwhile, reliability and maintainability depend on the structure. As the three performances are hard to improve, multicopters will outperform the other two at least in the consumer market for quite a long time. On the other hand, with the development in technologies of batteries, material, and motors, the time of endurance and the payload capacity will be both improved. As a consequence, multicopters will exhibit satisfactory behaviors in more aspects and will be unsurprisingly the choice of more and more customers.

1.2.3 Bottleneck

There also exist bottlenecks during the development of multicopters. The decoupled movements and simple structure depend on the rapid change of thrust and moments by the rapid change of the propeller angular speeds. Due to this principle, it is difficult to extend multicopters to a larger size.

Fig. 1.14 Volocopter VC200

(1) First, roughly, the longer the radius of a propeller is, the slower the dynamic response of rotors will be. Comparatively, helicopters increase or decrease overall lift by altering the angle of attack for all blades collectively by equal amounts at the same time, resulting in ascent and descent behavior.

(2) Secondly, the longer the radius of a propeller is, the more fatigue it is in the rotor hub due to the blade-flapping effects. *Blade flapping* is the upward-and-downward movement of a rotor blade, which will occur during flight of multicopters. Results caused by blade-flapping effects are similar to breaking a soft iron wire by bending it toward one direction and then the opposite repetitively. Multicopters are typically equipped with lightweight, fixed-pitch plastic propellers, because propellers that are too rigid can lead to transmission of these aerodynamic forces directly through to the rotor hub. This may cause a mechanical failure to the motor mounted or to the airframe itself. Therefore, large helicopters adopt flapping hinges. With flapping hinges, none of the bending forces or rolling moments are transferred to the body of the helicopter. Therefore, to provide a large payload capacity, each propeller of multicopters has to be modified to have turning joints, such as flapping hinges in helicopters or control vanes in ducted air vehicles [11]. This will make the modified multicopters complex and further reduce the intrinsic advantages on performance such as ease-of-use, reliability, and maintainability.

Nevertheless, there is a way to increase the payload capacity, which is to use more small propellers to replace large rotors. For example, a German company designed a multicopter VC200, as shown in Fig. 1.14, which can carry two people and achieves more than one hour flight time [12]. This solution circumvents the complex mechanical structure and has a good modularization. The cost paid is that the weight of aircraft is increased, and the time of endurance is sacrificed. Moreover, since it has eighteen motors, the failure rate of one motor is increased. However, there is much control redundancy. By some smart control allocation algorithms, the flight safety can be improved. Overall, compared with a quadcopter of a same size, VC200 does not have advantages on reliability, maintainability, time of endurance, and payload capacity, but it is a compromise as a full-scale multicopter. Therefore, it is unnecessary to have so many propellers for a small multicopter. For a manned aircraft, the design of VC200 is acceptable. However, the heavier an aircraft is, the safer it needs to be, especially for manned aircraft. This is related to the *airworthiness*,[7] which is a time-consuming and costly process. Therefore, people would rather choose small or micro-multicopters.

1.3   History of Multicopters

In the previous section, the reason that people choose small or micro-multicopters has been discussed. However, there is another question usually asked: Why are multicopters more popular now than they were before 2005? The answer can be drawn from the history of the development of multicopter technology. Generally, the history can be divided into five periods: dormancy period (before 1990), growing

---

[7]Airworthiness is the measure of an aircraft's suitability for safe flight. Certification of airworthiness is initially conferred by a certificate of airworthiness from a national aviation authority and is maintained by performing the required maintenance actions.

period (1990–2005), development period (2005–2010), activity period (2010–2013), and booming period (2013–).

1.3.1 The First Stage (Before 1990): Dormancy Period [13, pp. 1–2], [14]

As early as 1907 in France, guided by Charles Richet, the Breguet Brothers built their first human carrying helicopter—and they called it the Breguet-Richet Gyroplane No. 1, which was a quadcopter, as shown in Fig. 1.15a. The first attempt of flight was undertaken between August and September of 1907. The witnesses said they saw the quadcopter lifted 1.5 m into the air for a moments and landed soon after. This failure was caused by the impractical design. Despite ended up a failure, it might be the earliest attempt to build the multicopter [15]. Etienne Oemichen, another engineer, also began experimenting with rotating-wing designs in 1920. His first model failed to lift from the ground. However, after some calculations and redesigns, his second aircraft, the Oemichen No. 2, as shown in Fig. 1.15b, established a world record for helicopters in 1923 by remaining airborne for up to fourteen minutes. The army also had an interest in vertical lift machines. In 1921, George De Bothezat and Ivan Jerome were hired to develop one for the US Army Air Corps. Their quadcopter, as shown in Fig. 1.15c, was designed to take a payload of three people in addition to one pilot and was supposed to reach an altitude of one hundred meters, but the result was that it only managed to lift five meters. Since these early designs suffered from poor engine performance and could only reach a couple of meters high, not much improvement was done to the quadcopter design during the following three decades. It was not until the mid-1950s the first real quadcopter was flown, which was designed by Marc Adman Kaplan. His quadcopter design, Convertawings Model "A", as shown in Fig. 1.15d, was first flown in 1956 and proved to be a great success. The one ton heavy quadcopter was able to hover and maneuver with its two 90 horsepower engines. Controlling this quadcopter did not require additional propellers perpendicular to the main rotors, but was realized by varying the thrust of the main rotors instead. Even though this quadcopter flew successfully, people saw little interest in it because they expected heavy payload and this one could not compete with the conventional aircraft in terms of performance specifications such as speed, payload, flying range, and time of endurance. In the 1950s, the US Army continued to develop several VTOLs. In 1957, the Army contracted with Curtiss-Wright to develop VZ-7 as a prototype flying jeep, to carry small amounts of men and machinery over rough terrain, as

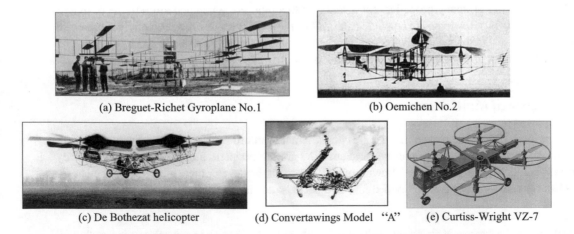

(a) Breguet-Richet Gyroplane No.1 (b) Oemichen No.2

(c) De Bothezat helicopter (d) Convertawings Model "A" (e) Curtiss-Wright VZ-7

Fig. 1.15 Some multicopters in the dormancy period (Photograph courtesy of http://www.aviastar.org)

shown in Fig. 1.15e. Curtiss-Wright produced two VZ-7 prototypes in 1958. The machines were proved to be very easy to hover and maneuver. However, they were not able to meet the altitude and speed requirements specified in the Army's contract. The VZ-7 program was canceled in the mid-1960s when it was realized that the job for which "flying jeeps" were designed could be performed more easily and efficiently by conventional helicopters. Before 1990, since all the components of a multicopter including motors and sensors were cumbersome, the multicopter was normally designed in a full-scale one. As analyzed in Sect. 1.2.2, all full-scale multicopters in those days used turning joints to change the lift to tune the attitude. They did not have advantages over full-scale single rotor helicopters. That is why US Army canceled these projects. Since then, multicopters were nearly abandoned. During the following 30 years, there was not much improvement and this type of aircraft attracted little attention.

1.3.2 The Second Stage (1990–2005): Growing Period

(1) Research
Until the 1990s, with the development of Micro-Electro-Mechanical System (MEMS), Inertial Measurement Unit (IMU) weighting several grams emerged. Although MEMS sensors have been designed, the low-cost MEMS IMUs produce big noise. Therefore, the measurement they took cannot be used directly. The research started to receive more and more attention on how to get rid of the noise in the attitude measurement of MEMS IMUs. The design of a small multicopter requires not only algorithms but also microcomputers on which these algorithms can run. At this stage, the computation speed of microcomputers such as Single Chip Microcomputers (SCMs) and Digital Signal Processor (DSP) had been improved significantly. These provided a possibility to design small multicopters. The researchers at universities started to build models and design control algorithms. Also, some pioneers started to build their own real multicopters [16–20].

(2) Products [13, 21]
As the concept of the quadcopter drifted away from military use only, it began to make its way to approach consumers via the RC toy market. During the early 1990s, a mini quadcopters, called Keyence Gyrosaucer (Fig. 1.16a), was marketed in Japan. It may be the earliest generation of mini quadcopter. Designed for indoor use, this mini quadcopter, with body and propellers made of polystyrene foam, used two gyroscopes for "posture and turning control." The user could get about three minutes of flight time with a single charge [22]. Back to the USA during the early 1990s, engineer Mike Dammar developed his own battery-powered quadcopter. Working at Spectrolutions Inc., he marketed the device called Roswell Flyer (Fig. 1.16b) in 1999, which was later adopted into a company called Draganfly. Under this new brand, there were a series of developments over the years [23]. In 2002, a quadcopter was invented and developed in the Jugend forscht (young researchers) competition in Germany [24]. It was called Silverlit X-UFO later (Fig. 1.16c).

1.3.3 The Third Stage (2005–2010): Development Period

(1) Research
Since 2005, more and more researchers paid attention to multicopters and published a large amount of papers. Moreover, some of them were not satisfied with simulation and started to build their own multicopters to test corresponding algorithms, especially the attitude control algorithms [25–31]. Although the algorithms were easy to design, it was not easy for a researcher to build a real multicopter, because E-commerce then was not mature and popular like it is today. Even if some multicopters were designed,

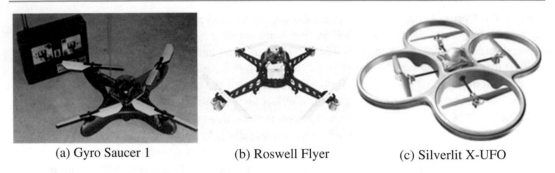

(a) Gyro Saucer 1   (b) Roswell Flyer   (c) Silverlit X-UFO

Fig. 1.16 Some multicopters in the growing period

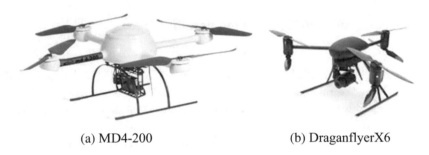

(a) MD4-200   (b) DraganflyerX6

Fig. 1.17 Some products in the development period

they were often not very reliable. Most of the time was spent on finding appropriate components, designing, and testing. Skipping the arduous work of building quadcopters themselves, some researchers utilized the existing reliable commercial quadcopters and the *optical motion-capture system* to build testing environments, such as "real-time indoor autonomous vehicle test environment" by the team of Jonathan P. How at Massachusetts Institute of Technology [32] and "the Grasp multiple micro-UAV test bed" by the team of Vijay Kumar at the University of Pennsylvania [33]. Based on the testing environments, some complex tasks were completed indoor.

(2) Products
Microdrones GmbH was founded in October 2005. The first product generation, namely MD4-200 as shown in Fig. 1.17a, was launched in April 2006. Over 250 units were sold within a short time. In 2010, MD4-1000 joined the family, which immediately became a benchmark [34]. These products had achieved a great success in the commercial market. In October 2006, a large community of MikroKopter autopilots was created by Holger Buss and Ingo Busker. They released an open source autopilot MikroKopter. As early as mid-2007, the aircraft equipped with MikroKopter could hover steadily. In a short time, other components were added. Therefore, it became possible to perform semi-autonomous flights [35]. In 2007, the two inventers of the Silverlit X-UFO teamed with other two prizewinners at Jugend forscht to start Ascending Technologies GmbH in Germany. This company designed multicopters for professional, civil, and research purposes [24]. In 2008, Draganflyer designed the Draganflyer X6 as shown in Fig. 1.17b. It featured carbon fiber construction, a folding airframe, autopilot technology, a wide variety of payloads, and a unique handheld controller. It was recognized as one of the "Best of What's New" awardees by Popular Science in 2008 [23]. At that time, since the commercial multicopters were expensive, they were far out of sight of mass customers. Similarly, consumer multicopters, such as Draganflyer IV with a camera or X-UFO without a camera, were also a little expensive as a toy. Moreover, they nearly cannot bear any payload and were not easy to control

because they did not have Global Positioning System (GPS) receivers installed. More importantly, since *smartphone devices*[8] had not been invented, it could be hardly used for entertainment purpose or taking pictures. Therefore, it was not worthwhile for consumers to buy such a toy. Meanwhile, in 2004, the Parrot Company initiated a project named "AR. Drone", aiming at producing a micro-UAV for the mass market of video games and home entertainment. The project was publicly presented at the Consumer Electronics Show (CES) 2010. On August 18, 2010, the AR. Drone was released. This project involved engineers from Parrot company with the technical support of SYSNAV and its academic partner MINES Paris-Tech for navigation and control design [36]. Multicopters were receiving more and more attention. In 2007, in the business column of *Nature*, an article was published to discuss the commercialization time line of drones [37].

1.3.4   The Fourth Stage (2010–2013): Activity Period

(1) Research
In February 2012, Vijay Kumar at the University of Pennsylvania made a speech at TED[9] conference. Several videos demonstrated that fleets of tiny flying robots performed a series of intricate maneuvers, working together on tasks [38]. This talk revealed the huge potential of multicopters. In 2012, IEEE Robotics and Automation Magazine published a special issue on aerial robotics and the quadcopter platform, which summarized and demonstrated some state-of-the-art technologies [39]. At this stage, many open source autopilots about multicopters emerged, which lowered the threshold of building multicopters for beginners. Table 1.3 shows the major projects of open source autopilots and their links.

(2) Related industries
The iPhone 4, released on June 24, 2010, featured full 9DoF motion sensing, including a three-axis accelerometer, three-axis gyroscope, and electronic compass (or three-axis magnetometer). These sensors facilitated much better rotational motion sensing, gaming, image stabilization, dead reckoning for GPS, gesture recognition, and other applications. With the advances in the smartphone technology, these MEMS sensors found many applications and received wide attention. The size, cost, and power consumption were reduced further. On the other hand, because of their wide applications, the companies had more motivation to make the GPS receivers minimized in its size and cheaper in price. At that time, the world's smallest GPS receiver was smaller than a penny and weighed only 0.3 g [40]. *Smart devices* could control and receive video signals by Wi-fi within a certain distance. Propelled by the requirement of smart devices, volumetric energy density of Lithium-ion battery was being increased. The development of related technologies in turn boosted the development of multicopters.

(3) Products
As shown in Fig. 1.18a, the quadcopter, called AR. Drone, was very successful in the toy market. Its technology and ideology were also very advanced. At first, it used a downward-looking camera to measure *optical flow* and two ultrasonic range finders to measure the altitude. Based on these, its velocity could be obtained by using estimation algorithms. This made AR. Drone able to hover indoor and easy to pilot. Therefore, the ease-of-use was improved a lot. It is ready-to-fly in minutes after unpacked. Secondly, it was light in weight and had an indoor hull made from foam. These safety considerations protected the users and also made it sturdy. Thirdly, AR. Drone could be controlled by a smartphone or a tablet by giving a point-of-view display for the drone's onboard cameras. That

---

[8]A smartphone device is an electronic device, generally connected to other devices or networks via different wireless protocols such as Bluetooth, NFC, Wi-fi, and 3G that can operate to some extent interactively and autonomously.

[9]TED is a nonprofit devoted to "ideas worth spreading"—through http://www.TED.com, the annual conferences, the annual TED Prize, and local TEDx events.

Table 1.3 Major open source projects

Open source projects	Web site
Ardupilot	http://ardupilot.com
Openpilot	http://www.openpilot.org
Paparazzi	http://paparazziuav.org
Pixhawk	https://pixhawk.ethz.ch
Mikrokopter	http://www.mikrokopter.de
KKmulticopter	http://www.kkmulticopter.kr
Multiwii	http://www.multiwii.com
Aeroquad	http://www.aeroquadstore.com
Crazyflie	https://www.bitcraze.io/category/crazyflie
CrazePony	http://www.crazepony.com
DR. R&D	http://www.etootle.com
ANO	http://www.anotc.com
Autoquad	http://autoquad.org
MegaPirate	http://megapiratex.com/index.php
Erlerobot	http://erlerobotics.com
MegaPirateNG	http://code.google.com/p/megapirateng
Taulabs	http://forum.taulabs.org
Flexbot	http://www.flexbot.cc
Dronecode (Operating System)	https://www.dronecode.org
Parrot API (SDK)	https://projects.ardrone.org/embedded/ardrone-api/index.html
3DR DRONEKIT (SDK)	http://www.dronekit.io
DJI DEVELOPER (SDK)	http://dev.dji.com/cn
DJI MATRICE 100+ DJI Guidance	https://developer.dji.com/cn/matrice-100
SDKfor XMission (SDK)	http://www.xaircraft.cn/en/xmission/developer
Ehang GHOST (SDK)	http://dev.ehang.com

(a) AR. Drone   (b) Phantom

Fig. 1.18 Some products in the activity period

means that it had strong entertainment functions. Furthermore, it offered a Software Development Kit (SDK) so that researchers could develop their own applications which meaned it was equipped with research potentials. As a result, the product spread rapidly in academia. At this stage, some companies, which only designed autopilots of multicopters before, also started to design ready-to-fly multicopters inspired by the success of AR. Drone. For example, at the end of 2012, DJI released an all-in-one solution, ready-to-fly "Phantom" quadcopter, as shown in Fig. 1.18b.

1.3.5   The Fifth Period (2013–): Booming Period

(1) Research

At this stage, the research on multicopters tended to make them more autonomous and cooperative. In June 2013, Raffaello D'Andrea at ETH Zurich made a speech at TED Global 2013 about "Machine Athleticism", i.e., the ability of machines to perform dynamic feats that fully exploit their physical capabilities, including play catch, balance and make decisions together, and a demo of Kinect-controlled quadcopter [41]. His TED talk on small aircraft later in 2016 was also impressive as well. In June 2015, a special section about Machine Intelligence of the magazine *Nature* published an article on "Science, technology and the future of small autonomous drones" [42]. This review paper summarized challenges in design and manufacturing, sensing and control, and future research trends in the field of small drones. To overview the development of multicopters in academia, the papers in databases "Engineering Village[10]" and "Web of Science[11]" with "quadrotor" and "multirotor" as the keywords are reviewed. The number of annually published papers in years between January 1990 and December 2015 was recorded as shown in Fig. 1.19. It can be seen that the amount of publications reached a climax in 2013 with the delay occurred in the review process considered. These preliminary research laid a solid foundation for the development of the multicopter industry.

(2) Products

At the end of 2012, DJI released an all-in-one solution, ready-to-fly "Phantom" quadcopter. Like AR. Drone, Phantom can be controlled easily and can hold its position by using GPS and altitude sensors. Compared with AR. Drone, it has a certain payload capacity and can resist wind. At that time, *sport cameras* were popular in the area of extreme sports. As a result, Phantom mounted with sport cameras spurted into popularity in the field of aerial photography. The enthusiasts shared their videos through the social networks or video networks. This made more people know about multicopters. By

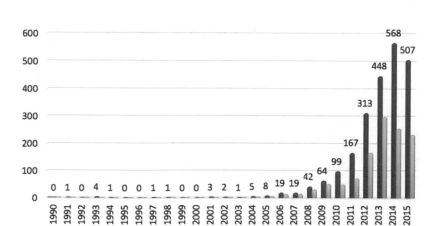

Fig. 1.19 The trend of the number of publications about multicopters

---

[10]http://www.engineeringvillage.com.

[11]http://www.webofknowledge.com.

Table 1.4 Some multicopter (Some of them are conception products) from August 2013 to September 2015

Name	Company	Released Time	Country	Features
Spiri	Patrick Edwards-Daugherty	2013.8	Canada	Autonomous, fully programmable
Stingray500	Curtis Youngblood	2013.12	USA	Full collective pitch 3D quadcopter
AR. Drone 2.0	Parrot	2013.12	France	APP controlled, extreme precise control, and automatic stabilization without GPS
AirDog	Helico Aerospace Industries	2014.6	Latvia	Rotor-arms able to fold away for easy storage, auto-follow
Rolling Spider	Parrot	2014.7	France	Small enough to fit in the palm of your hand, wheels enable the Rolling Spider to zip around the floor and up walls and ceilings
IRIS+	3D Robotics	2014.9	USA	The "Geofence," auto-follow
Nixie	Fly nexie	2014.11	USA	A small camera-equipped drone that can be worn as a wrist band
GHOST 1.0	Ehang	2014.11	China	Control is through one-one-click commands on your phone, auto-follow
Mind4	AirMind	2014.11	China	Precise and autonomous following and filming, Android-based smart drone
Inspire 1	DJI	2014.11	China	Vision positioning system, transforming design, 4K camera full-featured APP, optional dual-operator control
Bebop	Parrot	2014.12	France	A comprehensive update based on AR. Drone 2.0
Vertex VTOL	ComQuestVentures	2015.1	Puerto Rico	A hybrid aircraft that combines the hover and VTOL capabilities of a quadcopter
Skydio	Skydio	2015.1	United Sates	Autonomously navigate around obstacles, flew using intuitive gestures with a mobile device
Steadidrone Flare	Steadidrone	2015.1	Czech	Highest quality carbon fiber and aluminum constructed 'Rapid Deploy' folding airframe
Airborg H61500	Top Flight Technologies	2015.3	USA	A hybrid gas-electric multicopter UAV
Splash Drone	Urban Drones	2015.3	USA	Waterproof quadcopter
Solo	3D Robotics	2015.4	USA	Powered by twin computers intelligent flight and can define its own flight
Phantom 3	DJI	2015.4	China	Vision sensor for indoor flight, gimbal stabilized 4K camera
XPlanet	Xaircraft	2015.4	China	Agricultural UAV System, automatically set the spraying route
Phenox2	Hongo Aerospace	2015.4	Japan	Intelligent, interactive, and programmable drone
CyPhy LVL1	CyPhy Works	2015.4	USA	Six-rotor design with the exact angle of each rotor and the drone can fly completely horizontally even when making turns
Lily	Lily	2015.5	USA	Throw-and-shoot camera, waterproofing
PhoneDrone	xCraft	2015.5	New Zealand	Utilizes its user's smartphone as its brains
Yeair!	Airstier	2015.6	Germany	A quadcopter powered by a combustion engine
Tayzu	Tayzu Robotics	2015.7	United States	Fully-autonomous drones, large-scale data collection capability
Fotokite Phi	Perspective Robotics AG	2015.8	Switzerland	A quadcopter on a leash
UNICORN	FPV Style	2015.8	China	A racer drone, with a high forward flight speed
Micro Drone 3.0	Extreme Fliers	2015.8	UK	Small, smart, and streams HD footage to your phone
Feibot	Feibot	2015.9	China	Smartphone-based drone controller
Snap	Vantage Robotics	2015.9	USA	Airframe can be connected to the rotors and battery by pressing together magnetic strips and a tiny gimbal and camera that are housed inside the body of the drone
Flybi	Advance Robotix Corporation	2015.9	USA	Drone with virtual reality head-tracker goggles a kind rapid charging system. The sensor will detect any object in its flight path and re-route itself safely

2012, Chris Anderson, the former editor-in-chief of *Wired* magazine, joined 3D Robotics as CEO. Under 3D Robotics's support, global volunteers soon created a world-class universal flight code for APM [43, 44]. APM is comprised of several parts, vehicles and boards. The multicopter part of APM is headed by Randy Mackay. In July 2012, the PX4 team headed by Lorenz Meier at ETH Zurich announced availability of the PX4 autopilot platform, with hardware available immediately from 3D Robotics [45, 46]. In August 2013, the PX4 Project and 3D Robotics announced Pixhawk—an advanced open-hardware autopilot design for multicopters, fixed-wing aircraft, ground rovers, and amphibious vehicles. Pixhawk is designed to improve the ease-of-use and reliability while offering unprecedented safety features compared with existing solutions [47]. At the end of 2013, a video was released that Amazon hoped to deliver small packages home in just thirty minutes by "Prime Air"—a future delivery system consisted of quadcopters [48], within 5 years. This idea further shortened the distance between multicopters and mass customers. More and more multicopters were designed at this stage. Table 1.4 lists a part of them designed from 2013 to September 2015.

1.3.6 Conclusion Remark

From the history of the development of multicopters, two conclusions are summarized in the following.

(1) The multicopter is the product of the time. As stated at the beginning of this section, multicopters have existed for 100 years and small multicopters have also existed for over 25 years. The toys such as X-UFO and Draganflyer did not receive so much attention as they do today. However, as the smart devices are emerging, smart multicopters like AR. Drone appeared and grew rapidly in a short term. Similarly, as sport cameras and social networks are popular, multicopters are getting known by the world rapidly. The development of multicopters is also driven by the relevant techniques, like those of the sensors, motors, chips, and materials. It is expected that multicopters will dominate the mass market in quite a long time.

(2) Compared with pure autopilots, all-in-one solution and ready-to-fly multicopters are the future trend. As a package, most components are not exposed. Moreover, customers need not to consider assembling a multicopter or tuning its parameters. The performance can be tested intensively in advance. So, the ease-of-use and reliability are guaranteed. The trend is like that of autofocus cameras. The essential issue is to change the *user experience*—making the flight as simple as possible.

1.4 The Objective and Structure of the Book

1.4.1 Objective

As far as the author understands, there are a number of books about multicopters. The related master and doctoral dissertations are beyond count. The books [49, 50] are about drones, where many types of drones are introduced. The book [51] focuses on the visual navigation of multicopters. The books [52, 53] emphasize control and simulation issues, while the book [13] is about how to build a quadcopter. In spite of all these, it is still hard for a beginner to find an introductory book. This motivates the author to write this book. With the intention of covering most design phases on hardware and algorithms of multicopters, this book has fifteen chapters which is divided into five parts, including multicopter hardware design, modeling, perception, control, and decision. It aims to organize the design principles adopted in engineering practice into a discipline and to emphasize the power of fundamental concepts. This book is featured with four salient characteristics: basic, practical, comprehensive, and systematic.

(1) Basic and practical
The most contents related to multicopters are self-contained, aiming at making this book understandable to readers with the background of Electronic Engineering (EE). For such a purpose, the components

are introduced starting from their functions and key parameters. The introduction to the design process starts with the principle, while the modeling section starts with the theoretical mechanics. The state estimation section starts with the measurement principle of sensors. Before talking about the control, the notions of stability and controllability are introduced. In addition, most of basic and practical methods are presented. These methods are closely related to open source autopilots which are often used now.

(2) Comprehensive and systematic
This book aims to give a complete picture of multicopter systems rather than a single method or technique. Very often, the role of a single method or technique is often not sufficient to meet the user requirements or to solve a practical complex problem. On the other hand, improving other related methods or techniques will reduce the difficulty in a single method. For example, the improvement of the state estimation performance or the mechanical structure can avoid dealing with delay or vibration as a control problem. Through this, some complex controller design can be avoided. For an undergraduate student, basic knowledge has been obtained, such as mathematics, aerodynamics, materials, structures, electronics, filtering, and control algorithms, which correspond to numerous courses. This book is hoped to combine them together to lay foundations for *full stack developers*.

1.4.2   Structure

This book has five parts, fifteen chapters, as shown in Fig. 1.20.

Chapter 1. Introduction. It includes the basic conceptions, remote control and performance evaluation of a multicopter, history of technology development of multicopters, and the objective and structure of this book.

(1) Part I. Design Part
Through this part, readers can have a deeper understanding about the composition of a multicopter system, the configuration and structural design for the airframe, and the choice of the propulsion system. These correspond to Chaps. 2–4, respectively.

Chapter 2. Basic composition. This chapter includes three parts: the airframe, the propulsion system, and the command and control system. In the aspects of the function and key parameters, the fuselage, landing gear, duct, motor, ESC, propeller, battery, RC transmitter and receiver, autopilot, GCS, and radio telemetry are introduced.

Chapter 3. Airframe design. This chapter introduces the basic configuration and the brief consideration on how to attenuate vibration and to reduce the noise.

Chapter 4. Modeling and evaluation of propulsion system. The propulsion system of a multicopter consists of propellers, motors, ESCs, and batteries. The components are modeled with respect to energy. Then, based on these, the flight performance, such as the maximum flight time in hover mode and the maximum payload, is evaluated.

(2) Part II. Modeling Part
Through this part, readers can have a deeper understanding about the dynamic model of a multicopter, which will be further used for the state estimation and control. This part contains two chapters, including the coordinate system, attitude representation, dynamic model, and parameter measurement.

Chapter 5. Coordinate system and attitude representation. First, this chapter introduces the Earth-Fixed Coordinate Frame (EFCF) and Aircraft-Body Coordinate Frame (ABCF). Then, the attitude is represented in the form of three types: Euler angle, rotation matrix, and quaternion. This chapter is the basis for Chap. 6.

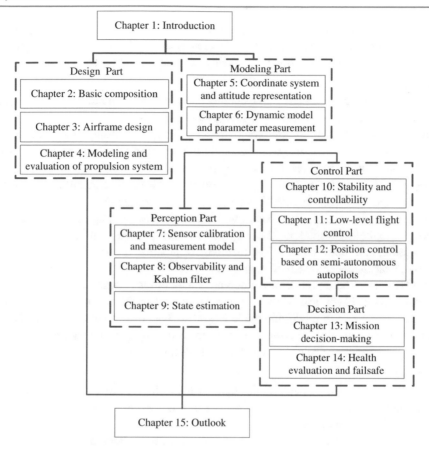

Fig. 1.20 Structure of this book

Chapter 6. Dynamic model and parameter measurement. First, the multicopter control model is introduced, which includes rigid body kinematics, rigid body dynamics, the control effectiveness model, and the propulsor model. Secondly, the aerodynamic drag model is further introduced, which will be used for the state estimation in Chap. 9. Finally, methods are proposed to identify the parameters of a multicopter model.

(3) Part III. Perception Part
Through this part, readers can have a deeper understanding about the state estimation of a multicopter, which will be further used for control. This part contains three chapters.

Chapter 7. Sensor calibration and measurement model. A multicopter is equipped with many sensors, such as the three-axis accelerometer, three-axis gyroscope, three-axis magnetometer, barometer, ultrasonic range finder, laser range finder, GPS receiver, and camera. Low-cost inertial sensors have bias, misalignment angles, and scale factors, which are not negligible and differ from one multicopter sensor board to another. If the sensors are not calibrated, then the position and attitude estimation are inaccurate before takeoff. This may cause flight accidents. In this chapter, the calibration methods are introduced first. Then, the measurement models of sensors are given.

Chapter 8. Observability and Kalman filter. In fact, with related sensors, there still exists a question whether the state information can be estimated in theory, that is, the observability problem. If the system is unobservable, then it makes no sense to design filters. After introducing the observability, a widely used filter, namely Kalman filter, is introduced in detail.

Chapter 9. State estimation. Sensors have noise and redundancy. Moreover, some states cannot be obtained from sensors directly. Therefore, multisensor data fusion is very important. This chapter will contain the attitude estimation, position estimation, velocity estimation, and obstacle estimation.

(4) Part IV. Control Part
Through this part, readers can have a deeper understanding about control of a multicopter, where the methods introduced are commonly used. There are three chapters in this part.

Chapter 10. Stability and controllability. Before controller design, the stability of a multicopter is defined and some simple stability criteria are given. Secondly, the controllability of a multicopter is also introduced. Without it, all fault-tolerant control methods are inapplicable. An interesting example of controllability analysis for hexacopter propulsor degradation and failure will be given. Controllability can be also used to evaluate the design of a multicopter or health of a multicopter.

Chapter 11. Low-level flight control. This chapter will introduce how to design a motor controller that drives a multicopter to fly to a desired position. The process includes the position control, attitude control, control allocation, and motor control. The methods for position control and attitude control are based on both the linear model and the nonlinear model.

Chapter 12. Position control based on Semi-Autonomous Autopilots (SAAs). In this chapter, the SAA will be considered as a "black box." This will be very helpful in the secondary development. Not only has it avoided the trouble of modifying the low-level source code of autopilots, but also can it utilize commercial reliable autopilots to achieve the targets. This simplifies the whole design. This chapter will introduce the system identification and controller design process.

(5) Part V. Decision Part
Through this part, readers can have a deeper understanding about high-level decision-making of a multicopter. Decision-making has two objectives: mission decision-making and *failsafe*,[12] which corresponds to Chaps. 13 and 14, respectively.

Chapter 13. Mission decision-making. For FAC, it consists of the task planning and path planning, where the former informs the multicopter where to go next and latter informs the multicopter how to go there step by step. In the path planning, two problems considered are how to follow a straight-line path and how to avoid an obstacle. For SAC, a switching logic between the autonomous hover and the remote control will be introduced.

Chapter 14. Health evaluation and failsafe. For civil applications, safety is often more important than fulfilling tasks. During a flight, the communication failure, sensor failure, and propulsion system failure all may happen. These accidents may ruin tasks. In this case, it is important to decide on the next action of multicopters. In this chapter, some failure problems are introduced first. Then, some methods of health evaluation are given. Moreover, failsafe suggestions are listed as well. Finally, a failsafe mechanism as a case is designed by using the Extended Finite State Machine (EFSM).

Chapter 15. Outlook. This chapter summarizes the potential technologies which will propel the development of multicopters in the first place. Secondly, innovation trends are proposed. Thirdly, potential risks are further analyzed. Finally, it provides the personal thoughts about the opportunities and challenges.

The structure shown in Fig. 1.20 has considered three types of readers. Readers who are only interested in design can read this chapter, Design Part, and Chap. 15. Readers who are only interested in perception can read the this chapter, Modeling Part, Perception Part, and Chap. 15. Readers who are only interested in control are suggested to read this chapter, Modeling Part, Control Part, Decision Part, and Chap. 15.

---

[12]A failsafe is that, in the event of a specific type of failure, responds in a way that will cause no harm, or at least a minimum of harm, to other devices or to personnel.

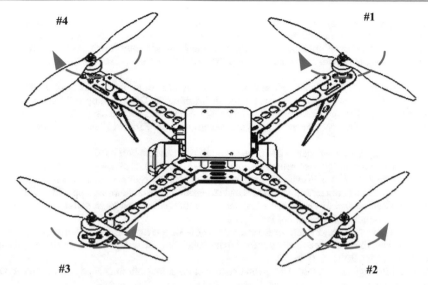

Fig. 1.21 A special quadcopter

Exercises

1.1 As shown in Fig. 1.21, if propellers of a quadcopter are spinning with propellers #1 and #4 rotating clockwise and propellers #2 and #3 counterclockwise, analyze whether it can hover and clarify the reasons.

1.2 Besides the advantages and disadvantages mentioned in this chapter, summarize more advantages and disadvantages of multicopters.

1.3 Choose one of your favorite open source projects and clarify its features.

1.4 Choose one of your favorite multicopters and clarify the reasons.

1.5 Analyze the development stage of multicopters and the trend of multicopters now.

References

1. Austin R (2010) Unmanned aircraft systems: UAVS design, development and deployment. Wiley, NY
2. FAA overview of small UAS notice of proposed rulemaking. http://www.faa.gov/regulations_policies/rulemaking/media/021515_sUAS_Summary.pdf. Accessed 25 Jan 2016
3. Quan Q (2015) Decryption of multirotor development. Robot Ind 2:72–83 (In Chinese)
4. Fan PH (2010) Design and control of multi-rotor aircraft. Dissertation, Beihang University (In Chinese)
5. Talay TA (1975) Introduction to the aerodynamics of flight, SP-367. Scientific and Technical Information Office, NASA, USA
6. Zhang RF (2011) A study on quadrotor compound helicopter oriented to reliable flight control. Dissertation, Beihang University (In Chinese)

7. Quan Q, Fu JS, Cai KY (2012) A compound multicopter. Chinese patent, ZL201220708839.7 (In Chinese)
8. F3-radio control soaring. http://www.fai.org/ciam-our-sport/f3-radio-control-soaring. Accessed 25 Jan 2016
9. Palmer D (2015) The FAA's interpretation of the special rule for model aircraft. Surg Radiol Anat 80:567–749
10. FAA Interpretation of the special rule for model aircraft. http://www.faa.gov/uas/media/model_aircraft_spec_rule. pdf. Accessed 25 Jan 2016
11. Lipera L, Colbourne JD, Patangui P et al (2001) The micro craft iSTAR micro air vehicle: control system design and testing. In: American Helicopter Society 57th Annual forum, Washington, DC, USA, pp 1998–2008
12. Volocopter. http://www.e-volo.com/ongoing-developement/vc-200. Accessed 25 Jan 2016
13. Norris D (2014) Build your own quadcopter: power up your designs with the Parallax Elev-8. McGraw-Hill Education, New York
14. All the world's rotorcraft. http://www.aviastar.org/helicopters.html. Accessed 28 Feb 2016
15. Leishman JG (2001) The breguet-richet quad-rotor helicopter of 1907. Vertiflite 47(3):58–60
16. Hamel T, Mahony R, Chriette A (2002) Visual servo trajectory tracking for a four rotor VTOL aerial vehicle. In: Proceedings of IEEE international conference on robotics and automation, Washington, DC, pp 2781–2786
17. Altug E (2003) Vision based control of unmanned aerial vehicles with applications to an autonomous four rotor helicopter, quadcopter. Dissertation, University of Pennsylvania
18. Kroo I, Printz F (1999) Mesicopter project. http://aero.stanford.edu/mesicopter. Accessed 25 Jan 2016
19. Borenstein J (1996) The hoverbot—an electrically powered flying robot. http://www-personal.umich.edu/johannb/hoverbot.html. Accessed 25 Jan 2016
20. Bouabdallah S, Murrieri P, Siegwart R (2004) Design and control of an indoor micro quadcopter. In: Proceedings of IEEE international conference on robotics and automation, New Orleans, USA, pp 4393–4398
21. Turi J (2014) Tracing the origins of the multicopter drone, for business and pleasure. http://www.engadget.com/2014/11/02/tracing-the-origins-of-the-multirotor-drone/. Accessed 25 Jan 2016
22. Keyence. Gyrosaucer by Keyence. http://www.oocities.org/bourbonstreet/3220/gyrosau.html. Accessed 25 Jan 2016
23. The story behind Draganfly Innovations, Innovative UAV Aircraft & Aerial Video Systems. http://www.draganfly.com/our-story/. Accessed 25 Jan 2016
24. Amazing Technology Company. http://www.asctec.de/en/ascending-technologies/company/. Accessed 28 Feb 2016
25. Tayebi A, Mcgilvray S (2006) Attitude stabilization of a VTOL quadrotor aircraft. IEEE Trans Control Syst T 14(3):562–571
26. Bouabdallah S, Siegwart R (2007) Full control of a quadcopter. In: Proceedings of international conference on intelligent robots and systems (IROS). San Diego, USA, pp 691–699
27. Pounds P, Mahony R, Corke P (2010) Modelling and control of a large quadrotor robot. Control Eng Pract 18(7):691–699
28. Huang H, Hoffmann GM, Waslanderet SL et al (2009) Aerodynamics and control of autonomous quadrotor helicopters in aggressive maneuvering. In: Proceedings of IEEE international conference on robotics and automation. Kobe, Japan, pp 3277–3282
29. Madani T, Benallegue A (2006) Backstepping sliding mode control applied to a miniature quadrotor flying robot. In: Proceedings of 32nd IEEE international conference on industrial electronics society. France, Paris, pp 700–705
30. Zhang RF, Wang XH, Cai KY (2009) Quadcopter aircraft control without velocity measurements. In: Joint 48th IEEE conference on decision and control and 28th Chinese control conference, pp 817–822
31. Soumelidis A, Gaspar P, Regula G (2008) Control of an experimental mini quad-rotor UAV. In: 16th IEEE mediterranean international conference on control and automation, pp 1252–1257
32. How JP, Bethke B, Frank A et al (2008) Real-time indoor autonomous vehicle test environment. IEEE Control Syst Mag 28(2):51–64
33. Michael N, Mellinger D, Lindsey Q et al (2010) The grasp multiple micro-uav testbed. IEEE Robot Autom Mag 17(3):56–65
34. Microdrones profile. https://www.microdrones.com/en/company/profile. Accessed 25 Jan 2016
35. MikroKopter history. http://wiki.mikrokopter.de/en/starting. Accessed 25 Jan 2016
36. Bristeau PJ, Callou F, Vissire D et al (2011) The navigation and control technology inside the AR. Drone micro UAV. In: 18th IFAC world congress, Milano, Italy, pp 1477–1484
37. Stafford N (2007) Spy in the sky. Nature 445(22):808–809
38. Kumar V (2012) Robots that fly and cooperate. https://www.ted.com/talks/vijay_kumar_robots_that_fly_and_cooperate. Accessed 25 Jan 2016
39. Mahony R, Kumar V (2012) Aerial robotics and the quadrotor [from the guest editors]. IEEE Robot Autom Mag 19(3):19–20
40. Mims C (2011) GPS receivers now small enough to attach to almost anything. http://www.technologyreview.com/view/425334/gps-receivers-now-small-enough-to-attach-to-almost-anything. Accessed 25 Jan 2016
41. D'Andrea R (2013) The astounding athletic power of quadcopters. https://www.ted.com/talks/raffaello_d_andrea_the_astounding_athletic_power_of_quadcopters. Accessed 25 Jan 2016
42. Floreano D, Wood RJ (2015) Science, technology and the future of small autonomous drones. Nature 521(7553):463–464

43. Ardupilot, APM. https://3drobotics.com/about/. Accessed 25 Jan 2016
44. History of Ardupilot. http://dev.ardupilot.com/wiki/history-of-ardupilot. Accessed 25 Jan 2016
45. Introducing the PX4 autopilot system. http://tech-insider.org/diy-drones/research/2012/0725-a.html. Accessed 25 Jan 2016
46. Autopilot hardware. https://pixhawk.org/start. Accessed 25 Jan 2016
47. PX4 and 3D robotics announce Pixhawk. http://3drobotics.com/px4-and-3d-robotics-announce-pixhawk. Accessed 25 Jan 2016
48. Prime air. http://www.amazon.com/b?node=8037720011. Accessed 25 Jan 2016
49. Lozano R (2013) Unmanned aerial vehicles: embedded control. Wiley, Hoboken
50. Nonami K, Kendoul F, Suzuki S et al (2010) Autonomous flying robots: Unmanned aerial vehicles and micro aerial vehicles. Springer, Japan
51. Carrillo LRG, Lpez AED, Lozano R et al (2012) Quad rotorcraft control: vision-based hovering and navigation. Springer-Verlag, London
52. Amir MY, Abbas V (2010) Modeling and neural control of quadcopter helicopter: MATLAB-SIMULINK based modeling, simulation and neural control of quadcopter helicopter. LAP LAMBERT Academic Publishing, Germany
53. Putro IE (2011) Modeling and control simulation for autonomous quadcopter: quadcopter nonlinear modeling and control simulation using Matlab/Simulink environment. LAP Lambert Academic Publishing

Part I
Design

Basic Composition

2

The compositions of multicopter systems are both simple yet complex. The compositions are thought to be simple because a multicopter system is generally composed of several well-modularized components such as the airframe, propulsion system, command and control system. These components for a multicopter can be described as the organs for a human, where the airframe corresponds to the body, carrying other hardware; the propulsion system would be the feet and hands powered by the heart and blood vessels, providing power for the multicopter; the command and control system would be the sense organs and brain, controlling the propulsion system to achieve tasks. On the other hand, the compositions are also complex, because each component is not independent, and they connect and constrain with others in a very complex way. Though there are countless combinations for a multicopter, only a few of them can really work. If designers are unaware of the principle of components and assemble multicopters blindly, then the assembled multicopters may have poor performance, or even could not work at all. Therefore, it is necessary to know the basic principle of each component and their relationship. This chapter aims to answer the question as below:

What are the basic compositions of a multicopter system?

This chapter consists of three parts, namely the airframe, propulsion system, and command and control system. The component of each part will be introduced from the corresponding function, working principle and key parameter, etc.

© Springer Nature Singapore Pte Ltd. 2017
Q. Quan, *Introduction to Multicopter Design and Control*, DOI 10.1007/978-981-10-3382-7_2

Playing Zither

The ancient Chinese had already recognized the interdependency between the entirety and the locality. The *"Emperor's Inner Canon"* is an ancient Chinese medical literature that was treated as the fundamental doctrinal source for Chinese traditional medicine for more than two millennia. The work is composed of two volumes, namely *"Suwen"* and *"Lingshu"*. According to this text, human body is composed of various organs related to each other organically. It recommends to study the etiology as a whole. Shi Su, a Chinese poet in the Song dynasty, wrote in *"Qinshi"*—one of his poems—"If the sound of a zither comes from the instrument itself, then why does not it sound when placed inside a case? If the sound of a zither comes from your fingers, then why do not we just listen to those fingers only? (The Chinese is "若言琴上有琴声,放在匣中何不鸣? 若言声在指头上,何不于君指上听", translated by Miao Guang, from http://www.hsilai.org/tc/ 365/0321.php)" This poem illustrates that the fingers, instrument, skills and emotion of the player constitute the music. These elements are interdependent, and none of them is dispensable.

2.1 Introduction

To give an intuitive understanding of multicopters, Figs. 2.1 and 2.2 show the combination and connection of a multicopter system. As shown in Fig. 2.1, Radio Controlled (RC) transmitter, RC receiver, autopilot (also known as the flight controller), Global Position System (GPS) receiver and Ground Control Station (GCS) belong to the command and control system. Moreover, a multicopter system includes the airframe and propulsion system. In Fig. 2.2, the relationship between the components of the propulsion system and the command and control system are shown clearly. Figure 2.3 is the structure diagram of a multicopter, and it is also the structure diagram of this chapter.

2.2 Airframe

Generally, a typical airframe only includes a fuselage and a landing gear. To cover overall types of multicopters, the duct is also taken as a part of the airframe in this chapter.

2.2.1 Fuselage

2.2.1.1 Function
The fuselage acts as the platform to carry all the equipment of a multicopter. The safety, durability, usability, and the performance of a multicopter are often highly dependent on the configuration of its fuselage. For a well-designed multicopter, all factors including the scale, shape, material, strength, and weight should be carefully taken into consideration.

2.2.1.2 Parameters

(1) Weight
The weight of the fuselage is mainly determined by its size and material. Under the same thrust (also referred to as lift in some papers), a smaller fuselage weight means a larger remaining payload capacity.

Fig. 2.1 Basic composition of a multicopter system

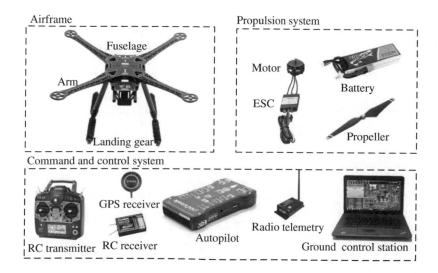

Fig. 2.2 Combination and
connection of a multicopter
system (photo by Jethro
Hazelhurst from http://
www.ardupilot.org)

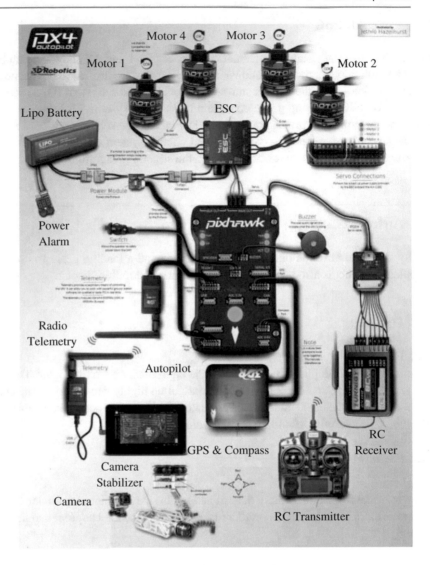

Fig. 2.3 Structure diagram
of this chapter

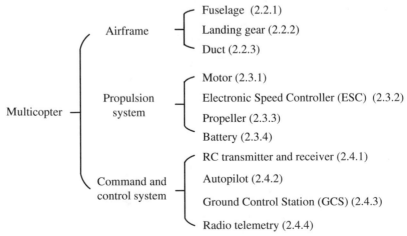

Therefore, in the premise of ensuring the performance of the fuselage, the weight is expected to be as small as possible.

(2) Configuration

The most common configurations include tricopter, quadcopter, hexacopter, and octocopter. Figure 2.4 shows some kinds of configurations that open source autopilots support.

(3) Diagonal size

Diagonal size is the diameter (usually in mm) of the circumcircle determined by the motor axes. In general, as shown in Fig. 2.5, it is the distance of motor axes in the diagonal line and it is used to indicate the size of an airframe. The diagonal size restricts the size of propeller, which determines the maximum thrust and then the payload capacity.

(4) Material

Properties of some kinds of materials are shown in Table 2.1.[1] It is observed that carbon fiber material has small density, high rigidity, and high strength, but it is expensive and hard to process. So, the carbon fiber propellers are widely applied to commercial multicopters that need to carry heavy payload. As a comparison, the acrylic plastic material is light, inexpensive and easy to process, but its rigidity and strength are small. So, the acrylic plastic material is often used in toys or small model airplanes.

Fig. 2.4 Some basic configurations of multicopters

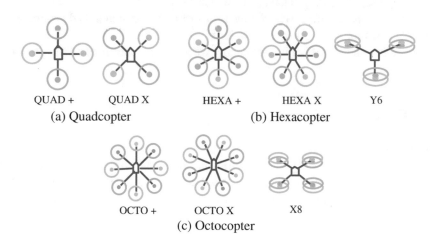

QUAD +   QUAD X   HEXA +   HEXA X   Y6
(a) Quadcopter   (b) Hexacopter

OCTO +   OCTO X   X8
(c) Octocopter

Fig. 2.5 Diagonal size of multicopters

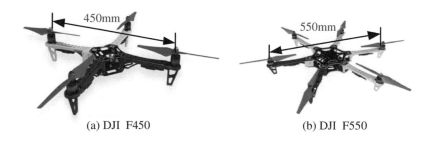

450mm   550mm

(a) DJI F450   (b) DJI F550

---

[1]Note: (a) Rigidity. Rigidity of material is a measure of the ability to overcome deformation when applied stress. It can be measured by Young's modulus (Msi). (b) Strength. Strength of material is measured by the maximum nominal tensile stress before the sample is broken when applied tensile force. It can be measured by tensile strength (Ksi).

Table 2.1 Properties of some materials [1]

	Carbon fiber	Fiberglass	Polycarbonate	Acrylic	Aluminium	Balsa
Density (lb/cuin)	0.05	0.07	0.05	0.04	0.1	0.0027–0.0081
Young's modulus (Msi)	9.3	2.7	0.75	0.38	10.3	0.16–0.9
Tensile strength (Ksi)	120	15–50	8–16	8–11	15–75	1–4.6
Cost (10:cheapest)	1	6	9	9	7	10
Producibility (10:simplest)	3	7	6	7	7	10

2.2.2 Landing Gear

Figure 2.6 shows a landing gear, functions of which include the follows:

(1) Supporting the whole multicopter when landing on the ground or taking off and keeping the level balance of the multicopter.
(2) Keeping propellers off ground at a safe distance.
(3) Weakening ground effect (the downwash stream hits the ground and generates some disturbance effect) when multicopters take off or land.
(4) Consuming and absorbing impact energy when multicopters land on the ground.

2.2.3 Duct

2.2.3.1 Function
In addition to protecting the blade and ensuring personal safety, the *duct* can also enhance the efficiency of thrust and reduce noise. The thrust of a multicopter with ducts is composed of two parts, i.e., the thrust of the propeller and the additional thrust induced by the duct, as shown in Fig. 2.7.

Fig. 2.6 A landing gear

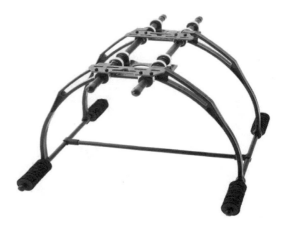

Fig. 2.7 Duct

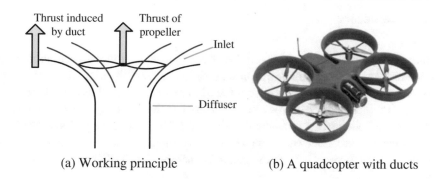

(a) Working principle   (b) A quadcopter with ducts

2.2.3.2 Working Principle
For a rotating propeller in the duct, airflow inside the inlet flows faster than the outside, while the pressure inside is lower than the outside according to *Bernoulli's Principle*, therefore an additional thrust is obtained. Beside, varying the cross section of the duct allows the designer to advantageously change the speed and pressure of the airflow. In addition, by adding a duct, the loss due to the tip vortex which occurs when a propeller operates in free space can be reduced. Therefore, the efficiency and maximum thrust can be increased, sometimes may be significantly.

2.2.3.3 Parameters
The diffuser length and propeller diameter are important parameters for the duct, and reader can refer to [2] for its detailed optimization design method. Although the duct may improve the efficiency to increase the hovering time, the duct itself is generally heavy which may significantly increase the weight and reduce the time. So, the final optimal design needs to achieve a trade-off.

2.3   Propulsion System

A propulsion system includes propellers, motors, ESCs, and often a battery. This system is the most important part of the multicopter, which determines the main performances such as the hovering time, the payload ability, and the flying speed and distance. Moreover, components of the propulsion system have to be compatible with each other, otherwise they cannot work properly or even fails in some extreme cases. For example, in some conditions, an aggressive maneuver may make the current exceed the safety threshold of the ESC, then make the motors stop working in the air, which is very dangerous. Performance evaluation of a propulsion system will be introduced in Chap. 4.

2.3.1   Propeller

2.3.1.1 Function
Propeller is a component that produces the thrust and torque to control a multicopter. The motor efficiency varies with the output torque (depends on the type, size, speed and other factors of a propeller). Therefore, a good match should make sure that the motor works in a high efficiency condition, which will guarantee less power consumed for the same thrust and then extend the time of endurance of the multicopter. Therefore, choosing appropriate propellers is a very direct way to improve the performance and efficiency of a multicopter.

2.3.1.2 Parameters

(1) Type

Generally, the propeller model is described by a four-digit number, such as a 1045 (or 10 × 45) propeller, among which the first two represents the diameter of the propeller (unit: inch), and the latter two represents the propeller pitch (also referred to as screw pitch, blade pitch or simplified as pitch, unit: inch). Therefore, the APC1045 propeller implies that the propeller belongs to APC series, and the diameter and pitch of the propeller is 10 in. and 4.5 in., respectively. The propeller pitch is defined as "the distance a propeller would move in one revolution if it were moving through a soft solid, like a screw through wood." For example, a 21-inch-pitch propeller would move forward 21 in. per revolution.

(2) Chord length

The definition of *chord length* of a propeller is shown in Fig. 2.8, which varies along the radius. Generally, the chord located at the 2/3 of the radius of the propeller is chosen as the nominal chord length.

(3) Moment of inertia

Moment of inertia is a quantity expressing the tendency of a body to resist angular acceleration, which is the sum of the products of the mass of each particle in the body with the square of its distance from the rotation axis. A smaller moment of inertia of the propeller can improve the response speed of the motor, which is important for the control effect and performance.

(4) Number of blades

Some typical propellers with different number of blades are shown in Fig. 2.9. Experiments indicated that the efficiency of two-blade propeller is better comparing with three-blade propeller [3, p. 65]. As shown in Fig. 2.10, the maximum thrust of the propeller is increased with the number of the blades (Fig. 2.10a), while the efficiency is decreased with the number of the blades (Fig. 2.10b). It can also be found that, to obtain the same thrust, a three-blade propeller has a less diameter compared with the corresponding two-blade propeller. Although the efficiency is reduced, the time of endurance may be improved to a certain extent by the reduction of the size and weight of the airframe.

(5) Safe rotation rate

Generally, the materials of propellers used on the multicopters are flexible. So, when rotation rate exceeds a certain value, the propellers may deform, which will reduce its efficiency. Therefore, when calculating the safety rotation rate limit, all the possible conditions should be considered. The APC

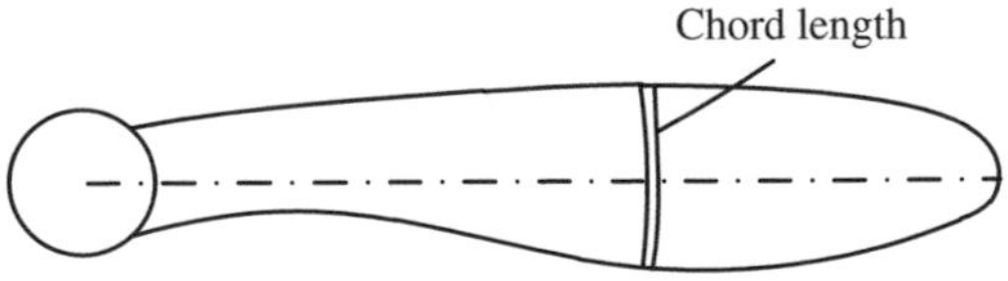

Fig. 2.8 Chord length of a propeller

Chord length

Fig. 2.9 Propellers with different number of blades

(a) Two-blade propeller   (b) Three-blade propeller   (c) Four-blade propeller

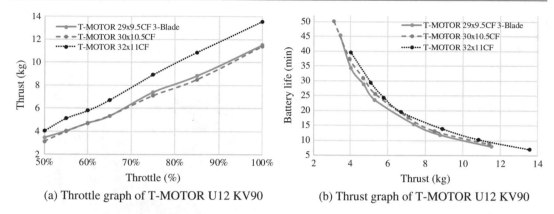

(a) Throttle graph of T-MOTOR U12 KV90   (b) Thrust graph of T-MOTOR U12 KV90

Fig. 2.10 Thrust and efficiency of two-blade propeller and three-blade propeller. Taking T-MOTOR 29 × 9.5CF 3-Blade for example, it indicates that the propeller has three blades, and it is made of Carbon Fiber (CF). The default number of blades is two

Web site[2] gives an empirical formula for the maximum speed of multicopter propellers, which is 105000 Revolution Per Minute (RPM)/prop diameter (inches). Taking the 10-in. propeller for example, its maximum speed is 10500RPM. By contrast, the maximum speed of Slow Flyer (SF) propellers is only 65000RPM/prop diameter (inches).

(6) Propeller specific thrust
Specific thrust, also referred to as efficiency (unit: g/W), is a very important parameter to measure the efficiency of energy transformation. The propeller specific thrust is defined as

$$\text{Mechanical power (unit: W)} = \text{Torque (unit: N} \cdot \text{m)} \times \text{Propeller speed (unit: rad/s)}$$
$$\text{Propeller specific thrust (unit: g/W)} = \frac{\text{Thrust (unit: g)}}{\text{Mechanical power (unit: W)}}.$$

(7) Material [4]
Material includes carbon fiber, plastic, and wood. Though the propellers made of carbon fiber cost almost twice as much as those made of plastic, they are more popular because propellers made of carbon fiber have the following advantages: 1) less vibration and noise because of its high rigidity; 2) lighter and stronger; and 3) more suitable for the motor with high KV. However, because of the high rigidity, the motor will absorb most of the impact when a crash occurs and the fiber blade can be treated as a high speed rotating razor which is too dangerous to human nearby. Propellers made of wood are much heavier and more expensive, which is suitable for multicopters with large payload capacity.

2.3.1.3 Static Balance and Dynamic Balance [5, 6]
The goal of static balance and dynamic balance is to reduce the vibration caused by the asymmetry centrifugal force which is further caused by the asymmetrical distribution of mass and shape. The imbalance of propeller is a major source of vibration for a multicopter. Propeller static imbalance occurs when the Center of Gravity (CoG) of the propeller does not coincide with the axis of rotation. Dynamic imbalance occurs when the CoG of the propeller does not coincide with the center of inertia.

---

[2]http://www.apcprop.com/Articles.asp?ID=255.

Fig. 2.11 A Du-Bro
propeller balancer

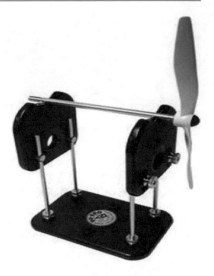

The imbalanced force not only affects the sensor measurement, but also makes the motor bearing wear down more quickly and increases the power consumption. These characteristics will shorten the lifetime of multicopters and increase the possibility of failure. Besides, an imbalanced propeller is far noisier than a balanced one. A balancer as shown in the Fig. 2.11 can be used to test the static balance of a propeller. The test of dynamic balance is often not an easy work as it needs sensors to record the data. An easy way without using sensors is introduced in a video [7]. If imbalance exists, some measures can be taken, such as pasting scotch tapes on the lighter blade or grinding the heavier one (not the edge) using sandpapers.

2.3.2 Motor [8, pp. 533–592]

2.3.2.1 Function
Motors of multicopters are mainly brushless DC motors for the various advantages such as high efficiency, potential to downsize, and low manufacturing costs. Brushless DC motors are used to convert electrical energy (stored in battery) into mechanical energy for propeller. Concretely, based on the position of rotors, brushless DC motors can be classified into the outer rotor type and inner rotor type as shown in Fig. 2.12. Considering that the motor of a multicopter is supposed to drive larger propellers to improve efficiency, the outer rotor type outperforms the inner rotor type as it can provide larger torques. Besides, compared with the inner rotor type, speed of the outer rotor type is more stable. Therefore, the outer rotor type is more popular in multicopters and most other aircraft.

2.3.2.2 Working Principle
As shown in Fig. 2.13, the control circuit generates a Pulse Width Modulated (PWM) signal to the ESC, where the signal is amplified by a driving circuit and sent to the power switch of the inverter. Then, it will control the motor winding to work in a certain sequence and generate a jump-type rotating magnetic field in the air gap of the motor. Common types of main circuits of brushless DC motors include the star-type three-phase half-bridge, star-type three-phase bridge, and angle-type three-phase bridge. Among them, the star-type three-phase bridge is used most widely. Three output signals of the position detector are controlled by a logic circuit to control the on and off states of the switch. There are two control modes: two-two conduction mode and three-three conduction mode. As shown

Fig. 2.12 Outer rotor type
and inner rotor type (photo
courtesy of www.nidec.
com)

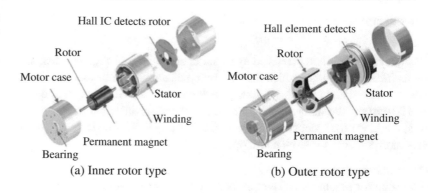

(a) Inner rotor type   (b) Outer rotor type

Fig. 2.13 Main circuit of
star-type three-phase
bridge

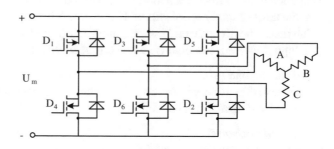

Fig. 2.14 Counter
electromotive force
waveform of three-phase
winding and its two-two
conduction mode

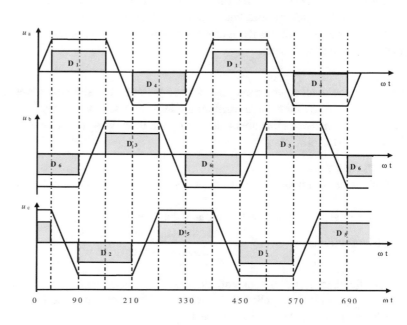

in Fig. 2.14, every 60° the rotor rotates, the switches of inverters commutate one time and the field of
the state of stator changes one time. There are six magnetic states and three phases for motors, each
phase conducts 120°.

2.3.2.3 Parameters

(1) Size

Size of motor is generally represented by its stator size with four-digit number, such as motor 2212 (or written as 22 × 12), among which the first two indicates its stator diameter (mm) and the latter two indicates its stator height (mm). For example, the motor 2212 indicates that the stator diameter of the motor is 22 mm and the stator height is 12 mm. That means, the larger the former two are, the wider the motor is; the larger the latter two are, the higher the motor is. A wide and high motor has high power, which is more suitable for large multicopters.

(2) KV value for motors

The KV value for brushless DC motors is the number of RPM that the motor will revolve when 1 V (Volt) is applied with no load attached to the motor. For example, 1000 KV just means that when 1 V is applied, the no-load motor speed will be 1000 RPM. A low KV motor has more windings of thinner wire, which means it will carry more power, produce a higher torque, and drive a bigger propeller. By contrast, a high KV motor can produce a low torque so that it can only drive a small propeller.

(3) No-load current and voltage

In the no-load test, the current passing through the three-phase winding of stator after applying nominal voltage (generally 10 or 24 V) is defined as the nominal no-load current.

(4) Maximum current/power

It is the maximum current or power the motor can undertake. For example, maximum continuous current "25A/30s" represents that the motor can work safely with continuous current up to 25A, beyond which for more than 30s the motor may be burnt out. The same definition can be applied for the maximum continuous power.

(5) Resistance

There is resistance in all motor armatures. It is very small but cannot be ignored because the current flowing through the resistance is tremendously large and sometimes reaches tens of Amperes. The existence of the resistance generates heat during the running of the motor, which may overheat the motor and reduce the efficiency.

(6) Motor efficiency

Motor efficiency is an important parameter to measure the performance. It is defined as follows:

$$\text{Electrical power (unit: W)} = \text{Input Voltage (unit: V)} \times \text{Effective current (unit: A)}$$
$$\text{Motor efficiency} = \frac{\text{Mechanical power (unit: W)}}{\text{Electrical power (unit: W)}}.$$

The motor efficiency is not a constant. In general, it varies with input voltage (throttle) and load (propeller). For the same propeller, the efficiency of the motor may be reduced as the input voltage (current) is increased. That is because the larger the current is, the more the heat (caused by the resistance) and other loss will be, which makes the ratio of the effective mechanical power reduced.

(7) Overall specific thrust

The overall performance of the propulsion system depends largely on a well-matched combination of motor and propeller. To evaluate the efficiency of the motor and propeller together, the overall specific thrust is calculated as

Motor model	Voltage (V)	Propeller Model	Throttle	Current (A)	Power (W)	Thrust (g)	Speed (RPM)	Efficiency (G/W)	Torque (N·m)	Temperature °C
T-MOTOR MN5212 KV340	24	T-MOTOR 15x5CF	50%	3.3	79	745	3821	9.44	0.142	38
			55%	4.2	99.8	910	4220	9.11	0.172	
			60%	5.2	123.6	1075	4576	8.7	0.198	
			65%	6.3	150.7	1254	4925	8.32	0.232	
			75%	9.1	217.2	1681	5663	7.74	0.31	
			85%	12.2	292.1	2115	6315	7.24	0.382	
			100%	17.8	426.7	2746	7167	6.44	0.498	
		T-MOTOR 18x6.1CF	50%	5.7	137.5	1318	3596	9.58	0.29	74
			55%	7.4	178.1	1612	3958	9.05	0.344	
			60%	9.3	222	1901	4310	8.56	0.411	
			65%	11.6	278.2	2259	4622	8.12	0.472	
			75%	16.5	395.5	2835	5226	7.17	0.605	
			85%	22.1	531.1	3477	5751	6.55	0.737	
			100%	31	744.7	4355	6358	5.85	0.918	

Fig. 2.15 Overall specific thrust of motor MN5212 KV420 (data from http://www.rctigermotor.com)

$$\text{Overall specific thrust (unit: g/W)} = \frac{\text{Thrust (unit: g)}}{\text{Electrical power(unit: W)}}$$
$$= \text{Propeller specific thrust} \times \text{Motor efficiency}.$$

Since both the propeller specific thrust and motor efficiency are not constant, the overall specific thrust changes with the working condition. The overall specific thrust is often given by the motor producers. Taking a motor for example, the overall specific thrust under different states is displayed in Fig. 2.15, where "Efficiency (G/W)" is in fact the overall specific thrust. This will help designers to choose combinations of a motor and a propeller according to their requirements.

2.3.3 Electronic Speed Controller

2.3.3.1 Function
The basic function of ESCs is to control the speed of motors based on the PWM signal that autopilots send, which is too weak to drive brushless DC motors directly. Furthermore, some ESCs act as a dynamic brake, or a power supply (battery elimination circuit module) for RC receiver or servo motors. Unlike a general ESC, the brushless ESC has a new function, i.e., it can act as an inverter transforming an onboard DC power input into a three-phase Alternating Current (AC) power that can be applied to brushless DC motors. Undoubtedly, there are some other auxiliary functions, such as battery protection and starting protection. The Fig. 2.16 shows the Xaircraft ESC-S20A for multicopters.

2.3.3.2 Parameters
(1) Maximum continuous/peak current
The most important parameter for brushless ESCs is current, which is usually represented by Ampere (A), such as 10 A, 20 A, and 30 A. Different motors need to be equipped with different ESCs. An inappropriate matching will burn ESCs or even cause motor failure. Concretely, there are two important parameters of the brushless ESC, namely the *maximum continuous current* and *peak current*. The former

Fig. 2.16 XAircraft
ESC-S20A for multicopters

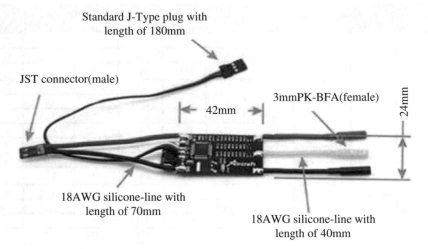

is the maximum continuous current in the normal working condition, while the latter is the maximum instantaneous current that the ESC can withstand. Each ESC will be labeled with a specified value, such as Hobbywing XRotor 15 A which indicates the maximum continuous current allowed. When choosing the type of ESCs, attention should be paid to the maximum continuous current, which needs to be checked whether it leaves a safety margin (20% for example) so as to efficiently avoid burning the power tube. Taking 50A ESC for example, 10A is often left as a safety margin.

(2) Voltage range
The range of voltage allowing the ESC to work properly is also an important parameter. Usually, the index like "3-4S LiPo" can be found on the ESC specification, which means that the voltage range of this ESC is 3-4 cells of LiPo battery, i.e., 11.1–14.8 V.

(3) Resistance
Since all ESCs have resistance, the heating power cannot be ignored because the current flowing through them can sometimes reach tens of Amperes. Considering the heat dissipation, the resistance of ESCs with high current is always designed to be small.

(4) Refresh rate
The motor response has a great relationship with the refresh rate of ESCs. Before the development of multicopters, ESCs were designed specifically for model airplanes or cars. The maximum operating frequency of servo motors was 50 Hz, therefore the refresh rate of ESCs was 50 Hz. Theoretically, the higher the refresh rate is, the faster the response will be. Since multicopters differ from other types of model airplanes in that the rapid thrust adjustment is realized by the rapid control of propeller angular speed, the refresh rate of multicopter ESCs is often faster. In order to ensure smooth outputs, low-pass filtering is often applied to the input or output of ESCs at the cost of reducing their response rate. This also implies the reduced control frequency.

(5) Programmability
The performance of ESCs can be optimized by tuning internal parameters. There are three ways to set the parameters of ESCs, i.e., programmable cards as shown in Fig. 2.17, computer software via the USB, and RC transmitters. The parameters that can be set up include: throttle range calibration, low voltage protection, power outage value, current limitation, brakes mode, throttle control mode, switch timing setting, starting mode, and PWM mode setting.

Fig. 2.17 Hobbywing
brushless ESC
programmable cards

(6) Compatibility
If the ESC and motor are incompatible, the motor is likely to be jammed, which may result in a fall
and crash for a multicopter in the air. Sometimes motors may get jammed in extreme cases, such as the
case that the control command for mode transitions changes sharply, generating large instantaneous
current, which are not easy to be detected.

2.3.3.3 Square-Wave Driver and Sinusoidal Driver [9]

(1) Square-wave driver
Square-wave driver type ESC outputs square wave. Its control elements work in the switch state, which
makes it simpler, cheaper, and easier to control.

(2) Sinusoidal driver
Sinusoidal driver-type ESC outputs sinusoidal wave, which uses Field Oriented Control (FOC). Therefore the sinusoidal driver performs better in the aspects of operation stability, speed range, efficiency,
and vibration reduction of noise. Now, the optical encoder, Hall sensor and observer-based method are
available to measure the angle of the rotor. For a multicopter, its motor rotors are working in a high
speed state, which means the FOC can be applied based on the observation of rotor electrical angle
with information including the motor model, current, and voltage. This is a good way to reduce the
cost.

2.3.4 Battery

2.3.4.1 Function
Battery is used to provide energy. A battery for small multicopters is shown in Fig. 2.18. A problem often
concerned on present multicopters is the time of endurance , which heavily depends on the capacity of
batteries. Now, there are many types of batteries, where the Lithium Polymer (LiPo) battery and Nickel
Metal Hydride (NiMH) battery are the most commonly used ones because of superior performance
and cheap price.

2.3.4.2 Parameters
The basic parameters of the battery include voltage, discharge capacity, internal resistance, and discharge rate. The nominal voltage of a single cell of LiPo battery is 3.7 V. When it is fully charged, the

Fig. 2.18 GENS ACE
Tattu UAV battery

Fig. 2.19 Connection
diagrams

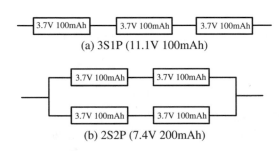

voltage can reach 4.2 V. In order to ensure that the total battery capacity and voltage are enough, several cells can be assembled together. In an actual process, the remaining voltage is decreased gradually with the discharge of the battery. Most research shows that in a certain range, the remaining voltage is in a linear relationship with the battery remaining capacity. However, in the late stage of discharge, the voltage may drop sharply, which may result in rapid thrust loss of multicopter. To ensure that a multicopter has enough power or capacity to return home before the carried battery runs out, it is necessary to set a safe voltage threshold for the battery. Besides, the output voltage will drop as the current of discharge is increased because of more voltage allocated to the internal resistance. It should be noted that the battery should not be completely discharged, otherwise it may have an irreversible damage.

(1) Connection
By combining battery cells in series, a higher voltage can be obtained, with capacity unchanged. On the other hand, by combining battery cells in parallel, a larger capacity can be obtained, with voltage unchanged. In some cases, there exist combinations of both series and parallel in battery packs. The letters S and P are used to represent for the series connection and parallel connection, respectively. For example, as shown in Fig. 2.19a, assuming that the voltage of one cell is 3.7 V and its capacity is 100 mAh, then 3S1P represents three cells in series connection (total voltage is 11.1 V, capacity is 100 mAh). As shown in Fig. 2.19b, for the 2S2P battery, its total voltage is 7.4 V and total capacity is 200 mAh.

(2) Capacity
The milliAmpere-hour (mAh) or Ampere-hour (Ah) is a technical index that how much electrical charge a particular battery has. The capacity of 5000 mAh for a LiPo battery means that the discharge of the battery will last for an hour with the current of 5000 mA when the voltage of a single cell is decreased from 4.2 to 3.0 V. However, the discharge ability will be decreased along with the process of discharge, and its output voltage will also be decreased slowly. As a result, the remaining capacity is not a linear function of the discharge time. In practice, there are two ways to detect whether the remaining capacity of a battery can support the flight. One way is to detect the voltage of the battery by sensors in real time, which is commonly used. The other way is to estimate the State of Charge (SoC) value of batteries, which will be shown in Chap. 14.

(3) Discharge Rate

Discharge rate is represented by

$$\text{Discharge Rate (unit: C)} = \frac{\text{Current of Discharge (unit: mA)}}{\text{Capacity (unit: mAh)}}.$$

For example, the discharge rate of a battery will be 0.2 C when its nominal capacity is 100 mAh and the discharge current is 20 mA. Obviously, the discharge rate measures the rate of discharge. When the maximum discharge rate of a battery with the nominal capacity of 5000 mAh is 20C, the maximum current of discharge is calculated as 5000 mA × 20C = 100 A. The total current of a multicopter cannot exceed its maximum current limit of the battery; otherwise, the battery may be burnt out. The battery having higher discharge rate can generate more current, which can be applied to multicopters demanding higher current because of heavier bodies and more motors.

(4) Resistance

Resistance of a battery is not a constant value, and it varies with the power status and service life. The resistance of a rechargeable battery is relatively small in the initial state. However, after a long period of use, because of the exhaustion of electrolyte and decrease in chemical substance activity of the battery, the internal resistance will be increased gradually until to a certain degree where the power in the battery cannot be released. Then, the battery can be regarded as being run out of.

(5) Energy Density

Energy density is the amount of energy stored in a given system or region of space per unit volume or mass, and the latter is more accurately termed specific energy. In general, the units for energy density and specific energy are (Watt × hour)/kg and (Watt × hour)/L, i.e., Wh/kg and Wh/L, respectively. Batteries with higher energy density are more popular due to the contradiction between volume (weight) and endurance for a product. Lithium-ion battery as a kind of clean energy is getting more and more attentions and is widely used in many applications. The energy density of Lithium-ion batteries varies from chemistry to chemistry and the energy density can range from 240 to 300 Wh/L (double of the NiCd, 1.5 times of NiMH).

2.4 Command and Control System

2.4.1 RC Transmitter and Receiver [10]

2.4.1.1 Function

An *RC transmitter*, as shown in Fig. 2.20a,[3] is used to transmit commands from remote pilots to the corresponding receiver. Then, the *receiver*, as shown in Fig. 2.20b, passes the commands to the autopilot after decoding them. Finally, the multicopter flies according to the commands. Some flight parameters can be set on the transmitter, such as the throttle direction, stick sensitivity, neutral position of RC servo motors, function definitions of channels, record and remind setting of flight time, and lever function setting. Advanced functions include battery voltage and current flight data of multicopters. At present, there are several open source transmitters. Readers interested in detailed information can refer to the Web site http://www.open-tx.org or http://www.reseau.org/arduinorc, based on which transmitters can be customized.

---

[3]For a multicopter, the *throttle control stick* is to control the upward-and-downward movement, and the *rudder control stick* is to control the yaw movement, while the *aileron stick* is to control the roll movement, and the *elevator stick* is to control the pitch movement.

Fig. 2.20 Futaba RC
transmitter and receiver

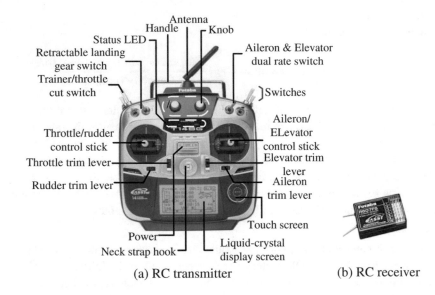

(a) RC transmitter   (b) RC receiver

2.4.1.2 Parameters

(1) Frequency

The RC transmitter and the receiver communicate by radio waves, and the commonly used radio frequency is 72 MHz and 2.4 GHz. Before the utilization of 2.4 GHz, the chance of co-channel interference was very high. The phenomenon might occur when one RC transmitter controls two model aircraft. With the development of model airplane, the safety becomes a serious issue. Therefore, 2.4 GHz transmitters emerged. The 2.4 GHz radio frequency falls in the microwave range, which was used in wireless audio transmission initially. The 2.4 GHz radio communication technology has the following advantages. 1) High frequency. The technology relies on microcomputers to plan frequencies automatically instead of setting the frequency by controlling the crystal. 2) Less chance of co-channel interference. When several transmitters work together, this technology allows frequency-hopping automatically to avoid mutual interference. 3) Low power consumption. As no crystal is used as frequency control parts, the power consumption is reduced significantly. 4) Smaller volume. Since the control wavelength is very short, transmitting and receiving antennas can be shortened greatly. 5) Rapid response and high control accuracy. Although the 2.4 GHz RC transmitter can deal with the co-channel interference, some problems still exist. For example, the 2.4 GHz microwave is of good straightness. In other words, the control signal has a bad performance when there exists an obstacle between the RC transmitter and the multicopter. As a result, the transmitting antenna and receiving antenna should be maintained line of sight; and the obstacles between them, such as houses and warehouse, should be avoided.

(2) Modulation [11, pp. 129–133], [12]

Pulse Code Modulation (PCM) implies the encoding of signal pulses, and Pulse Position Modulation (PPM) refers to the modulation of high-frequency signal. By operating sticks on the transmitter, the value of potentiometer varies accordingly. By the encoding circuit, it can be read and converted into a pulse coded signal, namely PPM or PCM, which will be further modulated through a high-frequency modulation circuit and sent by high-level circuit. The advantages of PCM are not only the strong anti-interference capacity, but also the convenience to be programmed by a computer. Compared with PCM, PPM is easier to realize and cheaper, but is more susceptible to interference.

(3) Channels

One channel corresponds to one separate operation, and generally there are six-channel transmitters, eight-channel transmitters, and ten-channel (or more) transmitters to control multicopters. The operations needed include: throttle control, yaw control, pitch control, and roll control. In this way, an RC transmitter requires four channels at least. Considering the mode transition and control of camera gimbal, transmitters with at least eight channels are recommended.

(4) Mode [13]

RC transmitter modes refer to the way how an RC transmitter is configured to control a multicopter, i.e., the relationship that sticks correspond to movements. For example shown in Fig. 2.21, "Mode 1": pitch/yaw on the left stick, throttle/roll on the right (also called right-hand mode, popular in Japan, more suitable for fixed-wing aircraft); "Mode 2": throttle/yaw on the left, pitch/roll on the right (also called left-hand mode, popular in the U.S. and other parts of the world including China, more suitable for multicopters).

(5) Throttle

In general, the throttle control stick in an RC transmitter is designed to be unable to recover back to its original position automatically. The *direct-type* RC transmitter in this book refers to the type that the total thrust has a positive correlation with the deflection of the throttle control stick. Furthermore, a connected motor will stop with the throttle control stick at the bottom and work at a full speed with the throttle control stick at the top. The transmitter can also be set to recover back to the midpoint automatically once it is released, which is called the *increment-type* RC transmitter in this book. Correspondingly, the motor speed will be increased when the stick is higher than the midpoint and decreased when lower than the midpoint. The mode transition can be realized by loosening and tightening the spring washers behind the stick and then using specified algorithms.

(6) Remote control distance

The control distance of an RC transmitter is restricted by its power. For example, the effective control distance of the "Md-200" is claimed to be 1000 m. In order to extend the control distance, power amplifiers and antennas can be used.

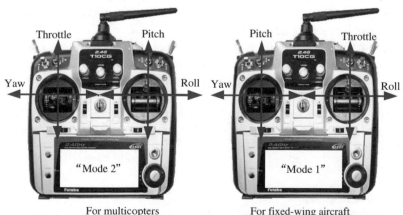

	For multicopters	For fixed-wing aircraft
Throttle :	control the upward-and-downward movement	corresponding to the throttle stick
Pitch :	control the forward-and-backward movement	corresponding to the elevator
Yaw :	control the yaw movement	corresponding to the rudder
Roll :	control the leftward-and-rightward movement	corresponding to the aileron

Fig. 2.21 Different control modes of an RC transmitter

2.4.2 Autopilot

A multicopter *autopilot* is a flight control system used to control the attitude, position, and trajectory of a multicopter. It can be semi-automatically (needs commands from remote pilot) or fully automatically. Autopilots have a control framework which is often based on Proportional-Integral-Derivative (PID) controllers, leaving parameters to be tuned for different multicopters.

2.4.2.1 Composition
A multicopter autopilot can be divided into the software part and hardware part. The software part is the brain of a multicopter and it is used to process and send information, while the hardware part generally includes the following components:

(1) GPS receiver. It is used to obtain the location information of multicopters.
(2) Inertial Measurement Unit (IMU). It includes: the three-axis accelerometer, three-axis gyroscope, and electronic compass (or three-axis magnetometer). It is used to obtain attitude information of a multicopter. In general, a six-axis IMU is the combination of a three-axis accelerometer and a three-axis gyroscope; a nine-axis IMU is the combination of a three-axis accelerometer, a three-axis gyroscope and a three-axis magnetometer; and a ten-axis is the combination of a nine-axis IMU and a barometer.
(3) Height sensor. The barometer and ultrasonic range finder are used to obtain the absolute height (altitude) and relative height (distance to the ground), respectively.
(4) Microcomputers. It acts as a platform to receive information and run algorithms to produce control command.
(5) Interface. It acts as a bridge between the microcomputer and the other devices, such as the sensors, ESC, and RC receiver.

2.4.2.2 Function
(1) Perception. It is used to solve the problem of "where the multicopter is." The GPS receiver, IMU, and height sensors all have a lot of noises, and their refresh rates are not the same. For example, the refresh rate of a GPS receiver is 5 Hz, while the refresh rate of an accelerometer may be 1000 Hz. One task of an autopilot is to fuse these information together to obtain accurate position and attitude. This mainly corresponds to Chaps. 7–9 in this book.
(2) Control. Control is to solve the problem of "how the multicopter flies to a desired position." Based on the position and attitude measured and given, the low-level flight control law is carried out, generating commands for ESCs to control the motors of a multicopter to achieve a desired position. This mainly corresponds to Chaps. 10–12 in this book.
(3) Decision. Decision is to solve the problem of "where the multicopter will go." The decision-making mainly includes the mission decision-making and failsafe. This mainly corresponds to Chaps. 13 and 14 in this book.

2.4.2.3 Open Source Autopilot
Currently, there are many free open source autopilots of multicopters. The Web sites can be found from Table 1.3 in Chap. 1. Open source autopilot flight control boards (hardware) are shown in Fig. 2.22. The types and performance of components of some autopilots are shown in Table 2.2.

Fig. 2.22 Open source
autopilots of multicopters

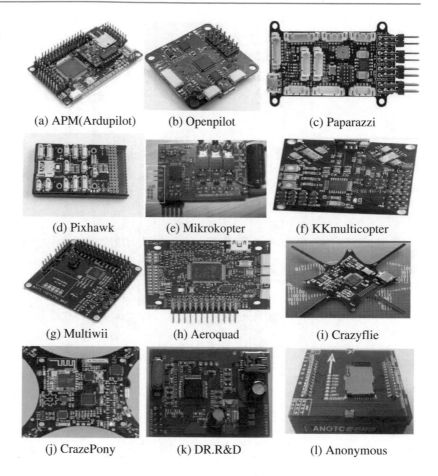

(a) APM(Ardupilot)   (b) Openpilot   (c) Paparazzi

(d) Pixhawk   (e) Mikrokopter   (f) KKmulticopter

(g) Multiwii   (h) Aeroquad   (i) Crazyflie

(j) CrazePony   (k) DR.R&D   (l) Anonymous

2.4.3   Ground Control Station

2.4.3.1 Function

An important part of a GCS is the software. Remote pilots can interact with the software using the mouse, keyboard, button, and joystick. So, way points can be planned by remote pilots for multicopters in advance. Furthermore, remote pilots can monitor the flight status in real time and set new missions to intervene flight. Besides, the software can record and playback flight for analysis.

2.4.3.2 Open Source GCS Software

There are a lot of free open source GCS software available for multicopters now. Figure 2.23 shows some screenshots of GCS software. Most of the GCS software can be downloaded from the corresponding autopilot Web sites, which are listed on the Table 1.3.

2.4.4   Radio Telemetry

2.4.4.1 Function

Radio telemetry refers to using Digital Signal Processing (DSP) technology, digital modulation and demodulation, radio technology to transmit data with high accuracy, and it is equipped with functions

Table 2.2 The parameters of open source autopilot flight control boards

Description	Size (mm)	Weight (g)	Processor	Process frequency (MHz)	Gyroscope	Accelerometer	Magnetometer	Barometer
Arducopter	66 × 40.5	23	ATmega2560	16	MPU-6000	MPU-6000	HMC5843	MS5611
Openpilot	36 × 36	8.5	STM32F103CB	72	ISZ/IDC-500	ADX330	HMC5843	BMP085
Paparazzi(Lisa/M)	51 × 25	10.8	STM32F105RCT6	60	MPU-6000	MPU-6000	HMC5843	MS5611
Pixhawk	40 × 30.2	8	LPC2148	60	ISZ/IDC-500	SCA3100-D04	HMC5843	BMP085
Mikrokopter	44.6 × 50	35	ATmega644	20	ADXRS610	LIS344ALH	KMZ51	MPX4115A
Kkmulticopter	49 × 49	11.7	ATmega168	20	ENC-03	–	–	–
Multiwii	N/A[a]	N/A[a]	Arduino[b]	8–20	ISZ/IDC-650	LIS3L02AL	HMC5883L	BMP085
Aeroquad	N/A[a]	N/A[a]	Arduino[b]	8–20	ITG3200	ADXL345	HMC5883L	BMP085
Crazyflie 2.0	90 × 90	19	STM32F405	168	MPU-9250	MPU-9250	MPU-9250	LPS25H
CrazePony-II(4)	38.9 × 39.55	20	STM32f103T8U6	72	MPU6050	MPU6050	HMC5883L	MS5611
Dr.R&D(2015)IV	33 × 33	300 (the whole)	STM32F103	72	MPU6050	MPU6050	HMC5883L	Ultrasound HC-SR04
Anonymous(V2)	75 × 45	40	STM32F407	168	MPU6050	MPU6050	AK8975	MS5611

Note a: uncertain. Because Multiwii and Aeroquad support dynamic hardware configuration, their sizes are related with construction. b: The flight control board is developed based on Arduino, so the processor can be changed. Most data is cited from [14]

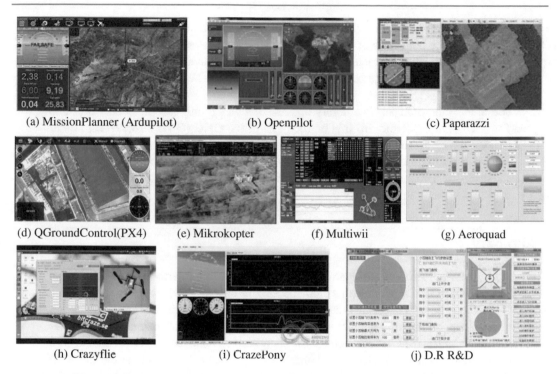

(a) MissionPlanner (Ardupilot)   (b) Openpilot   (c) Paparazzi

(d) QGroundControl(PX4)   (e) Mikrokopter   (f) Multiwii   (g) Aeroquad

(h) Crazyflie   (i) CrazePony   (j) D.R R&D

Fig. 2.23 Screenshots of some GCS softwares

of forward error correction and balanced soft decision. In contrast to analog radio telemetry which is made up of analog Frequency Modulation (FM) station and modem, digital radio telemetry provides transparent RS232 interface whose transmission rate is 19.2 Kbps. It is able to send and receive data in less than 10 ms and shows some parameters such as field intensity, temperature, voltage, state error statistics, alarm, and network management. As a medium of communication, radio telemetry has specific area of applications. In some special conditions, it can provide real-time and reliable data transmission for the monitoring signals in private network. Technology of radio telemetry is suitable for geographical environment which is scattered and complex as it features low cost, easy installation and maintenance, strong diffraction capability, flexible network structure, far coverage, etc. One end of radio telemetry is connected to the GCS software, and the other end is connected to the multicopter. Communication is performed using certain protocols to maintain the two-way communication of a multicopter and the corresponding GCS.

2.4.4.2 Parameter
(1) Frequency.
(2) Transmission distance.
(3) Transmission rate.

More parameter information on data links can be found in [15, pp. 191–246].

2.4.4.3 Communication Protocol
Communication protocol is also called *communication regulations*, referring to as the convention of the data transmission on both sides. The convention includes uniform rules of data format, synchronous

method, transmission rate, procedure, error checking, and correct on and definition of control characters, which should be recorded by both sides of communication. It is also called *link control regulations*. The formulation of communication protocol is advantageous to the separation of GCS and autopilot. As long as communication protocols are obeyed, the GCS software can be compatible with different autopilots.

MAVLink communication protocol is a library organization which only has the header files and it is designed for micro and small aircraft. It is on the basis of the GNU (GNU's Not Unix) Lesser General Public License (LGPL) and MAVLink can efficiently encapsulate C-data structure through a serial port, and send the data packet to the GCS. This communication protocol is widely tested by PX4, APM, and Parrot AR. Drone. There are other protocols. For example, Openpilot autopilot adopts UAVTalk protocol to communicate with GCS.

2.5   Summary

As the Chinese idiom says, "small as the sparrow is, it possesses all its internal organs (small, but complete!)." This proverb is also applicable to a multicopter. This chapter introduces each component of a multicopter mainly in the aspects of function and key parameters. The composition of a multicopter introduced above can support the fully-autonomous flight of a multicopter. For the multicopter under the Semi-Autonomous Control (SAC) mode, some components can be removed, such as GCS and GPS receiver. In order to choose better components and improve the performance or find out the causes of failure, it is necessary to have a deep and comprehensive understanding of multicopters. For example, by adjusting the ESC parameters and choosing appropriate propellers, the flight performance can be improved; by considering the compatibility of ESCs and motors in advance, some crashes can be avoided.

Exercises

2.1 Find a multicopter with the airframe, propeller, ESC, motor, battery, and then explain the meaning of the key parameters of each component in detail.

2.2 Give a method of checking propeller dynamic balance.

2.3 Explain the principle of ESC with sinusoidal driver.

2.4 Choose and compare two autopilots, then clarify their advantages and disadvantages.

2.5 In order to combine the advantage of airplanes and multicopters, there are some kinds of aircraft able to take off and land vertically being available online. Find out a product and analyze its flight principle, advantages and disadvantages.

References

1. Frame materials. http://aeroquad.com/showwiki.php?title=Frame-Materials. Accessed 7 Apr 2016
2. Hrishikeshavan V, Black J, Chopra I (2014) Design and performance of a quad-shrouded rotor micro air vehicle. J Aircr 51:779–791

3. Harrington AM (2011) Optimal propulsion system design for a micro quad rotor. Dissertation, University of Maryland

4. RC airplane propellers. http://www.rc-airplanes-simplified.com/rc-airplane-propellers.html. Accessed 29 Jan 2016

5. MacCamhaoil M (2012) Static and dynamic balancing of rigid rotors. Bruel & Kjaer application notes, BO: 0276–12

6. Wijerathne C (2015) Propeller balancing. http://okigihan.blogspot.com/p/propellerbalancing-propeller-unbalance.html. Accessed 29 Jan 2016

7. Laser balancing props. http://flitetest.com/articles/Laser_Balancing_Props. Accessed 29 Jan 2016

8. Chapman SJ (2005) Electric machinery fundamentals, 4th edn. McGraw-Hill Higher Education, Boston

9. Bertoluzzo M, Buja G, Keshri RK et al (2015) Sinusoidal versus square-wave current supply of PM brushless DC drives: a convenience analysis. IEEE Trans Ind Electron 62:7339–7349

10. Büchi R (2014) Radio control with 2.4 GHz. BoD–Books on Demand

11. Norris D (2014) Build your own quadcopter. McGraw-Hill Education, New York

12. Rother P (2000) PCM or PPM? Possibilities, performance? http://www.aerodesign.de/peter/2000/PCM/PCM_PPM_eng.html. Accessed 29 Jan 2016

13. RC transmitter modes for airplanes. http://www.rc-airplane-world.com/rc-transmitter-modes.html. Accessed 29 Jan 2016

14. Lim H, Park J, Lee D et al (2012) Build your own quadrotor: open-source projects on unmanned aerial vehicles. IEEE Robot Autom Mag 19:33–45

15. Fahlstrom P, Gleason T (2012) Introduction to UAV systems, 4th edn. Wiley, UK

Airframe Design

3

When given a multicopter, its configuration and structure come into our sight first, including the shape and size of the airframe, the choice of motors and propellers, and the distribution of the battery and the payload. All these designs not only influence the multicopter performance, but also help customers distinguish between different brands of multicopters. Compared with the configuration design, structural design to reduce vibration and noise is more important, especially for multicopters flying over residential areas. Considerable vibration will degrade the normal operation of a multicopter, consume more energy, accelerate component aging, and reduce the quality of aerial photography. Similarly, considerable noise will affect the nearby residents' life and make the multicopters not suitable for detective works. This chapter aims to answer the question below:

What should be taken into consideration when designing the airframe of a multicopter?

The answer to this question mainly involves three aspects including configuration design, anti-vibration design, and noise reduction.

© Springer Nature Singapore Pte Ltd. 2017
Q. Quan, *Introduction to Multicopter Design and Control*, DOI 10.1007/978-981-10-3382-7_3

Zhaozhou Bridge

The ancient Chinese had already realized that ingenious structure design could enhance the reliability. The Anji Bridge, also known as Zhaozhou Bridge, constructed in the years 605–618 during the Sui dynasty (581–618), was designed by a craftsman named Chun Li. This bridge has a long history of 1400 years. This magnificent bridge is about 50 m long with a central span of 37 m. It stands 7.3 m tall and has a width of 9 m. The whole bridge is made of limestone slabs. Instead of piers, this bridge has a central arch and two small side arches on each side of the main arch. Normally, the river can pass through the main arch. When the bridge is submerged in a flood, these side arches allow water to pass through. This kind of design is a great innovation in the bridge construction history. This structure can not only reduce the flood impact, but also save materials, thereby reducing the total weight of the bridge.

3.1 Configuration Design

The configuration design of a multicopter includes overall design of an airframe and layout of each components.

3.1.1 Airframe Configuration

3.1.1.1 Fuselage Configuration

(1) *Cross configuration*
Take quadcopters as an example. A quadcopter uses four propellers to produce thrust. Two booms cross at the center, and the propellers are symmetrically distributed at the end of four arms. The common structures are shown in Fig. 3.1. The space in the middle of the fuselage can be used to place the autopilot and other peripheral equipment. The cross configuration can be further divided into two types as shown in Fig. 3.1. Compared with plus-configuration quadcopters, X-configuration quadcopters have higher maneuverability (more rotors involved in pitch and roll control) and less occlusion of the forward field of view, so the X-configuration quadcopters are much more popular.

(2) *Ring configuration*
The ring configuration is like that shown in Fig. 3.2. The ring configuration can be treated as a whole, which is more rigid than the traditional cross fuselage and more helpful for reducing the vibration generated by motors and propellers. However, the cost is that its fuselage is heavier than the cross configuration, which may reduce the maneuverability in some degree.

Fig. 3.1 Cross configuration

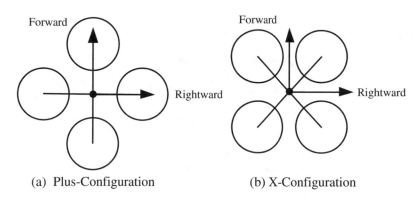

(a) Plus-Configuration (b) X-Configuration

Fig. 3.2 Ring configuration

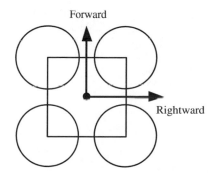

3.1.1.2   Motor-Propeller Mounted

(1) *Common form and co-axis form with two propellers*
Usually, one motor and one propeller are attached at the end of one arm, which is called the common form, as shown in Fig. 3.3a. But sometimes, in order to increase the loading capacity (by increasing the number of propellers) without increasing the size of the multicopter, designers mount two propellers at the end of one arm (one at the top and the other at the bottom), which is called the co-axis form, as shown in Fig. 3.3b. However, due to the interference in air flow, adopting this form will reduce the efficiency of the single propeller. The two propellers of the co-axis form are approximately equivalent to 1.6 propellers as the common form. In order to improve efficiency, it is necessary to further optimize the co-axis system. When combining with different motors and propellers, the co-axis efficiency can be improved to some extent.

Experiments have also indicated that the space between two propellers will affect the efficiency of the co-axis propeller system as well. According to the reference [1, p. 192], it is recommended that $h/r_p > 0.357$, where the definitions of h and r_p are shown in Fig. 3.4. In the mechanical design, for simplicity, two motors are connected together forming a co-axis system.

(2) *Angle of propeller disc*
Most propeller discs are horizontally assembled when the multicopter is placed on a levelled surface, because this kind of design is simple and easy to control. In this design, the thrust produced by the propellers is always perpendicular to the fuselage plane. As a result, the multicopter has to change its pitch angle to achieve a forward flight. At this moment, if a forward-looking camera is mounted with the multicopter, then, in order to make the optical axis of the camera point horizontal, the camera stabilizer needs to rotate as shown in Fig. 3.5a. On the other hand, there are some multicopters with non-horizontal propeller discs, forming a nonzero angle with the levelled surface when placed on it.

Fig. 3.3 Common form and co-axis form with two propellers

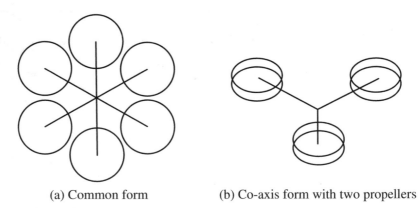

(a) Common form   (b) Co-axis form with two propellers

Fig. 3.4 A simple connection and its geometry parameters of a co-axis form with two propellers

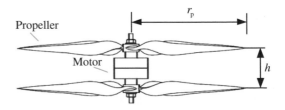

Fig. 3.5 Multicopters with horizontal disc and non-horizontal disc in forward flight

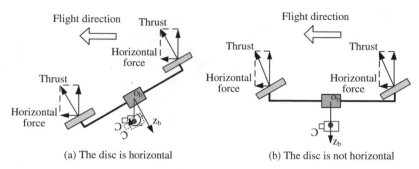

(a) The disc is horizontal   (b) The disc is not horizontal

This kind of design is beneficial because the produced thrust has components along the three body axes and there is no need to change the pitch of the multicopter in a forward flight, as shown in Fig. 3.5b. As a result, camera stabilizers can be removed. However, as a result, such a structure should have six propellers at least because there are six independent control variables. A multicopter with such a structure is shown in Fig. 3.6. Some researchers also designed the multicopter with changeable propeller discs to achieve flexible control [2].

(3) *Motor facing downward and upward*
For multicopters, there exist two ways installing motors: facing downward and facing upward. Propellers of a multicopter with motors facing downward produce push force, while propellers with motors facing upward produce pull force. Figure 3.7a shows a kind of multicopter with motors facing downward, advantages of which include:

1) Producing complete downwash, less interference from the fuselage.
2) Protecting motors against rain.
3) Ensuring a more accurate measurement of the barometer for that the downwash of air is lower than the position where the barometer places.

Figure 3.7b–d shows some other kinds of multicopters with motors facing upward, advantages of which include:

(1) Protecting propellers against crash in landing phase.
(2) Giving a wider view of the camera.

3.1.1.3   Propeller Radius and Airframe Radius
The size of a multicopter is closely related to the radius of the propeller. Assuming that the angle between two arms is θ, then for a Y6-configuration hexacopter (Fig. 3.8a), $\theta = 120°$; for a traditional quadcopter (Fig. 3.8b), $\theta = 90°$; for a traditional hexacopter (Fig. 3.8c), $\theta = 60°$. According to Fig. 3.8,

Fig. 3.6 CyPhyLVL1

Fig. 3.7 Multicopters with motor facing downward and upward

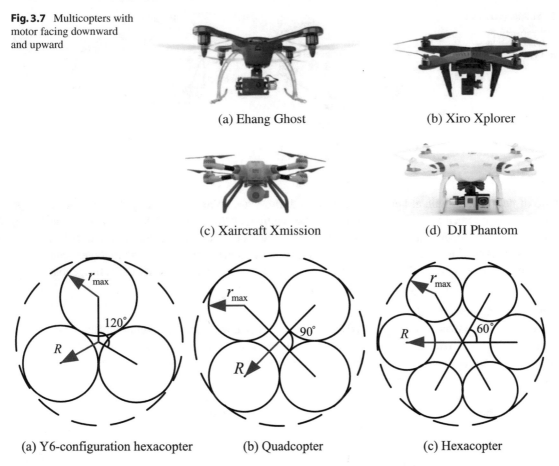

(a) Ehang Ghost

(b) Xiro Xplorer

(c) Xaircraft Xmission

(d) DJI Phantom

(a) Y6-configuration hexacopter

(b) Quadcopter

(c) Hexacopter

Fig. 3.8 Multicopters with different configurations and their geometry parameters

if the number of arms of multicopter is n, there is $\theta = 360°/n$, then the relationship between the airframe radius R and the maximum radius of a propeller r_{max} is

$$R = r_{max}/\sin\frac{\theta}{2} = r_{max}/\sin\frac{180°}{n}. \tag{3.1}$$

In [3], numerical simulation results show that the thrust is decreased because of the aerodynamic interference between propeller flow fields, and the cyclic fluctuation of the thrust appears. However, this influence is negligible when the rotors are not too close to each other. Additionally, as shown in [4, p. 81], by using smoke plume to visualize airflow, it can be clearly observed that when the space between propellers varies from $0.1r_{max}$ to r_{max}, each propeller has its independent vortex without affecting propeller performance. Hence, in order to make a multicopter more compact without losing much efficiency by the aerodynamic interference, the following rule of thumb can be considered

$$r_{max} = 1.05r_p \sim 1.2r_p \tag{3.2}$$

where the propeller radius r_p is determined by the weight and the maximum payload of a multicopter. So, once the propeller radius is determined, the airframe radius is determined according to Eq. (3.1).

3.1.1.4 Relationship Between Size and Maneuverability [5]

Maneuverability is the ability to change its own state, which is closely related to the maximum acceleration of a multicopter. Changing the size of a multicopter will change its weight and moment of inertia, and finally determine the maximum acceleration of translation and rotation. In this chapter, the acceleration model in the hover mode is built to analyze the influence of different sizes on the multicopter performance. The propeller thrust T_p and the propeller moment M_p can be expressed as (see Chap. 4 for detail)

$$T_p = (\frac{1}{2\pi})^2 C_T \rho \varpi^2 (2r_p)^4$$
$$M_p = (\frac{1}{2\pi})^2 C_M \rho \varpi^2 (2r_p)^5 \tag{3.3}$$

where C_T represents the thrust coefficient, C_M represents the torque coefficient, ρ is the air density, ϖ is the angular velocity of the propeller. Since as previously analyzed $r_p \sim R$ (here "$\sim$" means "in the same order of magnitude"), according to Eq. (3.3), the total force T and moment M on the multicopter satisfy

$$\begin{cases} T \sim T_p \\ M_{pitch}, M_{roll} \sim T_p \cdot R \\ M_{yaw} \sim M_p \end{cases} \Rightarrow \begin{cases} T \sim \varpi^2 R^4 \\ M \sim \varpi^2 R^5 \end{cases}$$

where $M_{pitch}, M_{roll}, M_{yaw}$ are the moments generating pitch, roll and yaw, respectively. Moreover, for a normal multicopter, the mass m and the moment of inertia J roughly satisfy the following relation with R that

$$m \sim R^3, J \sim R^5.$$

As a result, the acceleration of translation (also called linear acceleration) a is determined by the thrust and the mass acceleration of rotation (also called angular acceleration) α is determined by the moment and the moment of inertia, as follows:

$$a = \frac{T}{m} \sim \frac{\varpi^2 R^4}{R^3} = \varpi^2 R$$
$$\alpha = \frac{M}{J} \sim \frac{\varpi^2 R^5}{R^5} = \varpi^2. \tag{3.4}$$

In the following, to study the relationship between the propeller angular speed and the multicopter size, two widely used methods will be adopted.

(1) *Mach scaling*
According to [5], Mach scaling is used for compressible flows and essentially assumes that the blade tip speed is a constant, so

$$\varpi \sim 1/r_p.$$

According to this assumption, one has

$$a \sim \frac{1}{R}, \quad \alpha \sim \frac{1}{R^2}. \tag{3.5}$$

(2) *Froude scaling*
The Froude number is a dimensionless number defined as the ratio of the flow inertia to the external field [6]. According to [5], Froude scaling is used for incompressible flows and assumes that the Froude number is a constant, so

$$v_b^2 / Rg = \varpi^2 r_p^2 / Rg \sim 1$$

where v_b is the tip speed of propeller, and g is the acceleration of gravity. Since $r_p \sim R$,

$$\varpi \sim 1/\sqrt{r_{\mathrm{p}}}.$$

According to this assumption, one has

$$a \sim 1, \alpha \sim \frac{1}{R}. \tag{3.6}$$

From the analysis above, the body size has little relevance to the acceleration of translation ($a \sim 1$) according to the Froude number method. However, if the body size of a multicopter is small, the acceleration of rotation will be high (both $\alpha \sim \frac{1}{R^2}$ and $\alpha \sim \frac{1}{R}$), and the multicopter will have a high ability to change its attitude. Hence, the movement of micro multicopters is more flexible.

In addition to the principle mentioned above, the size and rigidity of a propeller should also be considered. The increase in propeller size (moment of inertia) will make its response slow, until the multicopter loses its controllability. Because of this, helicopters mainly change the effective pitch of the propeller rather than angular speed to change the thrust. Besides, the more rigid a propeller is, the greater the fatigue the blade-flapping effects bring to the rotor hub. The result is similar as bending a soft iron wire one way then the other repetitively. Therefore, it is important to have flexible propellers with appropriate sizes.

3.1.1.5 Position of the CoG

When designing a multicopter, it is necessary to put the CoG of the multicopter on the central axis, otherwise the extra moment will be caused. Another task is to determine whether the CoG should be above or below the propeller disc plane. In practice, the two structures both exist, as shown in Fig. 3.9. The commonly used way is shown in Fig. 3.9a. However, it is beneficial for upward shooting when placing the camera above the propeller disc, as shown in Fig. 3.9b.

Next, how the position of CoG affects the stability of a multicopter will be analyzed from the perspectives of the forward flight situation and the wind interference situation.

(1) *Forward flight*

As shown in Fig. 3.10a and b, when a multicopter flies forward, the induced airflow will derive the drag force on the propeller. If the CoG position is low (as shown in Fig. 3.10a), the drag torque (generated by the drag force with the CoG as the pivot) will make the pitch angle turn to zero. On the other hand, according to Fig. 3.10b, if the CoG is high, the drag torque will cause the increase in the pitch angle until the multicopter is turned over. Therefore, when a multicopter is in forward flight, the CoG under the propeller disc can make the multicopter stable.

(2) *Wind interference*

In this situation, the multicopters with different position of CoG are shown in Fig. 3.10c and d. When gust blows, the induced airflow will give rise to the drag. If the CoG is high (Fig. 3.10d), the drag torque will make the pitch angle turn to zero. On the other hand, if the CoG of the multicopter is low (Fig. 3.10c), then the drag torque will cause the increase in the pitch angle until the multicopter is turned over. Therefore, a multicopter with high CoG can reject the wind interference.

In the two cases, no matter where the CoG is, the multicopter cannot be completely stable. According to [7, 8], if the position of the CoG is far above the propeller disc, a certain dynamic mode of the multicopter will be very unstable according to the stability analysis, then more control effort has to be made. Under the comprehensive consideration, the multicopter with a low CoG position is better than that with a high CoG in stability. In practice, the CoG should be as close to the center as possible and a little under it if necessary.

Fig. 3.9 Two ways of camera installation of Freefly quadcopter

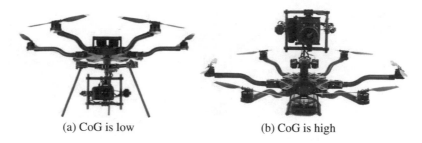

(a) CoG is low   (b) CoG is high

Fig. 3.10 Forces on multicopters

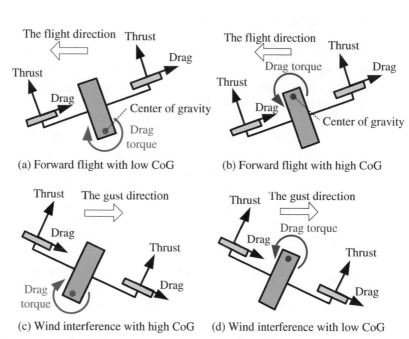

(a) Forward flight with low CoG   (b) Forward flight with high CoG

(c) Wind interference with high CoG   (d) Wind interference with low CoG

3.1.1.6   Autopilot Installation Position

Since there are a lot of sensors in the autopilot, the ideal installation position of the autopilot should coincide with the geometric center of a multicopter. Owing to the centrifugal acceleration and tangential acceleration, if the autopilot is far from the center, it would produce an error in the accelerometer measurement, which is also called *lever-arm effect*.

(1) *Standard installation position*

As shown in Fig. 3.11, there is a white arrow to indicate the forward direction of the autopilot. So, the autopilot is installed with the white arrow pointing toward the direction of forward flight and fixed by using a damping foam. The autopilot should be placed near the CoG of a multicopter with the assumption that the center of geometry and the CoG are coincident. In practice, the installation location is not so strict, but still should be as near as possible to the center of the aircraft and horizontal with motors.

(2) *Substitution* [9]

The substitution of position can be taken into consideration if it is difficult to install in the standard installation position. APM autopilot can be installed in the airframe at specific angle (such as 90°), so that the standard installation position is retrieved by using a corresponding calibration algorithm.

Fig. 3.11 Installation position of an autopilot (photo adapted from http://www.ardupilot.org)

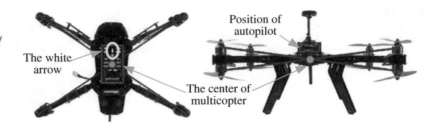

3.1.2 Aerodynamic Configuration

The aerodynamic design aims to reduce the drag of the fuselage of multicopter during flight. Drag (sometimes called air resistance) is a force acting opposite to the relative motion of any object moving with respect to a surrounding air. Drag force can be distinguished as four types: frictional drag, pressure drag, induced drag, and interference drag [10].

(1) Frictional drag. Frictional drag is caused by atmosphere viscidity. The friction will be increased with stronger air stickiness, coarser surface, and larger superficial area.

(2) Pressure drag. Pressure drag is the force caused by the pressure difference during flight. It is related to windward area, the larger the windward area, the greater the pressure drag. Also, the vehicle shape will influence the pressure drag. The three objects in Fig. 3.12 have the same windward area. However, the structure of streamline is subject to the least drag.

(3) Induced drag. The induced drag and the lift are formed simultaneously. If a multicopter flies at a high speed, the lift and the induced drag will be increased.

(4) Interference drag. The drag of a multicopter is not equal to the total drag produced by each part alone. The difference is the interference drag. If the relative position of each component is proper and the joint of components is smooth, the interference drag can be reduced.

In order to reduce the drag during flight and improve the flight efficiency, it is important to consider the aerodynamic design, especially for a multicopter with requirements of a high flight speed or a far flight distance. Overall, for a quadcopter, the pitch angle should be considered to decrease the maximum windward area, and a streamline body should be designed. Furthermore, the relative position of each component should be taken into account. The joint of components and the fuselage surface should

Fig. 3.12 Pressure drag of three objects

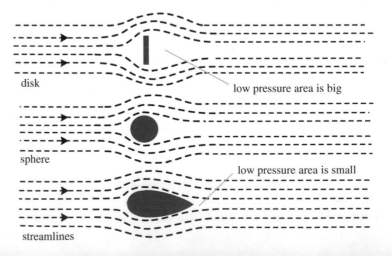

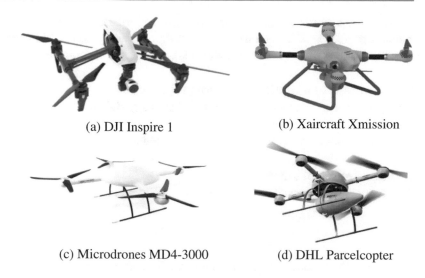

Fig. 3.13 Some commercial quadcopters with certain aerodynamic configuration

(a) DJI Inspire 1

(b) Xaircraft Xmission

(c) Microdrones MD4-3000

(d) DHL Parcelcopter

be as smooth as possible. In practice, the drag coefficient of a multicopter can be calculated through Computational Fluid Dynamics (CFD) software, and the aerodynamic configuration can be optimized repeatedly and over until the performance is satisfied. Some commercial quadcopters with certain aerodynamic configuration are shown in Fig. 3.13.

3.2 Structural Design

3.2.1 Design Principles of Airframe

(1) The structure of airframe is designed to support limited payloads without detrimental permanent deformation. At any payload up to limit payloads, the deformation may not interfere with safe operation. Moreover, the structure is able to support ultimate payloads[1] without failure.
(2) After other necessary requirements have been achieved, the total weight is as light as possible.
(3) The proportion of the length, the width, and the height of a multicopter is proper; and the structural layout is suitable.
(4) Esthetic and durable.

3.2.2 Anti-Vibration Consideration

3.2.2.1 Significance of Anti-Vibration
Accelerometers are used for the estimation of the position and the attitude, which is of great importance. However, it is very sensitive to vibration. Concretely,

(1) Acceleration is directly related to the estimation of attitude angles.
(2) The data of the acceleration is fused with the barometer and the GPS (Global Positioning System) to estimate the multicopter position.

---

[1]The ultimate payload is the limit payload multiplied by prescribed factors of safety. For example, a factor of safety of 1.5 can be used.

For such a purpose, prefilters (anti-aliasing filters) are inserted before acceleration signal sampling. Another important role of anti-vibration is to improve the quality of pictures. With perfect anti-vibration design, camera stabilizers can be removed, which is necessary for the miniaturization of multicopters.

3.2.2.2 Constraint of Vibration Strength [11]

(1) Generally, the lateral vibration strength is less than 0.3g, while the longitudinal vibration strength is less than 0.5g. (Note: g represents the acceleration of gravity here).
(2) In practice, the vibration strength along all axes is better bounded less than 0.1g.

3.2.2.3 Main Source of Vibration [9]

The body vibration mainly results from the airframe deformation and the asymmetry of motors and propellers.

(1) *Airframe*

  1) The airframe deformation especially the arm deformation will result in the asynchronous vibration. So the arms should be rigid.
  2) Common carbon fiber airframes are excellent in anti-torsion and anti-bending.
  3) Compared with others, an airframe made of aluminum is heavier but more rigid.

  Recently, the concept of wearable quadcopters has received more and more attention. One conceptual wearable quadcopter is shown in Fig. 3.14. However, the wearable quadcopter requires that its airframe be flexible and easily transformed. These problems should be considered in the design. In this case, the airframe can be blended unidirectionally, opposite to the direction of the thrust and moments. Moreover, the airframe is anti-torsion and anti-bending in the direction of the thrust.

(2) *Motor*

  1) Motors work smoothly.
  2) The rotor holder should be coaxial with the bearing and the center of the propeller, thus getting rid of the eccentric force produced when motor is rotating.
  3) All rotors are in balance.

(3) *Propeller*

  1) Propellers are in balance (refer to Sect. 2.3.1).
  2) Propellers should be matched with the airframe size and the multicopter weight, and there are similar toughness[2] when rotating both clockwise and counterclockwise.

Fig. 3.14 Nixie conceptual wearable quadcopter

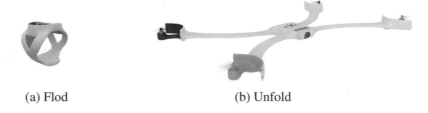

(a) Flod   (b) Unfold

---

[2]The ability of a material to absorb energy and plastically deform without fracturing.

3) Carbon fiber propellers are highly rigid, but the hidden danger to personnel exists when propellers are in rotation.

4) The low-speed large propeller is more efficient than the high-speed small one, but the large propeller will cause more vibration in amplitude.

After all factors above have been taken into account, the left problem is to consider methods for vibration damping and isolating.

3.2.2.4 Vibration Damping Between Autopilot and Airframe

(1) Traditionally, the double-side foam type and the nylon button are used to fix the autopilot.

(2) In many cases, as the autopilot has small mass, the double-side foam type and the nylon button are less effective. The tested feasible methods are to use the foam, the gal pad, the O-ring, and the earplug [11], as shown in Fig. 3.15.

At present, there also exist dampers specially for autopilots in market, shown in Fig. 3.16. It is made of two glass fiber frames, four damping balls, and two foam pads.

Fig. 3.15 Anti-vibration methods (photo adapted from http://www.ardupilot. org)

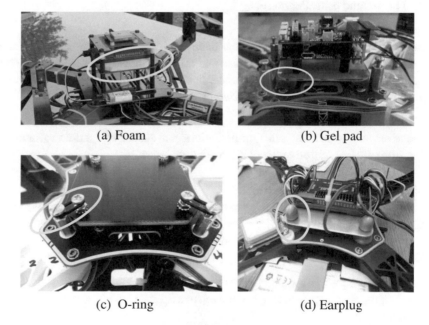

(a) Foam   (b) Gel pad

(c) O-ring   (d) Earplug

Fig. 3.16 An anti-vibration damper

3.2.3 Noise Reduction Consideration

The noise of the multicopter is mainly caused by the high-speed rotation. Here, by referring to [12], the sound generation principle and noise-reducing methods are introduced.

3.2.3.1 Main Harm of Propeller Noise

(1) If the body exists in the sound field produced by propellers, some sensors may become invalid because of the influence of noise.
(2) Noise will affect the surrounding flight environment, and cause pollution, especially when multi-copters are over residential areas.
(3) The body vibration and the acoustic fatigue induced by the noise may seriously influence the safety of aircraft.

3.2.3.2 Propeller Sound Generating Principle

For propellers and all kinds of turbo machines, they all generate sound by rotating blades. The high-speed rotating propeller will cause the high-frequency unsteady flow of the surrounding air, and then, the energy will be released out in the form of noise. Noise of propellers can be classified into tone noise and broadband propeller noise.

(1) *Tone noise*
Tone noise is caused by sources that repeat themselves exactly during each rotation of the propeller, which is further divided into loading noise and thickness noise. The loading noise is caused by the unsteady pressure (load) on the blade surface. In addition to unsteady loading noise, there is a contribution from the blade motion that is referred to as thickness noise.

(2) *Broadband rotor noise*
Broadband rotor noise is a random, non-periodic, signal caused by random variations in blade loading resulting from the interaction of the blades with turbulence.

3.2.3.3 Noise Reduction Method

To reduce the propeller noise, designers have to focus on reducing the blade tip speed, thickness, blade loading, unsteady flow, broadband, vibration, or other elements. Parameters including propeller pitch, propeller diameter, blade thickness, blade count, blade shape, and airfoil section are important for noise reduction of propellers [13, pp. 28–36].

(1) *Propeller pitch*. For the same thrust requirement, reasonably increasing the blade sweep may decrease the propeller angular speed and the relative tip speed, and therefore reduce the noise radiated by propeller.

(2) *Propeller diameter*. For the same thrust requirement, increasing the diameter of propeller may significantly decrease the revolving speed (according to Eq. (3.2)) and then reduce the noise in some degree. But from the compact design aspect, the propeller diameter is generally expected to be as small as possible.

(3) *Blade thickness*. Thickness noise is important at high tip Mach numbers. Reducing the total volume of blades can reduce the relative thickness and length of propeller blade profile. Then, thickness noise can be reduced.

(4) *Blade count*. For the same thrust requirement, increasing the number of propeller blades can decrease the revolving speed of propeller and then reduce the noise. In practice, a propeller with odd number of blades can break the symmetry and reduce the possibility of sympathetic vibration. This makes it quieter than even blade propeller. For instance, according to experimental results, three-blade

propellers are generally more stable and quieter than two-blade propellers for multicopters. However, along with the increase in blade count, the cost and the producible difficulty may be increased rapidly, and a poor quality propeller without guarantee of static and dynamic balance will produce even more noise.

(5) *Blade shape* and *airfoil section*. The sound power of blade tips is highest according to the diametric distribution of sound power. It is possible to reduce noise by designing a blade to move the peak of aerodynamic loads from span distribution to internal diameter direction. In general, these parameters have a much larger effect on the aerodynamic performance of the propeller than they have on the noise.

Except for propeller design shown above, there are also other noise reduction methods. For example, an audio sensor can detect the noise generated by one propeller. By feedback, another propeller can produce anti-noise sound that cancels the noise generated by the first one [14].

3.3   Summary

In general, after considering the basic design principles of airframe in Sect. 3.1.1, the design requirement should be further taken into consideration. For example, the hexacopter CyPhy LVL1 is quite distinctive for its non-horizontal propeller discs. Since the vibration is derived from the airframe transformation, and the asymmetry of motors and propellers, the airframe should be chosen as rigid as possible, and also high-quality motors and propellers should be chosen. In order to reduce noise, effective methods should be implemented. These can be achieved eventually by designing new propellers. In this chapter, some design principles without specific design methods are simply shown. The problems of how to design a multicopter minimizing the drag, vibration, and noise are deserved to research in the future.

Exercises

3.1 Show the reasons why X-configuration performs better than plus-configuration in maneuver.

3.2 In aircraft design, the Mach number and Froude number are always considered to be constant. Clarify the reasons.

3.3 Assume that the induced drag on propeller can be ignored. Based on that, build a simple model for the 2D multicopter as shown in Fig. 3.10 when it is hovering and analyze the influence on characteristic roots of open loop when CoG is high and low.

3.4 After inserting prefilters, the ability of anti-vibration can be improved. Clarify the reasons.

3.5 Except for the methods of anti-vibration introduced in this chapter, give some other methods.

3.6 Except for the methods of noise reduction introduced in this chapter, give some other methods.

References

1. Bohorquez F (2007) Rotor hover performance and system design of an efficient coaxial rotary wing micro air vehicle. Dissertation, University of Maryland College Park
2. Ryll M, Bulthoff HH, Giordano PR (2015) A novel overactuated quadrotor unmanned aerial vehicle: modeling, control, and experimental validation. IEEE Trans Control Syst Technol 23(2):540–556

3. Hwang JY, Jung MK, Kwon OJ (2014) Numerical study of aerodynamic performance of a multirotor unmanned-aerial-vehicle configuration. J Aircr 52(3):839–846
4. Harrington AM (2011) Optimal propulsion system design for a micro quadrotor. Dissertation, University of Maryland College Park
5. Mahony R, Kumar V, Corke P (2012) Multirotor aerial vehicles: modeling, estimation, and control of quadrotor. IEEE Robot Autom Mag 19(3):20–32
6. White FM (1999) Fluid mechanics. McGraw-Hill, New York
7. Bristeau PJ, Martin P, Salaun E et al (2009) The role of propeller aerodynamics in the model of a quadrotor UAV. In: European control conference (ECC), Budapest, August, pp 683–688
8. Pounds P, Mahony R, Corke P (2010) Modelling and control of a large quadrotor robot. Control Eng Pract 18(7):691–699
9. Mounting the flight controller. http://ardupilot.org/copter/docs/common-mounting-the-flight-controller.html. Accessed 10 Oct 2016
10. Thomas ASW (1984) Aircraft drag reduction technology. Lockheed-Georgia Co Marietta Flight Sciences Div
11. Vibration Damping. http://ardupilot.org/copter/docs/common-vibration-damping.html. Accessed 10 Oct 2016
12. Marte JE, Kurtz DW (1970) A review of aerodynamic noise from propellers, rotors and lift fans. Technical report 32–1462, Jet Propulsion Laboratory, California Institute of Technology, USA
13. Harper-Bourne M (1972) Model turbojet exhaust noise part 2. contract report no CR72/31. Institute of sound and vibration research, University of Southampton, Southampton
14. Beckman BC (2016) Vehicle noise control and communication. US patent 20,160,083,073

Modeling and Evaluation of Propulsion System

4

To design a multicopter, first of all, a designer has to select proper components to assemble a multicopter to meet the performance requirements, such as hover endurance, system efficiency, maximum payload, maximum pitch angle, and maximum flight distance. Multicopter performance is mainly determined by the chosen propulsion systems, consisting of propellers, brushless Direct Current (DC) motors, Electronic Speed Controllers (ESCs), and the batteries. Different components will lead to different flight performance in a very complex way. For instance, it seems that increasing the capacity of the battery can increase endurance, but it also increases the weight and power, thus the endurance of flight would probably even be decreased sometimes. As far as we know, in practice, many designers used to evaluate the performance of a multicopter through their experience or repeated experiments, which is an inefficient and costly process. Moreover, the optimal choice of components related to the performance requirements remains a problem, which is hard to solve by experiments and experience [1, 2]. This chapter aims to answer the question below:

How is the flight performance of a multicopter evaluated?

To answer this question, a practical modeling method is proposed for the propulsion system of multicopters to evaluate a series of performance indices. For a practical purpose, only technical specifications of components offered by manufacturers are required as the input to the proposed models. Testing examples are finally given to demonstrate the effectiveness of the proposed method. Furthermore, a Web site http://www.flyeval.com is established which can provide users with the performance evaluation mentioned in this chapter.

© Springer Nature Singapore Pte Ltd. 2017
Q. Quan, *Introduction to Multicopter Design and Control*, DOI 10.1007/978-981-10-3382-7_4

Chinese Noria

The ancient Chinese had known how to utilize the natural sources of energy since long. The Chinese noria, for example, was recorded to be introduced into the society during the Tang dynasty. The Chinese noria consists of wheel-shaped water lifts, which were called water wheels. Water wheel is a kind of water lift installed at the riverside, with water containers, which would change direction and turn downward when emptied and would be lifted up when filled with water. This amazing machine has aroused the enthusiasm among lots of writers and poems. Enormous poetries and books have mentioned this great work, such as *"Exploitation of the Works of Nature"*.

4.1   Problem Formulation

There are two major tasks in this chapter: propulsion system modeling and performance evaluation. The propulsion system modeling is divided into four parts: propeller modeling, motor modeling, ESC modeling, and battery modeling. To ensure commonality of the method, all input parameters of these models should be the basic parameters that can be easily found on the product descriptions, as shown in Table 4.1. The outputs are a series of performance indices. For simplicity, they are formulated into the following four problems.

Problem 1 In the hover mode,[1] according to the known parameters, estimate the hover endurance T_{hover}, the throttle command σ, the ESC input current I_e, the ESC input voltage U_e, the battery current I_b, and the motor speed N.

Problem 2 In the maximum throttle mode,[2] according to the known parameters, estimate the ESC input current I_e, the ESC input voltage U_e, the battery current I_b, the motor speed N, and the system efficiency[3] η.

Problem 3 In the forward flight mode, according to the known parameters, estimate the maximum load G_{maxload} and the maximum pitch θ_{max}.

Problem 4 In the forward flight mode, according to the known parameters, estimate the maximum flight speed V_{max}, and the maximum flight distance Z_{max}.

The modeling procedure is shown in Fig. 4.1.

4.2   Propulsion System Modeling

In the section, models for the propeller, the motor, the ESC, and the battery are established, respectively.

4.2.1   Propeller Modeling

For a multicopter, fixed-pitch propellers are often used. Propeller performance depends on its thrust T (unit: N) and torque M (unit: N·m). Referring to [3, 4], they are expressed as

Table 4.1 Propulsion system parameters

Component	Parameters
Propeller	$\Theta_p = \{$Diameter D_p, Pitch H_p, Blade Number B_p, Propeller Weight $G_p\}$
Motor	$\Theta_m = \{$KV Value K_{V0}, Maximum Continuous Current I_{mMax}, Nominal No-load Current I_{m0}, Nominal No-load Voltage U_{m0}, Resistance R_m, Motor Weight $G_m\}$
ESC	$\Theta_e = \{$Maximum Current I_{eMax}, Resistance R_e, ESC Weight $G_e\}$
Battery	$\Theta_b = \{$Capacity C_b, Resistance R_b, Total Voltage U_b, Maximum Discharge Rate K_b, Battery Weight $G_b\}$

---

[1]The multicopter stays fixed in the air, and relatively static to the ground.

[2]The maximum throttle case is an extreme case of a multicopter, in which motors are at full throttle state and the propellers have the maximum thrust.

[3]Here, system efficiency is the ratio of the propeller output power and the battery power at full throttle state.

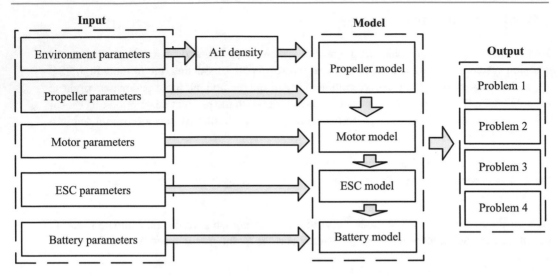

Fig. 4.1 Modeling procedure

$$T = C_T \rho \left(\frac{N}{60} \right)^2 D_p^4 \tag{4.1}$$

$$M = C_M \rho \left(\frac{N}{60} \right)^2 D_p^5 \tag{4.2}$$

where N (unit: RPM) denotes the propeller speed, D_p (unit: m) denotes the diameter of the propeller, C_T and C_M are the dimensionless thrust coefficient and torque coefficient, respectively. Here, the air density ρ (unit: kg/m^3), which varies with respect to the local altitude h (unit: m) and the temperature T_t (unit: °C), is written as

$$\rho = \frac{273 P_a}{101325(273 + T_t)} \rho_0 \tag{4.3}$$

where the standard air density $\rho_0 = 1.293$ (kg/m^3) (0 °C, 273 K). The atmospheric pressure P_a (unit: Pa) is obtained as [5, p. 5]

$$P_a = 101325(1 - 0.0065 \frac{h}{273 + T_t})^{5.2561}. \tag{4.4}$$

The height of a multicopter varies slightly while performing a task. So, h and T_t are treated as constants in the following performance evaluation. The remaining task in propeller modeling is to find the parameters C_T and C_M in Eqs. (4.1) and (4.2). To make things clearer, they are expressed as

$$\begin{aligned} C_T &= f_{C_T}(\Theta_p) \\ C_M &= f_{C_M}(\Theta_p) \end{aligned} \tag{4.5}$$

where Θ_p is the parameter set as in Table 4.1. The detailed procedure to obtain $f_{C_T}(\Theta_p)$ and $f_{C_M}(\Theta_p)$ is shown in Appendix 4.6.1, given as

$$\begin{aligned} C_T &= f_{C_T}(\Theta_p) \triangleq 0.25\pi^3 \lambda \zeta^2 B_p K_0 \frac{\varepsilon \arctan \frac{H_p}{\pi D_p} - \alpha_0}{\pi A + K_0} \\ C_M &= f_{C_M}(\Theta_p) \triangleq \frac{1}{8A}\pi^2 C_d \zeta^2 \lambda B_p^2 \end{aligned} \tag{4.6}$$

where

$$C_{\mathrm{d}} = C_{\mathrm{fd}} + \frac{\pi A K_0^2}{e} \frac{\left(\varepsilon \arctan \frac{H_{\mathrm{p}}}{\pi D_{\mathrm{p}}} - \alpha_0 \right)^2}{(\pi A + K_0)^2}. \tag{4.7}$$

The parameters $A, \varepsilon, \lambda, \zeta, e, C_{\mathrm{fd}}, \alpha_0$, which are not reflected in Table 4.1, are explained in Appendix 4.6.1. According to experimental results and the relevant literature [6, p. 150, p. 151, p. 174], [7, p. 62], [8, p.43], [9], it is suggested that the range of values of these parameters in Eq. (4.6) be taken as

$$\begin{aligned}
&A = 5 - 8, \varepsilon = 0.85 - 0.95, \lambda = 0.7 - 0.9 \\
&\zeta = 0.4 - 0.7, e = 0.7 - 0.9, C_{\mathrm{fd}} = 0.015 \\
&\alpha_0 = -\tfrac{\pi}{36} - 0, K_0 = 6.11.
\end{aligned} \tag{4.8}$$

Note that the above parameters may vary with the difference of types, material, and technology of propellers.

To verify the proposed modeling method, the experimental data provided at the Web site of APC propellers[4] is used for comparison. The parameters are chosen as $A = 5, \varepsilon = 0.85, \lambda = 0.75, \zeta = 0.55, e = 0.83, C_{\mathrm{fd}} = 0.015, \alpha_0 = 0, K_0 = 6.11$. The other parameters for propellers in Table 4.1 are used directly from the model parameters given by APC. The results are shown in Fig. 4.2. It can be clearly observed that the results of the theoretical model match well with the experimental data. After comprehensive consideration of various propellers from different manufacturers, a set of mean parameters $A = 5, \varepsilon = 0.85, \lambda = 0.75, \zeta = 0.5, e = 0.83, C_{\mathrm{fd}} = 0.015, \alpha_0 = 0, K_0 = 6.11$ is chosen to approximately evaluate the common propellers on the market. It should be noticed that $C_{\mathrm{T}}, C_{\mathrm{M}}$ can be obtained by experimental data directly. In this case, the method mentioned above not needed, and the experimental data $C_{\mathrm{T}}, C_{\mathrm{M}}$ can be used for performance evaluation directly.

4.2.2 Motor Modeling

Nowadays, the electric motors used in multicopters are often brushless DC motors. A brushless DC motor can be modeled as a permanent magnet DC motor [10]. Its equivalent circuit is shown in Fig. 4.3, where U_{m} (unit: V) is the equivalent motor voltage, I_{m} (unit: A) is the equivalent motor current, E_{a} (unit: V) is the back-electromotive force, R_{m} (unit: Ω) is the armature resistance, and L_{m} (unit: H) is the armature inductance of the motor. $\hat{I}_0$ (unit: A) is the no-load current required to overcome mechanical friction and air friction in the motor, as well as magnetic hysteresis and eddy current losses in the motor, and is approximately a constant at a certain motor speed [11]. And $I_{\mathrm{a}} = I_{\mathrm{m}} - \hat{I}_0$ helps produce the electromagnetic torque. Here, the armature inductance L_{m} and the transient process caused by switching elements are ignored.

The motor modeling aims to obtain U_{m} (unit: V) and I_{m} (unit: A) from the motor speed N (equal to the propeller speed), the motor load torque M (equal to the propeller torque), and the motor parameter class Θ_{m}. For clarity, they are expressed in functions as

$$\begin{aligned}
U_{\mathrm{m}} &= f_{U_{\mathrm{m}}} \left(\Theta_{\mathrm{m}}, M, N \right) \\
I_{\mathrm{m}} &= f_{I_{\mathrm{m}}} \left(\Theta_{\mathrm{m}}, M, N \right)
\end{aligned} \tag{4.9}$$

where N and M are determined according to Eqs. (4.1) and (4.2) in the above section, and Θ_{m} is shown in Table 4.1. The detailed procedure to obtain $f_{U_{\mathrm{m}}} \left(\Theta_{\mathrm{m}}, M, N \right)$ and $f_{I_{\mathrm{m}}} \left(\Theta_{\mathrm{m}}, M, N \right)$ is shown

---

[4]The data is from https://www.apcprop.com/.

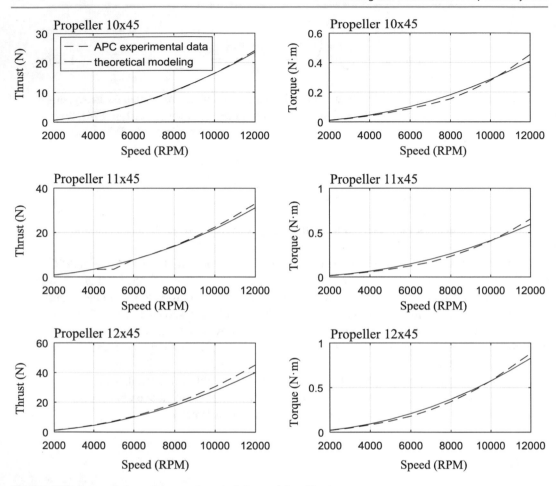

Fig. 4.2 Lift characteristics and torque characteristics model verification

Fig. 4.3 Equivalent motor
model

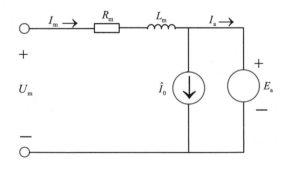

in Appendix 4.6.2, given as

$$
\begin{aligned}
U_m = f_{U_m}\left(\Theta_m, M, N\right) &\triangleq \left(\frac{MK_{V0}U_{m0}}{9.55(U_{m0}-I_{m0}R_m)}+I_{m0}\right)R_m + \frac{U_{m0}-I_{m0}R_m}{K_{V0}U_{m0}}N \\
I_m = f_{I_m}\left(\Theta_m, M, N\right) &\triangleq \frac{MK_{V0}U_{m0}}{9.55(U_{m0}-I_{m0}R_m)}+I_{m0}.
\end{aligned}
\tag{4.10}
$$

4.2.3   Electronic Speed Controller Modeling

An Electronic Speed Controller (ESC) is a device external to the motor that electronically performs the commutation achieved mechanically in brushed motors. The ESC converts the DC voltage of the battery to a three-phase alternating signal which is synchronized with the rotation of the rotor and applied to armature windings. ESCs regulate motor speed within a range depending on the motor load torque and battery voltage. An ESC equivalent circuit is shown in Fig. 4.4.

Given the motor modeling results U_m and I_m, the ESC parameter set Θ_e and the battery parameter set Θ_b (see Table 4.1 for detail), the ESC modeling aims to obtain the input throttle command σ (scale value, range from 0 to 1, no unit), the input voltage U_e (unit: V), and the input current I_e (unit: A). For clarity, they are expressed in functions as

$$
\begin{aligned}
\sigma &= f_\sigma \left(\Theta_e, U_m, I_m, U_b \right) \\
I_e &= f_{I_e} \left(\sigma, I_m \right) \\
U_e &= f_{U_e} \left(\Theta_b, I_e \right).
\end{aligned}
\tag{4.11}
$$

As shown in Fig. 4.4, U_{eo} is the equivalent DC voltage, given by

$$
U_{eo} = U_m + I_m R_e.
\tag{4.12}
$$

Based on Eq. (4.12), the throttle command σ is obtained as [12]

$$
\sigma = \frac{U_{eo}}{U_e} \approx \frac{U_{eo}}{U_b}.
\tag{4.13}
$$

Furthermore, since the input power of ESC is equal to the output power, the ESC input current is [12]

$$
I_e = \sigma I_m
\tag{4.14}
$$

which is limited by

$$
I_e \leq I_{eMax}.
\tag{4.15}
$$

where I_{eMax} (unit: A, from Θ_e) is the maximum current of ESC. The ESC voltage U_e is supplied by the battery voltage U_b (unit: V, from Θ_b), given by

$$
U_e = U_b - I_b R_b
\tag{4.16}
$$

where I_b (unit: A) is the battery current and R_b (unit: Ω, from Θ_b). For a multicopter, the number of ESCs is equal to the number of propulsors n_r, so

Fig. 4.4 ESC model

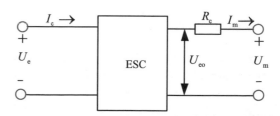

$$I_b = n_r I_e + I_{other} \tag{4.17}$$

where I_{other} (unit: A) includes the current from other devices (like autopilot and camera) and other current loss, usually, $I_{other} \approx 1$ A. From above, the functions in Eq. (4.11) are expressed as

$$\sigma = f_\sigma\left(\Theta_e, U_m, I_m, U_b\right) \triangleq \frac{U_m + I_m R_e}{U_b} \tag{4.18}$$

$$I_e = f_{I_e}\left(\sigma, I_m\right) \triangleq \sigma I_m \tag{4.19}$$

$$U_e = f_{U_e}\left(\Theta_b, I_b\right) \triangleq U_b - I_b R_b. \tag{4.20}$$

4.2.4 Battery Modeling

The battery model mainly aims to obtain the time of endurance T_b (unit: min) by using I_b, Θ_b, where I_b is expressed in Eq. (4.17) and Θ_b is shown in Table 4.1. For clarity, the model is expressed in an abstract function as

$$T_b = f_{T_b}\left(\Theta_b, I_b\right). \tag{4.21}$$

The battery discharge process is simplified so that the battery voltage remains constant and the battery capacity is decreased linearly.[5] Thus, the model $f_{T_b}\left(\Theta_b, I_b\right)$ is

$$T_b = f_{T_b}\left(\Theta_b, I_b\right) \triangleq \frac{C_b - C_{min}}{I_b} \cdot \frac{60}{1000} \tag{4.22}$$

where C_{min} (unit: mAh) is the battery minimum capacity set according to the safety margin. Generally, it can be chosen from $0.15C_b - 0.2C_b$. Finally, the battery current I_b cannot exceed the maximum current that the battery can withstand, which is expressed by the maximum discharge rate K_b (no unit, from Θ_b), so

$$I_b \leq \frac{C_b K_b}{1000}. \tag{4.23}$$

4.3 Performance Evaluation

So far, the propulsion system modeling is completed. In this section, Problems 1–4 are going to be investigated. Before evaluating the performance, a virtual multicopter is defined here. Its weight is G (unit: N) and it has n_r propulsors. This implies that n_r motors, ESCs, and propellers are required. The attitude of the multicopter is h, and the temperature is T_t. The parameters related to its propulsion system have the same notations shown in Table 4.1 for simplicity.

4.3.1 Solution to Problem 1

In the hover mode, the sum of thrust provided by n_r propellers should be equal to the weight of the multicopter G (unit: N). Thus, the thrust T (unit: N) provided by a single propeller is

---

[5]A more accurate model can be referred to [13, 14], which is nonlinear.

$$T = \frac{G}{n_\mathrm{r}}. \tag{4.24}$$

Then, according to Eqs. (4.1) and (4.2), the motor speed N and the propeller torque M can be obtained as

$$N = 60\sqrt{\frac{T}{\rho D_\mathrm{p}^4 C_\mathrm{T}}}$$
$$M = \rho D_\mathrm{p}^5 C_\mathrm{M} \left(\frac{N}{60}\right)^2 \tag{4.25}$$

where C_T and C_M are treated as the propeller parameters here, which should be estimated according to Eq. (4.6), or measured based on experiments. Then, substituting M, N into Eq. (4.10), U_m and I_m can be obtained

$$U_\mathrm{m} = f_{U_\mathrm{m}}(\Theta_\mathrm{m}, M, N)$$
$$I_\mathrm{m} = f_{I_\mathrm{m}}(\Theta_\mathrm{m}, M, N). \tag{4.26}$$

On the next stage, by using the given parameters of the ESC and the battery in Table 4.1, the throttle command σ, the input current I_e, and the voltage U_e of ESC are obtained according to Eqs. (4.18), (4.19) and (4.20) as

$$\sigma = f_\sigma(\Theta_\mathrm{e}, U_\mathrm{m}, I_\mathrm{m}, U_\mathrm{b})$$
$$I_\mathrm{e} = f_{I_\mathrm{e}}(\sigma, I_\mathrm{m})$$
$$U_\mathrm{e} = f_{U_\mathrm{e}}(\Theta_\mathrm{b}, I_\mathrm{b}). \tag{4.27}$$

Consequently, according to Eq. (4.17), the battery current is

$$I_\mathrm{b} = n_\mathrm{r} I_\mathrm{e} + I_\mathrm{other}. \tag{4.28}$$

Finally, the hover endurance T_hover (unit: min) is calculated by Eq. (4.22) as

$$T_\mathrm{hover} = f_{T_\mathrm{b}}(\Theta_\mathrm{b}, I_\mathrm{b}). \tag{4.29}$$

A flow chart providing the details of Problem 1 is plotted in Fig. 4.5. An example for Problem 1 is given in Table 4.2, where the adopted parameters and evaluation results are shown.

Table 4.2 Example for Problem 1

Environment	h=10 m, $T_\mathrm{t} = 25\,°\mathrm{C}$
Basic parameters	$G = 14.7\,\mathrm{N}$, $n_\mathrm{r} = 4$
Components	$\Theta_\mathrm{p} = \{D_\mathrm{p} = 10\,\mathrm{in}, H_\mathrm{p} = 4.5\,\mathrm{in}, B_\mathrm{p} = 2, G_\mathrm{p}\}$
	$\Theta_\mathrm{m} = \{K_{V0} = 890\,\mathrm{RPM/V}, I_\mathrm{mMax} = 19\,\mathrm{A}, I_{m0} = 0.5\,\mathrm{A},$
	$U_{m0}\,\Theta_\mathrm{e} = \{I_\mathrm{eMax} = 30\,\mathrm{A}, R_\mathrm{e} = 0.008\,\Omega, G_\mathrm{e}\}$
	$\Theta_\mathrm{b} = \{C_\mathrm{b} = 5000\,\mathrm{mAh}, R_\mathrm{b} = 0.01\,\Omega, U_\mathrm{b} = 12\,\mathrm{V},$
	$K_\mathrm{b} = 45\,\mathrm{C}, G_\mathrm{b}\}$
Other parameters	$A = 5, \varepsilon = 0.85, \lambda = 0.75, \zeta = 0.5, e = 0.83,$
	$C_\mathrm{fd} = 0.015, \alpha_0 = 0, K_0 = 6.11, C_1 = 3, C_2 - 1.5,$
	$C_\mathrm{min} = 0.2\,C_b$
Results	$T_\mathrm{loiter} = 15.8\,\mathrm{min}, \sigma = 54.6\%, I_\mathrm{e} = 3.6\,\mathrm{A}, U_e = 11.8\,\mathrm{V},$
	$I_\mathrm{b} = 15.2\,\mathrm{A}, N = 5223\,\mathrm{RPM}$

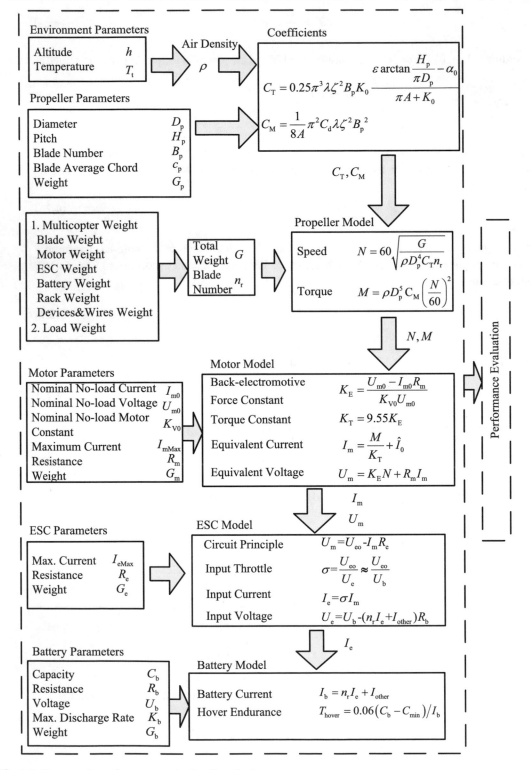

Fig. 4.5 Hover mode performance evaluation flow chart

4.3.2   Solution to Problem 2

In the maximum throttle mode, the throttle command $\sigma = 1$ which implies that the equivalent motor voltage U_m reaches its peak. First, I_m, U_m, M, N can be solved by using the following nonlinear equation group with the numerical iteration method

$$
\begin{aligned}
&f_\sigma \left(\Theta_e, U_m, I_m, U_b \right) = 1 \\
&M - \rho D_p^5 C_M \left(\tfrac{N}{60} \right)^2 = 0 \\
&U_m - f_{U_m} \left(\Theta_m, M, N \right) = 0 \\
&I_m - f_{I_m} \left(\Theta_m, M, N \right) = 0
\end{aligned}
\tag{4.30}
$$

which is derived from Eqs. (4.2), (4.10), and (4.18). This equation group can be solved through the numerical iteration method. Substituting I_m into Eq. (4.19) yields I_e. Then, based on Eqs. (4.17) and (4.20), U_e, I_b are obtained as

$$
\begin{aligned}
I_e &= f_{I_e} \left(1, I_m \right) \\
I_b &= n_r I_e + I_{other} \\
U_e &= f_{U_e} \left(\Theta_b, I_b \right).
\end{aligned}
\tag{4.31}
$$

The system efficiency is the ratio between the propeller output power and the battery power, which is calculated by

$$
\eta = \frac{\frac{2\pi}{60} n_r M N}{U_b I_b}.
\tag{4.32}
$$

An example for Problem 2 is given in Table 4.3, where the adopted parameters and evaluation results are shown.

4.3.3   Solution to Problem 3

The maximum payload $G_{maxload}$ (unit: N) and the maximum pitch angle of a multicopter θ_{max} (unit: rad) are also very important performance indices and are closely related to the safety of a multicopter. Generally, in order to ensure a sufficient control margin, the throttle command σ during a flight should be smaller than 0.9 (to satisfy the basic attitude control and wind resistance requirement). In

Table 4.3   Example for Problem 2

Environment	$h = 10\,\text{m}, T_t = 25\,°\text{C}$
Basic Parameters	$G = 14.7\,\text{N}, n_r = 4$
Components	$\Theta_p = \{D_p = 10\,\text{in}, H_p = 4.5\,\text{in}, B_p = 2, G_p\}$ $\Theta_m = \{K_{V0} = 890\,\text{RPM/V}, I_{mMax} = 19\,\text{A}, I_{m0} = 0.5\,\text{A},$ $U_{m0} = 10\,\text{V}, R_m = 0.101\,\Omega, G_m\}$ $\Theta_e = \{I_{eMax} = 30\,\text{A},$ $R_e = 0.008\,\Omega, G_e\}$ $\Theta_b = \{C_b = 5000\,\text{mAh},$ $R_b = 0.01\,\Omega, U_b = 12\,\text{V}, K_b = 45\,\text{C}, G_b\}$
Other parameters	$A = 5, \varepsilon = 0.85, \lambda = 0.75, \zeta = 0.5, e = 0.83,$ $C_{fd} = 0.015, \alpha_0 = 0, K_0 = 6.11, C_1 = 3, C_2 = 1.5,$ $C_{min} = 0.2\,C_b$
Results	$I_e = 16.5\,\text{A}, U_e = 11.3\,\text{V}, I_b = 66.2\,\text{A}, N = 8528\,\text{RPM},$ $\eta = 77.1\%$

Table 4.4 Example for Problem 3

Environment	$h = 10\,\text{m}, T_t = 25\,°\text{C}$
Basic Parameters	$G = 14.7\,\text{N}, n_r = 4$
Components	$\Theta_p = \{D_p = 10\,\text{in}, H_p = 4.5\,\text{in}, B_p = 2, G_p\}$ $\Theta_m = \{K_{V0} = 890\,\text{RPM/V}, I_{mMax} = 19\,\text{A}, I_{m0} = 0.5\,\text{A},$ $U_{m0} = 10\,\text{V}, R_m = 0.101\,\Omega, G_m\}$ $\Theta_e = \{I_{eMax} = 30\,\text{A},$ $R_e = 0.008\,\Omega, G_e\}$ $\Theta_b = \{C_b = 5000\,\text{mAh},$ $R_b = 0.01\,\Omega, U_b = 12\,\text{V}, K_b = 45\,\text{C}, G_b\}$
Other parameters	$A = 5, \varepsilon = 0.85, \lambda = 0.75, \zeta = 0.5, e = 0.83,$ $C_{fd} = 0.015, \alpha_0 = 0, K_0 = 6.11, C_1 = 3, C_2 = 1.5,$ $C_{min} = 0.2\,C_b$
Results	$G_{maxload} = 1.32\,\text{kg}, \theta_{max} = 1.01\,\text{rad}\,(57.9°)$

this chapter, $\sigma = 0.8$ is taken to estimate the maximum payload. Similar to Problem 2, according to Eqs. (4.2), (4.10), and (4.18), I_m, U_m, M, N are solved by the following equations

$$
\begin{aligned}
f_\sigma\left(\Theta_e, U_m, I_m, U_b\right) &= 0.8 \\
M - \rho D_p^5 C_M \left(\tfrac{N}{60}\right)^2 &= 0 \\
U_m - f_{U_m}\left(\Theta_m, M, N\right) &= 0 \\
I_m - f_{I_m}\left(\Theta_m, M, N\right) &= 0.
\end{aligned}
\tag{4.33}
$$

Then, according to Eq. (4.1), the produced propeller thrust T is calculated as

$$
T = \rho D_p^4 C_T \left(\frac{N}{60}\right)^2.
\tag{4.34}
$$

Therefore, the maximum payload is

$$
G_{maxload} = n_r T - G.
\tag{4.35}
$$

In addition, the maximum pitch angle is given by

$$
\theta_{max} = \arccos \frac{G}{n_r T}.
\tag{4.36}
$$

An example for Problem 3 is given in Table 4.4, where the adopted parameters and evaluation results are shown.

4.3.4   Solution to Problem 4

Maximum forward flight speed and maximum flight distance are the key performance indices, which both designers and users are concerned about. The solution to Problem 4 proceeds through the following three steps. First, the forward flight speed is solved. Secondly, the flight distance is obtained. Finally, the maximum flight speed and distance are solved as optimization problems.

4.3.4.1 Forward Flight Speed

The first step is to solve the forward flight speed under the given pitch angle θ (unit: rad) of a multicopter. The force diagram is shown in Fig. 4.6. According to Fig. 4.6, equations to describe the force equilibrium in forward flight state are established as follows:

$$
\begin{aligned}
F_{\text{drag}} &= G \tan \theta \\
T &= \frac{G}{n_r \cos \theta}
\end{aligned}
\tag{4.37}
$$

where F_{drag} is the drag acting on a multicopter. The drag is further expressed as [15]

$$
F_{\text{drag}} = \frac{1}{2} C_D \rho V^2 S
\tag{4.38}
$$

where V (unit: m/s) is the forward flight speed, S (unit: m^2) is the maximum cross-sectional area, and C_D is the whole vehicle drag coefficient related to θ. Considering that θ may be changed in a large range (up to 90°), C_D can be approximately expressed as[6]

$$
C_D = C_{D_1} \left(1 - \sin^3 \theta\right) + C_{D_2} \left(1 - \cos^3 \theta\right).
\tag{4.39}
$$

where C_{D_1} and C_{D_2} are the drag coefficient when θ equals to 0° and 90°, respectively. They are related to the aerodynamic configuration of a multicopter, which can be estimated by Computational Fluid Dynamics (CFD) simulation software. Finally, according to the equations above, the forward flight speed V can be written as the function of θ, as

$$
V(\theta) = \sqrt{\frac{2G \tan \theta}{\rho S [C_{D_1} \left(1 - \sin^3 \theta\right) + C_{D_2} \left(1 - \cos^3 \theta\right)]}}.
\tag{4.40}
$$

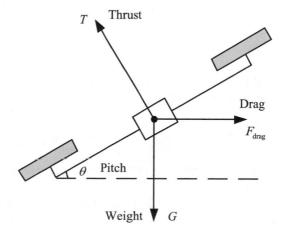

Fig. 4.6 Forces on a ulticopter in forward flight

---

[6]This equation is obtained by curve fitting from our CFD results. It can be changed to another form according to the relationship of C_D and θ.

4.3.4.2 Flight Distance

The second step is to solve for the flight distance under the given multicopter pitch angle θ. Considering that the flight speed of a multicopter is not very fast, for the use of estimation, the effects of incoming flow on the propeller thrust can be neglected. According to Eqs. (4.37) and (4.1), the motor speed N is given by

$$N = 60 \sqrt{\frac{G}{\rho C_T D_p^4 n_r \cos \theta}}. \tag{4.41}$$

Consequently, through Eq. (4.2), M is solved as

$$M = \frac{G C_M D_p}{C_T n_r \cos \theta}. \tag{4.42}$$

Finally, the total flight time T_{fly} is calculated through the equations group as

$$
\begin{aligned}
U_m &= f_{U_m} (\Theta_m, M, N) \\
I_m &= f_{I_m} (\Theta_m, M, N) \\
\sigma &= f_\sigma (\Theta_e, U_m, I_m, U_b) \\
I_e &= f_{I_e} (\sigma, I_m) \\
I_b &= n_r I_e + I_{\text{other}} \\
T_{\text{fly}} &= f_{T_b} (\Theta_b, I_b)
\end{aligned}
\tag{4.43}
$$

where Eqs. (4.10), (4.18), (4.19), (4.17), and (4.22) are utilized. In fact, T_{fly} is also the function of θ, so the flight distance Z under the given pitch angle θ can be written as the function form as

$$Z(\theta) = 60V(\theta) T_{\text{fly}}(\theta). \tag{4.44}$$

4.3.4.3 Maximum Flight Speed and Distance

The third step is to conduct the optimization solution of the previous two steps. In order to find the maximum forward flight speed V_{max}, an optimization problem is formulated as

$$V_{\text{max}} = \max_{\theta \in [0, \theta_{\text{max}}]} V(\theta) \tag{4.45}$$

where $V(\theta)$ is denoted by Eq. (4.40) and θ_{max} is solved in Problem 3. Furthermore, in order to find the maximum flight distance Z_{max}, an optimization problem is formulated as

$$Z_{\text{max}} = \max_{\theta \in [0, \theta_{\text{max}}]} Z(\theta) \tag{4.46}$$

where $Z(\theta)$ is obtained in Eq. (4.44). Both Eqs. (4.45) and (4.46) can be solved through the numerical traversal algorithm.

An example for Problem 4 is given in Table 4.5, where the adopted parameters and evaluation results are shown.

Table 4.5 Example for Problem 4

Environment	$h = 10\,\text{m}, T_t = 25\,^{\circ}\text{C}$
Basic parameters	$G = 14.7\,\text{N}, n_r = 4$
Components	$\Theta_p = \{D_p = 10\,\text{in}, H_p = 4.5\,\text{in}, B_p = 2, G_p\}$ $\Theta_m = \{K_{V0} = 890\,\text{RPM/V}, I_{mMax} = 19\,\text{A}, I_{m0} = 0.5\,\text{A},$ $U_{m0} = 10\,\text{V}, R_m = 0.101\,\Omega, G_m\}\ \Theta_e = \{I_{eMax} = 30\,\text{A},$ $R_e = 0.008\,\Omega, G_e\}\ \Theta_b = \{C_b = 5000\,\text{mAh},$ $R_b = 0.01\,\Omega, U_b = 12\,\text{V}, K_b = 45\,\text{C}, G_b\}$
Other parameters	$A = 5, \varepsilon = 0.85, \lambda = 0.75, \zeta = 0.5, e = 0.83,$ $C_{fd} = 0.015, \alpha_0 = 0,\ K_0 = 6.11, C_1 = 3, C_2 = 1.5,$ $C_{min} = 0.2\,C_b$
Results	$V_{max} = 11.2\,\text{m/s}, Z_{max} = 6021.4\,\text{m}$

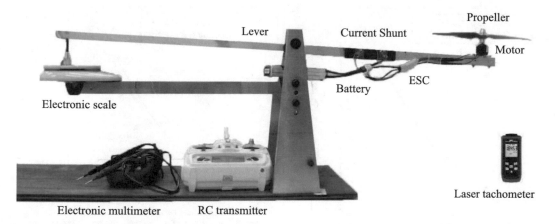

Fig. 4.7 Experiment apparatus

4.4 Test Case

To help the designer better evaluate the performance of a multicopter, an indoor measurement device for the propulsion system (as shown in Fig. 4.7) is introduced here. In that device, the thrust is measured through electronic scale, the propeller speed is measured through laser tachometer, the ESC current is measured through current shunt, the battery voltage is measured through electronic multimeter, and the throttle command is read from the RC transmitter directly. By using that device, an experiment is performed (parameters of components are given in Table 4.6), and the results are compared with the results of the proposed method, as shown in Fig. 4.8.

The device can also be used to measure the hover endurance of a multicopter. The method is that, the throttle command is tuned to make the propeller thrust equal to the hovering thrust (i.e., multicopter total weight divides propeller number), then the current and discharge time are measured to calculate the hovering time. The real experiment results of the hover endurance are compared with the value calculated by the proposed method, shown in Table 4.7.

Table 4.6 Parameters of experiment components

Environment	$h = 50\,\text{m}, T_{\text{t}} = 20\,°\text{C}$
Propeller	APC $10 \times 45\,\text{MR}$ ($D_{\text{p}} = 10\,\text{in}, H_{\text{p}} = 4.5\,\text{in}, B_{\text{p}} = 2$)
Motor	Sunnysky Angel A2212 ($K_{\text{V0}} = 980\,\text{RPM/V}, R_{\text{m}} = 0.12\Omega$, $I_{\text{mMax}} = 20\,\text{A}, I_{\text{m0}} = 0.5\,\text{A}, U_{\text{m0}} = 10\,\text{V}$)
ESC	$I_{\text{eMax}} = 30\,\text{A}, R_{\text{e}} = 0.008\,\Omega$
Battery	ACE ($C_{\text{b}} = 4000\,\text{mAh}, U_{\text{b}} = 12\,\text{V}, R_{\text{b}} = 0.016\,\Omega$, $K_{\text{b}} = 25\,\text{C}$)
Others	$A = 5, \varepsilon = 0.85, \lambda = 0.75, \zeta = 0.5, e = 0.83$, $C_{\text{fd}} = 0.015, \alpha_0 = 0, K_0 = 6.11, C_{\text{min}} = 0.2\,C_{\text{b}}$

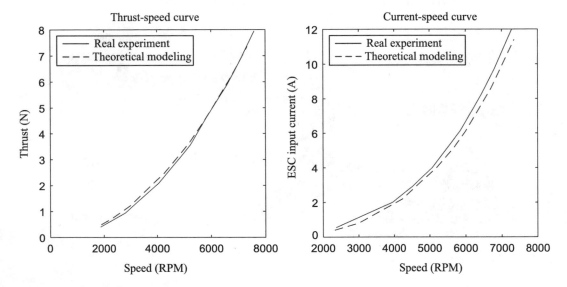

Fig. 4.8 Thrust-Speed curve and current-speed curve

Table 4.7 Results of hover endurance

Virtual multicopter	$G = 14.7\,\text{N}, n_{\text{r}} = 4$	
Environment	See Table 4.6	
Components	See Table 4.6	
Hover endurance	Experiment	12.5 min
	Proposed	12.2 min

4.5   Summary

In this chapter, a modeling method for the propulsion system is proposed, including the propeller, the motor, the ESC, and the battery. Based on these models, performances of a multicopter are evaluated in different modes, such as the hover mode, the maximum throttle mode, and the forward flight mode.

Based on the mathematical models described in this chapter, an online performance evaluation Web site

$$\text{http://www.flyeval.com}$$

for multicopters is established. Users can obtain the flight evaluations mentioned above by providing the airframe configuration parameters, the environment parameters, and the propulsion system parameters. Moreover, besides the evaluation function, this Web site has the design function that can recommend users with optimum configurations of the multicopter according to the flight requirement users provide. Also, readers can refer to another excellent Web site https://www.ecalc.ch to do similar evaluation which, however, does not have the design function so far.

4.6 Appendix

4.6.1 Procedure to Obtain Thrust Coefficient and Torque Coefficient

The task of this section is to give a concrete expression of $f_{C_T}(\Theta_p)$ as well as $f_{C_M}(\Theta_p)$, where Θ_p is defined in Table 4.1.

4.6.1.1 Blade Absolute Angle of Attack

The absolute angle of attack α_{ab} of the blade is a very important parameter. It is assumed that the blade angle θ_p (unit: rad) of a multicopter is constant along the direction of radius, given by [16, p. 40]

$$\theta_p = \arctan \frac{H_p}{\pi D_p} \tag{4.47}$$

where propeller pitch H_p (unit: m) and the diameter D_p (unit: m) are the parameters from Θ_p in Table 4.1. The effective angle of attack α (unit: rad) is

$$\alpha = \varepsilon(\theta_p - \phi_0) \tag{4.48}$$

where $\varepsilon \in \mathbb{R}_+$ is a correction factor that arises due to downwash, and ϕ_0 (unit: rad) is the helix angle. Considering flight characteristics of multicopters, it is assumed $\phi_0 \approx 0$. The absolute angle of attack α_{ab} (unit: rad) is [6, p. 136]

$$\alpha_{ab} = \alpha - \alpha_0 \tag{4.49}$$

where α_0 (unit: rad) is the zero-lift angle of attack. The relationship among α_{ab}, α and α_0 is shown in Fig. 4.9.

Fig. 4.9 Geometric and absolute angle of attack

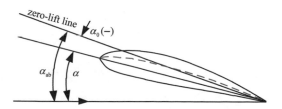

4.6.1.2 Lift Coefficient and Drag Coefficient of Blade Airfoil Section

The lift coefficient C_l of the blade airfoil section is related to the absolute angle of attack α_{ab}, which is given by [6, p. 164]

$$C_l = \frac{K_0 \alpha_{ab}}{1 + K_0/\pi A} \tag{4.50}$$

where K_0 is chosen as $K_0 \approx 6.11$ [6, p. 151], and $A = D_p/c_p$ is the aspect ratio, where c_p is the blade average chord length. Considering the downwash, A is in the range 5–8. Furthermore, the drag coefficient C_d of the blade airfoil section is given by [6, p. 174]

$$C_d = C_{fd} + \frac{1}{\pi A e} C_l^2 \tag{4.51}$$

where C_{fd} is the zero-lift drag coefficient which depends on the thickness of the blade, the *Reynolds number*, the angle of attack, etc. The *Oswald factor e* is selected between 0.7 and 0.9.

4.6.1.3 Thrust and Torque of Propeller

The blade airfoil lift is related to the airfoil and the rotational speed, given by [16, p. 117]

$$L = \frac{1}{2} C_l \rho S_{sa} W^2 \tag{4.52}$$

where W is the airspeed of a blade and $S_{sa} = B_p \lambda D_p c_p/2$ is the blade area, in which λ is the correction coefficient. The airspeed of a blade W in flight mainly has two components. One is from the average blade linear speed, and the other is from the multicopter flight speed. Generally, the former is much greater than the latter. Therefore, let W be the average blade linear speed so that

$$W \approx \pi \zeta D_p \frac{N}{60} \tag{4.53}$$

where N is the propeller speed and $\zeta = 0.4 - 0.7$. According to the *blade-element theory*, the blade element force diagram is shown in Fig. 4.10. In Fig. 4.10, it is shown that the propeller thrust T is not equal to its lift L. The thrust is

$$T = L \frac{\cos(\gamma + \phi_0)}{\cos(\gamma - \delta)} \tag{4.54}$$

where the correction angle γ (unit: rad) due to the downwash effect is expressed as [17, p. 50]

$$\gamma = \arctan \frac{dD}{dL} = \arctan \frac{\frac{1}{2} C_d \rho W^2 c dr}{\frac{1}{2} C_l \rho W^2 c dr} = \arctan \frac{C_d}{C_l}$$

and δ is neglected because the influence of downwash has been taken into consideration in ε. Moreover, since for multicopter $\phi_0 \approx 0$, so $T \approx L$. Based on the previous results, the propeller torque is approximated as

$$M = \frac{1}{4} \rho B_p C_d W^2 S_{sa} D_p. \tag{4.55}$$

Fig. 4.10 Forces on a
blade element

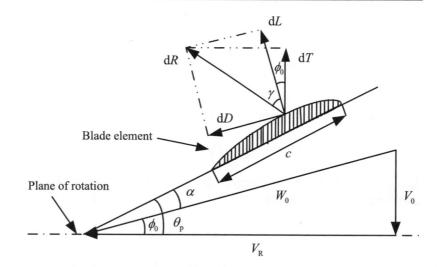

4.6.1.4 Results

According to Eqs. (4.47)–(4.55), $f_{C_T}(\Theta_p)$ and $f_{C_M}(\Theta_p)$ are obtained as follows:

$$f_{C_T}(\Theta_p) = C_T = 0.25\pi^3\lambda\zeta^2 B_p K_0 \frac{\varepsilon \arctan \frac{H_p}{\pi D_p} - \alpha_0}{\pi A + K_0}$$

$$f_{C_M}(\Theta_p) = C_M = \frac{1}{8A}\pi^2 C_d\zeta^2\lambda B_p^2 \tag{4.56}$$

where

$$C_d = C_{fd} + \frac{\pi A K_0^2}{e} \frac{\left(\varepsilon \arctan \frac{H_p}{\pi D_p} - \alpha_0\right)^2}{(\pi A + K_0)^2}. \tag{4.57}$$

4.6.2 Procedure to Obtain the Motor Equivalent Voltage and Current

The task of this section is to give a concrete expression of $U_m = f_{U_m}(\Theta_m, M, N)$ as well as $I_m = f_{I_m}(\Theta_m, M, N)$, where Θ_m is defined in Table 4.1, M is the propeller torque, and N is the motor speed.

4.6.2.1 Equivalent Motor Current

A brushless DC motor is modeled as a permanent magnet DC motor. According to *electric machine theory* [18, p. 516], the electromagnetic torque T_e (unit: N·m) is given by

$$T_e = K_T I_m \tag{4.58}$$

where K_T is the torque constant (unit: N·m/A) and I_m (unit: A) is the motor current. The output torque M is equal to the propeller torque, calculated as

$$M = T_e - T_0 = K_T(I_m - \hat{I}_0) \tag{4.59}$$

where $T_0 = K_T\hat{I}_0$ is the no-load torque and $\hat{I}_0$ (unit: A) is the no-load current in actual working state. By Eq. (4.59), I_m is obtained as

$$I_m = \frac{M}{K_T} + \hat{I}_0 \tag{4.60}$$

which is required that $I_m \leq I_{mMax}$.

4.6.2.2 Torque Constant
The back-electromotive force E_a (unit: V) is expressed as [18, p. 516]

$$E_a = K_E N \tag{4.61}$$

where K_E (unit: V/RPM) is the back-electromotive force constant. Under the nominal no-load running condition, the motor voltage satisfies [18, p. 539]

$$U_{m0} = K_E N_{m0} + I_{m0} R_m \tag{4.62}$$

where U_{m0} (unit: V) is the nominal no-load voltage, I_{m0} (unit: A) is the nominal no-load current, R_m (unit: Ω) is the resistance. They are all motor parameters as shown in Table 4.1. Furthermore, N_{m0} is the nominal no-load speed, given by

$$N_{m0} = K_{V0} U_{m0} \tag{4.63}$$

where K_{V0} (unit: RPM/V) is the KV value.

According to Eqs. (4.62) and (4.63)

$$K_E = \frac{U_{m0} - I_{m0} R_m}{K_{V0} U_{m0}}. \tag{4.64}$$

Considering that the torque constant K_T (unit: V/RPM) and the back-electromotive force constant K_E have the following relationship [18, p. 516]

$$K_T = 9.55 K_E \tag{4.65}$$

then

$$K_T = 9.55 \frac{U_{m0} - I_{m0} R_m}{K_{V0} U_{m0}}. \tag{4.66}$$

4.6.2.3 No-Load Current
So far, for Eq. (4.60), M, K_T have been derived, leaving the actual no-load current $\hat{I}_0$ to be calculated. By ignoring the mechanical loss, the nominal power consumption P_{Fe0} (unit: W) is attributed to the iron loss and copper loss, having the form as

$$P_{Fe0} = U_{m0} I_{m0} - I_{m0}^2 R_m. \tag{4.67}$$

In the same way as Eqs. (4.63) and (4.67), under the actual running condition, the actual input voltage U_m, the actual no-load current $\hat{I}_0$, the actual no-load speed $\hat{N}_0$, and actual power consumption P_{Fe} satisfy

$$\begin{aligned} \hat{N}_0 &= K_{V0} U_m \\ P_{Fe} &= U_m \hat{I}_0 - \hat{I}_0^2 R_m. \end{aligned} \tag{4.68}$$

Furthermore, according to [18, p. 516]

$$P_{\text{Fe}} = \left(\frac{\hat{N}_0}{N_{\text{m}0}}\right)^{1.3} P_{\text{Fe}0}$$
$$U_{\text{m}} = K_{\text{E}}\hat{N}_0 + \hat{I}_0 R_{\text{m}}. \tag{4.69}$$

On the basis of Eqs. (4.68) and (4.69), $\hat{I}_0$ can be expressed as

$$\hat{I}_0 = \frac{P_{\text{Fe}0}}{K_{\text{E}}\hat{N}_0}\left(\frac{\hat{N}_0}{N_{\text{m}0}}\right)^{1.3}$$
$$= \frac{P_{\text{Fe}0}}{K_{\text{E}}U_{\text{m}}K_{\text{V}0}}\left(\frac{U_{\text{m}}K_{\text{V}0}}{N_{\text{m}0}}\right)^{1.3}. \tag{4.70}$$

However, $\hat{I}_0$ and $I_{\text{m}0}$ are quite similar for multicopter motors through our calculation. For simplicity, it is reasonable to let $\hat{I}_0 = I_{\text{m}0}$ in approximate estimation.

4.6.2.4 Relationship Between KV Value and Propeller Size

A motor with higher $K_{\text{V}0}$ often matches smaller propellers, while a motor with lower $K_{\text{V}0}$ matches larger ones. Similar to $K_{\text{V}0}$, K_{V} is defined as the motor constant with load, obviously, $N = K_{\text{V}}E_{\text{a}}$. It will be shown that the product of K_{V} and K_{T} is constant approximately in the following. Motor electromagnetic power P_{em} is given by

$$P_{\text{em}} \approx \frac{2\pi}{60}NT_{\text{e}} \tag{4.71}$$

where T_{e} is the electromagnetic torque. Also, P_{em} can be expressed as

$$P_{\text{em}} = E_{\text{a}}I_{\text{m}}. \tag{4.72}$$

Combine with $N = K_{\text{V}}E_{\text{a}}$ and $T_{\text{e}} = K_{\text{T}}I_{\text{m}}$, one has

$$K_{\text{V}}K_{\text{T}} = \frac{N}{E_{\text{a}}}\frac{T_{\text{e}}}{I_{\text{m}}} \approx \frac{30}{\pi}. \tag{4.73}$$

Given a motor with input power P and input voltage U, the motor current is $I_{\text{m}} \approx P/U$. If the motor KV value is high, then the motor speed will be high. According to Eq. (4.73), K_{T} will be low and M will be small. Therefore, it is better to choose a small propeller. One the other hand, if the motor KV value is low, it is better to choose a big propeller.

4.6.2.5 Results

According to Eqs. (4.60)–(4.70), $f_{U_{\text{m}}}(\Theta_{\text{m}}, M, N)$ and $f_{I_{\text{m}}}(\Theta_{\text{m}}, M, N)$ are obtained as follows:

$$f_{U_{\text{m}}}(\Theta_{\text{m}}, M, N) = U_{\text{m}} = \left(\frac{MK_{\text{V}0}U_{\text{m}0}}{9.55(U_{\text{m}0}-I_{\text{m}0}R_{\text{m}})}+I_{\text{m}0}\right)R_{\text{m}} + \frac{U_{\text{m}0}-I_{\text{m}0}R_{\text{m}}}{K_{\text{V}0}U_{\text{m}0}}N$$
$$f_{I_{\text{m}}}(\Theta_{\text{m}}, M, N) = I_{\text{m}} = \frac{MK_{\text{V}0}U_{\text{m}0}}{9.55(U_{\text{m}0}-I_{\text{m}0}R_{\text{m}})} + I_{\text{m}0}. \tag{4.74}$$

Exercises

4.1 Given a quadcopter, where the total weight is 14.7 N, the altitude is 50 m, the local temperature is 25 °C and the parameters of components are shown in (Table 4.8). According to the given conditions, estimate the hover time of the quadcopter and give a detailed calculation process.

4.2 Assume the total weight of a hexacopter with payload is 98 N. Try to find a suitable set of hexacopter configurations (including the brands and models of propeller, motor, ESC, and battery) to make the hexacopter hover in the air for as long time as possible (at least more than 3 min), where the fundamental conditions are shown in Fig. 4.11. Use the evaluation Web site http://www.flyeval.com to select components and verify.

4.3 For a multicopter, suppose that the battery resistance R_b, ESC resistance R_e, motor resistance R_m, and other current consumption I_{other} need not to be considered. It can hover for T_{hover1} and T_{hover2} when the air density is ρ_1 and ρ_2, respectively (other conditions being equal). Show that

Table 4.8 Parameters of components

Component	Parameters
Propeller	APC1045 ($D_p = 10$ in, $H_p = 4.5$ in, $B_p = 2$), $C_T = 0.0984$, $C_M = 0.0068$
Motor	Sunnysky A2814–900 ($K_{V0} = 900$ RPM/V, $R_m = 0.08\,\Omega$,
	$W_{mMax} = 335$ W, $I_{m0} = 0.6$ A, $U_{m0} = 10$ V)
ESC	$I_{eMax} = 30$ A, $R_e = 0.008\,\Omega$
Battery	ACE ($C_b = 4000$ mAh, $U_b = 12$ V, $R_b = 0.0084\,\Omega$, $K_b = 65$ C)

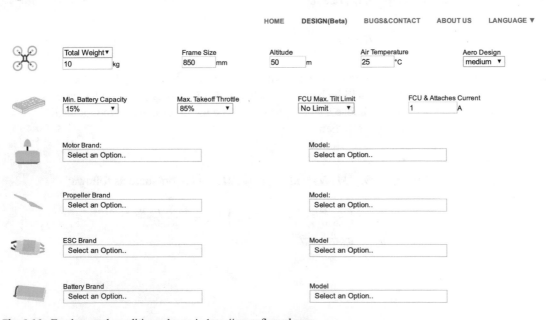

Fig. 4.11 Fundamental conditions shown in http://www.flyeval.com

$$\frac{T_{\text{hover1}}}{T_{\text{hover2}}} = \sqrt{\frac{\rho_1}{\rho_2}}.$$

4.4 For a multicopter, suppose that the battery resistance R_b, ESC resistance R_e, motor resistance R_m, and other current consumption I_{other} are not considered. The highest altitude it can hover at is h_{hover1} when the environment temperature is T_{t1}. Then, if the environment temperature changes to T_{t2} (other conditions being equal), give the highest altitude that the multicopter can hover at and express it by h_{hover1}, T_{t1}, T_{t2}, and other parameters if necessary.

4.5 In this chapter, the propulsion system is evaluated from the energy aspect. Analyze the disadvantages of this method and show your opinions.

References

1. Driessens S, Pounds P (2015) The triangular quadrotor: a more efficient quadrotor configuration. IEEE Trans Robot 31(6):1517–1526
2. Magnussen O, Hovland G, Ottestad M (2014) Multicopter UAV design optimization. In: Proceedings of the IEEE/ASME 10th international conference on mechatronic and embedded systems and applications (MESA), Senigallia, Italy, pp 1–6
3. Moffitt BA, Bradley TH, Parekh DE et al (2008) Validation of vortex propeller theory for UAV design with uncertainty analysis. In: 46th AIAA aerospace sciences meeting and exhibit, Reno, Nevada, AIAA 2008-406
4. Merchant MP, Miller LS (2006) Propeller performance measurement for low Reynolds number UAV applications. In: 44th AIAA Aerospace sciences meeting and exhibit, Reno, Nevada, AIAA 2006-1127
5. Cavcar M (2000) The international standard atmosphere (ISA). http://www.wxaviation.com/ISAweb-2.pdf. Accessed 7 Apr 2016
6. Torenbeek E, Wittenberg H (2009) Flight physics: essentials of aeronautical disciplines and technology, with historical notes. Springer, Netherlands
7. Liu PQ (2006) Air propeller theory and its application. Beihang University Press, Beijing. (In Chinese)
8. Zhu BL (2006) Unmanned aircraft aerodynamics. Aviation Industry Press, Beijing. (In Chinese)
9. Chen J, Yang SX, Mo L (2009) Modeling and experimental analysis of UAV electric propulsion system. J Aerosp Power 24(6):1339–1344. (In Chinese)
10. Bangura M, Lim H, Kim HJ et al (2014) Aerodynamic power control for multirotor aerial vehicles. In: Proceedings of the IEEE international conference on robotics and automation (ICRA), pp 529–536
11. Lawrence DA, Mohseni K (2005) Efficiency analysis for long-duration electric MAVs. In: Infotech@ Aerospace, Arlington, Virginia, AIAA 2005-7090
12. Lindahl P, Moog E, Shaw SR (2012) Simulation, design, and validation of an UAVSOFC propulsion system. IEEE Trans Aerosp Electron Syst 48(3):2582–2593
13. Doerffel D, Sharkh SA (2006) A critical review of using the Peukert equation for determining the remaining capacity of lead-acid and lithium-ion batteries. J Power Sources 155(2):395–400
14. Traub LW (2011) Range and endurance estimates for battery-powered aircraft. J Aircr 2(48):703–707
15. Orsag M, Bogdan S (2012) Influence of forward and descent flight on quadrotor dynamics. In: Recent Advances in Aircraft Technology, Zagreb, Croatia, pp 141–156
16. Hitchens F (2015) Propeller aerodynamics: the history, aerodynamics & operation of aircraft propellers. Andrews UK Limited
17. Johnson W (1980) Helicopter theory. Princeton University Press, Princeton
18. Chapman S (2005) Electric machinery fundamentals. McGraw-Hill, New York

Part II
Modeling

Coordinate System and Attitude Representation

5

In order to describe the attitude and position of a multicopter, it is necessary to establish appropriate coordinate frames. Coordinate frames are helpful to establish the relationship among dynamic variables, thus facilitating calculation. This chapter regards the multicopter as a rigid body whose attitude in space mainly describes the rotation between the Aircraft-Body Coordinate Frame (ABCF) and the Earth-Fixed Coordinate Frame (EFCF). There are many methods to represent the attitude. Each method has its advantages and disadvantages. This chapter will present the Euler angles, rotation matrix, and quaternions as well as the relationship between the corresponding attitude rate and the body's angular velocity. Different attitude representation methods correspond to different modeling methods, which are closely related to filtering methods and control methods in the following chapters. Deep understanding of the coordinate frames and attitude representation is helpful to understand the motions of a multicopter and then to design filters and controllers. This chapter aims to answer the question below:

What are the three attitude representation methods and the relationship between their derivatives and the aircraft body's angular velocity?

This chapter will start with the establishment of coordinate frames and give further details about the attitude representation as well as its derivation.

© Springer Nature Singapore Pte Ltd. 2017
Q. Quan, *Introduction to Multicopter Design and Control*, DOI 10.1007/978-981-10-3382-7_5

Huntianyi

Long time ago, the ancient Chinese had began the use of coordinate positioning systems for astronomical observations. *Huntianyi* is composed of *Hunxiang* and *Hunyi*. *Hunxiang* is a celestial sphere, which is carved with constellations, equator, ecliptic, etc. *Hunyi* is an armillary sphere with a scope inside, which is used to measure the coordinates of the equator and the ecliptic of Sun. The early model of *Hunyi* was composed of *Siyouyi*—an instrument to measure the angle between aster and the North Pole—and an equatorial ring. Then an ecliptical ring, ameridian ring and amore precise framework were added into the model during the period from the Han dynasty to the Northern Song dynasty. During the Yuan dynasty, Shoujing Guo—a Chinese astronomer—improved the structure of *Hunyi* by separating the device into two instruments: *Jianyi* and *Liyunyi*.

5.1 Coordinate Frame

5.1.1 Right-Hand Rule

Before defining the coordinate frames, the *right-hand rule* needs to be introduced. As shown in Fig. 5.1a, the thumb of the right hand points to the positive direction of the ox axis, the first finger points to the positive direction of the oy axis, and the middle finger points to the positive direction of the oz axis. Furthermore, as shown in Fig. 5.1b, in order to determine the positive direction of a rotation, the thumb of the right hand points the positive direction of the rotation axis and the direction of the bent fingers is the positive direction of rotation. All the coordinate frames used in this chapter and the positive direction of angles defined later obey the right-hand rule.

5.1.2 Earth-Fixed Coordinate Frame and Aircraft-Body Coordinate Frame

The EFCF $o_e x_e y_e z_e$ is used to study multicopter's dynamic states relative to the Earth's surface and to determine its three-dimensional (3D) position. The Earth's curvature is ignored, namely the Earth's surface is assumed to be flat. The initial position of the multicopter or the center of the Earth is often set as the coordinate origin o_e, the $o_e x_e$ axis points to a certain direction in the horizontal plane, and the $o_e z_e$ axis points perpendicularly to the ground. Then, the $o_e y_e$ axis is determined according to the right-hand rule.

   The ABCF $o_b x_b y_b z_b$ is fixed to the multicopter. The Center of Gravity (CoG) of the multicopter is chosen as the origin o_b of $o_b x_b y_b z_b$. The $o_b x_b$ axis points to the nose direction in the symmetric plane of the multicopter (nose direction is related to the plus-configuration multicopter or the X-configuration multicopter, readers can refer to Sect. 3.1.1 in Chap. 3 for more information). The $o_b z_b$ axis is in the symmetric plane of the multicopter, pointing downward, perpendicular to the $o_b x_b$ axis. The $o_b y_b$ axis is determined according to the right-hand rule. The relationship between the ABCF and the EFCF is shown in Fig. 5.2.

   Define the following unit vectors

$$\mathbf{e}_1 \triangleq \begin{bmatrix} 1 \\ 0 \\ 0 \end{bmatrix}, \mathbf{e}_2 \triangleq \begin{bmatrix} 0 \\ 1 \\ 0 \end{bmatrix}, \mathbf{e}_3 \triangleq \begin{bmatrix} 0 \\ 0 \\ 1 \end{bmatrix}.$$

Fig. 5.1 Coordinate axes and the positive direction of a rotation using the right-hand rule

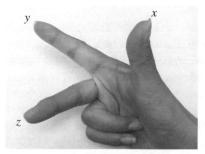

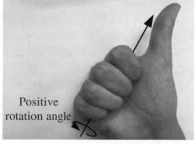

(a) Coordinate axes   (b) Positive direction of a rotation

Fig. 5.2 The relationship between the ABCF and the EFCF

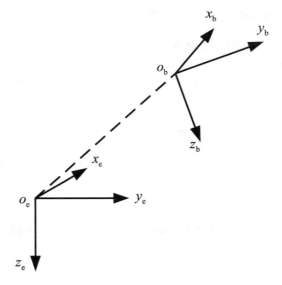

In the EFCF, the unit vectors along the $o_e x_e$ axis, $o_e y_e$ axis, and $o_e z_e$ axis are expressed as $\mathbf{e}_1, \mathbf{e}_2, \mathbf{e}_3$, respectively. In the ABCF, the unit vectors along the $o_b x_b$ axis, $o_b y_b$ axis, and $o_b z_b$ axis satisfy the following relationship

$$^{b}\mathbf{b}_1 = \mathbf{e}_1, \ ^{b}\mathbf{b}_2 = \mathbf{e}_2, \ ^{b}\mathbf{b}_3 = \mathbf{e}_3.$$

In the EFCF, the unit vectors along the $o_b x_b$ axis, $o_b y_b$ axis, and $o_b z_b$ axis are expressed as $^{e}\mathbf{b}_1, {}^{e}\mathbf{b}_2, {}^{e}\mathbf{b}_3$, respectively.

5.2 Attitude Representation

This section will present the Euler angles, rotation matrix, and quaternions.

5.2.1 Euler Angles

5.2.1.1 Definition of Euler Angles

The Euler angles are an intuitive way to represent the attitude. Their physical meanings are quite clear. Also, they are widely used in the attitude control. Based on Euler's theorem, the rotation of a rigid body around one fixed point can be regarded as the composition of several finite rotations around the fixed point [1, pp. 155–161]. The ABCF can be achieved by three elemental rotations of the EFCF around one fixed point. During these elemental rotations, each rotation axis is one of the coordinate axes of the rotating coordinate frame and each rotation angle is one of the Euler angles. Thus, the attitude matrix is closely related to the sequence of three elemental rotations, and it can be represented by the product of three elemental rotation matrices. Intuitively, let the EFCF align with the ABCF. Then, the yaw angle ψ, the pitch angle θ, and the roll angle ϕ are shown in Fig. 5.3 with their directions determined by the right-hand rule.

The Euler angle representation is complicated. As shown in Fig. 5.4, o_b' represents the projection of o_b in the plane $o_e x_e y_e$; o_{x_b} represents the intersection of the extension line of $x_b o_b$ and the plane

Fig. 5.3 Intuitive representation of Euler angles

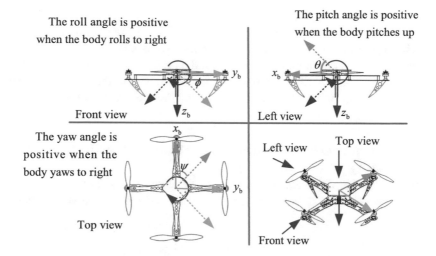

The roll angle is positive when the body rolls to right

The pitch angle is positive when the body pitches up

Front view   Left view

The yaw angle is positive when the body yaws to right

Left view   Top view

Top view

Front view

Fig. 5.4 Representation of Euler angles

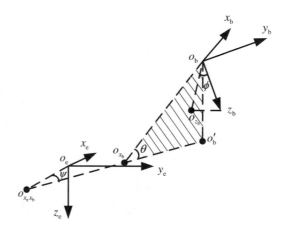

$o_e x_e y_e$; $o_b o_{z_b}$ represents the projection of $z_b o_b$ in the vertical plane $o_b o_{x_b} o'_b$ containing the $o_b x_b$ axis; $o_{x_e x_b}$ represents the intersection of the extension line of $x_e o_e$ and the extension line of $o'_b o_{x_b}$. The Euler angles are defined as follows [2, pp. 9–10].

(1) *Pitch angle* θ. It is the angle between the $o_b x_b$ axis and the plane $o_e x_e y_e$, as $\angle o_b o_{x_b} o'_b$ shown in Fig. 5.4. The pitch angle is positive when the $o_b x_b$ axis points upward and negative otherwise. The pitch angle shown in Fig. 5.4 is positive.

(2) *Roll angle* ϕ. It is the angle between the $o_b z_b$ axis and the plane $o_b o_{x_b} o'_b$, as $\angle z_b o_b o_{z_b}$ shown in Fig. 5.4. The roll angle is positive when the $o_b z_b$ axis points to the left of the plane $o_b o_{x_b} o'_b$ and negative otherwise. The roll angle shown in Fig. 5.4 is negative.

(3) *Yaw angle* ψ. It is the angle between the projection of $o_b x_b$ axis in the plane $o_e x_e y_e$ and the extension line of $o_e x_e$, as $\angle o_e o_{x_e x_b} o'_b$ shown in Fig. 5.4. The yaw angle is positive when the EFCF rotates clockwise and negative otherwise. The yaw angle shown in Fig. 5.4 is positive.

In the following, the rotation from the EFCF to the ABCF is presented. The rotation is composed of three elemental rotations about $\mathbf{e}_3$ axis, $\mathbf{k}_2$ axis, and $\mathbf{n}_1$ axis by ψ, θ, ϕ separately. More specifically,

(1) As shown in Fig. 5.5a, the rotation angle around $\mathbf{e}_3$ axis is the yaw angle ψ (provided that ψ is positive when the aircraft body turns to right and its value range is $[-\pi, \pi]$);

(2) As shown in Fig. 5.5b, the rotation angle around $\mathbf{k}_2$ axis is the pitch angle θ (provided that θ is positive when the aircraft nose pitches up and its value range is $[-\pi/2, \pi/2]$);

(3) As shown in Fig. 5.5c, the rotation angle around $\mathbf{n}_1$ axis is the roll angle ϕ (provided that ϕ is positive when the aircraft body rolls to right and its value range is $[-\pi, \pi]$).

5.2.1.2   Relationship Between the Attitude Rate and the Aircraft Body's Angular Velocity

As shown in Fig. 5.5, if the angular velocity of the aircraft body is $^{b}\boldsymbol{\omega} = [\,\omega_{x_b}\ \omega_{y_b}\ \omega_{z_b}\,]^{\mathrm{T}}$, then the relationship between the *attitude rate* and the *angular velocity* of the aircraft body is expressed as [3, pp. 35–36]

$$^{b}\boldsymbol{\omega} = \dot{\psi} \cdot {}^{b}\mathbf{k}_3 + \dot{\theta} \cdot {}^{b}\mathbf{n}_2 + \dot{\phi} \cdot {}^{b}\mathbf{b}_1. \tag{5.1}$$

Since

$$^{b}\mathbf{n}_2 = \mathbf{R}_n^b \cdot {}^{n}\mathbf{n}_2 = \mathbf{R}_n^b \cdot \mathbf{e}_2$$
$$^{b}\mathbf{k}_3 = \mathbf{R}_k^b \cdot {}^{k}\mathbf{k}_3 = \mathbf{R}_k^b \cdot \mathbf{e}_3$$

one has

$$\begin{aligned}
^{b}\mathbf{b}_1 &= \begin{bmatrix} 1 & 0 & 0 \end{bmatrix}^{\mathrm{T}} \\
^{b}\mathbf{n}_2 &= \mathbf{R}_n^b \cdot \mathbf{e}_2 = \begin{bmatrix} 0 & \cos\phi & -\sin\phi \end{bmatrix}^{\mathrm{T}} \\
^{b}\mathbf{k}_3 &= \mathbf{R}_k^b \cdot \mathbf{e}_3 = \begin{bmatrix} -\sin\theta & \cos\theta\sin\phi & \cos\theta\cos\phi \end{bmatrix}^{\mathrm{T}}
\end{aligned} \tag{5.2}$$

Here $\mathbf{R}_n^b = \mathbf{R}_x(\phi)$, $\mathbf{R}_k^b = \mathbf{R}_x(\phi)\mathbf{R}_y(\theta)$. The specific forms of $\mathbf{R}_x(\phi)$, $\mathbf{R}_y(\theta)$, $\mathbf{R}_z(\psi)$ are given as follows:

Fig. 5.5 Euler angles and frame transformation

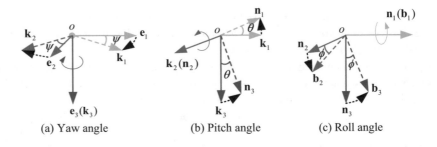

(a) Yaw angle   (b) Pitch angle   (c) Roll angle

$$\mathbf{R}_z\left(\psi\right) \triangleq \begin{bmatrix} \cos\psi & \sin\psi & 0 \\ -\sin\psi & \cos\psi & 0 \\ 0 & 0 & 1 \end{bmatrix}$$
$$\mathbf{R}_y\left(\theta\right) \triangleq \begin{bmatrix} \cos\theta & 0 & -\sin\theta \\ 0 & 1 & 0 \\ \sin\theta & 0 & \cos\theta \end{bmatrix} \qquad (5.3)$$
$$\mathbf{R}_x\left(\phi\right) \triangleq \begin{bmatrix} 1 & 0 & 0 \\ 0 & \cos\phi & \sin\phi \\ 0 & -\sin\phi & \cos\phi \end{bmatrix}.$$

Combining Eq. (5.1) with Eq. (5.2) yields

$$\begin{bmatrix} \omega_{x_b} \\ \omega_{y_b} \\ \omega_{z_b} \end{bmatrix} = \begin{bmatrix} 1 & 0 & -\sin\theta \\ 0 & \cos\phi & \cos\theta\sin\phi \\ 0 & -\sin\phi & \cos\theta\cos\phi \end{bmatrix} \begin{bmatrix} \dot{\phi} \\ \dot{\theta} \\ \dot{\psi} \end{bmatrix}. \qquad (5.4)$$

Furthermore

$$\dot{\boldsymbol{\Theta}} = \mathbf{W} \cdot {}^b\boldsymbol{\omega} \qquad (5.5)$$

where

$$\boldsymbol{\Theta} \triangleq \begin{bmatrix} \phi \\ \theta \\ \psi \end{bmatrix}, \mathbf{W} \triangleq \begin{bmatrix} 1 & \tan\theta\sin\phi & \tan\theta\cos\phi \\ 0 & \cos\phi & -\sin\phi \\ 0 & \sin\phi/\cos\theta & \cos\phi/\cos\theta \end{bmatrix}. \qquad (5.6)$$

Observed from Eq. (5.6), the denominators of some elements of matrix $\mathbf{W}$ are $\cos\theta$. In this case, the singularity problem, which should be avoided, may arise when $\cos\theta = 0$.

5.2.2   Rotation Matrix

5.2.2.1   Definition of Rotation Matrix

The rotation matrix satisfies

$$\begin{aligned} {}^e\mathbf{b}_1 &= \mathbf{R}_b^e \cdot {}^b\mathbf{b}_1 = \mathbf{R}_b^e \cdot \mathbf{e}_1 \\ {}^e\mathbf{b}_2 &= \mathbf{R}_b^e \cdot {}^b\mathbf{b}_2 = \mathbf{R}_b^e \cdot \mathbf{e}_2 \\ {}^e\mathbf{b}_3 &= \mathbf{R}_b^e \cdot {}^b\mathbf{b}_3 = \mathbf{R}_b^e \cdot \mathbf{e}_3. \end{aligned} \qquad (5.7)$$

Therefore, the rotation matrix is defined as

$$\mathbf{R}_b^e \triangleq \begin{bmatrix} {}^e\mathbf{b}_1 & {}^e\mathbf{b}_2 & {}^e\mathbf{b}_3 \end{bmatrix} \qquad (5.8)$$

where $\mathbf{R}_b^e \in SO(3)$, $SO(3) \triangleq \{\mathbf{R} | \mathbf{R}^T \mathbf{R} = \mathbf{I}_3, \det(\mathbf{R}) = 1, \mathbf{R} \in \mathbb{R}^{3\times3}\}$. According to the definition of Euler angles in Sect. 5.2.1 in this chapter, the rotation from the EFCF to the ABCF is composed of three elemental steps, which correspond to Fig. 5.5. The process in Fig. 5.5 is abstracted as follows [3, pp. 35–36]

$$
\begin{bmatrix} \mathbf{e}_1 \\ \mathbf{e}_2 \\ \mathbf{e}_3 \end{bmatrix} \xrightarrow{\mathbf{R}_z (\psi)} \begin{bmatrix} \mathbf{k}_1 \\ \mathbf{k}_2 \\ \mathbf{k}_3 = \mathbf{e}_3 \end{bmatrix} \xrightarrow{\mathbf{R}_y (\theta)} \begin{bmatrix} \mathbf{n}_1 \\ \mathbf{n}_2 = \mathbf{k}_2 \\ \mathbf{n}_3 \end{bmatrix} \xrightarrow{\mathbf{R}_x (\phi)} \begin{bmatrix} {}^e\mathbf{b}_1 = \mathbf{n}_1 \\ {}^e\mathbf{b}_2 \\ {}^e\mathbf{b}_3 \end{bmatrix}.
$$

Therefore, the rotation matrix $\mathbf{R}_b^e$, which represents the rotation from the ABCF to the EFCF, is expressed as

$$
\begin{aligned}
\mathbf{R}_b^e &= \left(\mathbf{R}_e^b \right)^{-1} \\
&= \mathbf{R}_z^{-1} (\psi) \, \mathbf{R}_y^{-1} (\theta) \, \mathbf{R}_x^{-1} (\phi) \\
&= \mathbf{R}_z^T (\psi) \, \mathbf{R}_y^T (\theta) \, \mathbf{R}_x^T (\phi) \\
&= \begin{bmatrix} \cos\theta\cos\psi & \cos\psi\sin\theta\sin\phi - \sin\psi\cos\phi & \cos\psi\sin\theta\cos\phi + \sin\psi\sin\phi \\ \cos\theta\sin\psi & \sin\psi\sin\theta\sin\phi + \cos\psi\cos\phi & \sin\psi\sin\theta\cos\phi - \cos\psi\sin\phi \\ -\sin\theta & \sin\phi\cos\theta & \cos\phi\cos\theta \end{bmatrix}.
\end{aligned} \tag{5.9}
$$

Conversely, Euler angles can be solved according to the rotation matrix. First, the rotation matrix $\mathbf{R}_b^e$ is defined as

$$
\mathbf{R}_b^e \triangleq \begin{bmatrix} r_{11} & r_{12} & r_{13} \\ r_{21} & r_{22} & r_{23} \\ r_{31} & r_{32} & r_{33} \end{bmatrix}.
$$

According to Eq. (5.9), one has

$$
\begin{aligned}
\tan\psi &= \frac{r_{21}}{r_{11}} \\
\sin\theta &= -r_{31} \\
\tan\phi &= \frac{r_{32}}{r_{33}}.
\end{aligned} \tag{5.10}
$$

By considering Euler angles' value ranges $\psi \in [-\pi, \pi]$, $\theta \in [-\pi/2, \pi/2]$, $\phi \in [-\pi, \pi]$, the solutions to Eq. (5.10) are

$$
\begin{aligned}
\psi &= \arctan \frac{r_{21}}{r_{11}} \\
\theta &= \arcsin (-r_{31}) \\
\phi &= \arctan \frac{r_{32}}{r_{33}}.
\end{aligned} \tag{5.11}
$$

Here, the value ranges for $\arctan (\cdot)$ and $\arcsin (\cdot)$ are $[-\pi/2, \pi/2]$. During an aggressive maneuver, the actual value range of pitch angle may be $\theta \in [-\pi, \pi]$. Therefore, the solutions above must be extended. Moreover, the Euler angle representation has the *singularity* problem, which will occur when $\theta = \pm\pi/2$, i.e., when $r_{11} = r_{21} = 0$.

When $\theta = \pm\pi/2$, $\mathbf{R}_b^e$ is rewritten as

$$
\mathbf{R}_b^e = \begin{bmatrix} 0 & -\sin (\psi \mp \phi) & \cos (\psi \mp \phi) \\ 0 & \cos (\psi \mp \phi) & \sin (\psi \mp \phi) \\ \mp 1 & 0 & 0 \end{bmatrix}. \tag{5.12}
$$

In this case, although there is an one-to-one correspondence between $\psi \mp \phi$ and $\mathbf{R}_b^e$, the concrete values of ψ, ϕ cannot be determined uniquely. There exist an infinite number of combinations. In other words, for $\theta = \pm\pi/2$, a unique solution for ψ, ϕ does not exist. In order to avoid the singularity problem, it is assumed that $\phi = 0$ in this case. In the following, a method is proposed to avoid the singularity problem and to obtain ϕ, θ, ψ. If $r_{11} = r_{21} = 0$, then

$$
\begin{aligned}
\phi &= 0 \\
\psi &= \arctan 2(-r_{12}, r_{22}) \\
\theta &= \text{sign}(-r_{31})\frac{\pi}{2}.
\end{aligned}
\tag{5.13}
$$

Otherwise

$$
(\phi, \theta, \psi) = \arg \min_{\phi_i, \theta_j, \psi_k, i, j, k \in \{0,1\}} \left\| \mathbf{R}_b^e - \mathbf{R}_z^{-1}(\psi_k)\,\mathbf{R}_y^{-1}(\theta_j)\,\mathbf{R}_x^{-1}(\phi_i) \right\|.
\tag{5.14}
$$

Here ϕ_i, θ_j, ψ_k are expressed as

$$
\begin{aligned}
\psi_0 &= \arctan 2(r_{21}, r_{11}), & \psi_1 &= \arctan 2(-r_{21}, -r_{11}) \\
\theta_0 &= \arcsin(-r_{31}), & \theta_1 &= \text{sign}(\theta_0)\pi - \theta_0 \\
\phi_0 &= \arctan 2(r_{32}, r_{33}), & \phi_1 &= \arctan 2(-r_{32}, -r_{33})
\end{aligned}
\tag{5.15}
$$

where the function arctan2 is defined as

$$
\arctan 2(y, x) \triangleq
\begin{cases}
\arctan(y/x) & x > 0 \\
\arctan(y/x) + \pi & y \geq 0, x < 0 \\
\arctan(y/x) - \pi & y < 0, x < 0 \\
+\pi/2 & y > 0, x = 0 \\
-\pi/2 & y < 0, x = 0 \\
\text{undefined} & y = 0, x = 0.
\end{cases}
\tag{5.16}
$$

Although each Euler angle has two optional values in (5.15), the true value is uniquely determined by the optimization (5.14) in most cases. However, there are also some exceptions. For example, when $r_{11} = r_{33} = 1, r_{21} = r_{32} = 0, r_{31} = 0$, the results consist of two cases: $\psi_0 = 0, \theta_0 = 0, \phi_0 = 0$; $\psi_1 = \pi, \theta_1 = \pi, \phi_1 = \pi$. For the two cases, the corresponding rotation matrices are the same, namely

$$
\mathbf{R}_b^e = \begin{bmatrix} 1 & 0 & 0 \\ 0 & 1 & 0 \\ 0 & 0 & 1 \end{bmatrix}.
$$

In practice, because of the continuity of rotation, the case which is close to the former value $(\psi(t - \Delta t), \theta(t - \Delta t), \phi(t - \Delta t), \Delta t \in \mathbb{R}_+$ is a small value) can be selected as the true one.

5.2.2.2   Relationship Between the Derivative of the Rotation Matrix and the Aircraft Body's Angular Velocity

If the rigid body's rotation (without translation) is only considered, then the derivative of a vector ${}^e\mathbf{r} \in \mathbb{R}^3$ satisfies

$$
\frac{d{}^e\mathbf{r}}{dt} = {}^e\boldsymbol{\omega} \times {}^e\mathbf{r}
\tag{5.17}
$$

where the symbol × represents the vector cross product. The circular motion in Fig. 5.6 can illustrate Eq. (5.17) intuitively.

The cross product of two vectors $\mathbf{a} \triangleq [\, a_x\ a_y\ a_z\,]^T$ and $\mathbf{b} \triangleq [\, b_x\ b_y\ b_z\,]^T$ is defined as [4, pp. 25–26]

$$\mathbf{a} \times \mathbf{b} = [\mathbf{a}]_\times \mathbf{b} \tag{5.18}$$

where

$$[\mathbf{a}]_\times \triangleq \begin{bmatrix} 0 & -a_z & a_y \\ a_z & 0 & -a_x \\ -a_y & a_x & 0 \end{bmatrix} \tag{5.19}$$

is a skew symmetric matrix.

According to Eq. (5.17), one has

$$\frac{d\left[\, ^e\mathbf{b}_1\ ^e\mathbf{b}_2\ ^e\mathbf{b}_3\,\right]}{dt} = \left[\, ^e\boldsymbol{\omega} \times\, ^e\mathbf{b}_1\ ^e\boldsymbol{\omega} \times^e \mathbf{b}_2\ ^e\boldsymbol{\omega} \times^e \mathbf{b}_3\,\right]. \tag{5.20}$$

Since $^e\boldsymbol{\omega} = \mathbf{R}_b^e \cdot {}^b\boldsymbol{\omega}$, by using the properties of the cross product, Eq. (5.20) is further written as

$$\frac{d\mathbf{R}_b^e}{dt} = \mathbf{R}_b^e \left[^b\boldsymbol{\omega} \right]_\times \tag{5.21}$$

where $\left[^b\boldsymbol{\omega} \right]_\times$ is the skew symmetric form of $^b\boldsymbol{\omega}$. The corresponding derivation is left as an exercise at the end of this chapter.

Fig. 5.6 The derivative of a vector presented by a circular motion

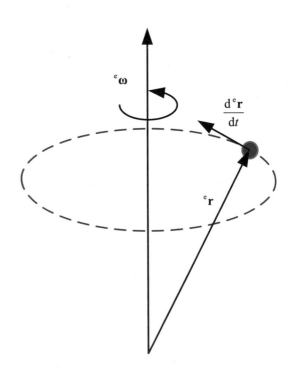

The use of the rotation matrix can avoid the singularity problem. However, since $\mathbf{R}_b^e$ has nine unknown variables, the computational burden of solving Eq. (5.21) is heavy. In the following section, another method known as quaternion representation is described.

5.2.3   Quaternions

So far, quaternion representation is one of the most widely used methods of attitude representation. Quaternion algebra was introduced by William Rowan Hamilton in 1843, and some related theorems were established, but neither of these theorems could be applied in real systems due to the limited computational resources at that time. In memory of Hamilton, people carved the formula for the quaternions into the stone of Brougham Bridge,[1] as shown in Fig. 5.7. Readers can refer to [5, 6] about this history. Recently, quaternions are getting more and more attention because of the wide application of high-performance computers and the rapid development of the aircraft's attitude control technologies.

5.2.3.1   Definition of Quaternions
Quaternions are normally written as

$$\mathbf{q} \triangleq \begin{bmatrix} q_0 \\ \mathbf{q}_v \end{bmatrix} \tag{5.22}$$

where $q_0 \in \mathbb{R}$ is the scalar part of $\mathbf{q} \in \mathbb{R}^4$ and $\mathbf{q}_v = [\, q_1 \ q_2 \ q_3 \,]^T \in \mathbb{R}^3$ is the vector part. For a real number $s \in \mathbb{R}$, the corresponding quaternion is defined as $\mathbf{q} = [\, s \ \mathbf{0}_{1\times3} \,]^T$. For a vector $\mathbf{v} \in \mathbb{R}^3$, the corresponding quaternion is $\mathbf{q} = [\, 0 \ \mathbf{v}^T \,]^T$.

5.2.3.2   Quaternions' Basic Operation Rules [7, 8]
(1) Addition and subtraction

$$\mathbf{p} \pm \mathbf{q} = \begin{bmatrix} p_0 \\ \mathbf{p}_v \end{bmatrix} \pm \begin{bmatrix} q_0 \\ \mathbf{q}_v \end{bmatrix} = \begin{bmatrix} p_0 \pm q_0 \\ \mathbf{p}_v \pm \mathbf{q}_v \end{bmatrix}.$$

Fig. 5.7 Quaternion plaque on Brougham (Broom) Bridge, Dublin (photo from https://en. wikipedia.org)

---

[1]It reads: "Here as he walked by on the 16th of October 1843 Sir William Rowan Hamilton in a flash of genius discovered the fundamental formula for quaternion multiplication $i^2 = j^2 = k^2 = ijk = -1$ & cut it on a stone of this bridge."

(2) Multiplication

$$\mathbf{p} \otimes \mathbf{q} = \begin{bmatrix} p_0 \\ \mathbf{p}_v \end{bmatrix} \otimes \begin{bmatrix} q_0 \\ \mathbf{q}_v \end{bmatrix} = \begin{bmatrix} p_0 q_0 - \mathbf{q}_v^T \mathbf{p}_v \\ \mathbf{p}_v \times \mathbf{q}_v + p_0 \mathbf{q}_v + q_0 \mathbf{p}_v \end{bmatrix}.$$

The product of two quaternions is expressed as two equivalent matrix products, namely

$$\mathbf{p} \otimes \mathbf{q} = \mathbf{p}^+ \mathbf{q} = \mathbf{q}^- \mathbf{p}$$

where

$$\mathbf{p}^+ = p_0 \mathbf{I}_4 + \begin{bmatrix} 0 & -\mathbf{p}_v^T \\ \mathbf{p}_v & [\mathbf{p}_v]_\times \end{bmatrix}, \mathbf{q}^- = q_0 \mathbf{I}_4 + \begin{bmatrix} 0 & -\mathbf{q}_v^T \\ \mathbf{q}_v & -[\mathbf{q}_v]_\times \end{bmatrix}.$$

Multiplication properties:

1) It is non-commutative

$$\mathbf{p} \otimes \mathbf{q} \neq \mathbf{q} \otimes \mathbf{p}.$$

2) It is distributive and associative

$$\mathbf{q} \otimes (\mathbf{r} + \mathbf{m}) = \mathbf{q} \otimes \mathbf{r} + \mathbf{q} \otimes \mathbf{m}$$
$$\mathbf{q} \otimes \mathbf{r} \otimes \mathbf{m} = (\mathbf{q} \otimes \mathbf{r}) \otimes \mathbf{m} = \mathbf{q} \otimes (\mathbf{r} \otimes \mathbf{m}).$$

3) Scalar multiplication

$$s\mathbf{q} = \mathbf{q}s = \begin{bmatrix} sq_0 \\ s\mathbf{q}_v \end{bmatrix}.$$

4) Multiplication between corresponding quaternions of two vectors

$$\mathbf{q_u} \otimes \mathbf{q_v} = \begin{bmatrix} 0 \\ \mathbf{u} \end{bmatrix} \otimes \begin{bmatrix} 0 \\ \mathbf{v} \end{bmatrix} = \begin{bmatrix} -\mathbf{u}^T \mathbf{v} \\ \mathbf{u} \times \mathbf{v} \end{bmatrix}.$$

(3) Conjugate
The conjugate of a quaternion is defined by

$$\mathbf{q}^* = \begin{bmatrix} q_0 \\ -\mathbf{q}_v \end{bmatrix} \tag{5.23}$$

with $\mathbf{q} = [\, q_0 \; \mathbf{q}_v^T \,]^T$. It has following properties

$$(\mathbf{q}^*)^* = \mathbf{q}$$
$$(\mathbf{p} \otimes \mathbf{q})^* = \mathbf{q}^* \otimes \mathbf{p}^*$$
$$(\mathbf{p} + \mathbf{q})^* = \mathbf{p}^* + \mathbf{q}^*.$$

(4) Norm
The norm of a quaternion is defined by

$$\begin{aligned}
\|\mathbf{q}\|^2 &= \|\mathbf{q} \otimes \mathbf{q}^*\| \\
&= \|\mathbf{q}^* \otimes \mathbf{q}\| \\
&= q_0^2 + \mathbf{q}_v^T \mathbf{q}_v \\
&= q_0^2 + q_1^2 + q_2^2 + q_3^2.
\end{aligned}$$

It has following properties

$$\begin{aligned}
\|\mathbf{p} \otimes \mathbf{q}\| &= \|\mathbf{p}\| \, \|\mathbf{q}\| \\
\|\mathbf{q}^*\| &= \|\mathbf{q}\| \, .
\end{aligned}$$

(5) Inverse
The inverse quaternion $\mathbf{q}^{-1}$ satisfies

$$\mathbf{q} \otimes \mathbf{q}^{-1} = \begin{bmatrix} 1 \\ \mathbf{0}_{3 \times 1} \end{bmatrix}$$

which can be computed by

$$\mathbf{q}^{-1} = \frac{\mathbf{q}^*}{\|\mathbf{q}\|}.$$

(6) Unit or normalized quaternion
For a unit quaternion, it satisfies $\|\mathbf{q}\| = 1$. Let $\|\mathbf{p}\| = \|\mathbf{q}\| = 1$. Then

$$\begin{aligned}
\|\mathbf{p} \otimes \mathbf{q}\| &= 1 \\
\mathbf{q}^{-1} &= \mathbf{q}^*.
\end{aligned}$$

(7) Division
If $\mathbf{r} \otimes \mathbf{p} = \mathbf{m}$, then

$$\begin{aligned}
\mathbf{r} \otimes \mathbf{p} \otimes \mathbf{p}^{-1} &= \mathbf{m} \otimes \mathbf{p}^{-1} \\
\mathbf{r} &= \mathbf{m} \otimes \mathbf{p}^{-1}.
\end{aligned}$$

If $\mathbf{p} \otimes \mathbf{r} = \mathbf{m}$, then

$$\begin{aligned}
\mathbf{p}^{-1} \otimes \mathbf{p} \otimes \mathbf{r} &= \mathbf{p}^{-1} \otimes \mathbf{m} \\
\mathbf{r} &= \mathbf{p}^{-1} \otimes \mathbf{m}.
\end{aligned}$$

5.2.3.3 Quaternions as Rotations

Based on the facts described above, the reason that quaternions are used to represent rotations is explained in the following. First, the vector rotation is considered. Assume that $\mathbf{q}$ represents a rotation process and $\mathbf{v}_1 \in \mathbb{R}^3$ represents a vector. Then, under the action of $\mathbf{q}$, the vector $\mathbf{v}_1$ is turned into $\mathbf{v}_1' \in \mathbb{R}^3$. This process is expressed as

$$\begin{bmatrix} 0 \\ \mathbf{v}_1' \end{bmatrix} = \mathbf{q} \otimes \begin{bmatrix} 0 \\ \mathbf{v}_1 \end{bmatrix} \otimes \mathbf{q}^{-1}. \tag{5.24}$$

Using the basic operation rules, it is easily proved that the first row of Eq. (5.24) always stands. The corresponding derivation is left as an exercise at the end of this chapter. As shown in Fig. 5.8, a unit quaternion can always be written in the form

$$\mathbf{q} = \begin{bmatrix} \cos \dfrac{\theta}{2} \\ \mathbf{v} \sin \dfrac{\theta}{2} \end{bmatrix} \tag{5.25}$$

where $\mathbf{v} \in \mathbb{R}^3$ represents the rotation axis satisfying $\|\mathbf{v}\| = 1$, $\theta \in \mathbb{R}$ represents the rotation angle, and $\mathbf{q} \in \mathbb{R}^4$ represents the rotation about $\mathbf{v}$ by θ (the positive direction of the rotation is also determined by the right-hand rule).

The reason why quaternions are able to represent a 3D rotation is stated in the following theorem.

Theorem 5.1 ([7]) *Let* $\mathbf{p}_v = [\, p_x \ p_y \ p_z \,]^\mathrm{T} \in \mathbb{R}^3$ *be a point in 3D space. It is represented as a quaternion using its homogeneous coordinates, namely* $\mathbf{p} = [\, 0 \ \mathbf{p}_v^\mathrm{T} \,]^\mathrm{T}$. *Let* $\mathbf{q}$ *be a unit quaternion. Then,*

(1) The product $\mathbf{q} \otimes \mathbf{p} \otimes \mathbf{q}^{-1}$ *takes* $\mathbf{p}$ *to* $\mathbf{p}' = [\, p_0' \ \mathbf{p}_v'^\mathrm{T} \,]^\mathrm{T}$, *with* $\|\mathbf{p}_v\| = \|\mathbf{p}_v'\|$.

(2) Any nonzero real multiple s *of* $\mathbf{q}$ *gives the same action.*

(3) The quaternion $\mathbf{q}$ *in Eq. (5.25) represents a rotation. It acts to make a vector* $\mathbf{v}_t \in \mathbb{R}^3$ *rotate around a unit axis* $\mathbf{v} \in \mathbb{R}^3$ *by* θ, *and achieve* $\mathbf{v}_t' \in \mathbb{R}^3$.

Proof Let us start with Part 1. For any unit quaternion $\mathbf{q}$, one has $\mathbf{q}^{-1} = \mathbf{q}^*$. The scalar part of $\mathbf{q}$, denoted by $S(\mathbf{q})$, can be extracted by using equation $2S(\mathbf{q}) = \mathbf{q} + \mathbf{q}^*$. Then

$$\begin{aligned}
2S(\mathbf{p}') &= 2S\left(\mathbf{q} \otimes \mathbf{p} \otimes \mathbf{q}^{-1}\right) \\
&= 2S\left(\mathbf{q} \otimes \mathbf{p} \otimes \mathbf{q}^*\right) \\
&= \mathbf{q} \otimes \mathbf{p} \otimes \mathbf{q}^* + (\mathbf{q} \otimes \mathbf{p} \otimes \mathbf{q}^*)^* \\
&= \mathbf{q} \otimes \mathbf{p} \otimes \mathbf{q}^* + \mathbf{q} \otimes \mathbf{p}^* \otimes \mathbf{q}^* \\
&= \mathbf{q} \otimes (\mathbf{p} + \mathbf{p}^*) \otimes \mathbf{q}^* \\
&= \mathbf{q} \otimes (2S(\mathbf{p})) \otimes \mathbf{q}^* \\
&= 2S(\mathbf{p})\, \mathbf{q} \otimes \mathbf{q}^* \\
&= 2S(\mathbf{p}).
\end{aligned} \tag{5.26}$$

Fig. 5.8 Physical meaning of unit quaternions

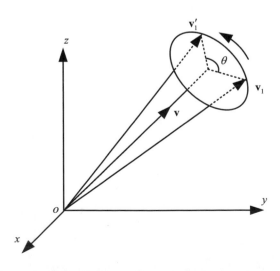

Hence, the action of the unit quaternion $\mathbf{q}$ on $\mathbf{p}$ gives $\mathbf{p}'$ and results in the same scalar part, namely $S\left(\mathbf{p}'\right) = p_0' = S\left(\mathbf{p}\right) = 0$. The value of p_0' can also be determined according to Eq. (5.24), which also equals to zero. Besides, since $\left\|\mathbf{p}'\right\| = \left\|\mathbf{q} \otimes \mathbf{p} \otimes \mathbf{q}^*\right\| = \left\|\mathbf{p}\right\|$, the norm of the quaternion is unchanged after the action, namely $\left\|\mathbf{p}_v\right\| = \left\|\mathbf{p}_v'\right\|$.

The process of Part 2 is simple. Since $(s\mathbf{q})^{-1} = \mathbf{q}^{-1}s^{-1}$, and scalar multiplication is commutative, one has

$$(s\mathbf{q}) \otimes \mathbf{p} \otimes (s\mathbf{q})^{-1} = s\mathbf{q} \otimes \mathbf{p} \otimes \mathbf{q}^{-1}s^{-1} = \mathbf{q} \otimes \mathbf{p} \otimes \mathbf{q}^{-1}s^{-1}s = \mathbf{q} \otimes \mathbf{p} \otimes \mathbf{q}^{-1}. \tag{5.27}$$

According to Eq. (5.27), any nonzero real multiple s of the unit quaternion $\mathbf{q}$ gives the same action. Furthermore, Part 1 is true for any nonzero quaternion.

Finally, let us turn to Part 3, the heart of the theorem. As shown in Fig. 5.9, two unit vectors $\mathbf{v}_0, \mathbf{v}_1$ ($\mathbf{v}_1 \neq \pm\mathbf{v}_0$) are defined with $\theta/2$ being the angle between them. Therefore, one has

$$\mathbf{v}_0^T\mathbf{v}_1 = \cos\frac{\theta}{2}. \tag{5.28}$$

A unit vector $\mathbf{v}$ is defined in the direction of the cross product of $\mathbf{v}_0$ and $\mathbf{v}_1$, namely

$$\mathbf{v} = \frac{\mathbf{v}_0 \times \mathbf{v}_1}{\left\|\mathbf{v}_0 \times \mathbf{v}_1\right\|} = \frac{\mathbf{v}_0 \times \mathbf{v}_1}{\left\|\mathbf{v}_0\right\| \left\|\mathbf{v}_1\right\| \sin\dfrac{\theta}{2}} = \frac{\mathbf{v}_0 \times \mathbf{v}_1}{\sin\dfrac{\theta}{2}}. \tag{5.29}$$

By the property of the cross product, the vector $\mathbf{v}$ is perpendicular to both $\mathbf{v}_0$ and $\mathbf{v}_1$. Furthermore,

$$\mathbf{v}_0 \times \mathbf{v}_1 = \mathbf{v}\sin\frac{\theta}{2}. \tag{5.30}$$

By using Eqs. (5.28) and (5.30), the unit quaternion $\mathbf{q}$ in Eq. (5.25) can be represented as

$$\mathbf{q} = \begin{bmatrix} \mathbf{v}_0^T\mathbf{v}_1 \\ \mathbf{v}_0 \times \mathbf{v}_1 \end{bmatrix} = \begin{bmatrix} 0 \\ \mathbf{v}_1 \end{bmatrix} \otimes \begin{bmatrix} 0 \\ \mathbf{v}_0 \end{bmatrix}^*. \tag{5.31}$$

In the following, it will be shown that the quaternion $\mathbf{q}$ is able to represent a rotation. Let

$$\begin{bmatrix} 0 \\ \mathbf{v}_2 \end{bmatrix} = \mathbf{q} \otimes \begin{bmatrix} 0 \\ \mathbf{v}_0 \end{bmatrix} \otimes \mathbf{q}^{-1}. \tag{5.32}$$

Fig. 5.9 Rotation represented by quaternions

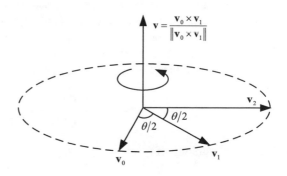

According to Part 1, $\|\mathbf{v}_2\| = \|\mathbf{v}_0\|$. Subsequently, the relative position between $\mathbf{v}_2$ and $\mathbf{v}_0$ is further studied. By multiplying Eq. (5.32) by $\begin{bmatrix} 0 \\ \mathbf{v}_1 \end{bmatrix}^*$ on both sides, and further using the properties of unit quaternions and conjugate quaternions, one has

$$
\begin{aligned}
\begin{bmatrix} 0 \\ \mathbf{v}_2 \end{bmatrix} \otimes \begin{bmatrix} 0 \\ \mathbf{v}_1 \end{bmatrix}^* &= \left(\mathbf{q} \otimes \begin{bmatrix} 0 \\ \mathbf{v}_0 \end{bmatrix} \otimes \mathbf{q}^{-1} \right) \otimes \begin{bmatrix} 0 \\ \mathbf{v}_1 \end{bmatrix}^* \\
&= \mathbf{q} \otimes \begin{bmatrix} 0 \\ \mathbf{v}_0 \end{bmatrix} \otimes \left(\begin{bmatrix} 0 \\ \mathbf{v}_0 \end{bmatrix} \otimes \begin{bmatrix} 0 \\ \mathbf{v}_1 \end{bmatrix}^* \right) \otimes \begin{bmatrix} 0 \\ \mathbf{v}_1 \end{bmatrix}^* \\
&= \mathbf{q} \otimes \left(\begin{bmatrix} 0 \\ \mathbf{v}_0 \end{bmatrix} \otimes \begin{bmatrix} 0 \\ \mathbf{v}_0 \end{bmatrix} \right) \otimes \left(\begin{bmatrix} 0 \\ \mathbf{v}_1 \end{bmatrix}^* \otimes \begin{bmatrix} 0 \\ \mathbf{v}_1 \end{bmatrix}^* \right) \\
&= \mathbf{q} \otimes \begin{bmatrix} -1 \\ \mathbf{0}_{3 \times 1} \end{bmatrix} \otimes \begin{bmatrix} -1 \\ \mathbf{0}_{3 \times 1} \end{bmatrix} \\
&= \mathbf{q} \\
&= \begin{bmatrix} 0 \\ \mathbf{v}_1 \end{bmatrix} \otimes \begin{bmatrix} 0 \\ \mathbf{v}_0 \end{bmatrix}^*.
\end{aligned}
\tag{5.33}
$$

From Eq. (5.33), one has

$$
\begin{aligned}
\mathbf{v}_2^{\mathrm{T}} \mathbf{v}_1 &= \mathbf{v}_1^{\mathrm{T}} \mathbf{v}_0 \\
\mathbf{v}_2 \times \mathbf{v}_1 &= \mathbf{v}_1 \times \mathbf{v}_0.
\end{aligned}
$$

So, $\mathbf{v}_2$ lies in the same plane as $\mathbf{v}_0$ and $\mathbf{v}_1$, as shown in Fig. 5.9, and also forms an angle $\theta/2$ with $\mathbf{v}_1$. The action of the unit quaternion $\mathbf{q}$ on $\mathbf{v}_0$ can be seen as the rotation of $\mathbf{v}_0$ about $\mathbf{v}$ by θ and achieves $\mathbf{v}_2$. Furthermore, $\mathbf{q}$ acts on $\mathbf{v}_1$ to produce a vector which is also in the same plane, at an angle of $\theta/2$ with $\mathbf{v}_2$. Define

$$
\begin{bmatrix} 0 \\ \mathbf{v}_3 \end{bmatrix} = \mathbf{q} \otimes \begin{bmatrix} 0 \\ \mathbf{v}_1 \end{bmatrix} \otimes \mathbf{q}^{-1}.
\tag{5.34}
$$

By multiplying Eq. (5.34) by $\begin{bmatrix} 0 \\ \mathbf{v}_2 \end{bmatrix}^*$ on both sides, and further using the properties of unit quaternions, conjugate quaternions and inverse quaternions, one has

$$
\begin{bmatrix} 0 \\ \mathbf{v}_3 \end{bmatrix} \otimes \begin{bmatrix} 0 \\ \mathbf{v}_2 \end{bmatrix}^* = \begin{bmatrix} 0 \\ \mathbf{v}_2 \end{bmatrix} \otimes \begin{bmatrix} 0 \\ \mathbf{v}_1 \end{bmatrix}^*.
\tag{5.35}
$$

The corresponding derivation is left as an exercise at the end of this chapter. Equations (5.33) and (5.35) show that the actions of $\mathbf{q}$ on $\mathbf{v}_0$ and $\mathbf{v}_1$ are, as required, to rotate around the axis $\mathbf{v}$ by an angle of θ. In fact, this quaternion can act on any vector $\mathbf{v}_t \in \mathbb{R}^3$, as is shown by splitting $\mathbf{v}_t$ into $s_0 \mathbf{v}_0 + s_1 \mathbf{v}_1 + s_2 \mathbf{v}$, where $s_0, s_1, s_2 \in \mathbb{R}$. Bilinearity allows to examine the action on $\mathbf{v}_0, \mathbf{v}_1, \mathbf{v}$, separately. As a result, it only needs to prove that Part 3 is true for the action on $\mathbf{v}$. It is easy to find

$$
\mathbf{q} \otimes \begin{bmatrix} 0 \\ \mathbf{v} \end{bmatrix} = \begin{bmatrix} 0 \\ \mathbf{v} \end{bmatrix} \otimes \mathbf{q}
\tag{5.36}
$$

according to the definition of $\mathbf{q}$ and $\mathbf{v}$. By multiplying Eq. (5.36) by $\mathbf{q}^*$ on both sides, one has

$$\mathbf{q} \otimes \begin{bmatrix} 0 \\ \mathbf{v} \end{bmatrix} \otimes \mathbf{q}^* = \begin{bmatrix} 0 \\ \mathbf{v} \end{bmatrix} \otimes \mathbf{q} \otimes \mathbf{q}^* = \begin{bmatrix} 0 \\ \mathbf{v} \end{bmatrix}$$

which is consistent with the interpretation of $\mathbf{v}$ as the axis of rotation. Thus, the action of $\mathbf{q}$ on every vector is a rotation around $\mathbf{v}$ by θ, which concludes the proof of Part 3. $\square$

By generalizing Theorem 5.1, it can be found that every 3D rotation corresponds to a unit quaternion. Furthermore, if $\mathbf{q}_1, \mathbf{q}_2$ are two quaternions, the action of $\mathbf{q}_1, \mathbf{q}_2$ on $\mathbf{p} = [\, p_0 \ \mathbf{p}_v^T \,]^T$ one after another is expressed as

$$\mathbf{p}' = \mathbf{q}_2 \otimes \left(\mathbf{q}_1 \otimes \mathbf{p} \otimes \mathbf{q}_1^{-1} \right) \otimes \mathbf{q}_2^{-1} = (\mathbf{q}_2 \otimes \mathbf{q}_1) \otimes \mathbf{p} \otimes (\mathbf{q}_2 \otimes \mathbf{q}_1)^{-1}$$

where $\mathbf{q}_2 \otimes \mathbf{q}_1$ is the composite rotation quaternion in the case of rotating vectors.

In the following, the coordinate frame rotation will be discussed. As shown in Fig. 5.10, a vector $\mathbf{v}_0 \in \mathbb{R}^3$ is fixed in the coordinate frame $oxyz$. The coordinate of $\mathbf{v}_0$ in $oxyz$ is $\mathbf{v}_1 = [x \ y \ z]^T \in \mathbb{R}^3$. The rotation of $oxyz$ about the axis $\mathbf{v} \in \mathbb{R}^3$ by the angle θ results in a new coordinate frame $ox'y'z'$. The coordinate of $\mathbf{v}_0$ in $ox'y'z'$ is $\mathbf{v}_1' = \left[x' \ y' \ z' \right]^T \in \mathbb{R}^3$. Then, the transformation of coordinates of the vector $\mathbf{v}_0$ between two frames can be written as

$$\begin{bmatrix} 0 \\ \mathbf{v}_1' \end{bmatrix} = \mathbf{q}^{-1} \otimes \begin{bmatrix} 0 \\ \mathbf{v}_1 \end{bmatrix} \otimes \mathbf{q} \tag{5.37}$$

where $\mathbf{q} = \left[\cos \theta/2 \ \mathbf{v}^T \sin \theta/2 \right]^T$ [9]. Readers may find Eq. (5.37) different from Eq. (5.24). This is because Eq. (5.24) represents a rotation of a vector in the same coordinate frame, whereas in Eq. (5.37), the coordinate frame is the one that rotates. Also, for this reason, Eq. (5.37) happens to be the inversion of Eq. (5.24). Furthermore, the composite rotation quaternion in the case of rotating coordinate frames, which is different from that of rotating vectors, is $\mathbf{q}_1 \otimes \mathbf{q}_2$, where $\mathbf{q}_1, \mathbf{q}_2$ are two quaternions that represent the sequential rotation of a coordinate frame, respectively.

Fig. 5.10 Coordinate frame rotation by a quaternion

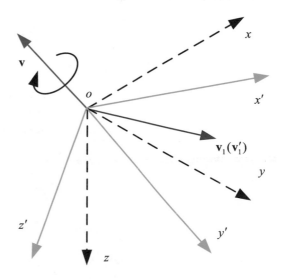

5.2.3.4 Quaternions and Rotation Matrix [10]

It is assumed that the rotation from the EFCF to the ABCF is represented by the quaternion $\mathbf{q}_e^b \triangleq [q_0 \ q_1 \ q_2 \ q_3]^T$. According to Eq. (5.37), one has

$$
\begin{aligned}
\begin{bmatrix} 0 \\ {}^e\mathbf{r} \end{bmatrix} &= (\mathbf{q}_b^e)^{-1} \otimes \begin{bmatrix} 0 \\ {}^b\mathbf{r} \end{bmatrix} \otimes \mathbf{q}_b^e \\
&= \mathbf{q}_e^b \otimes \begin{bmatrix} 0 \\ {}^b\mathbf{r} \end{bmatrix} \otimes (\mathbf{q}_e^b)^{-1}
\end{aligned}
\tag{5.38}
$$

where ${}^e\mathbf{r}, {}^b\mathbf{r} \in \mathbb{R}^3$ are the representations of the vector $\mathbf{r} \in \mathbb{R}^3$ in the two frames, respectively. The quaternions and the rotation matrix are related to each other. According to rules in Sect. 5.2.3.2, Eq. (5.38) is expressed as follows:

$$
\begin{bmatrix} 0 \\ {}^e\mathbf{r} \end{bmatrix} = \begin{bmatrix} q_0 & q_1 & q_2 & q_3 \\ -q_1 & q_0 & -q_3 & q_2 \\ -q_2 & q_3 & q_0 & -q_1 \\ -q_3 & -q_2 & q_1 & q_0 \end{bmatrix} \begin{bmatrix} q_0 & -q_1 & -q_2 & -q_3 \\ q_1 & q_0 & -q_3 & q_2 \\ q_2 & q_3 & q_0 & -q_1 \\ q_3 & -q_2 & q_1 & q_0 \end{bmatrix} \begin{bmatrix} 0 \\ {}^b\mathbf{r} \end{bmatrix}.
$$

Then,

$$
{}^e\mathbf{r} = \mathbf{C}(\mathbf{q}_e^b) \cdot {}^b\mathbf{r}
\tag{5.39}
$$

where

$$
\mathbf{C}(\mathbf{q}_e^b) \triangleq \begin{bmatrix} q_0^2 + q_1^2 - q_2^2 - q_3^2 & 2(q_1 q_2 - q_0 q_3) & 2(q_1 q_3 + q_0 q_2) \\ 2(q_1 q_2 + q_0 q_3) & q_0^2 - q_1^2 + q_2^2 - q_3^2 & 2(q_2 q_3 - q_0 q_1) \\ 2(q_1 q_3 - q_0 q_2) & 2(q_2 q_3 + q_0 q_1) & q_0^2 - q_1^2 - q_2^2 + q_3^2 \end{bmatrix}.
$$

Since $\mathbf{r}$ is an arbitrary vector, one has

$$
\mathbf{R}_b^e = \mathbf{C}(\mathbf{q}_e^b).
\tag{5.40}
$$

Furthermore, if $\mathbf{R}_b^e$ is assumed to be known as

$$
\mathbf{R}_b^e \triangleq \begin{bmatrix} r_{11} & r_{12} & r_{13} \\ r_{21} & r_{22} & r_{23} \\ r_{31} & r_{32} & r_{33} \end{bmatrix}
$$

then the corresponding quaternion is expressed as

$$
\begin{aligned}
q_0 &= \operatorname{sign}(q_0) \frac{1}{2} \sqrt{1 + r_{11} + r_{22} + r_{33}} \\
q_1 &= \operatorname{sign}(q_1) \frac{1}{2} \sqrt{1 + r_{11} - r_{22} - r_{33}} \\
q_2 &= \operatorname{sign}(q_2) \frac{1}{2} \sqrt{1 - r_{11} + r_{22} - r_{33}} \\
q_3 &= \operatorname{sign}(q_3) \frac{1}{2} \sqrt{1 - r_{11} - r_{22} + r_{33}}
\end{aligned}
\tag{5.41}
$$

where $\text{sign}(q_0)$ is 1 or -1. And

$$
\begin{aligned}
\text{sign}\,(q_1) &= \text{sign}\,(q_0)\,\text{sign}\,(r_{32} - r_{23}) \\
\text{sign}\,(q_2) &= \text{sign}\,(q_0)\,\text{sign}\,(r_{13} - r_{31}) \\
\text{sign}\,(q_3) &= \text{sign}\,(q_0)\,\text{sign}\,(r_{21} - r_{12})\,.
\end{aligned}
\tag{5.42}
$$

It should be noted that one rotation matrix corresponds to two quaternions, namely $\mathbf{q}_e^b$ and $-\mathbf{q}_e^b$.

5.2.3.5 Quaternions and Euler Angles

The rotation of a coordinate frame about $\mathbf{r} = [a\ b\ c]^T$, where $a^2 + b^2 + c^2 = 1$, by an angle α is expressed as a quaternion that

$$
\mathbf{q}\,(\alpha, \mathbf{r}) = \left[\cos\frac{\alpha}{2}\ \ a\sin\frac{\alpha}{2}\ \ b\sin\frac{\alpha}{2}\ \ c\sin\frac{\alpha}{2}\right]^T.
\tag{5.43}
$$

According to the rotation order in Sect. 5.2.1 in this chapter, one has

$$
\mathbf{q}_e^b = \mathbf{q}_z\,(\psi) \otimes \mathbf{q}_y\,(\theta) \otimes \mathbf{q}_x\,(\phi)
\tag{5.44}
$$

where

$$
\begin{aligned}
\mathbf{q}_x\,(\phi) &= \left[\cos\frac{\phi}{2}\ \ \sin\frac{\phi}{2}\ \ 0\ \ 0\right]^T \\
\mathbf{q}_y\,(\theta) &= \left[\cos\frac{\theta}{2}\ \ 0\ \ \sin\frac{\theta}{2}\ \ 0\right]^T \\
\mathbf{q}_z\,(\psi) &= \left[\cos\frac{\psi}{2}\ \ 0\ \ 0\ \ \sin\frac{\psi}{2}\right]^T.
\end{aligned}
$$

Furthermore, Eq. (5.44) is expressed as

$$
\mathbf{q}_e^b = \begin{bmatrix}
\cos\frac{\phi}{2}\cos\frac{\theta}{2}\cos\frac{\psi}{2} + \sin\frac{\phi}{2}\sin\frac{\theta}{2}\sin\frac{\psi}{2} \\
\sin\frac{\phi}{2}\cos\frac{\theta}{2}\cos\frac{\psi}{2} - \cos\frac{\phi}{2}\sin\frac{\theta}{2}\sin\frac{\psi}{2} \\
\cos\frac{\phi}{2}\sin\frac{\theta}{2}\cos\frac{\psi}{2} + \sin\frac{\phi}{2}\cos\frac{\theta}{2}\sin\frac{\psi}{2} \\
\cos\frac{\phi}{2}\cos\frac{\theta}{2}\sin\frac{\psi}{2} - \sin\frac{\phi}{2}\sin\frac{\theta}{2}\cos\frac{\psi}{2}
\end{bmatrix}.
\tag{5.45}
$$

According to Eq. (5.37), the rotation from $^e\mathbf{r} \in \mathbb{R}^3$ to $^b\mathbf{r}$ is written as

$$
\begin{aligned}
\begin{bmatrix} 0 \\ {}^b\mathbf{r} \end{bmatrix} &= \left(\mathbf{q}_e^b\right)^{-1} \otimes \begin{bmatrix} 0 \\ {}^e\mathbf{r} \end{bmatrix} \otimes \mathbf{q}_e^b \\
&= \left(\mathbf{q}_z\,(\psi) \otimes \mathbf{q}_y\,(\theta) \otimes \mathbf{q}_x\,(\phi)\right)^{-1} \otimes \begin{bmatrix} 0 \\ {}^e\mathbf{r} \end{bmatrix} \otimes \left(\mathbf{q}_z\,(\psi) \otimes \mathbf{q}_y\,(\theta) \otimes \mathbf{q}_x\,(\phi)\right) \\
&= (\mathbf{q}_x\,(\phi))^{-1} \otimes \left((\mathbf{q}_y\,(\theta))^{-1} \otimes \left((\mathbf{q}_z\,(\psi))^{-1} \otimes \begin{bmatrix} 0 \\ {}^e\mathbf{r} \end{bmatrix} \otimes \mathbf{q}_z\,(\psi)\right) \otimes \mathbf{q}_y\,(\theta)\right) \otimes \mathbf{q}_x\,(\phi)\,.
\end{aligned}
$$

It should be noted that the rotation order is $\psi \rightarrow \theta \rightarrow \phi$. This is consistent with the definition in Sect. 5.2.1 in this chapter (see Fig. 5.5).

In turn, Euler angles can be derived by using quaternions. According to Eq. (5.45), one has

$$
\begin{aligned}
\tan \phi &= \frac{2(q_0 q_1 + q_2 q_3)}{1 - 2\left(q_1^2 + q_2^2\right)} \\
\sin \theta &= 2\left(q_0 q_2 - q_1 q_3\right) \\
\tan \psi &= \frac{2\left(q_0 q_3 + q_1 q_2\right)}{1 - 2\left(q_2^2 + q_3^2\right)}.
\end{aligned}
\tag{5.46}
$$

By considering that Euler angles' value ranges are $\psi \in [-\pi, \pi], \theta \in [-\pi/2, \pi/2], \phi \in [-\pi, \pi]$, the solutions to Eq. (5.46) are

$$
\begin{aligned}
\phi &= \arctan \frac{2(q_0 q_1 + q_2 q_3)}{1 - 2\left(q_1^2 + q_2^2\right)} \\
\theta &= \arcsin \left(2\left(q_0 q_2 - q_1 q_3\right)\right) \\
\psi &= \arctan \left(\frac{2\left(q_0 q_3 + q_1 q_2\right)}{1 - 2\left(q_2^2 + q_3^2\right)}\right).
\end{aligned}
\tag{5.47}
$$

Here, the value ranges for $\arctan(\cdot)$ and $\arcsin(\cdot)$ are $[-\pi/2, \pi/2]$. During an aggressive maneuver, the actual value range of pitch angle may be $\theta \in [-\pi, \pi]$. Therefore, solutions above must be extended. Moreover, the Euler angle representation has the singularity problem. The problem occurs when $\theta = \pm \pi/2$, i.e., when $2\left(q_0 q_2 - q_1 q_3\right) = 1 || 2\left(q_0 q_2 - q_1 q_3\right) = -1$.

When $\theta = \pm \pi/2$, $\mathbf{q}_e^b$ in Eq. (5.45) is rewritten as

$$
\mathbf{q}_e^b = \frac{\sqrt{2}}{2}
\begin{bmatrix}
\cos\left(\dfrac{\psi}{2} \mp \dfrac{\phi}{2}\right) \\
\mp \sin\left(\dfrac{\psi}{2} \mp \dfrac{\phi}{2}\right) \\
\pm \cos\left(\dfrac{\psi}{2} \mp \dfrac{\phi}{2}\right) \\
\sin\left(\dfrac{\psi}{2} \mp \dfrac{\phi}{2}\right)
\end{bmatrix}.
\tag{5.48}
$$

Observed from Eq. (5.48), although there is an one-to-one correspondence between $\psi \mp \phi$ and $\mathbf{q}_e^b$, the concrete values of ψ, ϕ cannot be obtained. There exist an infinite number of combinations. In other words, given the quaternion $\mathbf{q}_e^b$, ψ and ϕ cannot be uniquely determined.

5.2.3.6 Relationship Between the Derivative of the Quaternions and the Aircraft Body's Angular Velocity

First, what should be known is the relationship between $\mathbf{q}_e^b(t + \Delta t)$ and $\mathbf{q}_e^b(t)$. According to Eq. (5.37), under the action of $\mathbf{q}_e^b(t)$, the representation of $^b\mathbf{v}(t)$ with respect to $^e\mathbf{v}(t)$ and $\mathbf{q}_e^b(t)$ is

$$
\begin{bmatrix} 0 \\ ^b\mathbf{v}(t) \end{bmatrix} = \mathbf{q}_e^b(t)^{-1} \otimes \begin{bmatrix} 0 \\ ^e\mathbf{v}(t) \end{bmatrix} \otimes \mathbf{q}_e^b(t).
\tag{5.49}
$$

During the rotation from the EFCF to the ABCF, it is assumed that the ABCF has a tiny perturbation. Here, $\Delta \mathbf{q}$ is used to represent the perturbation from t to $t + \Delta t$. According to Eq. (5.43), when Δt is small enough, $\Delta \mathbf{q}$ is expressed as

$$\Delta \mathbf{q} = \left[1 \ \frac{1}{2} {}^{b}\boldsymbol{\omega}^{T} \Delta t \right]^{T} \tag{5.50}$$

which can be written in the form of Eq. (5.43) with $\alpha = \left\| {}^{b}\boldsymbol{\omega} \right\| \Delta t$ and $\mathbf{r} = {}^{b}\boldsymbol{\omega} / \left\| {}^{b}\boldsymbol{\omega} \right\|$. Then, the derivative of $\mathbf{q}_{e}^{b}(t)$ is obtained as

$$\dot{\mathbf{q}}_{e}^{b} = \frac{1}{2} \begin{bmatrix} 0 & -{}^{b}\boldsymbol{\omega}^{T} \\ {}^{b}\boldsymbol{\omega} & -\left[{}^{b}\boldsymbol{\omega} \right]_{\times} \end{bmatrix} \mathbf{q}_{e}^{b}. \tag{5.51}$$

The corresponding derivation is left as an exercise at the end of this chapter. In practice, ${}^{b}\boldsymbol{\omega}$ can be measured approximately by a three-axis gyroscope and can be considered as a known value. In this sense, Eq. (5.51) is linear. With $\mathbf{q}_{e}^{b} = [\, q_{0} \ \mathbf{q}_{v}^{T} \,]^{T}$, Eq. (5.51) is further rewritten as

$$\begin{aligned} \dot{q}_{0} &= -\frac{1}{2} \mathbf{q}_{v}^{T} \cdot {}^{b}\boldsymbol{\omega} \\ \dot{\mathbf{q}}_{v} &= \frac{1}{2} \left(q_{0} \mathbf{I}_{3} + \left[\mathbf{q}_{v} \right]_{\times} \right) \cdot {}^{b}\boldsymbol{\omega} \end{aligned} \tag{5.52}$$

where $-\left[{}^{b}\boldsymbol{\omega} \right]_{\times} \mathbf{q}_{v} = \left[\mathbf{q}_{v} \right]_{\times} {}^{b}\boldsymbol{\omega}$ is used.

The attitude representation using Euler angles has a singularity problem in the case of large angles (e.g., $\theta = \pm \pi/2$), whereas quaternions can help to keep the linearity of the equation and avoid the singularity problem. Compared with Euler angles, quaternions require only simple calculation and work for all attitudes. Moreover, the formula (5.51) which represents the relationship between quaternions' derivative and the aircraft body's angular velocity has only four unknown variables, whereas the differential equation (5.21) expressed by using the rotation matrix has nine unknown variables. Therefore, compared with the rotation matrix, quaternions have a better numerical stability as well as a higher efficiency.

5.3 Summary

This chapter presents the coordinate frames and attitude representation. All the results shown are self-contained. Readers can learn about the three different methods of attitude representation, namely Euler angles, rotation matrix, and quaternions, as well as the transformation among them, as shown in Fig. 5.11. Moreover, this chapter derives the relationship between the derivative of these representations and the aircraft body's angular velocity, which are represented by Eqs. (5.5), (5.21), and (5.51), respectively. These will play an important part in the following chapters. For the coordinate frames and attitude representation, readers can also refer to an excellent book [11] for more details.

Exercises

5.1 In Sect. 5.2.2.1 in this chapter, the rotation matrix $\mathbf{R}_{b}^{e}$ is obtained by following the rotation order $\psi \to \theta \to \phi$. Write out the rotation matrix if the rotation order is $\theta \to \psi \to \phi$.

5.2 Fill in the steps of the derivation of Eq. (5.21). (Hint: Following property of the cross product is used: For a rotation matrix $\mathbf{R} \in \mathbb{R}^{3 \times 3}$ ($\det(\mathbf{R}) = 1$) and any two vectors $\mathbf{a}, \mathbf{b} \in \mathbb{R}^{3}$, one has $(\mathbf{R}\mathbf{a}) \times (\mathbf{R}\mathbf{b}) = \mathbf{R}(\mathbf{a} \times \mathbf{b})$ [12].)

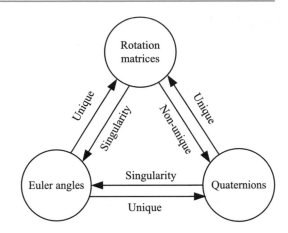

Fig. 5.11 Mutual transformations among three rotation expressions. Here, *Non-unique* implies that one rotation matrix corresponds to two quaternions; *Singularity* implies the infinite number of combinations for Euler angles. Readers can refer to Sects. 5.2.2 and 5.2.3 in this chapter for more details

5.3 Fill in the steps of the derivation of Eqs. (5.24) and (5.35).

5.4 In Sect. 5.2.3.5 in this chapter, the singularity problem of the Euler angle representation and the angle extension are mentioned. Propose a method to solve the singularity problem and to realize the angle extension. (Hint: Readers can refer to Sect. 5.2.2.1 in this chapter.)

5.5 Fill in the steps of the derivation of Eq. (5.51). (Hint: According to the composite rotation quaternion in the case of rotating coordinate frames, one has $\mathbf{q}_e^b(t + \Delta t) = \mathbf{q}_e^b(t) \otimes \Delta \mathbf{q}$.)

References

1. Goldstein H, Poole C, Safko J (2001) Classical mechanics, 3rd edn. Addison-Wesley, San Francisco
2. Wu ST, Fei YH (2005) Flight control system. Beihang University Press, Beijing (In Chinese)
3. Ducard GJ (2009) Fault-tolerant flight control and guidance systems: practical methods for small unmanned aerial vehicles. Springer-Verlag, London
4. Murray RM, Li Z, Sastry SS et al (1994) A mathematical introduction to robotic manipulation. CRC Press, Boca Raton
5. Hamilton WR (1866) Elements of quaternions. Longmans, Green & Company, London
6. Altmann SL (1989) Hamilton, Rodrigues, and the quaternion scandal. Math Mag 62(5): 291–308
7. Shoemake K (1994) Quaternions. Department of Computer and Information Science, University of Pennsylvania, USA. http://www.cs.ucr.edu/~vbz/resources/quatut.pdf. Accessed 06 July 2016
8. Sola J (2015) Quaternion kinematics for the error-state KF. Technical Report. Laboratoire d'Analyse et d'Architecture des Systemes-Centre national de la recherche scientifique (LAAS-CNRS), France. http://www.iri.upc.edu/people/jsola/JoanSola/objectes/notes/kinematics.pdf. Accessed 24 July 2016
9. Kuipers JB (1999) Quaternions and rotation sequences. Princeton University Press, Princeton
10. Euler angles, quaternions, and transformation matrices. NASA. http://ntrs.nasa.gov/archive/nasa/casi.ntrs.nasa.gov/19770024290.pdf. Accessed 06 July 2016
11. Corke P (2011) Robotics, vision and control: fundamental algorithms in MATLAB. Springer-Verlag, Berlin Heidelberg
12. Massey WS (1983) Cross products of vectors in higher dimensional Euclidean spaces. Am Math Mon 90(10): 697–701

Dynamic Model and Parameter Measurement

6

On the basis of the coordinate frames and attitude representations in Chap. 5, forces on a multicopter will be taken into account and dynamic models for filtering and control will be established. The forces and moments can change the acceleration and angular acceleration directly and further influence the velocity and angular velocity. This influence is related to most parameters of a multicopter. Furthermore, the velocity and angular velocity can change the position and the attitude of the multicopter. This influence is independent of parameters of the multicopter. For example, the position is calculated through the velocity without requiring the parameters of the multicopter. As pointed out in Chap. 1, multicopters are different from fixed-wing aircraft or single rotor blade helicopters. The differences appear mainly in its special dynamic model and control effectiveness model. A traditional multicopter has four independent control inputs: thrust, pitching moment, rolling moment, and yawing moment. The thrust is always perpendicular to the multicopter fuselage plane. Moreover, propellers of the multicopter can generate thrust to lift the multicopter directly. Therefore, the multicopter is simple and flexible in terms of the control allocation. Deep understanding of the multicopter dynamic model helps to understand its motion and further to design filters and controllers. This chapter aims to answer the questions below:

How is the multicopter dynamic model established and how are the model parameters measured?

The answer to these questions involves the multicopter control model, multicopter aerodynamic drag model, and multicopter model parameter measurement. The multicopter control model is mainly used for the control in the following chapters. It is also used for the state estimation. The multicopter aerodynamic drag model is mainly used for the state estimation as well. The last part of this chapter provides methods to measure the model parameters mentioned before.

© Springer Nature Singapore Pte Ltd. 2017
Q. Quan, *Introduction to Multicopter Design and Control*, DOI 10.1007/978-981-10-3382-7_6

Taijitu

The ancient Chinese had already understood that the development must follow certain rules called models. *Taijitu* is a Chinese symbol for the concept of *Yin* and *Yang*, also a model of the universe for the ancients in China. Sunzi, a Chinese military general, a strategist, and a philosopher, mentioned in his book "*Sun Tzu on the Art of War*" that "there are not more than five musical notes, yet the combinations of these five give rise to more melodies than can ever be heard. There are not more than five primary colors (blue, yellow, red, white, and black), yet in combination they produce more hues than can ever been seen (The Chinese is "声不过五，五声之变，不可胜听也;色不过五,五色之变,不可胜观也", translated by Lionel Giles. (1910). Sun Tzu on the art of war: The oldest military treatise in the world. Champaign, IL: Project Gutenberg)". This tells us that the adaptivity and flexibility come after the understanding of the principle and essence behind the change.

6.1   **Multicopter Control Model**

An accurate dynamic model is the foundation of analyzing and controlling a system. On the one hand, an over-complicated model leads to a complicated control algorithm design. On the other hand, an over-simple model separates the model from reality and brings negative influences into the control performance.

6.1.1   **General Description**

As shown in Fig. 6.1, the multicopter modeling mainly includes four parts.

(1) *Rigid-body kinematic model*. Kinematics are independent of the mass and force. It only studies variables such as position, velocity, attitude and angular velocity. For the multicopter kinematic model, the inputs are velocity and angular velocity, and the outputs are position and attitude.

(2) *Rigid-body dynamic model*. Dynamics involve both the movement and the force. They are related to the object's mass and moments of inertia. Equations such as Newton's second law, law of kinetic energy and law of momentum, are often used to study the mutual effect among different objects. For the multicopter dynamic model, the inputs are thrust and moments (pitching moment, rolling moment, and yawing moment), and the outputs are velocity and angular velocity. The rigid-body kinematic model and dynamic model constitute the general flight control rigid model of multicopters.

(3) *Control effectiveness model*. The inputs are propeller angular speeds, and the outputs are thrust and moments. For either a quadcopter or a hexacopter, the thrust and moments are all generated by propellers. Given the propeller angular speeds, the thrust and moments can be calculated by using control effectiveness model. The inversion of the control effectiveness model is called the *control allocation model*. When the thrust and moments are obtained by controller design, the propeller angular speeds can be calculated by using the control allocation model.

(4) *Propulsor model*. The propulsor model is a whole power mechanism that includes a brushless Direct Current (DC) motor, an Electronic Speed Controller (ESC), and a propeller. The input is a

Fig. 6.1 Architecture of the multicopter modeling

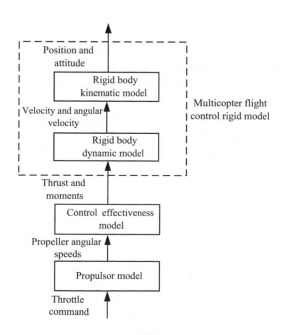

throttle command between 0 and 1 and the outputs are propeller angular speeds. In practice, the model with throttle command as input and propeller thrust as output can also be established.

6.1.2 Multicopter Flight Control Rigid Model

6.1.2.1 Assumptions
For convenience, when modeling a multicopter, the following assumptions are introduced.

Assumption 6.1 The multicopter is a rigid body.

Assumption 6.2 The mass and the moments of inertia are constant.

Assumption 6.3 The geometric center and the Center of Gravity (CoG) of the multicopter are the same.

Assumption 6.4 The multicopter is only under gravity and propeller thrust. Furthermore, the gravity points along the positive direction of the $o_e z_e$ axis and the propeller thrust points along the negative direction of the $o_b z_b$ axis.

Assumption 6.5 Propellers with odd indices rotate counterclockwise and propellers with even indices rotate clockwise.

The main difference between this model and other rigid-body dynamic models is that the thrust produced by the propellers is always perpendicular to the fuselage plane. In other words, the thrust direction is always consistent with the negative direction of the $o_b z_b$ axis.

6.1.2.2 Rigid-Body Kinematic Model
Let the vector of the multicopter's CoG be ${}^e\mathbf{p} \in \mathbb{R}^3$. Then

$$ {}^e\dot{\mathbf{p}} = {}^e\mathbf{v} \tag{6.1} $$

where ${}^e\mathbf{v} \in \mathbb{R}^3$ represents the velocity of the multicopter. In Chap. 5, the attitude kinematic model is divided into three types: Euler angle model, rotation matrix model, and quaternions model. Therefore, the following three rigid-body kinematic models are presented.

(1) *Euler angle model*

Based on *Assumption* 6.1, combining Eq. (5.5) in Chap. 5 with Eq. (6.1), one has

$$ \begin{aligned} {}^e\dot{\mathbf{p}} &= {}^e\mathbf{v} \\ \dot{\boldsymbol{\Theta}} &= \mathbf{W} \cdot {}^b\boldsymbol{\omega}. \end{aligned} \tag{6.2} $$

(2) *Rotation matrix model*

Based on *Assumption* 6.1, combining Eq. (5.21) in Chap. 5 with Eq. (6.1), one has

$$ \begin{aligned} {}^e\dot{\mathbf{p}} &= {}^e\mathbf{v} \\ \dot{\mathbf{R}} &= \mathbf{R}\left[{}^b\boldsymbol{\omega}\right]_{\times} \end{aligned} \tag{6.3} $$

where $\mathbf{R} \triangleq \mathbf{R}_b^e$ in order to simplify the descriptions.

(3) *Quaternions model*

Based on *Assumption* 6.1, combining Eq. (5.52) in Chap. 5 with Eq. (6.1), one has

$$
{}^e\dot{\mathbf{p}} = {}^e\mathbf{v}
$$
$$
\dot{q}_0 = -\frac{1}{2}\mathbf{q}_v^{\mathrm{T}} \cdot {}^b\boldsymbol{\omega}
$$
$$
\dot{\mathbf{q}}_v = \frac{1}{2}\left(q_0\mathbf{I}_3 + [\mathbf{q}_v]_\times\right){}^b\boldsymbol{\omega}.
$$
(6.4)

6.1.2.3 Rigid-Body Dynamic Model

(1) *Position dynamic model*

Assumption 6.4 implies that this chapter only considers the multicopter with levelled propeller disks (See Sect. 3.1.1.2 in Chap. 3). Based on *Assumption* 6.4, by analyzing the forces on a multicopter, one has

$$
{}^e\dot{\mathbf{v}} = g\mathbf{e}_3 - \frac{f}{m}{}^e\mathbf{b}_3
$$
(6.5)

where $f \in \mathbb{R}_+ \cup \{0\}$ represents the magnitude of the total propeller thrust and the thrust is unidirectional (the situation of negative thrust caused by variable-pitch propellers is not considered here) and $g \in \mathbb{R}_+$ represents the acceleration of gravity. Intuitively, the direction of the thrust points upward. In Sect. 5.1.2 in Chap. 5, the positive direction of the $o_b z_b$ axis points downward; therefore, $-f{}^e\mathbf{b}_3$ in Eq. (6.5) represents the thrust vector. Furthermore, one has

$$
{}^e\dot{\mathbf{v}} = g\mathbf{e}_3 - \frac{f}{m}\mathbf{R}\mathbf{e}_3 = g\mathbf{e}_3 + \mathbf{R}\frac{{}^e\mathbf{f}}{m}
$$
(6.6)

where ${}^e\mathbf{f} \triangleq -f\mathbf{e}_3$. Since

$$
{}^e\mathbf{v} = \mathbf{R} \cdot {}^b\mathbf{v}
$$
(6.7)

combining Eq. (6.6) with Eq. (6.7), one has

$$
{}^b\dot{\mathbf{v}} = -\left[{}^b\boldsymbol{\omega}\right]_\times {}^b\mathbf{v} + g\mathbf{R}^{\mathrm{T}}\mathbf{e}_3 - \frac{f}{m}\mathbf{e}_3.
$$
(6.8)

The derivation is left as an exercise at the end of this chapter.

(2) *Attitude dynamic model*

Based on *Assumptions* 6.1–6.3, the attitude dynamic equation in the ABCF is established as

$$
\mathbf{J} \cdot {}^b\dot{\boldsymbol{\omega}} = -{}^b\boldsymbol{\omega} \times \left(\mathbf{J} \cdot {}^b\boldsymbol{\omega}\right) + \mathbf{G}_a + \boldsymbol{\tau}
$$
(6.9)

where $\boldsymbol{\tau} \triangleq [\,\tau_x\ \tau_y\ \tau_z\,]^{\mathrm{T}} \in \mathbb{R}^3$ represents the moments generated by the propellers in the body axes; $\mathbf{J} \in \mathbb{R}^{3\times3}$ represents the multicopter moment of inertia. For a multicopter with n_r propellers, $\mathbf{G}_a \triangleq [\,G_{a,\phi}\ G_{a,\theta}\ G_{a,\psi}\,]^{\mathrm{T}} \in \mathbb{R}^3$ represents the *gyroscopic torques*. Their signs are related to the rotation direction of propellers. Based on *Assumption 6.5* and the definition of coordinate frames, a single propeller's angular velocity vector is $(-1)^k \varpi_k {}^b\mathbf{b}_3, k = 1, \ldots, n_r$, where $\varpi_k \in \mathbb{R}_+$ represents the angular speed (rad/s) of the kth propeller. Therefore, the gyroscopic torques caused by the rotation of a single propeller are expressed as

$$
\begin{aligned}
\mathbf{G}_{a,k} &= {}^b\mathbf{L}_k \times {}^b\boldsymbol{\omega} \\
&= J_{\mathrm{RP}}(-1)^k \varpi_k {}^b\mathbf{b}_3 \times {}^b\boldsymbol{\omega} \\
&= J_{\mathrm{RP}}\left({}^b\boldsymbol{\omega} \times \mathbf{e}_3\right)(-1)^{k+1}\varpi_k
\end{aligned}
$$
(6.10)

where $J_{RP} \in \mathbb{R}_+$ $(N \cdot m \cdot s^2)$ represents the total moments of inertia of the entire rotor and the propeller about the axis of rotation. Here, equations $^b\mathbf{b}_3 \times {}^b\boldsymbol{\omega} = -{}^b\boldsymbol{\omega} \times {}^b\mathbf{b}_3$, $^b\mathbf{b}_3 = \mathbf{e}_3$ are applied to derivation. For a multicopter with n_r propellers, one has

$$\mathbf{G}_a = \sum_{k=1}^{n_r} J_{RP}(^b\boldsymbol{\omega} \times \mathbf{e}_3)(-1)^{k+1}\varpi_k. \tag{6.11}$$

Furthermore, since

$$^b\boldsymbol{\omega} \times \mathbf{e}_3 = \begin{bmatrix} \omega_{y_b} \\ -\omega_{x_b} \\ 0 \end{bmatrix}$$

one has

$$\begin{aligned} G_{a,\phi} &= \sum_{k=1}^{n_r} J_{RP}\omega_{y_b}(-1)^{k+1}\varpi_k \\ G_{a,\theta} &= \sum_{k=1}^{n_r} J_{RP}\omega_{x_b}(-1)^{k}\varpi_k \\ G_{a,\psi} &= 0. \end{aligned} \tag{6.12}$$

As shown above, the gyroscopic torque does not exist in the yaw channel.

6.1.2.4 Multicopter Flight Control Rigid Model
By combining Eqs. (6.2), (6.3), (6.4), (6.6) and (6.9), the multicopter flight control rigid model is expressed as

$$\begin{aligned} ^e\dot{\mathbf{p}} &= {}^e\mathbf{v} \\ ^e\dot{\mathbf{v}} &= g\mathbf{e}_3 - \frac{f}{m}\mathbf{R}\mathbf{e}_3 \\ \dot{\boldsymbol{\Theta}} &= \mathbf{W} \cdot {}^b\boldsymbol{\omega} \\ \mathbf{J} \cdot {}^b\dot{\boldsymbol{\omega}} &= -{}^b\boldsymbol{\omega} \times (\mathbf{J} \cdot {}^b\boldsymbol{\omega}) + \mathbf{G}_a + \boldsymbol{\tau} \end{aligned} \tag{6.13}$$

or

$$\begin{aligned} ^e\dot{\mathbf{p}} &= {}^e\mathbf{v} \\ ^e\dot{\mathbf{v}} &= g\mathbf{e}_3 - \frac{f}{m}\mathbf{R}\mathbf{e}_3 \\ \dot{\mathbf{R}} &= \mathbf{R}[^b\omega]_\times \\ \mathbf{J} \cdot {}^b\dot{\boldsymbol{\omega}} &= -{}^b\boldsymbol{\omega} \times (\mathbf{J} \cdot {}^b\boldsymbol{\omega}) + \mathbf{G}_a + \boldsymbol{\tau} \end{aligned} \tag{6.14}$$

or

$$\begin{aligned} ^e\dot{\mathbf{p}} &= {}^e\mathbf{v} \\ ^e\dot{\mathbf{v}} &= g\mathbf{e}_3 - \frac{f}{m}\mathbf{R}\mathbf{e}_3 \\ \dot{q}_0 &= -\frac{1}{2}\mathbf{q}_v^T \cdot {}^b\boldsymbol{\omega} \\ \dot{\mathbf{q}}_v &= \frac{1}{2}\left(q_0\mathbf{I}_3 + [\mathbf{q}_v]_\times\right){}^b\boldsymbol{\omega} \\ \mathbf{J} \cdot {}^b\dot{\boldsymbol{\omega}} &= -{}^b\boldsymbol{\omega} \times (\mathbf{J} \cdot {}^b\boldsymbol{\omega}) + \mathbf{G}_a + \boldsymbol{\tau}. \end{aligned} \tag{6.15}$$

These models contain both ABCF and EFCF. On the one hand, the position and the velocity of the multicopter are expected to be described in the EFCF. This can help the remote pilots to determine the flight position and flight velocity. Besides, this presentation is consistent with GPS measurement.

On the other hand, in the ABCF, the presentations of thrust and moments appear intuitive, and the measurements by sensors are always expressed in the ABCF as well. The salient feature of the multicopter flight control rigid model is embedded in $-f/m \cdot \mathbf{Re}_3$. It means that the thrust direction is always consistent with the negative direction of the $o_b z_b$ axis. In many literatures, controllers were designed according to Eqs. (6.13)–(6.15) directly. In the following, the control effectiveness model will be considered in order to distinguish the quadcopter from the hexacopter.

6.1.3   Control Effectiveness Model

6.1.3.1 Single Propeller Thrust and Reaction Torque Model

According to Eq. (4.1) in Chap. 4, when a multicopter hovers without wind, the propeller thrust is expressed as

$$T_i = c_T \varpi_i^2 \tag{6.16}$$

where $c_T = 1/4\pi^2 \cdot \rho D_p^4 C_T$ is modeled as a constant that can be easily determined by experiments. Readers can refer to Chap. 4 for more details about its definition. The reaction torque is expressed as

$$M_i = c_M \varpi_i^2 \tag{6.17}$$

where M_i is the reaction torque of the ith propeller acting on the fuselage, $c_M = 1/4\pi^2 \cdot \rho D_p^5 C_M$ can also be determined by experiments. The parameters ρ, D_p, C_T, C_M are presented in Sect. 4.2.1 in Chap. 4.

Equation (6.17) is a static model about the reaction torque. Its dynamic model is

$$J_{RP} \dot{\varpi}_i = -c_M \varpi_i^2 + \tau_i \tag{6.18}$$

where $\tau_i \in \mathbb{R}$ is the torque acting on the ith propeller. According to Newton's third law, the reaction torque is as large as the torque acting on the ith propeller. Then,

$$M_i = c_M \varpi_i^2 + J_{RP} \dot{\varpi}_i \tag{6.19}$$

6.1.3.2 Thrust and Moments Model

The flight of the multicopter is driven by multiple propellers. The propeller angular speeds $\varpi_i, i = 1, 2, \ldots n_r$ will determine the total thrust f and moments $\boldsymbol{\tau}$. To make it easier to understand, this section starts with quadcopters.

(1) *Quadcopters*

As shown in Fig. 6.2a, the total thrust that acts on the quadcopter is

$$f = \sum_{i=1}^{4} T_i = c_T \left(\varpi_1^2 + \varpi_2^2 + \varpi_3^2 + \varpi_4^2 \right). \tag{6.20}$$

For a plus-configuration quadcopter, the moments produced by propellers are

$$\begin{aligned}
\tau_x &= d c_T \left(-\varpi_2^2 + \varpi_4^2 \right) \\
\tau_y &= d c_T \left(\varpi_1^2 - \varpi_3^2 \right) \\
\tau_z &= c_M \left(\varpi_1^2 - \varpi_2^2 + \varpi_3^2 - \varpi_4^2 \right)
\end{aligned} \tag{6.21}$$

Fig. 6.2 Two configurations of quadcopters

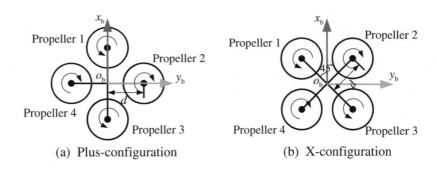

(a) Plus-configuration   (b) X-configuration

where $d \in \mathbb{R}_+$ represents the distance between the body center and any motor. According to Eqs. (6.20) and (6.21), one has the following matrix form

$$
\begin{bmatrix} f \\ \tau_x \\ \tau_y \\ \tau_z \end{bmatrix} = \begin{bmatrix} c_T & c_T & c_T & c_T \\ 0 & -dc_T & 0 & dc_T \\ dc_T & 0 & -dc_T & 0 \\ c_M & -c_M & c_M & -c_M \end{bmatrix} \begin{bmatrix} \varpi_1^2 \\ \varpi_2^2 \\ \varpi_3^2 \\ \varpi_4^2 \end{bmatrix}.
\tag{6.22}
$$

As shown in Fig. 6.2b, for an X-configuration quadcopter, the total thrust produced by propellers is still

$$
f = c_T \left(\varpi_1^2 + \varpi_2^2 + \varpi_3^2 + \varpi_4^2 \right).
\tag{6.23}
$$

But the moments are different, as shown in the following

$$
\begin{aligned}
\tau_x &= dc_T \left(\frac{\sqrt{2}}{2}\varpi_1^2 - \frac{\sqrt{2}}{2}\varpi_2^2 - \frac{\sqrt{2}}{2}\varpi_3^2 + \frac{\sqrt{2}}{2}\varpi_4^2 \right) \\
\tau_y &= dc_T \left(\frac{\sqrt{2}}{2}\varpi_1^2 + \frac{\sqrt{2}}{2}\varpi_2^2 - \frac{\sqrt{2}}{2}\varpi_3^2 - \frac{\sqrt{2}}{2}\varpi_4^2 \right) \\
\tau_z &= c_M \left(\varpi_1^2 - \varpi_2^2 + \varpi_3^2 - \varpi_4^2 \right).
\end{aligned}
\tag{6.24}
$$

(2) Multicopters

For multicopters, in order to implement the control allocation, the positions of all motors in the ABCF need to be determined. For a multicopter with n_r propellers, the propellers are marked in clockwise fashion from $i = 1$ to $i = n_r$, as shown in Fig. 6.3. The angle between the $o_b x_b$ axis and the supported arm of each motor is denoted by $\varphi_i \in \mathbb{R}_+ \cup \{0\}$. The distance between the body center and the ith motor is denoted by $d_i \in \mathbb{R}_+ \cup \{0\}$, $i = 1, 2, \ldots, n_r$.

Then, the thrust and moments produced by propellers are expressed as

$$
\begin{bmatrix} f \\ \tau_x \\ \tau_y \\ \tau_z \end{bmatrix} = \underbrace{\begin{bmatrix} c_T & c_T & \cdots & c_T \\ -d_1 c_T \sin\varphi_1 & -d_2 c_T \sin\varphi_2 & \cdots & -d_{n_r} c_T \sin\varphi_{n_r} \\ d_1 c_T \cos\varphi_1 & d_2 c_T \cos\varphi_2 & \cdots & d_{n_r} c_T \cos\varphi_{n_r} \\ c_M \delta_1 & c_M \delta_2 & \cdots & c_M \delta_{n_r} \end{bmatrix}}_{\mathbf{M}_{n_r}} \begin{bmatrix} \varpi_1^2 \\ \varpi_2^2 \\ \vdots \\ \varpi_{n_r}^2 \end{bmatrix}
\tag{6.25}
$$

where $\mathbf{M}_{n_r} \in \mathbb{R}^{4 \times n_r}$ represents the control effectiveness matrix and $\delta_i = (-1)^{i+1}$, $i = 1, \cdots, n_r$.

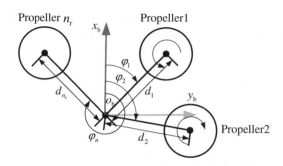

Fig. 6.3 Airframe configuration parameters of a multicopter

6.1.4 Propulsor Model

As shown in Fig. 6.4, the propulsor model here includes not only a brushless DC motor but also an ESC and a propeller. Throttle command σ is an input signal between 0 and 1, while the battery output voltage U_b cannot be controlled.

The ESC generates an equivalent average voltage $U_m = \sigma U_b$ after receiving the throttle command σ and the battery output voltage U_b. First, given a voltage signal, the motor can achieve a steady-state speed ϖ_{ss}. The relation is normally linear, which is expressed as

$$\varpi_{ss} = C_b U_m + \varpi_b = C_R \sigma + \varpi_b \tag{6.26}$$

where $C_R = C_b U_b$, C_b, and ϖ_b are constant parameters. The measurements of C_R and ϖ_b will be presented in Sect. 6.3 in this chapter. Secondly, when given a throttle command, the motor needs some time to achieve the steady-state speed ϖ_{ss}. This time constant denoted by T_m will determine the dynamic response. Generally, dynamics of a brushless DC motor can be simplified as a first-order low-pass filter. Its transfer function is expressed as

$$\varpi = \frac{1}{T_m s + 1} \varpi_{ss}. \tag{6.27}$$

In others words, when given a desired steady-state speed ϖ_{ss}, the motor speed cannot achieve ϖ_{ss} immediately. This process needs some time to adjust. Combining Eq. (6.26) with Eq. (6.27), one has a complete propulsor model as follows:

$$\varpi = \frac{1}{T_m s + 1}(C_R \sigma + \varpi_b) \tag{6.28}$$

where the input is the throttle command σ and the output is motor speed ϖ.

Now, the relationship between the throttle command perturbation and the reaction torque perturbation is studied further. When a multicopter is hovering, for the ith propeller, it is assumed that the

Fig. 6.4 Signal transmission of the propulsor model

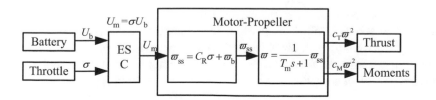

propeller angular speed in equilibrium is ϖ_i^*, the throttle command in equilibrium is σ_i^*, and the reaction torque in equilibrium is M_i^*. Then,

$$\varpi_i = \varpi_i^* + \Delta\varpi_i$$
$$\sigma_i = \sigma_i^* + \Delta\sigma_i$$
$$M_i = M_i^* + \Delta M_i$$

where $\Delta\varpi_i$, $\Delta\sigma_i$, and ΔM_i are the perturbations of propeller angular speed, throttle command, and reaction torque, respectively. Linearization of Eqs. (6.19) and (6.28) at equilibrium points are

$$\Delta M_i = 2c_{\mathrm{M}}\varpi_i^* \Delta\varpi_i + J_{\mathrm{RP}}\Delta\dot{\varpi}_i$$
$$\Delta\varpi_i = \frac{1}{T_{\mathrm{m}}s + 1} C_{\mathrm{R}} \Delta\sigma_i$$

Then, the transfer function from $\Delta\sigma_i$ to ΔM_i is further obtained as

$$\Delta M_i = \frac{C_{\mathrm{R}}\left(2c_{\mathrm{M}}\varpi_i^* + J_{\mathrm{RP}}s\right)}{T_{\mathrm{m}}s + 1} \Delta\sigma_i \qquad (6.29)$$

It is suggested that the dynamic model (6.29) be considered in the performance optimization for multicopters.

6.2   Multicopter Aerodynamic Drag Model

In the previous section, it is assumed that the multicopter is a rigid body. But in practice, multicopters are often equipped with lightweight, fixed-pitch plastic propellers. Such propellers are not rigid. Otherwise, they can easily produce fatigue in their roots and cause breakage. The aerodynamic and inertial forces acting on a propeller during flight are quite significant and can cause the propeller to flex. In the following section, blade-flapping effects, which are due to the flexibility of propellers, will be presented as well as the multicopter aerodynamic drag model. This model is quite helpful in the state estimation in Chap. 9.

6.2.1   Blade Flapping

Blade flapping is the upward-and-downward movement of a rotating blade [1]. As shown in Fig. 6.5a, the advancing blade, upon meeting the wind, gains a higher relative speed and responds to the increase of speed by producing larger thrust and generating the upward flapping velocity. As shown in Fig. 6.5b, c, the upward flapping velocity reduces the angle of attack and further the thrust.

As shown in Fig. 6.6a, the flight direction is right and the propeller rotates counterclockwise. The maximum relative velocity takes place at point A, while the maximum upward displacement takes place at point B. The point A lags behind the point B by $\pi/2$ phase, which is the same as the sinusoidal motion law, as shown in Fig. 6.6b. In Fig. 6.6c, the lower side is *advancing blade area* and the upper side is *retreating blade area*. Since the position lags behind the velocity by $\pi/2$ phase, the front side is known as the *upward displacement area* of a rotating blade and the back side is the *downward displacement area* of a rotating blade.

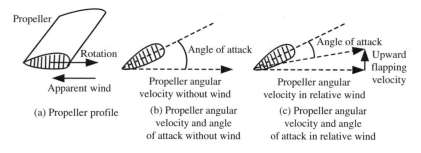

Fig. 6.5 Propeller angular velocity and the variation of angle of attack

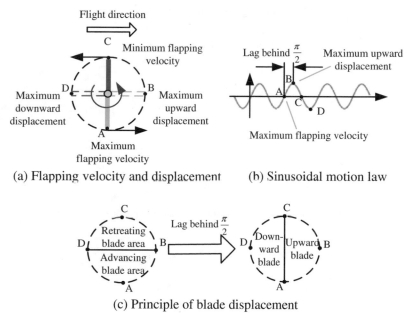

Fig. 6.6 Propeller flapping velocity and displacement

(a) Flapping velocity and displacement

(b) Sinusoidal motion law

(c) Principle of blade displacement

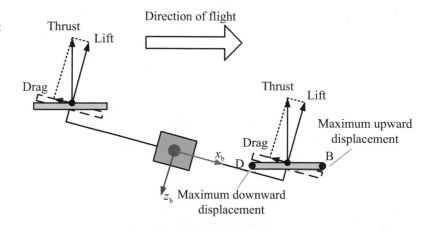

Fig. 6.7 Rotating blade thrust, lift, and drag during flight

The flapping effect can change the rotating blade direction and further the thrust direction. As shown in Fig. 6.7, thrust direction is no longer parallel to the $o_b z_b$ axis. There is an associated induced drag in the negative direction of the $o_b x_b$ axis. The induced drag is proportional to the thrust generated by the propeller, and it is also the main part of the multicopter's drag. The multicopter aerodynamic drag model is based on this induced drag. Readers can look for more details in [2–4].

6.2.2 Multicopter Aerodynamic Drag Model

According to Eq. (6.8), the position dynamic model is expressed as

$$\begin{aligned}
\dot{v}_{x_b} &= v_{y_b}\omega_{z_b} - v_{z_b}\omega_{y_b} - g\sin\theta \\
\dot{v}_{y_b} &= v_{z_b}\omega_{x_b} - v_{x_b}\omega_{z_b} + g\cos\theta\sin\phi
\end{aligned} \tag{6.30}$$

where $v_{x_b}, v_{y_b}, v_{z_b} \in \mathbb{R}$ are the multicopter velocities along the body axes $o_b x_b, o_b y_b, o_b z_b$, respectively. For the multicopter, drags applied to rotating blades are in the direction of the body axes. Due to the symmetry of the multicopter, the drags are simply expressed as

$$\begin{aligned}
f_{x_b} &= -k_{\text{drag}}v_{x_b} \\
f_{y_b} &= -k_{\text{drag}}v_{y_b}
\end{aligned} \tag{6.31}$$

where $f_{x_b}, f_{y_b} \in \mathbb{R}$ are drags along body axes $o_b x_b, o_b y_b$, respectively, and $k_{\text{drag}} \in \mathbb{R}_+$ is the drag coefficient. Consequently, the multicopter aerodynamic drag model (6.30) becomes

$$\begin{aligned}
\dot{v}_{x_b} &= v_{y_b}\omega_{z_b} - v_{z_b}\omega_{y_b} - g\sin\theta - \frac{k_{\text{drag}}}{m}v_{x_b} \\
\dot{v}_{y_b} &= v_{z_b}\omega_{x_b} - v_{x_b}\omega_{z_b} + g\cos\theta\sin\phi - \frac{k_{\text{drag}}}{m}v_{y_b}.
\end{aligned} \tag{6.32}$$

Since $\dot{v}_{x_b}$ and $\dot{v}_{y_b}$ are accelerations along the body axes, one has

$$\begin{aligned}
a_{x_b} &= \dot{v}_{x_b} + g\sin\theta = v_{y_b}\omega_{z_b} - v_{z_b}\omega_{y_b} - \frac{k_{\text{drag}}}{m}v_{x_b} \\
a_{y_b} &= \dot{v}_{y_b} - g\cos\theta\sin\phi = v_{z_b}\omega_{x_b} - v_{x_b}\omega_{z_b} - \frac{k_{\text{drag}}}{m}v_{y_b}
\end{aligned} \tag{6.33}$$

where $a_{x_b}, a_{y_b} \in \mathbb{R}$, which can be measured directly by accelerometers, are components of *specific force*. This will be detailed in Sect. 7.1.1 of Chap. 7 and Sect. 9.1.1 of Chap. 9. Drag coefficient k_{drag} can be estimated beforehand through a system identification process using higher accuracy calibration devices such as optical motion-capture systems. Alternatively, the coefficient k_{drag} can also be estimated as part of the states using the Extended Kalman filter (EKF). Further details are given in reference [5]. In this case, Eq. (6.32) is further expressed as

$$\begin{aligned}
\dot{v}_{x_b} &= v_{y_b}\omega_{z_b} - v_{z_b}\omega_{y_b} - g\sin\theta - \frac{k_{\text{drag}}}{m}v_{x_b} \\
\dot{v}_{y_b} &= v_{z_b}\omega_{x_b} - v_{x_b}\omega_{z_b} + g\cos\theta\sin\phi - \frac{k_{\text{drag}}}{m}v_{y_b} \\
\dot{k}_{\text{drag}} &= 0.
\end{aligned} \tag{6.34}$$

6.3   Multicopter Model Parameter Measurement

Model parameters need to be measured after the presentation of the multicopter control model.

6.3.1 Position of the Center of Gravity

The weight balancing needs to be done before flight. The objective is to make the geometric central axis pass through the CoG, achieving a better flight performance. When the propellers rotate at a constant speed, the resultant force is on the frame's central axis of symmetry. If the CoG has a position bias, the gravity and the thrust vectors are not going to be in the same straight line and the multicopter will tilt to one side. One of the most effective ways to determine the CoG is to use suspension wires. According to the principle of force balance, the gravitational force on the multicopter and the tension in the wires are in the same straight line after the multicopter is stable. As shown in Fig. 6.8, the concrete steps are as follows:

Step 1. Fetch a thin wire and tie a weight at one end of the wire. Then, tie one side of the boom arm in the middle of the wire and lift the other side of the wire. Record the wire position on the multicopter as the solid line shown in Fig. 6.8a. It will be the dotted line shown in Fig. 6.8b.
Step 2. Put the contact point in the other place and record the wire position in the same way as the solid line shown in Fig. 6.8b.
Step 3. The intersection of two recorded lines is the position of the CoG, as shown in Fig. 6.8b.
Step 4. Repeat the steps above several times in order to improve the measurement accuracy.

If a good balancing cannot be achieved, then the control gain of the motor needs to be adjusted. The motor speed should be increased in the heavier side in order to keep balance. This method is effective in correcting small bias of the CoG. If the bias of the CoG is large, the difference among control gains of the motors is big and this will influence the maneuvering performance of the multicopter. When the CoG is located at the geometric central axis, the matrix of moments of inertia is a diagonal matrix. The model is simpler and the coupling effect is smaller. Therefore, it is quite important to determine the CoG's position. One cannot perform the absolute balancing. If the balancing problem happens due to the aircraft structural characteristics, then the feedforward control needs to be added into the roll channel and pitch channel. In the Fully-Autonomous Control (FAC) mode, the integration in the PID control can eliminate the bias. In the Semi-Autonomous Control (SAC) mode, the remote pilots can observe and adjust manually the trim buttons of the roll channel and pitch channel in the RC transmitter, as shown in Fig. 6.9, to add a feedforward to compensate for the bias.

Fig. 6.8 The way to determinate the CoG's position

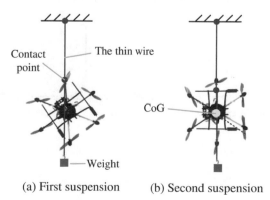

(a) First suspension   (b) Second suspension

Fig. 6.9 Trim button's position in the roll channel and pitch channel of RC transmitters

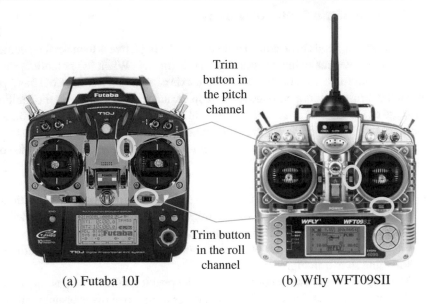

Trim button in the pitch channel

Trim button in the roll channel

(a) Futaba 10J   (b) Wfly WFT09SII

6.3.2 Weight

The normal electronic balance is used to measure the weight. It is noted that it is harder to put a large multicopter on the pan of the balance. The body can easily touch the ground which will make the measurement result erroneous. Meanwhile, the multicopter's CoG needs to be close to the center of the balance. A large bias between these two centers will lead to a moment effect in the balance and result in a force bias measured by pressure-sensitive sensors in the balance. This could influence the measurement accuracy.

6.3.3 Moment of Inertia [6]

6.3.3.1 Central Principal Moment of Inertia
A rigid body's moments of inertia are expressed as

$$\mathbf{J} = \begin{bmatrix} J_{xx} & -J_{xy} & -J_{xz} \\ -J_{yx} & J_{yy} & -J_{yz} \\ -J_{zx} & -J_{zy} & J_{zz} \end{bmatrix}. \tag{6.35}$$

Here $J_{xy} = J_{yx}$, $J_{xz} = J_{zx}$ and $J_{yz} = J_{zy}$. For objects with geometric symmetry like a standard multicopter, one has $J_{xy} = J_{xz} = J_{yz} = 0$. Therefore, Eq. (6.35) is simplified as

$$\mathbf{J} = \begin{bmatrix} J_{xx} & 0 & 0 \\ 0 & J_{yy} & 0 \\ 0 & 0 & J_{zz} \end{bmatrix} \tag{6.36}$$

where J_{xx}, J_{yy}, $J_{zz} \in \mathbb{R}_+$ are called *central principal moments of inertia*. Principal moments of inertia can be measured by using bifilar pendulum. As shown in Fig. 6.10, it is assumed that the mass of the object is m_0 and the local acceleration of gravity is g. The object is suspended by using two thin wires

Fig. 6.10 Measurement of principal moments of inertia by using a bifilar pendulum.

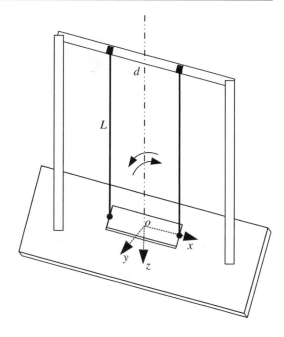

of the same length. The distance between the wires is d and the wire length is L. The object is level. Then, the oz axis of the object and the vertical central line of the wire coincide. The object is rotated manually around oz axis by a small angle. After the object is released, the periodic back-and-forth movement around the oz axis will proceed.

This swing is very similar to the simple pendulum. Its swing period is derived as

$$T_0 = 4\pi \sqrt{\frac{J_{zz}L}{m_0 g d^2}}. \tag{6.37}$$

Consequently,

$$J_{zz} = \frac{m_0 g d^2}{16\pi^2 L} T_0^2. \tag{6.38}$$

As shown in Fig. 6.10, change the direction of body axes and make the ox axis and oy axis be the rotation axis separately. Then, J_{xx}, J_{yy} can be derived in a similar way.

According to the principle above, a method to measure the three principal moments of inertia is proposed in the following. Figure 6.11 is the installation instruction for the measurement. Here, the measurement of J_{xx} is taken as an example. The concrete steps are as follows:

Step 1. According to the geometric symmetry, determine the rough position of the CoG by using hands and eyes to perceive the mass distribution.
Step 2. Fix two thin wires on the multicopter body. Adjust two contact points in order to ensure the equal altitude and the uniform distribution on two sides of the CoG, as shown in Fig. 6.11a. The moment of inertia about another axis can be measured by changing the position of two contact points. For some multicopters with special configurations, it is hard to keep the balance. A frame can be constructed to help the measurement by using light sticks, steel wires, plastic tapes, etc.

Fig. 6.11 Measurement of principal moments of inertia

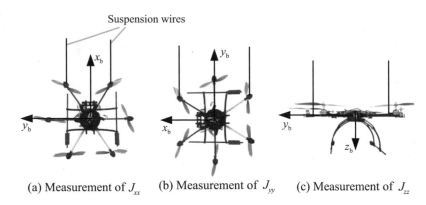

(a) Measurement of J_{xx} (b) Measurement of J_{yy} (c) Measurement of J_{zz}

Step 3. As shown in Fig. 6.11a, tie the top side of two thin wires in a horizontal stick and ensure the same length of two wires. The two wires generate a rectangle, which is similar to Fig. 6.10. Record the width d and the length L of this rectangle.

Step 4. Adjust the position of the contact points in order to keep a better balance.

Step 5. Rotate the multicopter around the suspension axis by a small angle, and the time that it takes for 50 swing periods is denoted by T. The period of the bifilar pendulum is further expressed as $T_0 = T/50$. The initial swing angle should not be too large. A rather high accuracy can be obtained by setting the swing angle smaller than 10 degrees. Under correct operations, the accuracy is the same as the counterpart of a simple pendulum.

Step 6. Calculate J_{xx} by substituting T_0 into Eq. (6.38).

Step 7. Change the suspension manner and repeat *Steps 1–6* in order to get J_{yy} and J_{zz}, as shown in Fig. 6.11b, c.

It is noted that $J_{xy} = J_{xz} = J_{yz} = 0$ holds only if the mass distribution of the multicopter is perfectly symmetric. The element J_{xz}, J_{xy}, J_{yz} and other elements are called *products of inertia*. Generally, they are ignored because of their small values compared with the principal moments of inertia. However, for some multicopters that cannot achieve a good balancing or some others with special configurations, the products of inertia cannot be ignored. Therefore, a method is proposed to measure the products of inertia in the following.

6.3.3.2 Products of Inertia

First of all, values of moments of inertia are different in different coordinate frames (the measurements of these values are often taken along the body axes). For example, a coordinate frame $\{H\}$, which is fixed to the multicopter body, is defined. One has

$$^{b}\boldsymbol{\omega}^{\mathrm{T}} \cdot \mathbf{J} \cdot {}^{b}\boldsymbol{\omega} = \left(\mathbf{R}_{\mathrm{h}}^{\mathrm{b}} \cdot {}^{\mathrm{h}}\boldsymbol{\omega}\right)^{\mathrm{T}} \mathbf{J} \left(\mathbf{R}_{\mathrm{h}}^{\mathrm{b}} \cdot {}^{\mathrm{h}}\boldsymbol{\omega}\right) = {}^{\mathrm{h}}\boldsymbol{\omega}^{\mathrm{T}} \cdot \mathbf{J}_{\mathrm{h}} \cdot {}^{\mathrm{h}}\boldsymbol{\omega} \tag{6.39}$$

where $^{b}\boldsymbol{\omega} \in \mathbb{R}^3$ is the angular velocity in the ABCF, $^{\mathrm{h}}\boldsymbol{\omega} \in \mathbb{R}^3$ is the angular velocity in the coordinate frame $\{H\}$, $\mathbf{R}_{\mathrm{h}}^{\mathrm{b}} \in \mathbb{R}^{3\times3}$ is the rotation matrix from $\{H\}$ to the ABCF, and $\mathbf{J}_{\mathrm{h}} \in \mathbb{R}^{3\times3}$ represents the multicopter moments of inertia in the coordinate frame $\{H\}$. Then, one has

$$\mathbf{J}_{\mathrm{h}} = \left(\mathbf{R}_{\mathrm{h}}^{\mathrm{b}}\right)^{\mathrm{T}} \cdot \mathbf{J} \cdot \mathbf{R}_{\mathrm{h}}^{\mathrm{b}}. \tag{6.40}$$

A situation is considered, as shown in Fig. 6.12. The vertical rotation axis V is perpendicular to the $o_b y_b$ axis and forms an angle α with $o_b x_b$ axis and an angle $90° - \alpha$ with $o_b z_b$ axis. Therefore, one has

$$\mathbf{R}_h^b = \begin{bmatrix} \cos\alpha & 0 & -\sin\alpha \\ 0 & 1 & 0 \\ \sin\alpha & 0 & \cos\alpha \end{bmatrix}. \tag{6.41}$$

By combining Eqs. (6.35), (6.40) and (6.41), the following equation is obtained

$$J_V = (\mathbf{J}_h)_{xx} = J_{xx}\cos^2\alpha + J_{zz}\sin^2\alpha - J_{xz}\sin 2\alpha \tag{6.42}$$

where $J_V \in \mathbb{R}_+$ is the multicopter moment of inertia about the vertical rotation axis V (V is the ox axis of $\{H\}$). This value can be obtained by using the suspension manner shown in Fig. 6.12 and the measurement processes mentioned above. Observed from equation (6.42), the axis V and $o_b x_b$ axis coincide when $\alpha = 0$, then $J_V = J_{xx}$. Similarly, $J_V = J_{zz}$, when $\alpha = 90°$. When α is determined and J_{xx}, J_{zz}, and J_V are known, J_{xz} is expressed as

$$J_{xz} = \frac{J_{xx}\cos^2\alpha + J_{zz}\sin^2\alpha - J_V}{\sin 2\alpha}. \tag{6.43}$$

The concrete steps are summarized in the following.

Step 8. Measure the principal moments of inertia J_{xx}, J_{yy}, and J_{zz} according to *Steps 1–7* mentioned before.

Step 9. Suspend the multicopter as shown in Fig. 6.12. Record the angle α and measure the moment of inertia J_V.

Step 10. Calculate J_{xz} according to Eq. (6.43).

Step 11. Repeat *Steps 9–10* several times by choosing different values of α. Calculate the average value in order to reduce the error.

Step 12. Change the suspension manner and repeat *Steps 9-11* in order to get J_{xy} and J_{yz}.

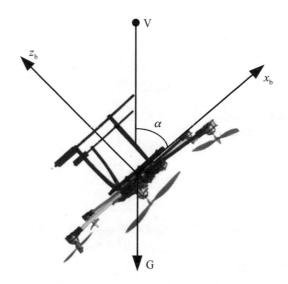

Fig. 6.12 Side view of the measurement of J_{xz}. The view is along the $o_b y_b$ axis direction, and the two suspension wires are fixed to the landing gear

6.3.4 Propulsor Model's Parameter Measurement

In practice, the corresponding propeller angular speeds, the final thrust and moments are unknown expect for the known throttle command given by pushing the throttle control stick. Thus, the relationship between the throttle command and propeller angular speed, and then the relationship between the propeller angular speed and the thrust (moment) need to be determined by using the following series of experiments.

6.3.4.1 Composition of a Propulsor

The propulsor model is shown in Fig. 6.13. Here, an RC transmitter is used because the propulsor needs to be started by a corresponding control signal generation system. In actual tests, the signal line of the ESC is connected to the throttle channel of the RC transmitter. The input signal between 0 and 1, denoted by σ, is obtained by pushing the throttle control stick.

The model from the throttle command σ to the motor speed ϖ is given in Eq. (6.28). A method used to measure system parameters is discussed in the following. As shown in Fig. 6.14b, the unit step response of the first-order system at $t = T_m$ reaches a value of 0.632. Thus, the time constant T_m can be obtained by the corresponding time of $\varpi = 0.632\varpi_{ss}$. For some RC transmitters, the value of σ can be obtained directly from the electronic screen. However, for those without a screen or a display function of the throttle command, the value of σ can be read from the dial in the throttle control stick. It should be emphasized that the measurement of the propulsion system is realized on the basis of the throttle range calibration for ESC. The throttle range calibration aims to establish a corresponding relationship between the throttle command and the output of ESC. The ESC can only output correctly the maximum and minimum speeds after it recorded the maximum and minimum values of the throttle control stick by the throttle range calibration.

6.3.4.2 Measurement Processes

Generally, it is hard to measure the reaction torque of propulsors because the propeller angular speed is very fast and the reaction torque is very small due to the efficiency of the blades and the small drag. Generally speaking, the order of magnitude of the reaction torque is only $0.1\,\mathrm{N\cdot m}$. In other words, the force generated by a one-meter stick is only $0.1\,\mathrm{N}$, which is far out of the sensitivity range of general

Fig. 6.13 Signal transmission from RC transmitter to thrust and torque

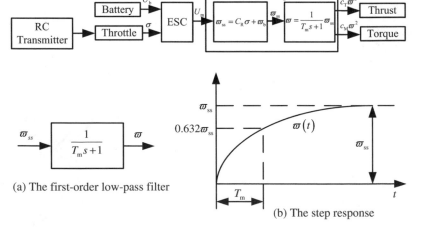

Fig. 6.14 Step response of motor-propeller model

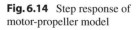

(a) The first-order low-pass filter

(b) The step response

Fig. 6.15 A device for the measurement of propulsion system's parameters

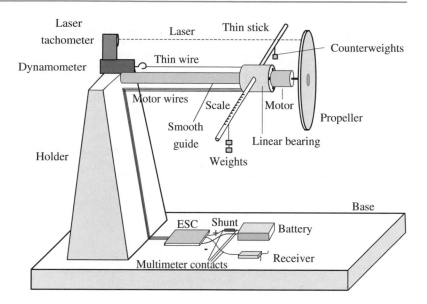

dynamometers. Here, a device, as shown in Fig. 6.15, is designed to solve this problem. It can measure all parameters of a propulsor model.

As shown in Fig. 6.15, the blade thrust is measured directly by a dynamometer, the motor speed is measured directly by a non-contact tachometer, and the value of the throttle command is read directly from an RC transmitter or calculated by a multimeter. The concrete steps for the torque measurement are summarized in the following.

Step 1. When the control signal is 0 and the propeller is static, adjust the position of weights and counterweights in order to keep the thin stick level. Record the initial position of the weights p_1.

Step 2. Make the motor rotate by giving a specified throttle command. When the motor speed is stable, the reaction torque will tilt the thin stick. Adjust finely the position of the weights in order to keep the thin stick level. Record the position of the weights p_2.

Step 3. Assume that the mass of the weights is m_f. The reaction torque generated by the propulsor is expressed as

$$M = (p_2 - p_1) m_f g. \tag{6.44}$$

Observed from Eq. (6.44), it is assumed that $m_f = 0.01$ kg. Then, a torque of 0.05 N·m results in a difference of force arm $\Delta p = p_2 - p_1 = 0.5$ m. This value can be measured precisely by a ruler. The linear bearing is used because it can move and rotate smoothly along the guided direction (in order to measure the blade thrust) and the rotation direction (in order to measure the torque). Meanwhile, the drag and the torque generated by the movement and the rotation are small enough to be ignored. Experiments show that this method can detect the tiny change of the moments with a high accuracy.

6.3.4.3 Measurement Results

The lines and devices are established and connected according to Fig. 6.15. A common motor XXD 2212, whose KV value is 1000 RPM/V, is used in this test. The battery is ACE 3S1P. The RC system is Walkera DEVO-10 Suite. The propellers are APC1047 and the ESC is Hobbywing 30A. It is considered that the throttle range calibration for ESC has already been performed before the experiments.

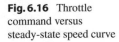

Fig. 6.16 Throttle command versus steady-state speed curve

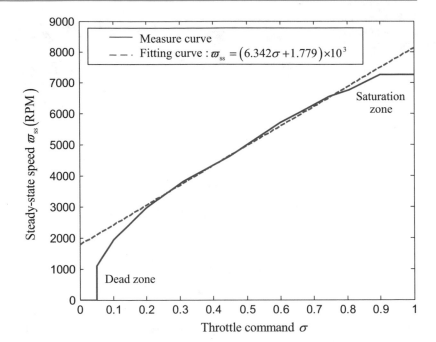

The complete measurement scheme is shown as follows. Set a throttle command σ and use the system described before to measure the motor speed ϖ_{ss}, the thrust T, and the moment M. After a considerable amount of test data has been obtained, one can draw the throttle command versus steady-state speed curve, the steady-state speed versus thrust curve, the steady-state speed versus moment curve, and the motor response curve. Then, the parameter fitting is used to get C_R, ϖ_b, c_T, and c_M. A step change to the throttle command is undertaken (the ESC input line is connected directly to the dial switch of the RC transmitter) in order to activate the dynamometer, which records the data and is connected to a personal computer, to get the thrust variation curve. The time constant T_m is obtained by analyzing the settling time.

As shown in Fig. 6.16, the throttle command versus steady-state speed curve is not linear in the whole interval. This is determined by the ESC safety mechanism. A dead zone is set by the ESC. If $\sigma < 0.05$, then the output voltage of the ESC is 0 and the motor speed is 0, which means that the motor stops rotating. The safety is guaranteed when the throttle control stick is nearly pulled to the bottom. If $\sigma > 0.9$, then the motor speed is no longer increased. The speed is now close to the motor limit and the output speed (the thrust) is ensured to be in a safe range, which is denoted as the saturation zone. When $0.2 < \sigma < 0.8$, the throttle command versus steady-state speed curve is nearly linear. After the linear fitting is applied to this curve, one has

$$\varpi_{ss} = 6342\sigma + 1779. \tag{6.45}$$

Observed from Eq. (6.45), one has $C_R = 6432$ RPM and $\varpi_b = 1779$ RPM. It should be noted that the speed unit is RPM.

As shown in Fig. 6.17, the quadratic curve fitting is used. Observed from the fitting result, the thrust and the speed do have a quadratic relationship. Their relationship is expressed as

$$T = 0.1984(\varpi_{ss}/1000)^2. \tag{6.46}$$

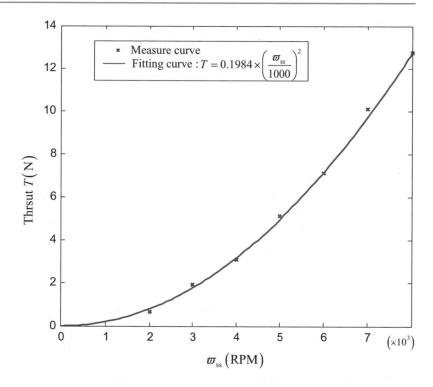

Fig. 6.17 Steady-state speed versus thrust curve

Therefore, one has $c_T = 1.984 \times 10^{-7}$ N/RPM^2.

The curve shown in Fig. 6.18 is the steady-state speed versus moment curve. The following equation is obtained by using the quadratic curve fitting

$$M = 0.003733(\varpi_{ss}/1000)^2. \qquad (6.47)$$

Therefore, one has $c_M = 3.733 \times 10^{-9}$ N · m/RPM^2.

The Fig. 6.19 is the motor-propeller response curve. The ESC input line is connected to the three-position switch of the RC transmitter (the output jumps among 0, 0.5, and 1). The dynamometer is used to record the thrust variation after the switch position is changed quickly. A high-end dynamometer can be connected to a personal computer via a data cable. The curve is displayed in real time, and the data can be exported. It is assumed that the Motor-propeller model is a first-order low-pass filter. The time constant T_m of the first-order low-pass filter determines the step response. Its value is the time taken to achieve $0.632\varpi_{ss}$ (ϖ_{ss} is the steady-state speed) from zero. Due to the quadratic relation between the thrust and the motor speed, the value of T_m is actually the time needed to achieve a thrust value of $0.4T_{max}$ (T_{max} is the steady-state thrust) from zero-thrust. The steady-state thrust T_{max} needs to be determined first. In the case of Fig. 6.19, $T_{max} = 8.55$ N. Then, the corresponding time of $0.4T_{max}$, denoted by t_e, is determined. Here, one has $t_e = 0.318$ s. When the initial time for the motor rotation, denoted by t_s, is determined ($t_s = 0.22$ s in Fig. 6.19), the time constant is expressed as

$$T_m = \Delta t = t_e - t_s = 0.098 \text{ s}. \qquad (6.48)$$

Fig. 6.18 Steady-state speed versus moment curve

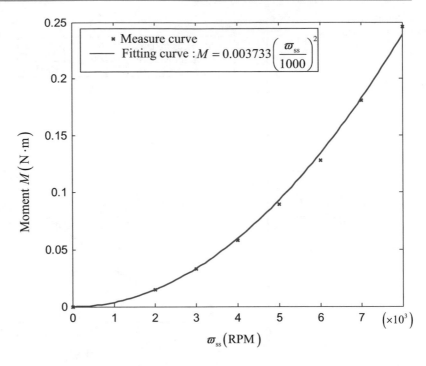

Fig. 6.19 Motor-propeller response curve

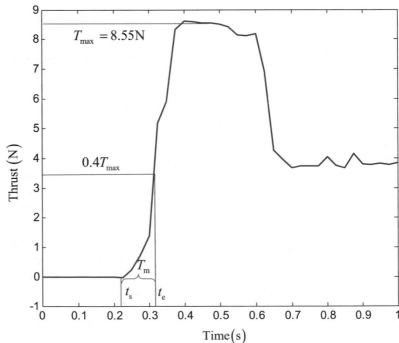

6.4 Summary

In this chapter, the multicopter flight control rigid model (6.13) or (6.14) or (6.15), the control effectiveness model (6.25), and the propulsor model (6.28) are given. These models constitute the multicopter control model. Readers can also refer to [7–9] for more information. Furthermore, the multicopter aerodynamic drag model is established. For real systems, the model parameters need to be determined. Therefore, the last section of this chapter presents methods to obtain the model parameters. In general, the multicopter modeling process is relatively mature as well as important in practice. However, for high-accuracy control, models with higher accuracy are required, such as dynamic model involving disturbances induced by wind or the ground effect.

Exercises

6.1 Fill in the steps of the derivation of Eq. (6.8).

6.2 Based on Sect. 6.1.3.2 in this chapter, write out the control effectiveness models for the multicopters in Figs. 1.3b and 1.4 in Chap. 1.

6.3 Establish a multicopter flight control rigid model for the CyPhy LVL 1 drone shown in Fig. 3.6 in Chap. 3 and write out the control effectiveness model. (Hint: The main feature of CyPhy LVL 1 is that the propellers are not parallel to the fuselage, which means that the thrust direction is no longer parallel to the body axis.)

6.4 Measure the moments of inertia of your cell phone.

6.5 Given the measurement data from Web site http://rfly.buaa.edu.cn/course, identify the parameters using the parameter fitting methods.

References

1. Blade flapping. http://www.dynamicflight.com/aerodynamics/flapping/. Accessed 22 Jan 2016
2. Bristeau PJ, Callou F, Vissiere D, Petit N (2011) The navigation and control technology inside the ar. drone micro UAV. In: Proceedings of the 18th IFAC world congress, Milano, Italy, pp 1477–1484
3. Mahony R, Kumar V, Corke P (2012) Multirotor aerial vehicles: modeling, estimation, and control of quadrotor. IEEE Robot Autom Mag 19(3):20–32
4. Abeywardena D, Kodagoda S, Dissanayake G, Munasinghe R (2013) Improved state estimation in quadrotor MAVs: a novel drift-free velocity estimator. IEEE Robot Autom Mag 20(4):32–39
5. Leishman RC, Macdonald JC, Beard RW, McLain TW (2014) Quadrotors and accelerometers: state estimation with an improved dynamic model. IEEE Control Syst Mag 34(1):28–41
6. Quan Q, Dai XH, Wei ZB, Wang J, Cai KY (2013) A method to measure the moments of inertia and products of inertia of small aircraft. Chinese patent, ZL201310479270.0. (In Chinese)
7. Bresciani T (2008) Modelling, identification and control of a quadrotor helicopter. Dissertation, Lund University, Sweden
8. De Oliveira M (2011) Modeling, identification and control of a quadrotor aircraft. Dissertation, Czech Technical University, Czech
9. Pounds P, Mahony R, Corke P (2010) Modelling and control of a large quadrotor robot. Control Eng Pract 18(7): 691–699

Part III
Perception

Sensor Calibration and Measurement Model

7

There are many sensors mounted in a multicopter, such as the three-axis accelerometer, three-axis gyroscope, three-axis magnetometer, ultrasonic range finder, 2D laser range finder, Global Positioning System (GPS) receiver and camera. These sensors in a multicopter are like sensory organs of a human being. A multicopter can obtain its position and attitude with these sensors. However, these sensors based on Micro-Electro-Mechanical System (MEMS) are inaccurate. For example, the devices on the circuit board can affect the measurement of a magnetometer. So it is impossible to obtain an accurate orientation, not to mention making flight reliably control. On the other hand, unsteady takeoff of multicopters can also be related to uncalibrated sensors. Uncalibrated sensors could make multicopters unstable once they take off. Thus, the calibration should be performed as a key step of manufacturing. Some calibration methods that do not need any additional equipment are introduced. This chapter aims to answer the question below:

How are sensors calibrated and what are the measurement models?

The answer to this question involves fundamental principles, calibration methods, and measurement models of sensors.

© Springer Nature Singapore Pte Ltd. 2017
Q. Quan, *Introduction to Multicopter Design and Control*, DOI 10.1007/978-981-10-3382-7_7

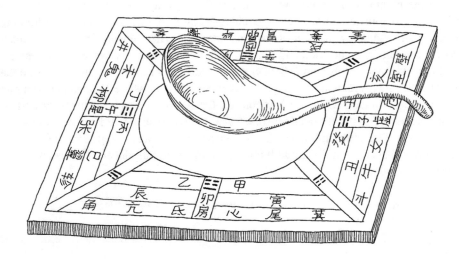

Compass

The Compass was invented by the ancient Chinese ages ago and was initially used in the expansion of the territory. According to historical records, the use of compass dates back to the Warring States Period (475–221 BC). This is recorded in the book "*Han Feizi*", an ancient Chinese text attributed to a legalist philosopher during the third century BC. Furthermore, the compass was explained more explicitly in the book "*Brush Talks from Dream Brook*", written by Kuo Shen during the Song dynasty. Kuo Shen also discovered the phenomenon of magnetic declination. This is the first record of magnetic declination in the world, about 400 years earlier than the discovery made by Christopher Columbus.

7.1 Three-Axis Accelerometer

7.1.1 Fundamental Principle

The three-axis accelerometer is a kind of inertial sensor which measures the *specific force*, namely the total acceleration eliminating gravity or the non-gravitational force per unit mass [1, p. 10], [2]. When an accelerometer stays still, the acceleration of gravity can be sensed by the accelerometer, but the total acceleration is zero. In free-fall, the total acceleration is the acceleration of gravity, but the three-axis accelerometer will output zero. The principle of three-axis accelerometers can be used to measure angles. Intuitively, as shown in Fig. 7.1, the amount of spring compression is decided by the angle between the accelerometer and the ground. The specific force can be measured by the length the spring compressed. So, in the absence of external forces, the accurate pitch angle and roll angle can be measured without accumulative error.

Accelerometers used in multicopters are often based on MEMS technology so that the size of accelerometers can be as small as a fingernail. Also, the energy consumption is low. Nowadays, the MEMS technology is widely applied in producing accelerometers mounted in smartphones [3, pp. 31–32]. The MEMS accelerometers are based on piezoresistive effect, piezoelectric effect, or capacitive principle; the specific force of those effects is proportional to resistance, voltage, or capacitance, respectively. These values can be collected through the amplification circuit and filter circuit. The output error induced by vibration is large, which is the shortcoming of this kind of sensor [4, pp. 67–75].

7.1.2 Calibration

This section is mainly adapted from [5]. The given method is only one of calibration methods.

(1) Error Model
Let $^b\mathbf{a}_m \in \mathbb{R}^3$ be the calibrated specific force and $^b\mathbf{a}'_m \in \mathbb{R}^3$ be the specific force without calibration. They have the following relationship

$$^b\mathbf{a}_m = \mathbf{T}_a \mathbf{K}_a \left(^b\mathbf{a}'_m + \mathbf{b}'_a \right). \tag{7.1}$$

Here,

$$\mathbf{T}_a = \begin{bmatrix} 1 & \Delta\psi_a & -\Delta\theta_a \\ -\Delta\psi_a & 1 & \Delta\phi_a \\ \Delta\theta_a & -\Delta\phi_a & 1 \end{bmatrix}, \mathbf{K}_a = \begin{bmatrix} s_{ax} & 0 & 0 \\ 0 & s_{ay} & 0 \\ 0 & 0 & s_{az} \end{bmatrix}, \mathbf{b}'_a = \begin{bmatrix} b'_{ax} \\ b'_{ay} \\ b'_{az} \end{bmatrix} \tag{7.2}$$

Fig. 7.1 Measuring principle of MEMS accelerometers

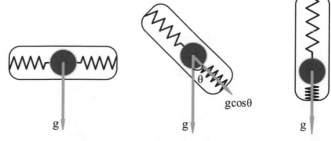

(a) Indicating value is 0   (b) Indicating value is gcosθ   (c) Indicating value is g

where $\mathbf{T}_a \in \mathbb{R}^{3\times3}$ denotes the tiny tilt in the process of mounting sensors, $\mathbf{K}_a \in \mathbb{R}^{3\times3}$ denotes the scale factor, and $\mathbf{b}'_a \in \mathbb{R}^3$ denotes the bias.

(2) Calibration Principle
In order to calibrate an accelerometer, the following unknown parameters

$$\mathbf{\Theta}_a \triangleq \begin{bmatrix} \Delta\psi_a & \Delta\theta_a & \Delta\phi_a & s_{ax} & s_{ay} & s_{az} & b'_{ax} & b'_{ay} & b'_{az} \end{bmatrix}^{\mathrm{T}} \tag{7.3}$$

need to be estimated. Then, the right side of the model (7.1) can be written as a function as follows:

$$\mathbf{h}_a\left(\mathbf{\Theta}_a, {}^b\mathbf{a}'_m\right) \triangleq \mathbf{T}_a\mathbf{K}_a\left({}^b\mathbf{a}'_m + \mathbf{b}'_a\right). \tag{7.4}$$

Here, the measurement noise is eliminated because the specific force is measured many times in a single attitude. Concretely, the accelerometer is rotated at different angles for $M \in \mathbb{Z}_+$ times. At each angle, the accelerometer will stay still for a short time. The average specific force in the kth interval is denoted as ${}^b\mathbf{a}'_{m,k} \in \mathbb{R}^3$ where $k = 1, 2, \ldots, M$. The calibration principle is that *the magnitude of specific force keeps constant with different attitude of accelerometers, which is the local gravity, denoted as g.* According to this principle, the following optimization is given to estimate the unknown parameters

$$\mathbf{\Theta}^*_a = \arg \min_{\mathbf{\Theta}_a \in \mathbb{R}^9} \sum_{k=1}^{M} \left(\left\|\mathbf{h}_a\left(\mathbf{\Theta}_a, {}^b\mathbf{a}'_{m,k}\right)\right\| - g\right)^2. \tag{7.5}$$

The Levenberg–Marquardt algorithm can be used to find the optimal solution $\mathbf{\Theta}^*_a$.

(3) Calibration Results
The experimental data is collected from a Pixhawk, an advanced autopilot for building Unmanned Aerial Vehicles (UAVs), based on the PX4 open-hardware project. The Pixhawk is connected to a computer and the data is read via a series bus. According to the method mentioned in the previous part, the following calibration results are obtained

$$\mathbf{T}_a = \begin{bmatrix} 1 & 0.0093 & -0.0136 \\ -0.0093 & 1 & 0.0265 \\ 0.0136 & -0.0265 & 1 \end{bmatrix}$$

$$\mathbf{K}_a = \begin{bmatrix} 1.0203 & 0 & 0 \\ 0 & 1.0201 & 0 \\ 0 & 0 & 1.0201 \end{bmatrix}$$

$$\mathbf{b}'_a = \begin{bmatrix} -2.755 \\ 1.565 \\ -9.942 \end{bmatrix} \times 10^{-5}.$$

The index

$$Dist_a \triangleq \left(\left\|\mathbf{h}_a\left(\mathbf{\Theta}_a, {}^b\mathbf{a}'_{m,k}\right)\right\| - g\right)^2 \tag{7.6}$$

is defined to describe the optimization result, as shown in Fig. 7.2. It is observed that the index is decreased.

Fig. 7.2 Optimization results of calibrated and uncalibrated accelerometers

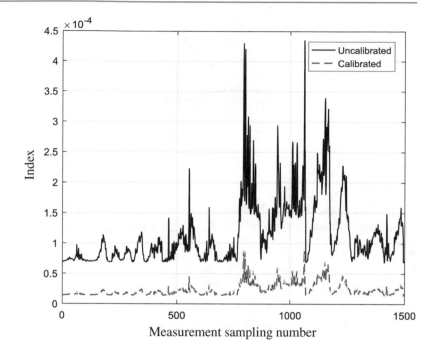

7.1.3   Measurement Model

MEMS accelerometers are fixed to the Aircraft-Body Coordinate Frame (ABCF), which can measure specific forces along different axes. The measurement is denoted as

$$^{b}\mathbf{a}_m = {}^{b}\mathbf{a} + \mathbf{b}_a + \mathbf{n}_a. \tag{7.7}$$

Here, $^{b}\mathbf{a} \in \mathbb{R}^3$ denotes the true value of the specific force and $\mathbf{b}_a \in \mathbb{R}^3$ denotes the drift error. The noise $\mathbf{n}_a \in \mathbb{R}^3$ is the accelerometer measurement noise vector, which is often considered as a Gaussian White Noise (GWN) vector. Furthermore, the drift error $\mathbf{b}_a$ is considered to have the following model

$$\dot{\mathbf{b}}_a = \mathbf{n}_{\mathbf{b}_a} \tag{7.8}$$

where $\mathbf{n}_{\mathbf{b}_a} \in \mathbb{R}^3$ is often considered to be a GWN vector.

7.2   Three-Axis Gyroscope

7.2.1   Fundamental Principle

As shown in Fig. 7.3a, the MEMS gyroscope is based on *Coriolis force* that is the apparent deflection of moving objects from a straight path when viewed from a rotating frame of reference. According to the change in the amount of capacity, the angular velocity can be calculated because the Coriolis force is proportional to the angular velocity. Usually, MEMS gyroscopes are designed in the form as shown in Fig. 7.3b. The directions of accelerations of two masses are opposite, but the magnitudes are equal. As a result of the two different forces, the two capacitor plates are forced to generate the capacitance

Fig. 7.3 Measuring
principle of MEMS
gyroscopes

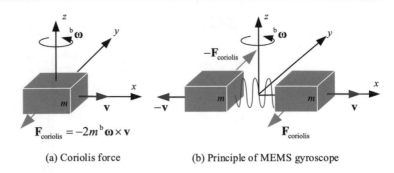

(a) Coriolis force   (b) Principle of MEMS gyroscope

difference which is proportional to the angular velocity. On the other hand, the change of acceleration will have the same influence on the two capacitor plates, so the angular velocity measurement will not be affected.

7.2.2   Calibration

This section is mainly adapted from [5]. The given method is only one of calibration methods.

(1) Error Model
Similar to accelerometers, there exist some errors in gyroscopes. Let $^b\boldsymbol{\omega}_m \in \mathbb{R}^3$ be the angular velocity after calibration and $^b\boldsymbol{\omega}'_m \in \mathbb{R}^3$ be the angular velocity uncalibrated. They have the following relationship

$$^b\boldsymbol{\omega}_m = \mathbf{T}_g\mathbf{K}_g\left(^b\boldsymbol{\omega}'_m + \mathbf{b}'_g\right). \tag{7.9}$$

Here,

$$\mathbf{T}_g = \begin{bmatrix} 1 & \Delta\psi_g & -\Delta\theta_g \\ -\Delta\psi_g & 1 & \Delta\phi_g \\ \Delta\theta_g & -\Delta\phi_g & 1 \end{bmatrix}, \mathbf{K}_g = \begin{bmatrix} s_{gx} & 0 & 0 \\ 0 & s_{gy} & 0 \\ 0 & 0 & s_{gz} \end{bmatrix}, \mathbf{b}'_g = \begin{bmatrix} b'_{gx} \\ b'_{gy} \\ b'_{gz} \end{bmatrix} \tag{7.10}$$

where $\mathbf{T}_g \in \mathbb{R}^{3\times 3}$ denotes the tiny tilt in the process of mounting sensors, $\mathbf{K}_g \in \mathbb{R}^{3\times 3}$ denotes the scale factor, and $\mathbf{b}'_g \in \mathbb{R}^3$ denotes the bias.

(2) Calibration Principle
In order to calibrate a gyroscope, the following unknown parameters

$$\boldsymbol{\Theta}_g \triangleq \begin{bmatrix} \Delta\psi_g & \Delta\theta_g & \Delta\phi_g & s_{gx} & s_{gy} & s_{gz} & b'_{gx} & b'_{gy} & b'_{gz} \end{bmatrix}^T \tag{7.11}$$

need to be estimated. The calibration for gyroscopes is based on calibrated accelerometers. The calibration principle is that *the integration of angular velocity by a gyroscope can be used to calculate the angle, which should have the same accuracy as the angle calculated by a calibrated accelerometer.* Let

$$^b\mathbf{a}_{m,k} \triangleq \begin{bmatrix} a_{x_bm,k} & a_{y_bm,k} & a_{z_bm,k} \end{bmatrix}^T \tag{7.12}$$

be the kth calibrated specific force from a calibrated accelerometer. Then, the kth Euler angles are

$$\theta_k = \arcsin\left(\frac{a_{x_b m,k}}{g}\right)$$
$$\phi_k = -\arcsin\left(\frac{a_{y_b m,k}}{g\cos\theta_k}\right) \qquad (7.13)$$
$$\psi_k = 0$$

where the yaw angle is set to 0. According to Eq. (5.45) in Chap. 5, the corresponding quaternion is

$$\mathbf{q}_{a,k} = \begin{bmatrix} \cos\frac{\phi_k}{2}\cos\frac{\theta_k}{2}\cos\frac{\psi_k}{2} + \sin\frac{\phi_k}{2}\sin\frac{\theta_k}{2}\sin\frac{\psi_k}{2} \\ \sin\frac{\phi_k}{2}\cos\frac{\theta_k}{2}\cos\frac{\psi_k}{2} - \cos\frac{\phi_k}{2}\sin\frac{\theta_k}{2}\sin\frac{\psi_k}{2} \\ \cos\frac{\phi_k}{2}\sin\frac{\theta_k}{2}\cos\frac{\psi_k}{2} + \sin\frac{\phi_k}{2}\cos\frac{\theta_k}{2}\sin\frac{\psi_k}{2} \\ \cos\frac{\phi_k}{2}\cos\frac{\theta_k}{2}\sin\frac{\psi_k}{2} - \sin\frac{\phi_k}{2}\sin\frac{\theta_k}{2}\cos\frac{\psi_k}{2} \end{bmatrix}. \qquad (7.14)$$

In order to clarify the calibration process for the gyroscope, a function $\boldsymbol{\Psi}$ is defined as follows:

$$\mathbf{q}'_{a,k+1} = \boldsymbol{\Psi}\left(\boldsymbol{\Theta}_g, {}^b\boldsymbol{\omega}'_{m,k:k+1}, \mathbf{q}_{a,k}\right) \qquad (7.15)$$

where ${}^b\boldsymbol{\omega}'_{m,k:k+1} \in \mathbb{R}^{3\times N}$ denotes the N angular velocities between the kth and $(k+1)$th acceleration measurements, $\mathbf{q}_{a,k}$ is the kth measurement by an accelerometer according to Eqs. (7.13) and (7.14), and $\mathbf{q}'_{a,k+1}$ denotes the $(k+1)$th estimate based on ${}^b\boldsymbol{\omega}'_{m,k:k+1}$ and $\mathbf{q}_{a,k}$. According to Eq. (5.47) in Chap. 5, the estimated Euler angles $\theta'_{k+1}, \phi'_{k+1}, \psi'_{k+1}$ are obtained. Furthermore, the estimated specific force $\mathbf{a}'_{k+1}$ is

$$\mathbf{a}'_{k+1} = g\begin{bmatrix} -\sin\theta'_{k+1} \\ \cos\theta'_{k+1}\sin\phi'_{k+1} \\ \cos\theta'_{k+1}\cos\phi'_{k+1} \end{bmatrix}. \qquad (7.16)$$

The process is described as

$$\mathbf{a}'_k = \mathbf{h}_g\left(\boldsymbol{\Theta}_g, {}^b\boldsymbol{\omega}'_{m,k-1:k}, {}^b\mathbf{a}_{m,k-1}\right) \qquad (7.17)$$

where $\mathbf{h}_g$ is the function to estimate the specific force based only on the last specific force and angular velocities. The calibration principle is further expressed as follows *the calibrated gyroscope should make the estimation $\mathbf{a}'_k$ and measurement ${}^b\mathbf{a}_{m,k}$ by the calibrated accelerometer as close as possible.* According to this principle, the following optimization is given as

$$\boldsymbol{\Theta}^*_g = \arg\min_{\boldsymbol{\Theta}_g \in \mathbb{R}^9} \sum_{k=1}^M \left(\mathbf{h}_g\left(\boldsymbol{\Theta}_g, {}^b\boldsymbol{\omega}'_{m,k-1:k}, {}^b\mathbf{a}_{m,k-1}\right) - {}^b\mathbf{a}_{m,k}\right)^2. \qquad (7.18)$$

In the following, the function $\boldsymbol{\Psi}$ between the angular velocity and quaternion is clarified. According to Eq. (5.51) in Chap. 5, one has

$$\dot{\mathbf{q}} = \underbrace{\frac{1}{2}\begin{bmatrix} 0 & -\boldsymbol{\omega}^T(t) \\ \boldsymbol{\omega}(t) & -[\boldsymbol{\omega}(t)]_\times \end{bmatrix}\mathbf{q}}_{\mathbf{f}(\mathbf{q},t)}. \qquad (7.19)$$

Based on this relationship, $\mathbf{q}'_{a,k+1}$ can be estimated based on ${}^b\boldsymbol{\omega}'_{m,k:k+1}$ and $\mathbf{q}_{a,k}$. Here, the function $\boldsymbol{\Psi}$ can be optimized by using the four-order Runge–Kutta method. Concretely, the model is written as

$$q'_{a,k+1} = \frac{q_{a,k} + \frac{1}{6}\Delta t \left(k_1 + 2k_2 + 2k_3 + k_4\right)}{\left\| q_{a,k} + \frac{1}{6}\Delta t \left(k_1 + 2k_2 + 2k_3 + k_4\right)\right\|}. \tag{7.20}$$

Here,

$$\begin{aligned} k_i &= f\left(q^{(i)}, t_k + c_i \Delta t\right), i = 1, 2, 3, 4 \\ q^{(1)} &= q_k \\ q^{(i)} &= q_k + \Delta t \sum_{j=1}^{i-1} a_{ij} k_j, i = 2, 3, 4 \end{aligned} \tag{7.21}$$

where these parameters used are

$$c_1 = 0, c_2 = \frac{1}{2}, c_3 = \frac{1}{2}, c_4 = 1, a_{21} = \frac{1}{2}, a_{31} = 0, a_{41} = 0, a_{32} = \frac{1}{2}, a_{42} = 0, a_{43} = 1$$

Δt denotes the step size of the acceleration measurement and t_k is the time of the kth measurement.

(3) Calibration Results
The experimental data is collected from a Pixhawk. The Pixhawk is connected to a computer and the data is read via a series bus. According to the method mentioned in the previous part, the following calibration results are obtained

$$T_g = \begin{bmatrix} 1 & 0.1001 & -0.1090 \\ -0.1001 & 1 & 0.1002 \\ 0.1090 & -0.1002 & 1 \end{bmatrix}$$

$$K_g = \begin{bmatrix} 1 & 0 & 0 \\ 0 & 1 & 0 \\ 0 & 0 & 1 \end{bmatrix}$$

$$b'_g = \begin{bmatrix} 0.2001 \\ 0.2002 \\ 0.2004 \end{bmatrix} \times 10^{-3}.$$

The index

$$Dist_g \triangleq \left(h_g\left(\Theta_g, {}^b\omega'_{m,k-1:k}, {}^b a_{k-1}\right) - {}^b a_k\right)^2 \tag{7.22}$$

is defined to describe the optimization result, as shown in Fig. 7.4. It is observed that the index is decreased.

7.2.3   Measurement Model

MEMS gyroscopes are fixed to the ABCF, which can measure angular velocities along different axes. So, the measurement is denoted as

$${}^b\omega_m = {}^b\omega + b_g + n_g \tag{7.23}$$

where ${}^b\omega \in \mathbb{R}^3$ denotes the true value of angular velocity, $b_g \in \mathbb{R}^3$ denotes the drift error, and $n_g \in \mathbb{R}^3$ is the angular velocity measurement noise vector which is often considered as a GWN vector. Furthermore, the drift error is considered to have the following model

$$\dot{b}_g = n_{b_g} \tag{7.24}$$

where $n_{b_g} \subset \mathbb{R}^3$ is often considered as a GWN vector

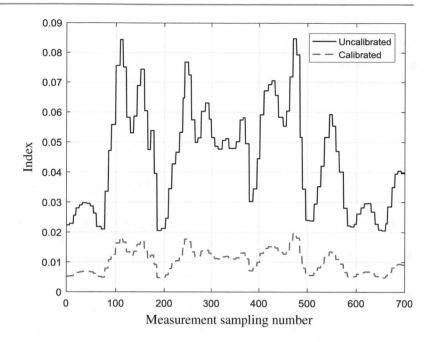

Fig. 7.4 Optimization results of calibrated and uncalibrated gyroscopes

---

7.3   Three-Axis Magnetometer

7.3.1   Fundamental Principle

Magnetometers are essential components for providing navigation and location-based services. Magnetometers use *anisotropic magnetoresistance* or *Hall effect* to detect the strength of the space magnetic induction [6, pp. 130–144]. Besides, *Lorentz-force* magnetometers have also been studied and developed [7]. Based on the Lorenz principle, the electric and magnetic fields arise a force to change the capacitance in the circuit.

7.3.2   Calibration

This section is mainly adapted from [5]. The given method is only one of calibration methods.

(1) Error Model
Similar to accelerometers, there exist some errors in magnetometers. Assume that $^b\mathbf{m}_m \in \mathbb{R}^3$ denotes the magnetic induction value after calibration and $^b\mathbf{m}'_m \in \mathbb{R}^3$ denotes the magnetic induction value without calibration. They have the following relationship

$$^b\mathbf{m}_m = \mathbf{T}_m \mathbf{K}_m \left(^b\mathbf{m}'_m + \mathbf{b}'_m \right). \tag{7.25}$$

Here,

$$\mathbf{T}_m = \begin{bmatrix} 1 & \Delta\psi_m & -\Delta\theta_m \\ -\Delta\psi_m & 1 & \Delta\phi_m \\ \Delta\theta_m & -\Delta\phi_m & 1 \end{bmatrix}, \quad \mathbf{K}_m = \begin{bmatrix} s_{mx} & 0 & 0 \\ 0 & s_{my} & 0 \\ 0 & 0 & s_{mz} \end{bmatrix}, \quad \mathbf{b}'_m = \begin{bmatrix} b'_{mx} \\ b'_{my} \\ b'_{mz} \end{bmatrix}$$

where $\mathbf{T}_m \in \mathbb{R}^{3\times3}$ denotes the tiny tilt in the process of mounting sensors, $\mathbf{K}_m \in \mathbb{R}^{3\times3}$ denotes the scale factor, and $\mathbf{b}'_m \in \mathbb{R}^3$ denotes the bias.

(2) Calibration Principle
In order to calibrate a magnetometer, the following unknown parameters

$$\boldsymbol{\Theta}_m \triangleq \begin{bmatrix} \Delta\psi_m & \Delta\theta_m & \Delta\phi_m & s_{mx} & s_{my} & s_{mz} & b'_{mx} & b'_{my} & b'_{mz} \end{bmatrix}^{\mathrm{T}} \tag{7.26}$$

need to be estimated. Then, the right side of the model (7.25) can be written as a function as follows:

$$\mathbf{h}_m \left(\boldsymbol{\Theta}_m, {}^b\mathbf{m}'_m \right) \triangleq \mathbf{T}_m \mathbf{K}_m \left({}^b\mathbf{m}'_m + \mathbf{b}'_m \right). \tag{7.27}$$

Here, the measurement noise is eliminated because the magnetic induction value is measured for many times at one attitude. To be specific, the magnetometer is rotated at different angles for $M \in \mathbb{Z}_+$ times. At each angle, the magnetic induction will stay still for a short time. In normal, the magnetic induction keeps constant with different attitude of magnetometer. For simplicity, let $\left\| {}^b\mathbf{m}_{m,k} \right\|^2 = 1, k = 1, 2, \ldots, M$. Here, the magnetic induction value is normalized. So, ${}^b\mathbf{m}_{m,k}, k = 1, 2, \ldots, M$, are all on the unit circle. The calibration principle is that *the value of* $\boldsymbol{\Theta}_m$ *should make the value of* $\left\| \mathbf{T}_m \mathbf{K}_m \left({}^b\mathbf{m}'_m + \mathbf{b}'_m \right) \right\|$ *as close to 1 as possible* [8]. According to this principle, the following optimization is given

$$\boldsymbol{\Theta}_m^* = \arg \min_{\boldsymbol{\Theta}_m \in \mathbb{R}^9} \sum_{k=1}^{M} \left(\left\| \mathbf{h}_m \left(\boldsymbol{\Theta}_m, {}^b\mathbf{m}'_{m,k} \right) \right\| - 1 \right)^2. \tag{7.28}$$

(3) Calibration Results
The experimental data is collected from a Pixhawk. The Pixhawk is connected to a computer and the data is read via a series bus. According to the method mentioned in the previous part, the calibration result can be obtained as follows:

$$\mathbf{T}_m = \begin{bmatrix} 1 & -0.0026 & 0.0516 \\ 0.0026 & 1 & -0.0156 \\ -0.0516 & 0.0156 & 1 \end{bmatrix}$$

$$\mathbf{K}_m = \begin{bmatrix} 0.9999 & 0 & 0 \\ 0 & 1 & 0 \\ 0 & 0 & 0.9999 \end{bmatrix}$$

$$\mathbf{b}'_m = \begin{bmatrix} -0.3223 \\ -0.1280 \\ 0.1589 \end{bmatrix} \times 10^{-5}.$$

The index

$$Dist_m \triangleq \left(\left\| \mathbf{h}_m \left(\boldsymbol{\Theta}_m, {}^b\mathbf{m}'_{m,k} \right) \right\| - 1 \right)^2 \tag{7.29}$$

is defined to describe the optimization result, as shown in Fig. 7.5. It is observed that the index is decreased.

Fig. 7.5 Optimization results of calibrated and uncalibrated magnetometers

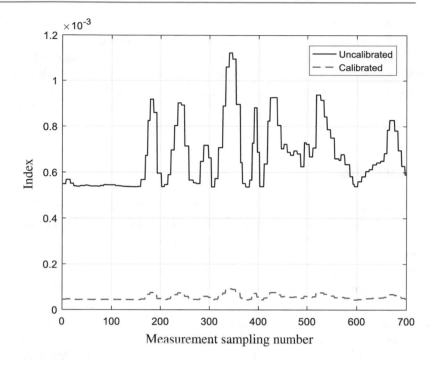

7.3.3 Measurement Model

The magnetic field is denoted as ${}^e\mathbf{m} \in \mathbb{R}^3$. The magnetometer is aligned with the ABCF, which can measure magnetic field intensity along different axes, denoted ${}^b\mathbf{m}_m \in \mathbb{R}^3$. When the sensor is placed horizontally, the vector sum of the two vector components in the horizontal direction is always pointing to the north after compensating for the *magnetic declination* (refer to Sect. 9.1.1.2 in Chap. 9). The measurement is denoted as

$$ {}^b\mathbf{m}_m = \mathbf{R}_e^b \cdot {}^e\mathbf{m} + \mathbf{b}_m + \mathbf{n}_m \tag{7.30} $$

where $\mathbf{R}_e^b$ denotes rotation matrix from the Earth-Fixed Coordinate Frame (EFCF) to the ABCF, $\mathbf{b}_m \in \mathbb{R}^3$ denotes the drift error, and $\mathbf{n}_m \in \mathbb{R}^3$ is the magnetic field intensity measurement noise vector which is often considered as a GWN vector. Furthermore, the drift error $\mathbf{b}_m$ can be considered as

$$ \dot{\mathbf{b}}_m = \mathbf{n}_{\mathbf{b}_m} \tag{7.31} $$

where $\mathbf{n}_{\mathbf{b}_m} \in \mathbb{R}^3$ is often considered as a GWN vector. The magnetometer is used to measure the yaw angel.

7.4 Ultrasonic Range Finder

7.4.1 Fundamental Principle

Ultrasound is sound with a frequency greater than the upper limit of human hearing (greater than 20 kHz). With good directivity and powerful penetrability, ultrasonic is widely used in ranging speed or scale, etc. The ultrasound signal is transmitted by an ultrasonic transducer, reflected by an obstacle and received by another transducer where the signal is detected. The distance of an object is given by multiplying the half time delay of the position of the peak by the speeds of acoustic waves in air. Taking the different temperatures into consideration, the speeds of acoustic waves in the air are also different. Thus, the method of calculating the distance needs to be adjusted. Besides, there exist some disadvantages about this sensor. First, it only can measure a short distance. Moreover, as shown in Fig. 7.6, little or no echo may be reflected by a soft object or the object at an angle relative to the transducer.

7.4.2 Calibration

These measurements are used for position control. The slight deviations will not cause significant performance degradation to multicopters. Therefore, the deviation of these sensors can be corrected by online state estimation when multicopters are in the air.

7.4.3 Measurement Model

Ultrasonic range finders are typically used to measure the relative height. It is often mounted at the bottom of a multicopter and faces downward. The distance measured by the sensor is $d_{\mathrm{sonar}} \in \mathbb{R}_+ \cup \{0\}$, and the true height from the ground is denoted as $-p_{z_e} \in \mathbb{R}_+ \cup \{0\}$. Then, they have the following relation

$$d_{\mathrm{sonar}} = \frac{-1}{\cos\theta\,\cos\phi} p_{z_e} + n_{d_{\mathrm{sonar}}} \tag{7.32}$$

where $\theta, \phi \in \mathbb{R}$ denotes the pitch angel and the roll angle, and $n_{d_{\mathrm{sonar}}} \in \mathbb{R}$ is often considered as a GWN. If the ultrasonic range finder has a drift error, then its model can be extended similar to the models mentioned in Sects. 7.1.1–7.1.3.

7.5 Barometer

7.5.1 Fundamental Principle

If a multicopter is hovering far from the ground, ultrasonic range finders may be unavailable. In this condition, barometers should be used. The piezoelectric barometers are often used in multicopters. Barometers are intended to measure the atmospheric pressure, the corresponding altitude, or relative altitude by subtracting the two altitudes.

Fig. 7.6 Little or no echo cases when ultrasonic range finders are used

7.5.2 Calibration

These measurements are used for controlling position. The slight deviation will not cause significant performance degradation to multicopters. Therefore, deviation of barometers can be corrected by online state estimation when multicopters are flying.

7.5.3 Measurement Model

Barometers are used to measure the absolute height or the relative altitude. The altitude is denoted as

$$d_{\text{baro}} = -p_{z_e} + b_{d_{\text{baro}}} + n_{d_{\text{baro}}} \tag{7.33}$$

where $d_{\text{baro}} \in \mathbb{R}_+ \cup \{0\}$ denotes the height measured by the barometer, $b_{d_{\text{baro}}} \in \mathbb{R}$ denotes the drift error, and $n_{d_{\text{baro}}} \in \mathbb{R}$ is often considered as a GWN. Furthermore, $b_{d_{\text{baro}}} \in \mathbb{R}$ can be considered as

$$\dot{b}_{d_{\text{baro}}} = n_{b_{d_{\text{baro}}}} \tag{7.34}$$

where $n_{b_{d_{\text{baro}}}} \in \mathbb{R}$ is often considered as a GWN.

7.6   2D Laser Range Finder

7.6.1 Fundamental Principle

The 2D laser scanner can be used as a range finder which uses a laser beam to determine the distance. The most common form of laser range finders operates based on the Time of Flight (ToF). A laser pulse is sent in a narrow beam toward the object, and the time is measured by the pulse to be reflected off the target and returned to the transmitter. The principle is shown in Fig. 7.7.

7.6.2 Calibration

Because 2D laser scanners are often used for height measurement and obstacle avoidance, the slight deviation of these will not cause significant performance degradation of the aircraft. Thus, these sensors are generally accurate.

7.6.3 Measurement Model

The height from ground measured by the 2D laser range finder is $d_{\text{laser}} \in \mathbb{R}_+ \cup \{0\}$ shown as follows:

$$d_{\text{laser}} = \frac{1}{M} \sum_{i=1}^{M} \rho_i \cos\varphi_i = \frac{-1}{\cos\theta \cos\phi} p_{z_e} + n_{d_{\text{laser}}} \tag{7.35}$$

where θ, ϕ denote pitch angle and roll angle; $\rho_i \in \mathbb{R}_+ \cup \{0\}$ and $\varphi_i \in [-\varphi_{\max}, \varphi_{\max}]$ are the values of distance measured by the range finder and the corresponding angle, respectively; $M \in \mathbb{Z}_+$ denotes the number of samples, and $n_{d_{\text{laser}}} \in \mathbb{R}$ is considered as a GWN.

Fig. 7.7 The principle of
2D laser range finders

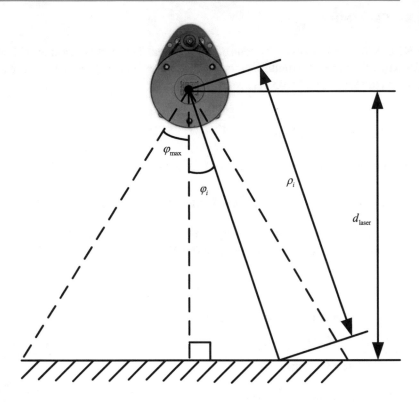

7.6.4   Supplement: LiDAR

LiDAR (Light Detection and Ranging) is a remote sensing way by using light in the form of a pulsed laser to measure ranges (varying distances). When an airborne laser is pointed at a targeted object, the reflected light is then recorded by a sensor to measure the range. As shown in Fig. 7.8, "when laser ranges are combined with position and orientation data generated from integrated GPS receivers and Inertial Measurement Unit (IMU) systems, scan angles, and calibration data, the result is a dense, detail-rich group of elevation points, called a *point cloud* [9]". LiDAR can produce more accurate shoreline maps to assist in emergency response operations and in many other applications. Without GPS receiver and IMU, supported by the data only from LiDAR, the 3D space model can also be built, as shown in Fig. 7.9.

The performance of LiDAR depends on weight, scale, power, horizontal viewing angle and vertical viewing angle, the number of scanning dot per second, scanning frequency, identification accuracy, and number of channels. For example, the weight of VLP-16 from Velodyne company is only 0.83 kg. Scanning radius scale is 100 m. The power is 8 W. Horizontal viewing angle is 360°. Vertical viewing angle is ±15°. The number of scanning dots is up to 300,000. The scale of scanning frequency is 5–20 Hz. Thanks to the miniaturization of LiDAR, it can be carried by multicopters, shown in Fig. 7.10.

Fig. 7.8 Process of producing point cloud

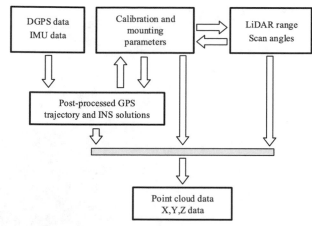

Fig. 7.9 3D space model by the LiDAR without GPS receiver and IMU. (photograph from http://www.prweb.com by Wolfgang Juchmann)

Fig. 7.10 LiDAR of Velodyne company and multicopter with LiDAR

7.7   Global Positioning System

7.7.1   Fundamental Principle [10]

The Global Positioning System (GPS), shown as in Fig. 7.11, is a Global Navigation Satellite System (GNSS)[1] that uses satellites to provide locations and time. The satellites use atomic clocks which are synchronized to each other, and the time of clocks is corrected by the true time of the ground clocks. GPS satellites' locations are monitored precisely. GPS receivers have clocks which are less stable since they are not synchronized with the true time. While GPS satellites broadcast their position

---

[1] There are other similar GNSS systems such as the BeiDou (China), GLONASS (Russia), and Galileo (European Union).

Fig. 7.11 GPS schematic
diagram

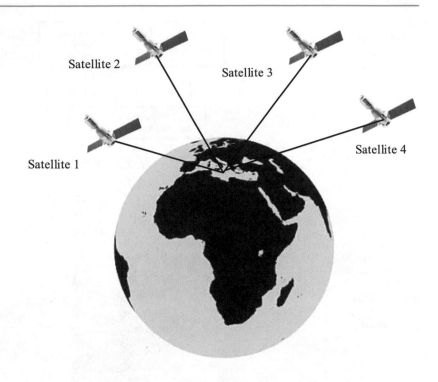

and time, GPS receivers can get signals of multiple satellites. Thus, GPS receivers can calculate the
exact position by solving equations, and the deviation can be eliminated as well. The accuracy of GPS
is generally in meters. The GPS observations are affected by different factors: (1) satellite related,
including orbital errors and satellite clock errors; (2) propagation errors, including the ionospheric
delay error, tropospheric refraction error, and multipath errors; (3) the receiver error, including the
receiver clock error and observation errors. The distances from a receiver to satellites can be obtained
by measuring the ranging code by the GPS receiver. Because satellite related errors and propagation
errors exist, the distance is called *pseudorange*. For Coarse/Acquisition (C/A) code, the C/A code
pseudorange has an accuracy of 20 m. Taking the influence of the ionosphere, troposphere and the
clocks into consideration, the basic observation equation of pseudorange positioning is expressed as

$$\rho = \rho' + c\left(\delta_t + \delta_T\right) + \delta_I \tag{7.36}$$

where $\rho \in \mathbb{R}_+$ denotes the true distance from the satellite to the GPS receiver, $\rho' \in \mathbb{R}_+$ denotes
the pseudorange from the satellite to the GPS receiver, $c \in \mathbb{R}_+$ denotes the speed of light in the air,
$\delta_t \in \mathbb{R}$ denotes the satellite clock error correction given by the satellite navigation message, $\delta_I \in \mathbb{R}$
denotes signal propagation delay correction in the atmosphere which can be calculated with the suitable
mathematical model, $\delta_T \in \mathbb{R}$ denotes error correction of the receiver clock relative to the GPS time,
which is unknown. Assuming that the positions of satellites are $\mathbf{p}_{s,k} \in \mathbb{R}^3$, $k = 1, \ldots, n_s$ and the
receiver position is $\mathbf{p}_r \in \mathbb{R}^3$, according to Eq. (7.36), one has

$$\left\| \mathbf{p}_{s,k} - \mathbf{p}_r \right\| - \delta_I = \rho'_k + c\left(\delta_{t,k} + \delta_T\right), k = 1, \ldots, n_s. \tag{7.37}$$

Since there are four parameters, namely $\mathbf{p}_r$, δ_T, need to be calculated, the receiver needs to be linked to four satellites at least so that the receiver position can be determined.

Differential GPS (DGPS) can improve location accuracy by eliminating the common error. In DGPS system, the fixed ground-based reference stations are necessary. Stations broadcast the difference between the position given by GPS systems and the known fixed position. In fact, the difference between the measured satellite pseudorange and actual pseudorange is broadcast. By using this difference, the pseudorange can be corrected in real time. Furthermore, the correction signal is broadcast by ground-based transmitters locally. Using differential technology, errors in the common part of GPS can be eliminated, but the inherent error of the receiver cannot be eliminated. Real-Time Kinematic (RTK) shares the similar concept that it uses measurements of the phase of the signal's carrier wave rather than the information content of the signal. This is a significant improvement. For example, The C/A code is broadcast on the 1575.42 MHz L1 carrier as a 1.023 MHz signal. The frequency of the carrier corresponds to a wavelength of 19 cm. As a result, a $\pm 1\%$ error in L1 carrier phase measurement corresponds to a ± 1.9 mm error. In practical situation, an RTK system includes a single base station with a number of mobile units. The mobile unit receives the phase measurements from the base station, and compare them with its own phase measurement. And there are many ways to transmit the correction signal from the base station, one of the most popular ways is to use a radio modem, because that is a real-time, low-cost signal transmission in the Ultra High Frequency (UHF) band. Most countries allocate certain frequencies for RTK specifically, which makes the system more accurate. The typical accuracy of this system is 1 cm ± 2 Parts-Per-Million (PPM) horizontally and 2 cm ± 2 PPM vertically.

7.7.2   Calibration

These measurements are used for position control. The slight deviation of these will not cause significant performance degradation to the multicopter. Therefore, deviation of these sensors can be corrected by online state estimation when the multicopter is flying.

7.7.3   Measurement Model

A GPS receiver is mounted on a multicopter to measure its position ${}^e\mathbf{p} \in \mathbb{R}^3$ in the EFCF. It can be described as

$$^e\mathbf{p}_{\text{GPS}} = {}^e\mathbf{p} + \mathbf{b}_p + \mathbf{n}_p \tag{7.38}$$

where ${}^e\mathbf{p}_{\text{GPS}} \in \mathbb{R}^3$ denotes the position measured, $\mathbf{b}_p \in \mathbb{R}^3$ denotes the drift error, and $\mathbf{n}_p \in \mathbb{R}^3$ is often considered as a GWN vector. Furthermore, $\mathbf{b}_p$ is described as

$$\dot{\mathbf{b}}_p = \mathbf{n}_{\mathbf{b}_p} \tag{7.39}$$

where $\mathbf{n}_{\mathbf{b}_p} \in \mathbb{R}^3$ is often considered as a GWN vector. In general, this model is suitable for GPS and DGPS, but the accuracy of them, namely $\mathbf{b}_p$ and $\mathbf{n}_p$, are different. Moreover, the frequencies of a GPS and a DGPS may be also different. Besides position information, the velocity of the receiver can be calculated according to carrier phase Doppler measurements from GPS satellite signals [11, pp. 140–141]. Since the speeds of multicopters are often low, the velocity the receiver offers is often inaccurate compared with the real one.

7.7.4   Supplement: Latitude-and-Longitude Distance and Heading Calculation

The latitude and longitude are combined to form the geographic coordinate system, see Fig. 7.12a. Geographic coordinate system is a coordinate system that enables every location on the Earth to be specified by a set of numbers or letters, or symbols. There are two ways to describe it:

(1) DMS (Degree-Minute-Second) form, such as (40° 44′55″N, 73° 59′11″W).
(2) Unsigned decimal form, such as (40.7486° −73.9864°).

Here, the north and the east are assumed to be positive, while the west and the south are negative. In the following section, the calculation of distance between two known points is introduced [12]. The position of point A and point B is (ϕ_A, λ_A) and (ϕ_B, λ_B), respectively, where ϕ is latitude and λ is longitude. The radius of the Earth is R_E. By using the haversine formula, the distance is calculated by

$$a_{AB} = \sin^2\left(\frac{\phi_B-\phi_A}{2}\right) + \cos\phi_A \cos\phi_B \sin^2\left(\frac{\lambda_B-\lambda_A}{2}\right)$$
$$c_{AB} = 2\cdot\text{atan2}\left(\sqrt{a_{AB}}, \sqrt{1-a_{AB}}\right) \tag{7.40}$$
$$d_{AB} = R_E\cdot c_{AB}$$

and

$$\psi_{AB} = \text{atan2}\left(\sin(\lambda_B - \lambda_A)\cos\phi_B, \cos\phi_A \sin\phi_B - \sin\phi_A \cos\phi_B \cos(\lambda_B - \lambda_A)\right) \tag{7.41}$$

where $d_{AB} \in \mathbb{R}_+\cup\{0\}$ denotes the as-the-crow-flies distance between the two points, and it is the shorter distance on the surface of the Earth between A and B; ψ_{AB} represents an initial azimuth relative to the true north. It is positive for clockwise deflection. The function atan2(y, x) is the arctangent function with two arguments. The purpose of using two arguments in function atan2(y, x) instead of one as tan(y/x) is to gather information on the signs of the inputs in order to return the appropriate quadrant of the computed angle. There are some other methods to measure the distance between two points, but this method using the haversine formula is the most accurate. In general, the current heading will vary as a multicopter is following a great circle path (orthodrome). The final heading will differ from the initial heading by varying degrees according to distance and latitude. If you want to go from Baghdad (35°N, 45°E) to Osaka (35°N, 135°E), then you will start on a heading of 60° and end up on a heading of 120°, as shown in the Fig. 7.12b. For multicopters, the flight distance is often short, so the change of the heading angle can be omitted. Let us consider an inverse problem. Given the point A (ϕ_A, λ_A),

Fig. 7.12 Latitude and longitude, and great circle path

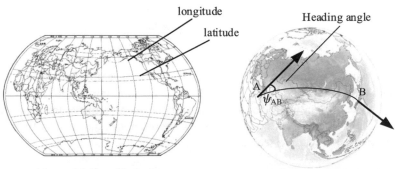

(a) Longitude and latitude   (b) Great circle path from A to B

the distance d_{AB} from point B and the initial heading angle ψ_{AB}, the problem is to find the position of B. It can be calculated by the following equations

$$\phi_B = \text{asin} (\sin \phi_A \cos \delta + \sin \phi_A \sin \delta \cos \psi_{AB})$$
$$\lambda_B = \lambda_A + \text{atan2} (\sin \psi_{AB} \sin \delta \sin \phi_A, \cos \delta - \sin \phi_A \sin \phi_B) . \tag{7.42}$$

7.8   Camera

Today, many multicopters are using cameras as sensors to measure speed and track targets, avoid obstacles and so on. In this section, coordinate frames will be introduced at first, and then, the linear geometric camera models will be introduced [13, pp. 3–31].

7.8.1   Fundamental Principle

A camera makes the three-dimensional (3D) scene project on a two-dimensional (2D) image plane. The following frames are used to describe it.

(1) Image Coordinate Frame (ICF)
Image captured by cameras in the form of a standard television signal is transmitted to the computer through the high-speed transmission system. Each piece of digital images (in black and white picture) in the computer is stored as an array. As shown in Fig. 7.13, a coordinate frame $o_p uv$ is defined in the image, where the position of pixel (u, v) is the column and row of a pixel in the array, respectively. So, (u, v) is the position of pixel in the ICF. The absolute position in the image cannot be determined in the ICF. So, another coordinate frame needs to be added in the form of physical location in the image. In this coordinate frame $o_i x_i y_i$, the origin point is o_i. The $o_i x_i$ axis and $o_i y_i$ axis are parallel to the $o_p u$ axis and $o_p v$ axis, respectively, as shown in Fig. 7.13. The position of the origin point is usually the center of the image.

(2) Camera Coordinate Frame (CCF)
Geometric camera models are shown in Fig. 7.14. The coordinate frame fixed to a camera $o_c x_c y_c z_c$ is called CCF, where the point o_c is the *optical center* of the camera. The $o_c x_c$ axis and $o_c y_c$ axis are parallel to the $o_i x_i$ axis and $o_i y_i$ axis, respectively. The $o_c z_c$ axis is the optical axis of the camera, which is perpendicular to the image plane. The length of the line segment $o_c o_i$ is the *focal length* of the camera.

(3) Earth-Fixed Coordinate Frame (EFCF)
Because a camera can be placed anywhere, it is important to choose a suitable coordinate frame to describe the position of the camera and other objects. This coordinate frame $o_e x_e y_e z_e$ is called EFCF. In computer vision, the transformation from the 3D scene to the 2D image by a camera is defined as the

Fig. 7.13 Image coordinate frame

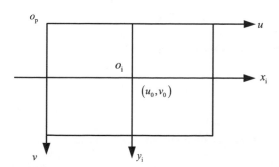

Fig. 7.14 Geometric
camera models

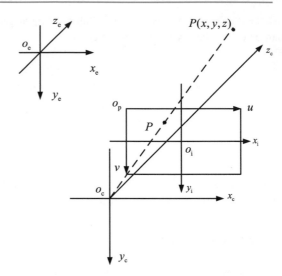

geometric camera model. Through this model, the point in the image can be linked to the position of
this point in the real world. This model is also called the *center photography* or *perspective projection*
model.

7.8.2   Measurement Model

As shown in Fig. 7.14, $\left(p_{x_e}, p_{y_e}, p_{z_e}\right)$ is the position in the EFCF, while $(p_{x_c}, p_{y_c}, p_{z_c})$ is the corresponding position in the CCF. Furthermore, $\left(p_{x_i}, p_{y_i}\right)$ and (u, v) are the physical positions in the image
and the pixel coordinates, respectively. The linear camera model consists of several parts.

(1) Conversion from EFCF to CCF
The process of conversion can be described by the rotation matrix $\mathbf{R}_e^c$ and translation vector $\mathbf{T} \in \mathbb{R}^3$.
So, the position P can be expressed as $\left(p_{x_e}, p_{y_e}, p_{z_e}, 1\right)$ in the EFCF and $\left(p_{x_c}, p_{y_c}, p_{z_c}, 1\right)$ in the CCF.
The conversion is described as

$$
\begin{bmatrix} p_{x_c} \\ p_{y_c} \\ p_{z_c} \\ 1 \end{bmatrix} = \underbrace{\begin{bmatrix} \mathbf{R}_e^c & \mathbf{T} \\ \mathbf{0}_{1\times 3} & 1 \end{bmatrix}}_{\mathbf{M}_2} \begin{bmatrix} p_{x_e} \\ p_{y_e} \\ p_{z_e} \\ 1 \end{bmatrix}
\tag{7.43}
$$

where $\mathbf{R}_e^c \in \mathbb{R}^{3\times 3}$ denotes the rotation matrix from EFCF to CCF, $\mathbf{T} \in \mathbb{R}^3$ denotes the translation
vector, and $\mathbf{M}_2 \in \mathbb{R}^{4\times 4}$.

(2) Conversion from CCF to ICF
The conversion from CCF to ICF is a perspective projection model relationship. Based on the principle
of similar triangles, the point coordinate in the CCF is

$$
\begin{aligned}
p_{x_i} &= \frac{f\,p_{x_c}}{p_{z_c}} \\
p_{y_i} &= \frac{f\,p_{y_c}}{p_{z_c}}
\end{aligned}
\tag{7.44}
$$

where $f \in \mathbb{R}_+$ is the *focal length* of the camera. Combining Eq. (7.43) with Eq. (7.44) gives

$$s \begin{bmatrix} p_{x_i} \\ p_{y_i} \\ 1 \end{bmatrix} = \underbrace{\begin{bmatrix} f & 0 & 0 & 0 \\ 0 & f & 0 & 0 \\ 0 & 0 & 1 & 0 \end{bmatrix}}_{\mathbf{P}} \begin{bmatrix} p_{x_c} \\ p_{y_c} \\ p_{z_c} \\ 1 \end{bmatrix} \tag{7.45}$$

where $s = p_{z_c}$ is the *scale factor*, and $\mathbf{P} \in \mathbb{R}^{3 \times 4}$ is the *perspective projection matrix*. As shown in Fig. 7.13, let (u_0, v_0) be the position of o_i in the ICF, the distances between two neighborhood pixels are expressed as dx_i and dy_i along the axes $o_i x_i$ and $o_i y_i$, respectively. The position conversion between the ICF and the physical ICF is

$$\begin{aligned} u &= \frac{p_{x_i}}{dx_i} + u_0 \\ v &= \frac{p_{y_i}}{dy_i} + v_0. \end{aligned} \tag{7.46}$$

Its homogeneous coordinate is expressed as

$$\begin{bmatrix} u \\ v \\ 1 \end{bmatrix} = \begin{bmatrix} \frac{1}{dx_i} & 0 & u_0 \\ 0 & \frac{1}{dy_i} & v_0 \\ 0 & 0 & 1 \end{bmatrix} \begin{bmatrix} p_{x_i} \\ p_{y_i} \\ 1 \end{bmatrix}. \tag{7.47}$$

By combining Eqs. (7.43) and (7.45) with (7.47), the relationship between EFCF and the ICF is written as

$$s \begin{bmatrix} u \\ v \\ 1 \end{bmatrix} = \begin{bmatrix} \frac{1}{dx_i} & 0 & u_0 \\ 0 & \frac{1}{dy_i} & v_0 \\ 0 & 0 & 1 \end{bmatrix} \begin{bmatrix} f & 0 & 0 & 0 \\ 0 & f & 0 & 0 \\ 0 & 0 & 1 & 0 \end{bmatrix} \begin{bmatrix} \mathbf{R}_e^c & \mathbf{T} \\ \mathbf{0}_{1 \times 3} & 1 \end{bmatrix} \begin{bmatrix} p_{x_e} \\ p_{y_e} \\ p_{z_e} \\ 1 \end{bmatrix}$$

$$= \underbrace{\begin{bmatrix} \mathbf{A} & \mathbf{0}_{3 \times 1} \end{bmatrix}}_{\mathbf{M}_1} \underbrace{\begin{bmatrix} \mathbf{R}_e^c & \mathbf{T} \\ \mathbf{0}_{1 \times 3} & 1 \end{bmatrix}}_{\mathbf{M}_2} \begin{bmatrix} p_{x_e} \\ p_{y_e} \\ p_{z_e} \\ 1 \end{bmatrix} = \mathbf{M}_1 \mathbf{M}_2 \mathbf{X}_e = \mathbf{M} \mathbf{X}_e \tag{7.48}$$

where $\alpha_x \triangleq f/dx_i$ and $\alpha_y \triangleq f/dy_i$ are the *scale factor* in the axis $o_i x_i$ and $o_i y_i$, respectively. And, $\mathbf{M} \in \mathbb{R}^{3 \times 4}$ is called *projection matrix*, $\mathbf{M}_1 \in \mathbb{R}^{4 \times 4}$ is determined by α_x, α_y, u_0, and v_0. Since α_x, α_y, u_0, and v_0 are related to the inner structure of a camera, they are classified as *intrinsic parameters*. On the other hand, $\mathbf{M}_2 \in \mathbb{R}^{4 \times 4}$ is determined by the pose of a camera, namely *extrinsic parameters*.

As for the solution to camera calibration parameter, many methods have been proposed. Camera calibration is performed based on a calibration object whose geometry in 3D space is known with a very good precision. In the next section, a classical method will be introduced in detail [14]. It only requires the camera to observe a planar object shown in Fig. 7.15 at a few (at least two) different orientations. Besides the introduction to a special method, some toolboxes are also attached in order to facilitate the calibration practice.

7.8.3   Calibration

(1) Notation

As shown in Fig. 7.15, a 3D point is denoted by $\mathbf{M} \triangleq [\, p_{x_e} \; p_{y_e} \; p_{z_e} \,]^T$. A 2D feature corner point is denoted by $\mathbf{m} \triangleq [\, u \; v \,]^T$. The notation $\bar{\mathbf{x}}$ is used to denote the augmented vector by adding a one as the

Fig. 7.15 A planar object

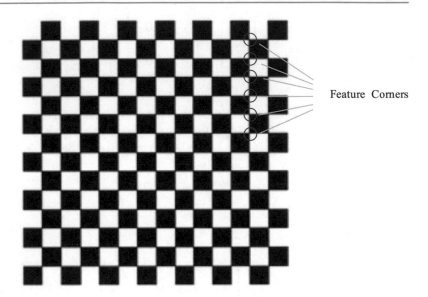

Feature Corners

last element, namely $\bar{\mathbf{M}} \triangleq [\, p_{x_e} \; p_{y_e} \; p_{z_e} \; 1 \,]^T$ and $\bar{\mathbf{m}} \triangleq [\, u \; v \; 1 \,]^T$. Based on Eq. (7.48), the relationship between a 3D point $\mathbf{M}$ and its image projection $\mathbf{m}$ is given by

$$ s\bar{\mathbf{m}} = \mathbf{A} \left[\, \mathbf{R}_e^c \; \mathbf{T} \, \right] \bar{\mathbf{M}}. \tag{7.49} $$

Without loss of generality, the planar object is assumed on $o_e x_e y_e$, or say $p_{z_e} = 0$. The ith column of the rotation matrix $\mathbf{R}_e^c$ is denoted by $\mathbf{r}_i$. Then,

$$
s \begin{bmatrix} u \\ v \\ 1 \end{bmatrix} = \mathbf{A} \left[\, \mathbf{r}_1 \; \mathbf{r}_2 \; \mathbf{r}_3 \; \mathbf{T} \, \right] \begin{bmatrix} p_{x_e} \\ p_{y_e} \\ 0 \\ 1 \end{bmatrix}
$$

$$
= \mathbf{A} \left[\, \mathbf{r}_1 \; \mathbf{r}_2 \; \mathbf{T} \, \right] \begin{bmatrix} p_{x_e} \\ p_{y_e} \\ 1 \end{bmatrix}. \tag{7.50}
$$

Let

$$ \tilde{\mathbf{M}} \triangleq \left[\, p_{x_e} \; p_{y_e} \; 1 \, \right]^T. \tag{7.51} $$

Then

$$ \bar{\mathbf{m}} = \mathbf{H}\tilde{\mathbf{M}} \tag{7.52} $$

where

$$ \mathbf{H} = \lambda \mathbf{A} \left[\, \mathbf{r}_1 \; \mathbf{r}_2 \; \mathbf{T} \, \right] \in \mathbb{R}^{3 \times 3} \tag{7.53} $$

and $\lambda = 1/s$ is a constant. Define

$$\hat{\mathbf{H}} \triangleq \begin{bmatrix} \hat{\mathbf{h}}_1^T \\ \hat{\mathbf{h}}_2^T \\ \hat{\mathbf{h}}_3^T \end{bmatrix} \tag{7.54}$$

up to a scale factor. So the points projected by $\hat{\mathbf{H}}$ satisfy

$$\hat{\mathbf{m}}_i = \frac{1}{\hat{\mathbf{h}}_3^T \tilde{\mathbf{M}}_i} \begin{bmatrix} \hat{\mathbf{h}}_1^T \tilde{\mathbf{M}}_i \\ \hat{\mathbf{h}}_2^T \tilde{\mathbf{M}}_i \end{bmatrix} \tag{7.55}$$

where $\hat{\mathbf{m}}_i, \tilde{\mathbf{M}}_i, i = 1, 2, \cdots$ correspond to the ith feature corners, respectively. It is expected that the real coordinate is close to the projected coordinate. According to this principle, the following optimization can be formulated

$$\mathbf{H} = \arg \min_{\hat{\mathbf{H}} \in \mathbb{R}^{3\times 3}} \sum_i \left\| \mathbf{m}_i - \hat{\mathbf{m}}_i \right\|^2 \tag{7.56}$$

where $i \in \mathbb{Z}_+$ represents the ith feature corner.

(2) Intrinsic parameter calibration
Given an image of the model plane, a homograph can be estimated. According to Eq. (7.50) and using the knowledge that $\mathbf{r}_1$ and $\mathbf{r}_2$ are orthogonal, namely $\mathbf{r}_1^T \mathbf{r}_2 = 0$, one has

$$\begin{aligned} \mathbf{h}_1^T \mathbf{B} \mathbf{h}_2 &= 0 \\ \mathbf{h}_1^T \mathbf{B} \mathbf{h}_1 &= \mathbf{h}_2^T \mathbf{B} \mathbf{h}_2 \end{aligned} \tag{7.57}$$

where

$$\mathbf{B} = \mathbf{A}^{-T} \mathbf{A}^{-1} = \begin{bmatrix} B_{11} & B_{12} & B_{13} \\ B_{21} & B_{22} & B_{23} \\ B_{31} & B_{32} & B_{33} \end{bmatrix} = \begin{bmatrix} \frac{1}{\alpha_x^2} & 0 & \frac{-u_0}{\alpha_x^2} \\ 0 & \frac{1}{\alpha_y^2} & -\frac{v_0}{\alpha_y^2} \\ \frac{-u_0}{\alpha_x^2} & -\frac{v_0}{\alpha_y^2} & \frac{u_0^2}{\alpha_x^2} + \frac{v_0^2}{\alpha_y^2} + 1 \end{bmatrix}. \tag{7.58}$$

These are the two basic constraints on the intrinsic parameters, given one homograph. Because homograph has eight degrees of freedom and there are six extrinsic parameters (three for rotation and three for translation), only two constraints on the intrinsic parameters can be obtained. Define

$$\mathbf{b} \triangleq \begin{bmatrix} B_{11} & B_{12} & B_{22} & B_{13} & B_{23} & B_{33} \end{bmatrix}^T \tag{7.59}$$

which satisfies

$$\begin{bmatrix} 0 & 1 & 0 & 0 & 0 & 0 \end{bmatrix} \mathbf{b} = 0 \tag{7.60}$$

where $B_{12} = 0$ is used. Let the ith column in $\mathbf{H}$ be $\mathbf{h}_i \triangleq [h_{i1} \ h_{i2} \ h_{i3}]^T$. Then,

$$\mathbf{h}_i^T \mathbf{B} \mathbf{h}_j = \mathbf{v}_{ij}^T \mathbf{b} \tag{7.61}$$

where

$$\mathbf{v}_{ij} = [h_{i1}h_{j1} \ h_{i1}h_{j2} + h_{i2}h_{j1} \ h_{i2}h_{j2} \ h_{i3}h_{j1} + h_{i1}h_{j3} \ h_{i3}h_{j2} + h_{i2}h_{j3} \ h_{i3}h_{j3}]^T. \tag{7.62}$$

According to Eq. (7.61), Eq. (7.57) is rewritten as

$$
\begin{bmatrix} \mathbf{v}_{12}^{\mathrm{T}} \\ (\mathbf{v}_{11} - \mathbf{v}_{22})^{\mathrm{T}} \end{bmatrix} \mathbf{b} = \mathbf{0}_{2\times1}.
\tag{7.63}
$$

If $N \in \mathbb{Z}_+$ images of the planar object are observed, then, by stacking N such equations, one has

$$
\mathbf{V}\mathbf{b} = \mathbf{0}_{2N\times1}
\tag{7.64}
$$

where $\mathbf{V} \in \mathbb{R}^{2N\times6}$. If $N \geq 2$, then a unique solution $\mathbf{b}$ can be determined. The solution to Eq. (7.64) is well known as the eigenvector of $\mathbf{V}^{\mathrm{T}}\mathbf{V}$ associated with the smallest eigenvalue. Once $\mathbf{b}$ is estimated, all camera intrinsic parameters are obtained, including $\mathbf{A}$. Furthermore, the extrinsic parameters for each image are readily computed. According to the definition of $\mathbf{H}$ in Eq. (7.53), one has

$$
\begin{aligned}
\mathbf{r}_1 &= \lambda^{-1}\mathbf{A}^{-1}\mathbf{h}_1 \\
\mathbf{r}_2 &= \lambda^{-1}\mathbf{A}^{-1}\mathbf{h}_2 \\
\mathbf{r}_3 &= \mathbf{r}_1 \times \mathbf{r}_2 \\
\mathbf{T} &= \lambda^{-1}\mathbf{A}^{-1}\mathbf{h}_3
\end{aligned}
\tag{7.65}
$$

where $\lambda = \left\| \mathbf{A}^{-1}\mathbf{h}_1 \right\| = \left\| \mathbf{A}^{-1}\mathbf{h}_2 \right\|$. However, because of the noise in data, the solution obtained is inaccurate to some degree. It can be refined through an optimization. Given N images of a planar object with $M \in \mathbb{Z}_+$ points, the estimation can be obtained by the following optimization

$$
\left(\mathbf{A}^*, \mathbf{R}_i^*, \mathbf{T}_i^*\right) = \arg \min_{\mathbf{A},\mathbf{R}_i,\mathbf{T}_i} \sum_{i=1}^{N} \sum_{j=1}^{M} \left\| \mathbf{m}_{ij} - \hat{\mathbf{m}}\left(\mathbf{A}, \mathbf{R}_i, \mathbf{T}_i, \mathbf{M}_j\right) \right\|
\tag{7.66}
$$

where $\hat{\mathbf{m}}\left(\mathbf{A}, \mathbf{R}_i, \mathbf{T}_i, \mathbf{M}_j\right)$ is the projection of the jth with corner point position $\mathbf{M}_j \in \mathbb{R}^3$ in the ith image.

7.8.4 Some Toolboxes

There are some toolboxes used to calibrate cameras, as shown in Table 7.1.

7.9   Summary

In this chapter, fundamental principles of sensors are introduced. Book [11, pp. 120–142] also provides an excellent overview on major sensors in this chapter. Errors always exist in measurement, but the state estimation is only performed when multicopters are in the air, i.e., it cannot be done before multicopters take off. As a result, it is important to eliminate the error offline. Through calibration offline, unknown parameters of measurement models can be determined. This chapter introduces the principles and methods to calibrate three-axis accelerometers, three-axis gyroscopes, and three-axis magnetometers automatically. Since these methods can work without any additional equipment, they are very practical. These methods are used to calibrate the intrinsic parameters of sensors. Due to installation problems, it is impossible to place sensors on the Center of Gravity (CoG). So the methods of calibrating the

Table 7.1 Toolboxes for calibration

Toolbox	Description	Web site
Computer vision system toolbox	The system toolbox supports camera calibration, stereo vision, 3D reconstruction, and 3D point cloud processing	http://cn.mathworks.com/help/vision/index.html
Camera calibration toolbox for MATLAB	This is a release of a camera calibration toolbox for MATLAB with a complete documentation, such as fish-eye lens cameras and catadioptric cameras	http://www.vision.caltech.edu/bouguetj/calib_doc
Camera calibration toolbox for generic lenses	This is a camera calibration toolbox for MATLAB which can be used for calibrating several different kinds of cameras	http://www.ee.oulu.fi/~jkannala/calibration/calibration.html
The DLR camera calibration toolbox	It is followed from the strategic purpose of both upgrading the former CalLab package and at the same time developing a platform independent application	http://dlr.de/rmc/rm/en/desktopdefault.aspx/tabid-3925/6084_read-9201
Fully automatic camera and hand to eye calibration	The first part covers a fully automatic calibration procedure and the second covers the calibration of the camera to a robot-arm or an external marker (known as hand-eye calibration)	http://www2.vision.ee.ethz.ch/software/calibration_toolbox/calibration_toolbox.php
Camera calibration toolbox	This toolbox is a Windows application designed to streamline the camera calibration process. It can be used to capture calibration images from a camera attached to your PC, detect the calibration object and calculate the intrinsic and extrinsic camera parameters	http://www0.cs.ucl.ac.uk/staff/Dan.Stoyanov/calib
Omnidirectional calibration toolbox	This toolbox can be used to calibrate the camera such as hyperbolic, parabolic, folded mirror, spherical, wide-angle cameras	http://www.robots.ox.ac.uk/~cmei/Toolbox.html
Camera calibration toolbox for generic multiple cameras	This is a camera calibration toolbox for generic multiple cameras using 1D target. The toolbox can be used to calibrate the following: (1) two conventional cameras; (2) two fish-eye cameras; (3) two mixed cameras; and (4) multiple cameras	http://quanquan.buaa.edu.cn/CalibrationToolbox.html

extrinsic parameters need to be further studied. Besides the introduction to calibration, measurement models are further given, which will be applied to filter design in the following chapters.

Exercises

7.1 In the calibration of accelerometers, the local gravity g in Eq. (7.5) is set to 1. What will be changed to the calibrated specific force by adopting optimization (7.5)?

7.2 By using the methods in Sects. 7.1–7.3, calibrate an IMU using data from the Web site http://rfly.buaa.edu.cn/course.

Fig. 7.16 Positions of a
receiver and four satellites

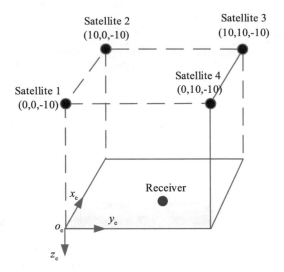

7.3 Discuss advantages and disadvantages of altitude measuring by GPS and by barometer, respectively.

7.4 Suppose that the positions of four satellites are shown in Fig. 7.16 and $\delta_t = \delta_T = \delta_I = 0$. The receiver is near the plane, the distances to the four satellites are $10\sqrt{2} + \varepsilon_1(t)$, $10\sqrt{3} + \varepsilon_2(t)$, $10\sqrt{2} + \varepsilon_3(t)$ and $10 + \varepsilon_4(t)$, respectively, where noise $\varepsilon_i(t) \sim \mathcal{N}(0, 0.01)$, $i = 1, 2, 3, 4$, and the sampling period is 0.1s. (1) Solve the receiver position without noises. (2) By considering the noise, solve and plot the receiver position with respect to time.

7.5 Calibrate the camera of your cell phone using a toolbox in Table 7.1.

References

1. Titterton DH, Weston JL (2004) Strapdown inertial navigation technology, 2nd edn. The Institution of Engineering and Technology, Stevenage, UK
2. Leishman RC, Macdonald JC, Beard RW, Timothy WM (2014) Quadrotors and accelerometers: state estimation with an improved dynamic model. IEEE Control Syst Mag 34(1):28–41
3. Hol JD (2008) Pose estimation and calibration algorithms for vision and inertial sensors. Dissertation, Linköping University
4. Beeby S, Ensell G, Kraft M, White N (2004) MEMS mechanical sensors. Artech House, London
5. Tedaldi D, Pretto A, Menegatti E (2014) A robust and easy to implement method for IMU calibration without external equipments. In: Proceedings of IEEE International Conference on Robotics and Automation, Hong Kong, pp 3042–3049
6. Ripka P (2001) Magnetic sensors and magnetometers. Artech House, Norwood
7. Rouf VT, Li M, Horsley DA (2013) Area-efficient three axis MEMS Lorentz force magnetometer. IEEE Sens J 13(11):4474–4481
8. Dorveaux E, Vissière D, Martin A, Petit N (2009) Iterative calibration method for inertial and magnetic sensors. In: Proceedings of the 48th IEEE Conference on Decision and Control Joint with the 28th Chinese Control Conference (CDC/CCC), Shanghai, pp 8296–8303
9. What is LiDAR. http://oceanservice.noaa.gov/facts/lidar.html. Accessed 5 Dec 2016
10. Misra P, Enge P (2006) Global positioning system: signals, measurements, and performance. Ganga-Jamuna Press, Lincoln, Massachusetts
11. Beard RW, McLain TW (2012) Small unmanned aircraft: theory and practice. Princeton University Press, Princeton
12. Movable Type Scripts (2014) Calculate distance, bearing and more between latitude/longitude points. http://www.movable-type.co.uk/scripts/latlong.html. Accessed 2 Mar 2016
13. Forsyth DA, Ponce J (2016) Computer vision: a modern approach. Pearson Education Limited, New York
14. Zhang ZY. A flexible new technique for camera calibration. IEEE Trans Pattern Anal Mach Intell 22(11):1330–1334

Observability and Kalman Filter

8

The state of a multicopter may not be measured directly by existing sensors. For example, since the speed of a multicopter is very low, its accurate value is difficult to be measured directly through speed sensors, such as Pitot tubes or a Global Positioning System (GPS) receiver. For such a purpose, a question that arises is what kind of sensor or the combination of sensors is necessary to estimate a given state. For example, given sensors such as an accelerometer, a GPS receiver, and a monocular camera, which sensor can give stable estimate for the speed of a multicopter? This involves the *observability* of a system. Observability is a measure for how well a system can be observed, i.e., to determine the initial states by measuring the outputs. Formally, a system is observable if, from the sequence of outputs and control vectors, the current state can be uniquely determined in a finite time interval. The concept of the observability can also be extended to the social survey. In the era of big data, huge reliable, easy-to-access, and local data can be collected online. Based on these data, the question is whether a local index related to the society can be observed online. If so, it will not only avoid false statistics, but also can save much labor and a lot of resources. Back to the topic of this chapter, after guaranteeing the observability, an observer design, such as the Kalman filter design, will make sense. This chapter aims to answer the questions below:

What is the observability and how is a Kalman filter derived?

The answer to these questions involves the concept of observability and its criteria for the systems such as the continuous-time linear system, discrete-time linear system, and continuous-time nonlinear system. Furthermore, the classical Kalman filter is derived in detail, and its extensions, namely multirate Kalman filter, Extended Kalman Filter (EKF) and Implicit Extended Kalman Filter (IEKF), are introduced.

The Blind Men and the Elephant

The philosophy under observability is whether an observation is comprehensive and objective. Shi Su, a great poet in the Northern Song dynasty, wrote in one of his poems named *"Written on the Wall of Xilin Temple"*: "It is a mountain range when viewed horizontally, a peak when looked at vertically; near, far, high or low—how its appearance varies. One cannot know the true nature of Mountain Lu; unless one is on that very mountain itself (The Chinese is "横看成岭侧成峰, 远近高低各不同；不识庐山真面目，只缘身在此山中" translated by John Balcon (2011). After Many Autumns: A collection of Chinese Buddhist LIterature. Buddha's Light Publishing, p. 131)". This poem described the magnificent view of Mount Lushan, and told us that the observation of a problem should be comprehensive and objective. Otherwise, we would jump to an incorrect conclusion. There was a parable of blind men and an elephant. All of these blind men believed that they got the sense of the real elephant. However, what they were getting was only a part of the elephant.

8.1 Observability

The observability and its criteria for the continuous-time linear system, discrete-time linear system, and continuous-time nonlinear system are introduced in the following.

8.1.1 Linear System

8.1.1.1 Continuous-Time Linear System

Consider the continuous-time linear system as follows:

$$\dot{\mathbf{x}} = \mathbf{A}\mathbf{x} + \mathbf{B}\mathbf{u}$$
$$\mathbf{y} = \mathbf{C}^{\mathrm{T}}\mathbf{x} \tag{8.1}$$

where $\mathbf{x} \in \mathbb{R}^n$, $\mathbf{y} \in \mathbb{R}^m$, $\mathbf{u} \in \mathbb{R}^m$ represent the state, output, and control input of the system, respectively; and $\mathbf{A} \in \mathbb{R}^{n \times n}$, $\mathbf{B} \in \mathbb{R}^{n \times m}$, $\mathbf{C} \in \mathbb{R}^{n \times m}$ are constant matrices or vectors. For system (8.1), the observability is defined in the following.

Definition 8.1 ([1, p. 153]) System (8.1) is observable on $[t_0, t_f]$ if and only if given the input $\mathbf{u}_{[t_0,t_f]}$ and the corresponding output $\mathbf{y}_{[t_0,t_f]}$, the initial state $\mathbf{x}(t_0)$ is uniquely determined.

For system (8.1), assume that the initial time $t_0 = 0$ without loss of generality. Then, the state and output of the system are calculated as

$$\mathbf{x}(t) = e^{\mathbf{A}t}\mathbf{x}(0) + \int_0^t e^{\mathbf{A}(t-\tau)}\mathbf{B}\mathbf{u}(\tau)\,\mathrm{d}\tau$$
$$\mathbf{y}(t) = \mathbf{C}^{\mathrm{T}}e^{\mathbf{A}t}\mathbf{x}(0) + \int_0^t \mathbf{C}^{\mathrm{T}}e^{\mathbf{A}(t-\tau)}\mathbf{B}\mathbf{u}(\tau)\,\mathrm{d}\tau. \tag{8.2}$$

Since the matrices $\mathbf{A}$, $\mathbf{B}$, $\mathbf{C}^{\mathrm{T}}$ and the input $\mathbf{u}(t)$ are known, the integral term in Eq. (8.2) is completely known. Let

$$\mathbf{y}'(t) = \mathbf{y}(t) - \int_0^t \mathbf{C}^{\mathrm{T}}e^{\mathbf{A}(t-\tau)}\mathbf{B}\mathbf{u}(\tau)\,\mathrm{d}\tau.$$

Then, Eq. (8.2) becomes

$$\mathbf{y}'(t) = \mathbf{C}^{\mathrm{T}}e^{\mathbf{A}t}\mathbf{x}(0). \tag{8.3}$$

According to the definition of observability, a question is formulated whether the initial state $\mathbf{x}(0)$ can be uniquely determined based on the outputs $\mathbf{y}'(t)$ in finite time $t_0 \leq t \leq t_f$. Multiplying $e^{\mathbf{A}^{\mathrm{T}}t}\mathbf{C}$ and then integrating on both sides of Eq. (8.3) results in

$$\int_0^{t_1} e^{\mathbf{A}^{\mathrm{T}}t}\mathbf{C}\mathbf{y}'(t)\,\mathrm{d}t = \mathbf{V}(0, t_1)\mathbf{x}(0) \tag{8.4}$$

where $\mathbf{V}(0, t_1) = \int_0^{t_1} e^{\mathbf{A}^{\mathrm{T}}t}\mathbf{C}\mathbf{C}^{\mathrm{T}}e^{\mathbf{A}t}\,\mathrm{d}t$. If $\mathbf{V}(0, t_1)$ is non-singular, $\mathbf{V}^{-1}(0, t_1)$ exists. Then, the initial state $\mathbf{x}(0)$ is uniquely determined as

$$\mathbf{x}(0) = \mathbf{V}^{-1}(0, t_1)\int_0^{t_1} e^{\mathbf{A}^{\mathrm{T}}t}\mathbf{C}\mathbf{y}'(t)\,\mathrm{d}t \tag{8.5}$$

so the system is observable. On the contrary, if $\mathbf{V}(0, t_1)$ is singular, it means that there exist infinite solutions to Eq. (8.4). So the system state cannot be uniquely determined, i.e., the system is unobservable. It is proved that there exists $t_1 > 0$ to make $\mathbf{V}(0, t_1)$ non-singular if and only if

$$\text{rank}\mathscr{O}\left(\mathbf{A}, \mathbf{C}^{\mathrm{T}}\right) = n \tag{8.6}$$

where

$$\mathscr{O}\left(\mathbf{A}, \mathbf{C}^{\mathrm{T}}\right) \triangleq \begin{bmatrix} \mathbf{C}^{\mathrm{T}} \\ \mathbf{C}^{\mathrm{T}}\mathbf{A} \\ \vdots \\ \mathbf{C}^{\mathrm{T}}\mathbf{A}^{n-1} \end{bmatrix}. \tag{8.7}$$

This result is summarized as the following theorem.

Theorem 8.2 ([1, p. 156]) *The continuous-time linear system (8.1) is observable if and only if the observability matrix is of full rank, i.e.,* $\text{rank}\mathscr{O}\left(\mathbf{A}, \mathbf{C}^{\mathrm{T}}\right) = n$.

An intuitive way to explain the condition $\text{rank}\mathscr{O}\left(\mathbf{A}, \mathbf{C}^{\mathrm{T}}\right) = n$ is given in the following. Taking successive derivatives of the output equation in system (8.1) gives

$$\begin{aligned} \mathbf{y} &= \mathbf{C}^{\mathrm{T}}\mathbf{x} \\ \dot{\mathbf{y}} &= \mathbf{C}^{\mathrm{T}}\mathbf{A}\mathbf{x} + \mathbf{C}^{\mathrm{T}}\mathbf{B}\mathbf{u} \Rightarrow \dot{\mathbf{y}} - \mathbf{C}^{\mathrm{T}}\mathbf{B}\mathbf{u} = \mathbf{C}^{\mathrm{T}}\mathbf{A}\mathbf{x} \\ \ddot{\mathbf{y}} &= \mathbf{C}^{\mathrm{T}}\mathbf{A}^2\mathbf{x} + \mathbf{C}^{\mathrm{T}}\mathbf{A}\mathbf{B}\mathbf{u} + \mathbf{C}^{\mathrm{T}}\mathbf{B}\dot{\mathbf{u}} \Rightarrow \ddot{\mathbf{y}} - \mathbf{C}^{\mathrm{T}}\mathbf{A}\mathbf{B}\mathbf{u} - \mathbf{C}^{\mathrm{T}}\mathbf{B}\dot{\mathbf{u}} = \mathbf{C}^{\mathrm{T}}\mathbf{A}^2\mathbf{x} \\ &\vdots \end{aligned}$$

$$\mathbf{y}^{(n-1)} - \sum_{k=0}^{n-2}\mathbf{C}^{\mathrm{T}}\mathbf{A}^{n-2-k}\mathbf{B}\mathbf{u}^{(k)} = \mathbf{C}^{\mathrm{T}}\mathbf{A}^{n-1}\mathbf{x}$$

where $(\cdot)^{(k)}$ denotes the kth derivative. Arranging the equations above results in

$$\begin{bmatrix} \mathbf{y} \\ \dot{\mathbf{y}} - \mathbf{C}^{\mathrm{T}}\mathbf{B}\mathbf{u} \\ \ddot{\mathbf{y}} - \mathbf{C}^{\mathrm{T}}\mathbf{A}\mathbf{B}\mathbf{u} - \mathbf{C}^{\mathrm{T}}\mathbf{B}\dot{\mathbf{u}} \\ \vdots \\ \mathbf{y}^{(n-1)} - \sum_{k=0}^{n-2}\mathbf{C}^{\mathrm{T}}\mathbf{A}^{n-2-k}\mathbf{B}\mathbf{u}^{(k)} \end{bmatrix} = \mathscr{O}\left(\mathbf{A}, \mathbf{C}^{\mathrm{T}}\right)\mathbf{x} \tag{8.8}$$

where $\mathscr{O}\left(\mathbf{A}, \mathbf{C}^{\mathrm{T}}\right)$ is the observability matrix already defined in Eq. (8.7). Since the left-hand side of Eq. (8.8) is known, the state $\mathbf{x}$ in Eq. (8.8) can be uniquely determined if and only if the observability matrix is of full rank, i.e., $\text{rank}\mathscr{O}\left(\mathbf{A}, \mathbf{C}^{\mathrm{T}}\right) = n$. And it is noticed that adding higher-order derivatives to Eq. (8.8) such as

$$\mathscr{O}' = \begin{bmatrix} \mathscr{O}\left(\mathbf{A}, \mathbf{C}^{\mathrm{T}}\right) \\ \mathbf{C}^{\mathrm{T}}\mathbf{A}^n \end{bmatrix}$$

does not increase the rank of the observability matrix. That is because the additional term $\mathbf{C}^{\mathrm{T}}\mathbf{A}^n$ is the linear combination of components of $\mathscr{O}\left(\mathbf{A}, \mathbf{C}^{\mathrm{T}}\right)$. According to the Cayley–Hamilton theorem [2], one has

$$\mathbf{C}^{\mathrm{T}}\mathbf{A}^n = \alpha_0\mathbf{C}^{\mathrm{T}} + \alpha_1\mathbf{C}^{\mathrm{T}}\mathbf{A} + \cdots + \alpha_{n-1}\mathbf{C}^{\mathrm{T}}\mathbf{A}^{n-1} \tag{8.9}$$

where $\alpha_0, \alpha_1, \ldots, \alpha_{n-1} \in \mathbb{R}$. Thus, $\mathrm{rank}\mathscr{O}' = \mathrm{rank}\mathscr{O}\left(\mathbf{A}, \mathbf{C}^{\mathrm{T}}\right)$. Next, an example is given to improve the understanding of the concept of observability.

Example 8.3 Given two sensors, a GPS receiver and an accelerometer, the problem is that which one can give stable estimate for the speed of a car moving in a one-dimensional straight line.

First, consider the use of a GPS receiver to estimate the speed. The simplified model is built as

$$\begin{bmatrix} \dot{p} \\ \dot{v} \end{bmatrix} = \underbrace{\begin{bmatrix} 0 & 1 \\ 0 & 0 \end{bmatrix}}_{\mathbf{A}}\begin{bmatrix} p \\ v \end{bmatrix} + \begin{bmatrix} 0 \\ \varepsilon_1 \end{bmatrix}$$
$$y = \underbrace{\begin{bmatrix} 1 & 0 \end{bmatrix}}_{\mathbf{C}^{\mathrm{T}}}\begin{bmatrix} p \\ v \end{bmatrix}$$

where $p, v \in \mathbb{R}$ represent the position and the speed of the car, respectively; $\varepsilon_1 \in \mathbb{R}$ is the noise; $y \in \mathbb{R}$ is the measurement of the GPS receiver. According to Eq. (8.7), the observability matrix is

$$\mathscr{O}\left(\mathbf{A}, \mathbf{C}^{\mathrm{T}}\right) = \begin{bmatrix} 1 & 0 \\ 0 & 1 \end{bmatrix}.$$

Since $\mathrm{rank}\mathscr{O}\left(\mathbf{A}, \mathbf{C}^{\mathrm{T}}\right) = 2$, according to Theorem 8.2, the speed is observable. Therefore, the GPS receiver can be utilized to estimate the speed of the car. The result is reasonable because the speed is the derivative of the position. Similarly, consider the use of an accelerometer to estimate the speed. The simplified model is as follows:

$$\begin{bmatrix} \dot{v} \\ \dot{a} \end{bmatrix} = \underbrace{\begin{bmatrix} 0 & 1 \\ 0 & 0 \end{bmatrix}}_{\mathbf{A}}\begin{bmatrix} v \\ a \end{bmatrix} + \begin{bmatrix} 0 \\ \varepsilon_2 \end{bmatrix}$$
$$y = \underbrace{\begin{bmatrix} 0 & 1 \end{bmatrix}}_{\mathbf{C}^{\mathrm{T}}}\begin{bmatrix} v \\ a \end{bmatrix} \tag{8.10}$$

where $v, a \in \mathbb{R}$ represent the speed and the acceleration of the car, respectively; $\varepsilon_2 \in \mathbb{R}$ is the noise; $y \in \mathbb{R}$ is the acceleration based on the measurement of the accelerometer for simplicity. According to Eq. (8.7), the observability matrix is

$$\mathscr{O}\left(\mathbf{A}, \mathbf{C}^{\mathrm{T}}\right) = \begin{bmatrix} 0 & 1 \\ 0 & 0 \end{bmatrix}.$$

Since $\mathrm{rank}\mathscr{O}\left(\mathbf{A}, \mathbf{C}^{\mathrm{T}}\right) \neq 2$, according to Theorem 8.2, the speed is unobservable. This may be inconsistent with common sense because the speed is the integral of the acceleration.

Now consider two objects falling down, where the initial speed of one object is $v_1 = 0$ m/s, while that of the other is $v_2 = 100$ m/s. Their accelerations are both $a = 9.8$ m/s^2. Therefore, the initial speed cannot be identified only from the information of acceleration. So, the speed is unobservable from the definition. Surely, the speed can be obtained by integrating the measured acceleration when the initial state is known. However, the error will be accumulated due to the measurement noise.

Thus, the estimate error of the speed will be increased with time. In particular, for most multicopters, Micro-Electro-Mechanical System (MEMS) accelerometers are used, where the measurement noise is relatively large. Therefore, the speed cannot be estimated stably based on system like (8.10).

8.1.1.2 Discrete-Time Linear System

For the continuous-time linear system (8.1), it can be transformed into a discrete-time linear system through the sampling period $T_s \in \mathbb{R}_+$, expressed as

$$\begin{aligned} \mathbf{x}_k &= \mathbf{\Phi}\mathbf{x}_{k-1} + \mathbf{u}_{k-1} \\ \mathbf{y}_k &= \mathbf{C}^\mathrm{T}\mathbf{x}_k \end{aligned} \tag{8.11}$$

where $\mathbf{\Phi} = e^{\mathbf{A}T_s}$, $\mathbf{u}(t) = \mathbf{u}'_{k-1}, t \in [(k-1)T_s, kT_s]$, $\mathbf{u}_{k-1} = \int_0^{T_s} e^{\mathbf{A}\tau}\mathbf{B}\mathrm{d}\tau \mathbf{u}'_{k-1}$. The matrices $\mathbf{\Phi} \in \mathbb{R}^{n \times n}$, $\mathbf{C} \in \mathbb{R}^{n \times m}$ are constant. The subscript k denotes the kth discrete time. For system (8.11), the observability is defined in the following.

Definition 8.4 ([1, p. 170]) System (8.11) is observable in a finite time interval $\left[k_0, k_f\right]$ if and only if given the inputs $\mathbf{u}_{[k_0, k_f]}$ and corresponding outputs $\mathbf{y}_{[k_0, k_f]}$, the initial state $\mathbf{x}_0$ at k_0 is uniquely determined.

According to Eq. (8.11), one has

$$\begin{aligned} \mathbf{y}_0 &= \mathbf{C}^\mathrm{T}\mathbf{x}_0 \\ \mathbf{y}_1 &= \mathbf{C}^\mathrm{T}\mathbf{x}_1 = \mathbf{C}^\mathrm{T}\mathbf{\Phi}\mathbf{x}_0 + \mathbf{C}^\mathrm{T}\mathbf{u}_0 \Rightarrow \mathbf{y}_1 - \mathbf{C}^\mathrm{T}\mathbf{u}_0 = \mathbf{C}^\mathrm{T}\mathbf{\Phi}\mathbf{x}_0 \\ \mathbf{y}_2 &= \mathbf{C}^\mathrm{T}\mathbf{x}_2 = \mathbf{C}^\mathrm{T}\mathbf{\Phi}\mathbf{x}_1 + \mathbf{C}^\mathrm{T}\mathbf{u}_1 \Rightarrow \mathbf{y}_2 - \mathbf{C}^\mathrm{T}\mathbf{u}_1 - \mathbf{C}^\mathrm{T}\mathbf{\Phi}\mathbf{u}_0 = \mathbf{C}^\mathrm{T}\mathbf{\Phi}^2\mathbf{x}_0 \\ &\vdots \\ \mathbf{y}_n &- \sum_{k=0}^{n-1}\mathbf{C}^\mathrm{T}\mathbf{\Phi}^{n-1-k}\mathbf{u}_k = \mathbf{C}^\mathrm{T}\mathbf{\Phi}^n\mathbf{x}_0. \end{aligned}$$

Arranging the equations above results in

$$\begin{bmatrix} \mathbf{y}_0 \\ \mathbf{y}_1 \\ \vdots \\ \mathbf{y}_{n-1} - \sum_{k=0}^{n-2}\mathbf{C}^\mathrm{T}\mathbf{\Phi}^{n-2-k}\mathbf{u}_k \end{bmatrix} = \mathcal{O}\left(\mathbf{\Phi}, \mathbf{C}^\mathrm{T}\right)\mathbf{x}_0. \tag{8.12}$$

Similarly with the continuous-time linear system, the observability criterion is stated in the following theorem.

Theorem 8.5 ([1, p. 171]) *The discrete-time linear system (8.11) is observable if and only if*

$$\mathrm{rank}\,\mathcal{O}\left(\mathbf{\Phi}, \mathbf{C}^\mathrm{T}\right) = n. \tag{8.13}$$

8.1.2 Continuous-Time Nonlinear System

8.1.2.1 Basic Definition

For nonlinear systems, the observability becomes more complex compared with linear systems. Also, the system is always locally observable. Consider a class of continuous-time nonlinear systems

$$\dot{\mathbf{x}} = \mathbf{f}(\mathbf{x}) + \sum_{k=1}^{m} u_k \mathbf{g}_k(\mathbf{x})$$

$$\mathbf{y} = \mathbf{h}(\mathbf{x}) \tag{8.14}$$

where $\mathbf{x} \in \mathbb{R}^n$, $\mathbf{y} \in \mathbb{R}^m$ represent the state and output, respectively; $\mathbf{f}(\mathbf{x})$, $\mathbf{g}_1(\mathbf{x})$, ..., $\mathbf{g}_m(\mathbf{x})$ are given vector fields defined in the open set $\mathscr{X} \subset \mathbb{R}^n$; the control input $\mathbf{u} = [u_1 \cdots u_m]^{\mathrm{T}} \in \mathbb{R}^m$. The question remains as follows: Are there simple necessary and sufficient conditions for system (8.14) to be locally observable? First, a few concepts are introduced.

Definition 8.6 ([3, p. 414]) Consider the continuous-time nonlinear system (8.14), two states $\mathbf{x}_0$ and $\mathbf{x}_1$ are *distinguishable* if there exists a control input $\mathbf{u}$ such that

$$\mathbf{y}(\cdot, \mathbf{x}_0, \mathbf{u}) \neq \mathbf{y}(\cdot, \mathbf{x}_1, \mathbf{u})$$

where $\mathbf{y}(\cdot, \mathbf{x}_i, \mathbf{u})$, $i = 0, 1$ are the output functions of system (8.14) corresponding to the initial state $\mathbf{x}(0) = \mathbf{x}_i$, $i = 0, 1$ and the input $\mathbf{u}$. The system is (locally) *observable* at the state $\mathbf{x}_0 \in \mathscr{X}$ if there exists a neighborhood $\mathscr{N}$ of $\mathbf{x}_0$ such that every state $\mathbf{x} \in \mathscr{N}$ other than $\mathbf{x}_0$ is distinguishable from $\mathbf{x}_0$. Finally, the system is (locally) *observable* if it is locally observable at each state $\mathbf{x} \in \mathscr{X}$.

Unlike linear systems, there are several unique properties in the observability of nonlinear systems. According to Definition 8.6, it is not necessary for all inputs $\mathbf{u}$ to hold the condition. In order to illustrate these, an example in [3] is used.

Example 8.7 ([3, p. 415]) For an SISO bilinear system

$$\dot{\mathbf{x}} = \mathbf{A}\mathbf{x} + u\mathbf{B}\mathbf{x}$$

$$y = \mathbf{C}^{\mathrm{T}}\mathbf{x} \tag{8.15}$$

where

$$\mathbf{A} = \begin{bmatrix} 0 & 1 & 0 \\ 0 & 0 & 1 \\ 0 & 0 & 0 \end{bmatrix}, \mathbf{B} = \begin{bmatrix} 0 & 0 & 0 \\ 1 & 0 & 0 \\ 0 & 0 & 0 \end{bmatrix}, \mathbf{C} = \begin{bmatrix} 0 \\ 1 \\ 0 \end{bmatrix} \tag{8.16}$$

consider the following two situations:

(1) Suppose that the control input $u(t) \equiv 0$. Then, system (8.15) becomes

$$\dot{\mathbf{x}} = \mathbf{A}\mathbf{x}$$

$$y = \mathbf{C}^{\mathrm{T}}\mathbf{x}.$$

It can be easily found that the system is unobservable as rank$\mathscr{O}(\mathbf{A}, \mathbf{C}^{\mathrm{T}}) = 2$. If $\mathbf{x}_0 = \mathbf{0}_{3 \times 1}$, then $y(t) = 0$. And, if $\mathbf{x}_1 = [1\ 0\ 0]^{\mathrm{T}}$, then $y(t) = \mathbf{C}^{\mathrm{T}} e^{\mathbf{A}t} \mathbf{x}_1 = 0$. According to Definition 8.6, the states $\mathbf{x}_0$ and $\mathbf{x}_1$ cannot be distinguishable from zero input, namely $y(\cdot, \mathbf{x}_0, 0) \equiv y(\cdot, \mathbf{x}_1, 0)$.

(2) Suppose a constant control input $u(t) \equiv 1$. Then, system (8.15) becomes

$$\dot{\mathbf{x}} = (\mathbf{A} + \mathbf{B})\,\mathbf{x}$$
$$y = \mathbf{C}^{\mathrm{T}}\mathbf{x}.$$

Similarly, the system is unobservable either. But the output $y(t) \equiv 0$ with the state $\mathbf{x}_0 = \mathbf{0}_{3\times 1}$, whereas $y \neq 0$ if $\mathbf{x}_1 = [\,1\ 0\ 0\,]^{\mathrm{T}}$. Hence, the states $\mathbf{x}_0$ and $\mathbf{x}_1$ are distinguishable from the constant input $u(t) \equiv 1$, namely $y(\cdot, \mathbf{x}_0, 1) \neq y(\cdot, \mathbf{x}_1, 1)$.

From the analysis above, the observability of the nonlinear system is related to the control input. In general, for a constant input $\forall u(t) \equiv k$, the state equation will become

$$\dot{\mathbf{x}} = (\mathbf{A} + k\mathbf{B})\,\mathbf{x}.$$

And the observability matrix is

$$\mathcal{O}\left(\mathbf{A} + k\mathbf{B}, \mathbf{C}^{\mathrm{T}}\right) = \begin{bmatrix} \mathbf{C}^{\mathrm{T}} \\ \mathbf{C}^{\mathrm{T}}\,(\mathbf{A} + k\mathbf{B}) \\ \mathbf{C}^{\mathrm{T}}\,(\mathbf{A} + k\mathbf{B})^2 \end{bmatrix} = \begin{bmatrix} 0 & 1 & 0 \\ k & 0 & 1 \\ 0 & k & 0 \end{bmatrix}.$$

Clearly, the first row and the third row of the matrix $\mathcal{O}\left(\mathbf{A} + k\mathbf{B}, \mathbf{C}^{\mathrm{T}}\right)$ are linearly dependent for any k. Thus, $\operatorname{rank}\mathcal{O}\left(\mathbf{A} + k\mathbf{B}, \mathbf{C}^{\mathrm{T}}\right) < 3$. Consequently, system (8.15) is unobservable for each fixed k. On the other hand, according to Definition 8.6, there exist nonzero states that produce an identically zero output, i.e., the states in the set

$$\left\{\mathbf{x} \in \mathbb{R}^3 \,\middle|\, \mathcal{O}\left(\mathbf{A} + k\mathbf{B}, \mathbf{C}^{\mathrm{T}}\right)\mathbf{x} = 0\right\} = \left\{\mathbf{x} = \alpha[1\ 0\ -k]^{\mathrm{T}}, \alpha \neq 0\right\}$$

are not distinguishable from $\mathbf{x}_0 = \mathbf{0}_{3\times 1}$ with any input $u(t) \equiv k$. However, given any nonzero initial state $\mathbf{x}$, there exists a choice of k such that $\mathbf{x} \neq \alpha[\,1\ 0\ -k\,]^{\mathrm{T}}, \alpha \neq 0$. Hence, for any state $\mathbf{x} \neq \mathbf{0}_{3\times 1}$, some input $u(t) \equiv k$ can be found such that $y(\cdot, \mathbf{0}_{3\times 1}, \mathbf{u}) \neq y(\cdot, \mathbf{x}, \mathbf{u})$. So the states $\mathbf{x}_0 = \mathbf{0}_{3\times 1}$ and $\mathbf{x} \neq \mathbf{0}_{3\times 1}$ are distinguishable, i.e., system (8.15) is (locally) observable. This example shows that only through linearization, the observability of a system may not be determined based on the observability matrix sometimes.

8.1.2.2 Observability Criterion

In this section, the sufficient conditions of local observability for the continuous-time nonlinear system (8.14) are discussed. Consider a smooth scalar function $h(\mathbf{x})$, whose gradient is defined as (see footnote 1 in Sect. 8.2.1)

$$\nabla h \triangleq \left(\frac{\partial h}{\partial \mathbf{x}}\right)^{\mathrm{T}}. \tag{8.17}$$

The gradient is a row vector, and the jth component is denoted by $(\nabla h)_j = \partial h / \partial x_j$. Similarly, for a vector field $\mathbf{f}(\mathbf{x})$, the Jacobian matrix of $\mathbf{f}$ is defined as

$$\nabla \mathbf{f} \triangleq \frac{\partial \mathbf{f}}{\partial \mathbf{x}}. \tag{8.18}$$

The Jacobian matrix is an $n \times n$ matrix, and the ijth component is denoted by $(\nabla \mathbf{f})_{ij} = \partial f_i / \partial x_j$.

Definition 8.8 ([3, p. 381]) Consider a smooth scalar function $h : \mathbb{R}^n \to \mathbb{R}$ and a vector field $\mathbf{f} : \mathbb{R}^n \to \mathbb{R}^n$. The map $\mathbf{x} \mapsto \nabla h(\mathbf{x}) \cdot \mathbf{f}(\mathbf{x})$ is called the *Lie derivative* of the function h with respect to the vector field $\mathbf{f}$, and is denoted by $L_{\mathbf{f}} h$. And $L_{\mathbf{f}} h$ is thought of as the directional derivative of the function h along the integral curves of $\mathbf{f}$.

Furthermore, the higher-order Lie derivative is derived as

$$L_{\mathbf{f}}^0 h = h$$
$$L_{\mathbf{f}}^i h = L_{\mathbf{f}} \left(L_{\mathbf{f}}^{i-1} h \right) = \nabla \left(L_{\mathbf{f}}^{i-1} h \right) \mathbf{f} \tag{8.19}$$

where $L_{\mathbf{f}}^i h$ denotes the ith order derivative of $L_{\mathbf{f}} h$. If $\mathbf{g}$ is another vector field, then one has the scalar function

$$L_{\mathbf{g}} L_{\mathbf{f}} h = \nabla (L_{\mathbf{f}} h) \mathbf{g}. \tag{8.20}$$

First, the derivative of the output of system (8.14) is

$$y_j = h_j(\mathbf{x})$$
$$\dot{y}_j = (\nabla h_j)(\mathbf{x}) \dot{\mathbf{x}} = (L_{\mathbf{f}} h_j)(\mathbf{x}) + \sum_{k=1}^{m} u_k (L_{\mathbf{g}_k} h_j)(\mathbf{x}) \tag{8.21}$$

where the Lie derivatives $L_{\mathbf{f}} h_j$ and $L_{\mathbf{g}_k} h_j$ are defined in Definition 8.8, and $y_j, h_j(\mathbf{x})$ are the jth components of $\mathbf{y}$ and $\mathbf{h}(\mathbf{x})$ in system (8.14), respectively. Taking one more derivative gives

$$\ddot{y}_j = (L_{\mathbf{f}}^2 h_j)(\mathbf{x}) + \sum_{k=1}^{m} u_k (L_{\mathbf{g}_k} L_{\mathbf{f}} h_j)(\mathbf{x}) + \sum_{k=1}^{m} u_k (L_{\mathbf{g}_k} h_j)(\mathbf{x})$$
$$+ \sum_{k=1}^{m} u_k (L_{\mathbf{f}} L_{\mathbf{g}_k} h_j)(\mathbf{x}) + \sum_{k=1}^{m} \sum_{s=1}^{m} u_k u_s (L_{\mathbf{g}_s} L_{\mathbf{g}_k} h_j)(\mathbf{x}). \tag{8.22}$$

The expressions for higher-order derivatives of y_j become extremely complex, but the format is clear. The quantity $y_j^{(k)}$ is a linear combination of terms of the form $(L_{\mathbf{z}_s} L_{\mathbf{z}_{s-1}} \cdots L_{\mathbf{z}_1} h_j)(\mathbf{x})$, where $1 \le s \le k$, and each of the vector fields $\mathbf{z}_1, \ldots, \mathbf{z}_s$ comes from the set $\{\mathbf{f}, \mathbf{g}_1, \ldots, \mathbf{g}_m\}$.

Theorem 8.9 ([3, p. 418]) *For the continuous-time nonlinear system (8.14), suppose $\mathbf{x}_0 \in \mathcal{X}$ is given, and the forms*

$$\left(\nabla L_{\mathbf{z}_s} L_{\mathbf{z}_{s-1}} \cdots L_{\mathbf{z}_1} h_j \right)(\mathbf{x}_0), \, s \geqslant 0, \mathbf{z}_i \in \{\mathbf{f}, \mathbf{g}_1, \ldots, \mathbf{g}_m\} \tag{8.23}$$

are evaluated at state $\mathbf{x}_0$. Then, the system is locally observable around $\mathbf{x}_0$ if there are n linearly independent row vectors in this set.

Detailed proofs can be found in [3, p. 418]. Examples as follows are given to understand the theorem better.

Example 8.10 ([3, p. 418]) **Observability of a linear system**

If $\mathbf{f}(\mathbf{x}) = \mathbf{A}\mathbf{x}$, $\mathbf{g}_i(\mathbf{x}) = \mathbf{b}_i$, $h_j(\mathbf{x}) = \mathbf{c}_j \mathbf{x}$, where $\mathbf{b}_i$ and $\mathbf{c}_j$ denote the ith column of the matrix $\mathbf{B}$ and the jth row of the matrix $\mathbf{C}^{\mathrm{T}}$, respectively. Hence, if $s = 0$, Eq. (8.23) becomes

$$(\nabla h_j)(\mathbf{x}) = \mathbf{c}_j, \, j = 1, \ldots, l.$$

Furthermore, if $s = 1$, then

$$\left(L_{\mathbf{f}} h_j\right)(\mathbf{x}) = \mathbf{c}_j \mathbf{A} \mathbf{x}, \ \left(L_{\mathbf{g}_i} h_j\right)(\mathbf{x}) = \mathbf{c}_j \mathbf{b}_i.$$

Then, the nonzero vectors of the form (8.23) with $s = 1$ are

$$\left(\nabla L_{\mathbf{f}} h_j\right)(\mathbf{x}) = \mathbf{c}_j \mathbf{A}, \ j = 1, \ldots, l.$$

When taking repeated Lie derivatives as in Eq. (8.23), the constant functions do not contribute anything. Hence, the only nonzero vectors of the form (8.23) are

$$\mathbf{c}_j \mathbf{A}^k, \ k \geqslant 0, \ j = 1, \ldots, l.$$

Theorem 8.9 states that the system is locally observable if this set contains n linearly independent row vectors. According to the Cayley–Hamilton theorem, the span of the set

$$\mathbf{c}_j \mathbf{A}^k, \ k = 0, \ldots, n - 1, \ j = 1, \ldots, l$$

needs to be considered. Then, the linear system is observable if

$$\text{rank} \begin{bmatrix} \mathbf{C}^{\mathrm{T}} \\ \mathbf{C}^{\mathrm{T}} \mathbf{A} \\ \vdots \\ \mathbf{C}^{\mathrm{T}} \mathbf{A}^{n-1} \end{bmatrix} = n.$$

This conclusion is consistent with the conclusion in Theorem 8.2.

Example 8.11 ([5]) **Observability of a nonlinear system**

According to Eqs. (6.33)–(6.34) in Chap. 6, a simplified multicopter aerodynamic drag model along the $o_{\mathrm{b}} x_{\mathrm{b}}$ axis of a multicopter is

$$\underbrace{\begin{bmatrix} \dot{\theta} \\ \dot{v}_{x_{\mathrm{b}}} \\ \dot{k}_{\mathrm{drag}} \end{bmatrix}}_{\dot{\mathbf{x}}} = \underbrace{\begin{bmatrix} 0 \\ -g \sin \theta - \frac{k_{\mathrm{drag}}}{m} v_{x_{\mathrm{b}}} \\ 0 \end{bmatrix}}_{\mathbf{f}(\mathbf{x})} + \omega_{y_{\mathrm{b}}} \underbrace{\begin{bmatrix} 1 \\ 0 \\ 0 \end{bmatrix}}_{\mathbf{g}_1} \tag{8.24}$$

where $\theta, v_{x_{\mathrm{b}}}, k_{\mathrm{drag}}, \omega_{y_{\mathrm{b}}}, m, g$ are the pitch angle, velocity along the $o_{\mathrm{b}} x_{\mathrm{b}}$ axis, the drag coefficient, the angular velocity about the $o_{\mathrm{b}} y_{\mathrm{b}}$ axis, the mass of the multicopter, and the gravity, respectively. The measurement model is

$$y_1 = h_1(\mathbf{x}) = -\frac{k_{\mathrm{drag}}}{m} v_{x_{\mathrm{b}}}. \tag{8.25}$$

The gradient of $h_1(\mathbf{x})$ is

$$(\nabla h_1)(\mathbf{x}) = \begin{bmatrix} 0 & -\frac{k_{\mathrm{drag}}}{m} & -\frac{v_{x_{\mathrm{b}}}}{m} \end{bmatrix}. \tag{8.26}$$

Since $\nabla L_{\mathbf{g}_1} h_1 (\mathbf{x}) = [0\ 0\ 0]$, then

$$(\nabla L_{\mathbf{f}} h_1)(\mathbf{x}) = \nabla (\nabla h_1 \mathbf{f})(\mathbf{x})$$
$$= \left[g\cos\theta \frac{k_{\text{drag}}}{m} \quad \left(\frac{k_{\text{drag}}}{m}\right)^2 \quad \left(\frac{g\sin\theta}{m} + \frac{2k_{\text{drag}}v_{x_b}}{m^2}\right) \right]. \tag{8.27}$$

Further, the second-order Lie derivative of $\mathbf{f}(\mathbf{x})$ is

$$(\nabla L_{\mathbf{f}} L_{\mathbf{f}} h_1)(\mathbf{x}) = \left[-g\cos\theta \left(\frac{k_{\text{drag}}}{m}\right)^2 \quad -\left(\frac{k_{\text{drag}}}{m}\right)^3 \quad \left(-\frac{2k_{\text{drag}}g\sin\theta}{m^2} - \frac{3k_{\text{drag}}^2 v_{x_b}}{m^3}\right) \right]. \tag{8.28}$$

According to Eqs. (8.26)–(8.28), combined with Therome 8.9, the observability matrix can be obtained as

$$\mathcal{O} = \begin{bmatrix} 0 & -\dfrac{k_{\text{drag}}}{m} & -\dfrac{v_{x_b}}{m} \\[2mm] g\cos\theta \dfrac{k_{\text{drag}}}{m} & \left(\dfrac{k_{\text{drag}}}{m}\right)^2 & \left(\dfrac{g\sin\theta}{m} + \dfrac{2k_{\text{drag}}v_{x_b}}{m^2}\right) \\[2mm] -g\cos\theta \left(\dfrac{k_{\text{drag}}}{m}\right)^2 & -\left(\dfrac{k_{\text{drag}}}{m}\right)^3 & \left(-\dfrac{2k_{\text{drag}}g\sin\theta}{m^2} - \dfrac{3k_{\text{drag}}^2 v_{x_b}}{m^3}\right) \end{bmatrix}. \tag{8.29}$$

Its determinant is

$$\det(\mathcal{O}) = \frac{k_{\text{drag}}^3 mg\cos\theta \left(-g\sin\theta - \dfrac{k_{\text{drag}}}{m} v_{x_b}\right)}{m^5}.$$

Since $k_{\text{drag}} \neq 0$, $\cos\theta \neq 0$, if and only if

$$-g\sin\theta - \frac{k_{\text{drag}}}{m} v_{x_b} = 0 \tag{8.30}$$

then $\det(\mathcal{O}) = 0$. So the matrix $\mathcal{O}$ is singular, i.e., the system is unobservable. Except from this condition, the system is locally observable. Since

$$\dot{v}_{x_b} = -g\sin\theta - \frac{k_{\text{drag}}}{m} v_{x_b}$$

the condition (8.30) implies $\dot{v}_{x_b} = 0$, namely the system is non-accelerated. In this case, the system may be unobservable. On the other hand, the system becomes observable when $\dot{v}_{x_b} \neq 0$. In this case, the variables θ, v_{x_b}, k_{drag} can be observed by using the accelerometer.

8.2 Kalman Filter

Kalman filter, also known as Linear Quadratic Estimate (LQE), is an algorithm that uses a series of measurements observed over time, containing statistical noise and other inaccuracies, and produces estimates of unknown variables that tend to be more precise than those based on a single measurement alone. The filter is named after Rudolf E. Kalman, one of the primary developers of this theory. The Kalman filter yields the exact conditional probability estimate in the special case that all errors are Gaussian-distributed. The algorithm works in a two-step process. In the prediction stage, the Kalman filter produces estimates of the current state variables, along with their uncertainties. Once the next measurement (necessarily corrupted with some amount of error, including random noise) is observed, these estimates are updated using a weighted average, with higher weighting given to estimates with higher certainty. The algorithm is a recursive method and can operate in real time, using only the present input measurements and the previously calculated state and its uncertainty matrix; no additional past information is required [6].

8.2.1 Objective

Before proceeding to derive the Kalman filter, it is useful to review some basic concepts for optimum estimate. Denote $\hat{\vartheta}\,(X_1, X_2, \ldots, X_n)$ as the estimate of the parameter $\vartheta \in \mathbb{R}$, where random components, $X_1, X_2, \ldots, X_n$, are the measured or empirical data. Then, $\hat{\vartheta}$ is a random variable. The following definitions are introduced.

(1) Unbiased estimate: The estimator is said to be *unbiased* if and only if the expected value of the estimation equals the true value, namely $\mathrm{E}\left(\hat{\vartheta}\right) = \vartheta$.

(2) Minimum-variance estimate: The variance of a set of observed values is

$$\mathrm{Var}\left(\hat{\vartheta}\right) \triangleq \mathrm{E}[\left(\hat{\vartheta} - \mathrm{E}\left(\hat{\vartheta}\right)\right)^2].$$

In probability theory and statistics, variance measures how far a set of numbers is spread out. A small variance indicates that the data points tend to be very close to the expected value and hence to each other, while a high variance indicates that the data points are very spread out around the mean and from each other. If $\mathrm{Var}\left(\hat{\vartheta}\right) \leq \mathrm{Var}\left(\tilde{\vartheta}\right)$ holds for any estimator $\tilde{\vartheta}\,(X_1, X_2, \ldots, X_n)$, then $\hat{\vartheta}$ is the *minimum-variance estimate*.

(3) Minimum-variance unbiased estimate (MVUE): If an estimate $\hat{\vartheta}$ is both the unbiased estimate and minimum-variance estimate of the parameter ϑ, then $\hat{\vartheta}$ is MVUE.

(4) If the parameter $\boldsymbol{\vartheta} \in \mathbb{R}^n$ is a vector, the concept of covariance other than variance is used. The covariance of the estimation $\hat{\boldsymbol{\vartheta}}$ is defined as

$$\mathrm{Cov}\left(\hat{\boldsymbol{\vartheta}}\right) \triangleq \mathrm{E}\left[\left(\hat{\boldsymbol{\vartheta}} - \boldsymbol{\vartheta}\right)\left(\hat{\boldsymbol{\vartheta}} - \boldsymbol{\vartheta}\right)^{\mathrm{T}}\right].$$

Based on the introduction above, the Kalman filter, a kind of recursive linear minimum-variance unbiased estimation algorithm, has three properties: unbiased, minimum-variance, and real-time performance.

8.2.2 Preliminary [7, pp. 21–23]

This section contains some scale derivatives with respect to a vector and matrix. Denote a vector $\mathbf{x} \triangleq [\,x_1\ x_2\ \cdots\ x_n\,]^{\mathrm{T}} \in \mathbb{R}^n$ and a matrix $\mathbf{X} \triangleq [\,\mathbf{x}_1\ \mathbf{x}_2\ \cdots\ \mathbf{x}_m\,] \in \mathbb{R}^{n \times m}$, $\mathbf{x}_i \in \mathbb{R}^n$, $i = 1, \ldots, m$. The derivative of a scalar function a, with respect to the vector $\mathbf{x}$, is the vector denoted as[1]

$$\frac{\partial a}{\partial \mathbf{x}} \triangleq \left[\,\frac{\partial a}{\partial x_1}\ \frac{\partial a}{\partial x_2}\ \cdots\ \frac{\partial a}{\partial x_n}\,\right]^{\mathrm{T}} \in \mathbb{R}^n. \tag{8.31}$$

---

[1]The derivative is a column vector rather than a row vector. This is consistent with [7, pp. 21–23]. On the other hand, the vector derivative with respect to a vector is a Jacobian matrix (See Eq. (8.18)).

Similarly, the derivative of a scalar a with respect to a matrix $\mathbf{X}$ is defined as

$$\frac{\partial a}{\partial \mathbf{X}} \triangleq \left[\frac{\partial a}{\partial \mathbf{x}_1} \ \frac{\partial a}{\partial \mathbf{x}_2} \ \cdots \ \frac{\partial a}{\partial \mathbf{x}_m} \right] \in \mathbb{R}^{n \times m}. \tag{8.32}$$

Some conclusions are listed directly as follows:

$$\frac{\partial \mathbf{x}^T \mathbf{y}}{\partial \mathbf{x}} = \frac{\partial \mathbf{y}^T \mathbf{x}}{\partial \mathbf{x}} = \mathbf{y} \tag{8.33}$$

$$\frac{\partial \mathbf{x}^T \mathbf{N} \mathbf{x}}{\partial \mathbf{x}} = 2\mathbf{N}\mathbf{x}, \mathbf{N} = \mathbf{N}^T \tag{8.34}$$

$$\frac{\partial}{\partial \mathbf{A}} \text{tr} (\mathbf{A}\mathbf{C}) = \mathbf{C}^T \tag{8.35}$$

$$\frac{\partial}{\partial \mathbf{A}} \text{tr} \left(\mathbf{A}\mathbf{B}\mathbf{A}^T \right) = 2\mathbf{A}\mathbf{B}, \mathbf{B} = \mathbf{B}^T \tag{8.36}$$

$$\frac{\partial}{\partial \mathbf{x}} (\mathbf{H}\mathbf{x} - \mathbf{z})^T \mathbf{W} (\mathbf{H}\mathbf{x} - \mathbf{z}) = 2\mathbf{H}^T \mathbf{W} (\mathbf{H}\mathbf{x} - \mathbf{z}), \mathbf{W} = \mathbf{W}^T \tag{8.37}$$

where $\mathbf{x}, \mathbf{y}, \mathbf{z}, \mathbf{b}$ are vectors and $\mathbf{A}, \mathbf{B}, \mathbf{C}, \mathbf{N}, \mathbf{H}, \mathbf{W}$ are matrices.

8.2.3   Theoretical Derivation

The derivation of the discrete-time Kalman filter is introduced first under the assumption that both the model and measurements are available in a discrete-time form [8, pp. 143–148]. Suppose that the discrete-time model and measurements are corrupted by noise.

8.2.3.1 Model Description
The "truth" model for discrete-time cases is given by

$$\begin{aligned} \mathbf{x}_k &= \mathbf{\Phi}_{k-1}\mathbf{x}_{k-1} + \mathbf{u}_{k-1} + \mathbf{\Gamma}_{k-1}\mathbf{w}_{k-1} \\ \mathbf{z}_k &= \mathbf{H}_k\mathbf{x}_k + \mathbf{v}_k \end{aligned} \tag{8.38}$$

where $\mathbf{x}_k \in \mathbb{R}^n$ is the system state vector at the discrete time k; $\mathbf{z}_k \in \mathbb{R}^m$ is the measurement vector at the time k; $\mathbf{u}_{k-1} \in \mathbb{R}^n$ is the control vector at the time $k-1$; $\mathbf{\Phi}_{k-1} \in \mathbb{R}^{n \times n}$ is the transition matrix taking the state $\mathbf{x}_k$ from time $k-1$ to time k, and it will remain constant for system (8.11) when the matrix $\mathbf{\Phi}_{k-1}$ is time-invariant; $\mathbf{w}_{k-1} \in \mathbb{R}^n$ is the process noise at the time $k-1$; $\mathbf{\Gamma}_{k-1} \in \mathbb{R}^{n \times n}$ is the system noise matrix to reflect the influence each noise signal made to the system state; $\mathbf{H}_k \in \mathbb{R}^{m \times n}$ is the measurement matrix at the time k; $\mathbf{v}_k \in \mathbb{R}^m$ is the measurement noise at the time k. Also, $\{\mathbf{w}_k\}, \{\mathbf{v}_k\}$ are assumed to be zero-mean Gaussian white noise processes. This means that the errors are uncorrelated forward or backward in time so that

$$\begin{aligned} \text{E}(\mathbf{w}_{k-1}) &= \mathbf{0}_{n \times 1}, \text{E}(\mathbf{v}_k) = \mathbf{0}_{m \times 1}, \mathbf{R}_{\mathbf{wv}}(k, j) = \mathbf{0}_{n \times m} \\ \mathbf{R}_{\mathbf{ww}}(k, j) &= \text{E}\left(\mathbf{w}_k \mathbf{w}_j^T \right) = \mathbf{Q}_k \delta_{kj} = \begin{cases} \mathbf{Q}_k, & k = j \\ \mathbf{0}_{n \times n}, & k \neq j \end{cases} \\ \mathbf{R}_{\mathbf{vv}}(k, j) &= \text{E}\left(\mathbf{v}_k \mathbf{v}_j^T \right) = \mathbf{R}_k \delta_{kj} = \begin{cases} \mathbf{R}_k, & k = j \\ \mathbf{0}_{m \times m}, & k \neq j \end{cases} \end{aligned} \tag{8.39}$$

where $\mathbf{Q}_k \in \mathbb{R}^{n \times n}$ is the system noise covariance matrix and $\mathbf{R}_k \in \mathbb{R}^{m \times m}$ is the measurement noise covariance matrix. In a Kalman filter, they are *positive semi-definite* matrices at least, represented by

$$\mathbf{Q}_k \geqslant \mathbf{0}_{n \times n}, \mathbf{R}_k > \mathbf{0}_{m \times m} \tag{8.40}$$

where $\mathbf{Q}_k \geqslant \mathbf{0}_{n \times n}$ means that there may be no noise in some system states, while $\mathbf{R}_k > \mathbf{0}_{m \times m}$ is necessary, implying that each measurement must contain noise; δ_{kj} is the Kronecker δ function, thus

$$\delta_{kj} = \begin{cases} 1, \, k = j \\ 0, \, k \neq j. \end{cases} \tag{8.41}$$

Suppose that the initial condition of a state $\mathbf{x}_0$ satisfies

$$\mathrm{E}\,(\mathbf{x}_0) = \hat{\mathbf{x}}_0, \mathrm{Cov}\,(\mathbf{x}_0) = \mathbf{P}_0 \tag{8.42}$$

where $\hat{\mathbf{x}}_0$ and $\mathbf{P}_0$ are known as prior knowledge. Besides, $\mathbf{x}_0$, $\mathbf{u}_k$ and $\mathbf{w}_{k-1}$, $\mathbf{v}_k$, $k \geqslant 1$ are uncorrelated that

$$\begin{aligned} \mathbf{R}_{\mathbf{xw}}\,(0, k) &= \mathrm{E}\,\left(\mathbf{x}_0 \mathbf{w}_k^\mathrm{T}\right) = \mathbf{0}_{n \times n} \\ \mathbf{R}_{\mathbf{xv}}\,(0, k) &= \mathrm{E}\,\left(\mathbf{x}_0 \mathbf{v}_k^\mathrm{T}\right) = \mathbf{0}_{n \times m} \\ \mathbf{R}_{\mathbf{uw}}\,(k, j) &= \mathrm{E}\,\left(\mathbf{u}_k \mathbf{w}_j^\mathrm{T}\right) = \mathbf{0}_{n \times n}. \end{aligned} \tag{8.43}$$

Suppose that the measurements $\mathbf{z}_1, \mathbf{z}_2, \ldots, \mathbf{z}_{k-1}$ are given, and the optimal estimate $\hat{\mathbf{x}}_{k-1|k-1}$ of the state $\mathbf{x}_{k-1}$ is the MVUE, implying

$$\mathrm{E}\,\left(\mathbf{x}_{k-1} - \hat{\mathbf{x}}_{k-1|k-1}\right) = \mathbf{0}_{n \times 1}. \tag{8.44}$$

The goal of the Kalman filter is to find an MVUE $\hat{\mathbf{x}}_{k|k}$ at the time k using the information contained in the new measurement $\mathbf{z}_k$. With a linear estimator as the objective, the estimate $\hat{\mathbf{x}}_{k|k}$ is expressed as a linear combination of the priori estimate and the new measurement, as shown by

$$\hat{\mathbf{x}}_{k|k} = \mathbf{K}_k' \hat{\mathbf{x}}_{k-1|k-1} + \mathbf{K}_k \mathbf{z}_k + \mathbf{K}_k'' \mathbf{u}_{k-1}. \tag{8.45}$$

Next, the matrices $\mathbf{K}_k' \in \mathbb{R}^{n \times n}$, $\mathbf{K}_k \in \mathbb{R}^{n \times m}$, $\mathbf{K}_k'' \in \mathbb{R}^{n \times n}$ will be determined in the following. The process is also the derivation of a discrete-time Kalman filter.

8.2.3.2 Derivation of $\mathbf{K}_k'$ and $\mathbf{K}_k''$

Based on Eq. (8.44), the goal is to determine $\mathbf{K}_k'$ and $\mathbf{K}_k''$ so that $\hat{\mathbf{x}}_{k|k}$ is the *unbiased estimate* of the state $\mathbf{x}_k$, namely

$$\mathrm{E}\,\left(\tilde{\mathbf{x}}_{k|k}\right) = \mathbf{0}_{n \times 1} \tag{8.46}$$

where the state error is defined by

$$\tilde{\mathbf{x}}_{k|k} \triangleq \mathbf{x}_k - \hat{\mathbf{x}}_{k|k}. \tag{8.47}$$

According to Eqs. (8.45) and (8.47), $\tilde{\mathbf{x}}_{k|k}$ becomes

$$\tilde{\mathbf{x}}_{k|k} = \mathbf{x}_k - \left(\mathbf{K}_k' \hat{\mathbf{x}}_{k-1|k-1} + \mathbf{K}_k \mathbf{z}_k + \mathbf{K}_k'' \mathbf{u}_{k-1}\right). \tag{8.48}$$

Substituting Eq. (8.38) into Eq. (8.48) results in

$$
\begin{aligned}
\tilde{\mathbf{x}}_{k|k} = {} & (\boldsymbol{\Phi}_{k-1}\mathbf{x}_{k-1} + \mathbf{u}_{k-1} + \boldsymbol{\Gamma}_{k-1}\mathbf{w}_{k-1}) - \mathbf{K}'_k\hat{\mathbf{x}}_{k-1|k-1} \\
& - \mathbf{K}_k \left[\mathbf{H}_k \left(\boldsymbol{\Phi}_{k-1}\mathbf{x}_{k-1} + \mathbf{u}_{k-1} + \boldsymbol{\Gamma}_{k-1}\mathbf{w}_{k-1}\right) + \mathbf{v}_k\right] - \mathbf{K}''_k\mathbf{u}_{k-1}.
\end{aligned}
\tag{8.49}
$$

Consequently, Eq. (8.49) is rearranged as

$$
\begin{aligned}
\tilde{\mathbf{x}}_{k|k} = {} & (\boldsymbol{\Phi}_{k-1} - \mathbf{K}_k\mathbf{H}_k\boldsymbol{\Phi}_{k-1})\,\mathbf{x}_{k-1} - \mathbf{K}'_k\hat{\mathbf{x}}_{k-1|k-1} \\
& + (\boldsymbol{\Gamma}_{k-1} - \mathbf{K}_k\mathbf{H}_k\boldsymbol{\Gamma}_{k-1})\,\mathbf{w}_{k-1} + (\mathbf{I}_n - \mathbf{K}_k\mathbf{H}_k - \mathbf{K}''_k)\,\mathbf{u}_{k-1} - \mathbf{K}_k\mathbf{v}_k.
\end{aligned}
\tag{8.50}
$$

By introducing a zero term $\mathbf{K}'_k\mathbf{x}_{k-1} - \mathbf{K}'_k\mathbf{x}_{k-1} = \mathbf{0}_{n\times1}$, Eq. (8.50) further becomes

$$
\begin{aligned}
\tilde{\mathbf{x}}_{k|k} = {} & \left(\boldsymbol{\Phi}_{k-1} - \mathbf{K}_k\mathbf{H}_k\boldsymbol{\Phi}_{k-1} - \mathbf{K}'_k\right)\mathbf{x}_{k-1} + \mathbf{K}'_k\left(\mathbf{x}_{k-1} - \hat{\mathbf{x}}_{k-1|k-1}\right) \\
& + (\boldsymbol{\Gamma}_{k-1} - \mathbf{K}_k\mathbf{H}_k\boldsymbol{\Gamma}_{k-1})\,\mathbf{w}_{k-1} + (\mathbf{I}_n - \mathbf{K}_k\mathbf{H}_k - \mathbf{K}''_k)\,\mathbf{u}_{k-1} - \mathbf{K}_k\mathbf{v}_k.
\end{aligned}
\tag{8.51}
$$

The expected value of $\tilde{\mathbf{x}}_{k|k}$ in Eq. (8.51) is

$$
\begin{aligned}
\mathrm{E}\left(\tilde{\mathbf{x}}_{k|k}\right) = {} & \left(\boldsymbol{\Phi}_{k-1} - \mathbf{K}_k\mathbf{H}_k\boldsymbol{\Phi}_{k-1} - \mathbf{K}'_k\right)\mathrm{E}\left(\mathbf{x}_{k-1}\right) \\
& + \left(\mathbf{I}_n - \mathbf{K}_k\mathbf{H}_k - \mathbf{K}''_k\right)\mathbf{u}_{k-1} \\
& + \mathbf{K}'_k\mathrm{E}\left(\mathbf{x}_{k-1} - \hat{\mathbf{x}}_{k-1|k-1}\right) \\
& + (\boldsymbol{\Gamma}_{k-1} - \mathbf{K}_k\mathbf{H}_k\boldsymbol{\Gamma}_{k-1})\,\mathrm{E}\left(\mathbf{w}_{k-1}\right) - \mathbf{K}_k\mathrm{E}\left(\mathbf{v}_k\right).
\end{aligned}
\tag{8.52}
$$

Since $\mathrm{E}\left(\mathbf{x}_{k-1} - \hat{\mathbf{x}}_{k-1|k-1}\right) = \mathbf{0}_{n\times1}, \mathrm{E}(\mathbf{v}_k) = \mathbf{0}_{m\times1}, \mathrm{E}(\mathbf{w}_{k-1}) = \mathbf{0}_{n\times1}$ in Eq. (8.52), equation $\mathrm{E}\left(\tilde{\mathbf{x}}_{k|k}\right) = \mathbf{0}_{n\times1}$ holds if and only if

$$
\left(\boldsymbol{\Phi}_{k-1} - \mathbf{K}_k\mathbf{H}_k\boldsymbol{\Phi}_{k-1} - \mathbf{K}'_k\right)\mathrm{E}\left(\mathbf{x}_{k-1}\right) + \left(\mathbf{I}_n - \mathbf{K}_k\mathbf{H}_k - \mathbf{K}''_k\right)\mathbf{u}_{k-1} = \mathbf{0}_{n\times1}.
\tag{8.53}
$$

Furthermore, since $\mathrm{E}(\mathbf{x}_{k-1})$, $\mathbf{u}_{k-1}$ are nonzero all the time, it is obvious that Eq. (8.53) holds when

$$
\begin{aligned}
\boldsymbol{\Phi}_{k-1} - \mathbf{K}_k\mathbf{H}_k\boldsymbol{\Phi}_{k-1} - \mathbf{K}'_k &= \mathbf{0}_{n\times n} \\
\mathbf{I}_n - \mathbf{K}_k\mathbf{H}_k - \mathbf{K}''_k &= \mathbf{0}_{n\times n}.
\end{aligned}
\tag{8.54}
$$

Based on Eq. (8.54), the expression of $\mathbf{K}'_k$ and $\mathbf{K}''_k$ is obtained as

$$
\begin{aligned}
\mathbf{K}'_k &= \boldsymbol{\Phi}_{k-1} - \mathbf{K}_k\mathbf{H}_k\boldsymbol{\Phi}_{k-1} \\
\mathbf{K}''_k &= \mathbf{I}_n - \mathbf{K}_k\mathbf{H}_k.
\end{aligned}
\tag{8.55}
$$

In other words, if $\mathbf{K}'_k$ and $\mathbf{K}''_k$ are given as in Eq. (8.55), then

$$
\mathrm{E}\left(\tilde{\mathbf{x}}_{k|k}\right) = \mathbf{0}_{n\times1}.
$$

This means that the goal of the unbiased estimate of the state $\mathbf{x}_k$ is achieved by $\mathbf{K}'_k$ and $\mathbf{K}''_k$ chosen by Eq. (8.55). Substituting Eq. (8.55) into Eq. (8.45) yields

$$
\hat{\mathbf{x}}_{k|k} = \hat{\mathbf{x}}_{k|k-1} + \mathbf{K}_k\left(\mathbf{z}_k - \hat{\mathbf{z}}_{k|k-1}\right)
\tag{8.56}
$$

where

$$
\begin{aligned}
\hat{\mathbf{x}}_{k|k-1} &\triangleq \boldsymbol{\Phi}_{k-1}\hat{\mathbf{x}}_{k-1|k-1} + \mathbf{u}_{k-1} \\
\hat{\mathbf{z}}_{k|k-1} &\triangleq \mathbf{H}_k\hat{\mathbf{x}}_{k|k-1}.
\end{aligned}
\tag{8.57}
$$

The following summaries are made: (1) $\hat{\mathbf{x}}_{k|k-1}$ is the one-step prediction of the state $\mathbf{x}_k$ based on $\hat{\mathbf{x}}_{k-1|k-1}$; (2) $\hat{\mathbf{z}}_{k|k-1}$ is the one-step prediction of the measurement $\mathbf{z}_k$ on the basis of state prediction $\hat{\mathbf{x}}_{k|k-1}$. However, the prediction will diverge, although the expected value is unbiased in the sense of probability. So, in the following, the error between observed output $\mathbf{z}_k$ and the predicted output $\hat{\mathbf{z}}_{k|k-1}$ is utilized to correct the estimate.

8.2.3.3 Derivation of $\mathbf{K}_k$

So far, $\mathbf{K}_k'$ and $\mathbf{K}_k''$ in Eq. (8.45) have been determined as shown in Eq. (8.55), which is related to the parameter $\mathbf{K}_k$. In this section, based on Eqs. (8.56) and (8.57), the goal is to determine $\mathbf{K}_k$ such that $\hat{\mathbf{x}}_{k|k}$ is the *minimum-variance estimate* of the system state $\mathbf{x}_k$.

The *trace* of the error covariance matrix $\text{tr}(\mathbf{P}_{k|k})$ is used to measure the covariance, where

$$
\mathbf{P}_{k|k} \triangleq \text{E}\left(\tilde{\mathbf{x}}_{k|k}\tilde{\mathbf{x}}_{k|k}^{\text{T}}\right).
\tag{8.58}
$$

It is known that

$$
\text{tr}\left(\mathbf{P}_{k|k}\right) = \text{E}\left(\text{tr}\left(\tilde{\mathbf{x}}_{k|k}\tilde{\mathbf{x}}_{k|k}^{\text{T}}\right)\right) = \text{E}\left(\tilde{\mathbf{x}}_{k|k}^{\text{T}}\tilde{\mathbf{x}}_{k|k}\right).
$$

Our goal is to find a matrix $\mathbf{K}_k$ to minimize $\text{tr}(\mathbf{P}_{k|k})$. Three steps are proposed as follows:

(1) In order to derive the expression of $\mathbf{K}_k$, the matrix $\mathbf{P}_{k|k}$ with respect to $\mathbf{K}_k$ and $\mathbf{P}_{k|k-1}$ is obtained. First, the estimate error $\tilde{\mathbf{x}}_{k|k}$ is

$$
\begin{aligned}
\tilde{\mathbf{x}}_{k|k} &= \mathbf{x}_k - \hat{\mathbf{x}}_{k|k} \\
&= \mathbf{x}_k - \hat{\mathbf{x}}_{k|k-1} - \mathbf{K}_k\left(\mathbf{z}_k - \mathbf{H}_k\hat{\mathbf{x}}_{k|k-1}\right) \\
&= \tilde{\mathbf{x}}_{k|k-1} - \mathbf{K}_k\left(\mathbf{H}_k\left(\mathbf{x}_k - \hat{\mathbf{x}}_{k|k-1}\right) + \mathbf{v}_k\right) \\
&= \left(\mathbf{I}_n - \mathbf{K}_k\mathbf{H}_k\right)\tilde{\mathbf{x}}_{k|k-1} - \mathbf{K}_k\mathbf{v}_k.
\end{aligned}
\tag{8.59}
$$

According to Eq. (8.59), the expression of $\tilde{\mathbf{x}}_{k|k}\tilde{\mathbf{x}}_{k|k}^{\text{T}}$ is expanded as

$$
\begin{aligned}
\tilde{\mathbf{x}}_{k|k}\tilde{\mathbf{x}}_{k|k}^{\text{T}} &= \left(\left(\mathbf{I}_n - \mathbf{K}_k\mathbf{H}_k\right)\tilde{\mathbf{x}}_{k|k-1} - \mathbf{K}_k\mathbf{v}_k\right)\left(\tilde{\mathbf{x}}_{k|k-1}^{\text{T}}(\mathbf{I}_n - \mathbf{K}_k\mathbf{H}_k)^{\text{T}} - \mathbf{v}_k^{\text{T}}\mathbf{K}_k^{\text{T}}\right) \\
&= \left(\mathbf{I}_n - \mathbf{K}_k\mathbf{H}_k\right)\tilde{\mathbf{x}}_{k|k-1}\tilde{\mathbf{x}}_{k|k-1}^{\text{T}}(\mathbf{I}_n - \mathbf{K}_k\mathbf{H}_k)^{\text{T}} - \left(\mathbf{I}_n - \mathbf{K}_k\mathbf{H}_k\right)\tilde{\mathbf{x}}_{k|k-1}\mathbf{v}_k^{\text{T}}\mathbf{K}_k^{\text{T}} \\
&\quad - \mathbf{K}_k\mathbf{v}_k\tilde{\mathbf{x}}_{k|k-1}^{\text{T}}(\mathbf{I}_n - \mathbf{K}_k\mathbf{H}_k)^{\text{T}} + \mathbf{K}_k\mathbf{v}_k\mathbf{v}_k^{\text{T}}\mathbf{K}_k^{\text{T}}.
\end{aligned}
\tag{8.60}
$$

Because the state prediction $\tilde{\mathbf{x}}_{k|k-1}$ is the linear function of measurements $\mathbf{z}_1, \mathbf{z}_2, \ldots, \mathbf{z}_{k-1}$ and is uncorrelated with measurement noise $\mathbf{v}_k$, one has

$$
\begin{aligned}
\text{E}\left(\tilde{\mathbf{x}}_{k|k-1}\mathbf{v}_k^{\text{T}}\right) &= \mathbf{0}_{n\times m} \\
\text{E}\left(\mathbf{v}_k\tilde{\mathbf{x}}_{k|k-1}^{\text{T}}\right) &= \mathbf{0}_{m\times n}.
\end{aligned}
\tag{8.61}
$$

Then, according to Eqs. (8.60) and (8.61), the matrix $\mathbf{P}_{k|k}$ is

$$
\begin{aligned}
\mathbf{P}_{k|k} &= \mathrm{E}\left(\tilde{\mathbf{x}}_{k|k}\tilde{\mathbf{x}}_{k|k}^{\mathrm{T}}\right) \\
&= (\mathbf{I}_n - \mathbf{K}_k\mathbf{H}_k)\,\mathbf{P}_{k|k-1}(\mathbf{I}_n - \mathbf{K}_k\mathbf{H}_k)^{\mathrm{T}} + \mathbf{K}_k\mathbf{R}_k\mathbf{K}_k^{\mathrm{T}}
\end{aligned}
\tag{8.62}
$$

where $\mathbf{P}_{k|k-1} \triangleq \mathrm{E}\left(\tilde{\mathbf{x}}_{k|k-1}\tilde{\mathbf{x}}_{k|k-1}^{\mathrm{T}}\right)$, $\mathbf{R}_k = \mathrm{E}\left(\mathbf{v}_k\mathbf{v}_k^{\mathrm{T}}\right)$.

(2) The expression of $\mathbf{P}_{k|k-1}$ in Eq. (8.60) is further determined, and then, the matrix $\mathbf{P}_{k|k}$ with respect to $\mathbf{K}_k$ and $\mathbf{P}_{k-1|k-1}$ is obtained. The state error $\tilde{\mathbf{x}}_{k|k-1}$ is derived as

$$
\begin{aligned}
\tilde{\mathbf{x}}_{k|k-1} &= \mathbf{x}_k - \hat{\mathbf{x}}_{k|k-1} \\
&= \boldsymbol{\Phi}_{k-1}\mathbf{x}_{k-1} + \mathbf{u}_{k-1} + \boldsymbol{\Gamma}_{k-1}\mathbf{w}_{k-1} - \left(\boldsymbol{\Phi}_{k-1}\hat{\mathbf{x}}_{k-1|k-1} + \mathbf{u}_{k-1}\right) \\
&= \boldsymbol{\Phi}_{k-1}\tilde{\mathbf{x}}_{k-1|k-1} + \boldsymbol{\Gamma}_{k-1}\mathbf{w}_{k-1}.
\end{aligned}
\tag{8.63}
$$

Then, the expression of $\tilde{\mathbf{x}}_{k|k-1}\tilde{\mathbf{x}}_{k|k-1}^{\mathrm{T}}$ is derived as

$$
\begin{aligned}
\tilde{\mathbf{x}}_{k|k-1}\tilde{\mathbf{x}}_{k|k-1}^{\mathrm{T}} &= \left(\boldsymbol{\Phi}_{k-1}\tilde{\mathbf{x}}_{k-1|k-1} + \boldsymbol{\Gamma}_{k-1}\mathbf{w}_{k-1}\right)\left(\tilde{\mathbf{x}}_{k-1|k-1}^{\mathrm{T}}\boldsymbol{\Phi}_{k-1}^{\mathrm{T}} + \mathbf{w}_{k-1}^{\mathrm{T}}\boldsymbol{\Gamma}_{k-1}^{\mathrm{T}}\right) \\
&= \boldsymbol{\Phi}_{k-1}\tilde{\mathbf{x}}_{k-1|k-1}\tilde{\mathbf{x}}_{k-1|k-1}^{\mathrm{T}}\boldsymbol{\Phi}_{k-1}^{\mathrm{T}} + \boldsymbol{\Phi}_{k-1}\tilde{\mathbf{x}}_{k-1|k-1}\mathbf{w}_{k-1}^{\mathrm{T}}\boldsymbol{\Gamma}_{k-1}^{\mathrm{T}} \\
&\quad + \boldsymbol{\Gamma}_{k-1}\mathbf{w}_{k-1}\tilde{\mathbf{x}}_{k-1|k-1}^{\mathrm{T}}\boldsymbol{\Phi}_{k-1}^{\mathrm{T}} + \boldsymbol{\Gamma}_{k-1}\mathbf{w}_{k-1}\mathbf{w}_{k-1}^{\mathrm{T}}\boldsymbol{\Gamma}_{k-1}^{\mathrm{T}}.
\end{aligned}
\tag{8.64}
$$

Although $\tilde{\mathbf{x}}_{k-1|k-1}$ is correlated with the noise $\mathbf{w}_0, \ldots \mathbf{w}_{k-2}$, it is uncorrelated with $\mathbf{w}_{k-1}$. Thus,

$$
\begin{aligned}
\mathbf{P}_{k|k-1} &= \mathrm{E}\left(\tilde{\mathbf{x}}_{k|k-1}\tilde{\mathbf{x}}_{k|k-1}^{\mathrm{T}}\right) \\
&= \boldsymbol{\Phi}_{k-1}\mathbf{P}_{k-1|k-1}\boldsymbol{\Phi}_{k-1}^{\mathrm{T}} + \boldsymbol{\Gamma}_{k-1}\mathbf{Q}_{k-1}\boldsymbol{\Gamma}_{k-1}^{\mathrm{T}}
\end{aligned}
\tag{8.65}
$$

where $\mathbf{P}_{k-1|k-1} = \mathrm{E}\left(\tilde{\mathbf{x}}_{k-1|k-1}\tilde{\mathbf{x}}_{k-1|k-1}^{\mathrm{T}}\right)$ and $\mathbf{Q}_{k-1} = \mathrm{E}\left(\mathbf{w}_{k-1}\mathbf{w}_{k-1}^{\mathrm{T}}\right)$. Substituting Eq. (8.65) into Eq. (8.62) leads to

$$
\mathbf{P}_{k|k} = (\mathbf{I}_n - \mathbf{K}_k\mathbf{H}_k)\left(\boldsymbol{\Phi}_{k-1}\mathbf{P}_{k-1|k-1}\boldsymbol{\Phi}_{k-1}^{\mathrm{T}} + \boldsymbol{\Gamma}_{k-1}\mathbf{Q}_{k-1}\boldsymbol{\Gamma}_{k-1}^{\mathrm{T}}\right)(\mathbf{I}_n - \mathbf{K}_k\mathbf{H}_k)^{\mathrm{T}} + \mathbf{K}_k\mathbf{R}_k\mathbf{K}_k^{\mathrm{T}}.
\tag{8.66}
$$

(3) The matrix $\mathbf{K}_k$ in Eq. (8.66) is determined. The principle of choosing $\mathbf{K}_k$ is to minimize $\mathrm{tr}\left(\mathbf{P}_{k|k}\right)$. The expression of $\mathrm{tr}\left(\mathbf{P}_{k|k}\right)$ is

$$
\begin{aligned}
\mathrm{tr}\left(\mathbf{P}_{k|k}\right) &= \mathrm{tr}\left[(\mathbf{I}_n - \mathbf{K}_k\mathbf{H}_k)\,\mathbf{P}_{k|k-1}(\mathbf{I}_n - \mathbf{K}_k\mathbf{H}_k)^{\mathrm{T}} + \mathbf{K}_k\mathbf{R}_k\mathbf{K}_k^{\mathrm{T}}\right] \\
&= \mathrm{tr}\left(\mathbf{P}_{k|k-1}\right) - 2\mathrm{tr}\left(\mathbf{K}_k\mathbf{H}_k\mathbf{P}_{k|k-1}\right) + \mathrm{tr}\left(\mathbf{K}_k\left(\mathbf{H}_k\mathbf{P}_{k|k-1}\mathbf{H}_k^{\mathrm{T}} + \mathbf{R}_k\right)\mathbf{K}_k^{\mathrm{T}}\right).
\end{aligned}
$$

Based on the preliminary given in Sect. 8.2.2, the derivative with respect to $\mathbf{K}_k$ is

$$
\frac{\partial}{\partial \mathbf{K}_k}\mathrm{tr}\left(\mathbf{P}_{k|k}\right) = -2\mathbf{P}_{k|k-1}^{\mathrm{T}}\mathbf{H}_k^{\mathrm{T}} + 2\mathbf{K}_k\left(\mathbf{H}_k\mathbf{P}_{k|k-1}\mathbf{H}_k^{\mathrm{T}} + \mathbf{R}_k\right).
\tag{8.67}
$$

Since $\mathbf{H}_k\mathbf{P}_{k|k-1}\mathbf{H}_k^{\mathrm{T}} + \mathbf{R}_k$ is positive definite matrix (recall that $\mathbf{R}_k > \mathbf{0}_{m\times m}$ is required in Eq. (8.40)), the optimal solution to $\mathbf{K}_k$ will make

$$
\frac{\partial}{\partial \mathbf{K}_k}\mathrm{tr}\left(\mathbf{P}_{k|k}\right) = \mathbf{0}_{n\times m}.
$$

Then,

$$\mathbf{K}_k = \mathbf{P}_{k|k-1}\mathbf{H}_k^{\mathrm{T}}\left(\mathbf{H}_k\mathbf{P}_{k|k-1}\mathbf{H}_k^{\mathrm{T}} + \mathbf{R}_k\right)^{-1} \tag{8.68}$$

where $\mathbf{K}_k$ is also called the *Kalman gain*. Substituting Eq. (8.68) into Eq. (8.62) results in

$$
\begin{aligned}
\mathbf{P}_{k|k} &= (\mathbf{I}_n - \mathbf{K}_k\mathbf{H}_k)\,\mathbf{P}_{k|k-1}(\mathbf{I}_n - \mathbf{K}_k\mathbf{H}_k)^{\mathrm{T}} + \mathbf{K}_k\mathbf{R}_k\mathbf{K}_k^{\mathrm{T}} \\
&= \mathbf{P}_{k|k-1} - \mathbf{K}_k\mathbf{H}_k\mathbf{P}_{k|k-1} - \mathbf{P}_{k|k-1}\mathbf{H}_k^{\mathrm{T}}\mathbf{K}_k^{\mathrm{T}} + \mathbf{K}_k\left(\mathbf{H}_k\mathbf{P}_{k|k-1}\mathbf{H}_k^{\mathrm{T}} + \mathbf{R}_k\right)\mathbf{K}_k^{\mathrm{T}} \\
&= (\mathbf{I}_n - \mathbf{K}_k\mathbf{H}_k)\,\mathbf{P}_{k|k-1} - \mathbf{P}_{k|k-1}\mathbf{H}_k^{\mathrm{T}}\mathbf{K}_k^{\mathrm{T}} \\
&\quad + \mathbf{P}_{k|k-1}\mathbf{H}_k^{\mathrm{T}}\underbrace{\left(\mathbf{H}_k\mathbf{P}_{k|k-1}\mathbf{H}_k^{\mathrm{T}} + \mathbf{R}_k\right)^{-1}\left(\mathbf{H}_k\mathbf{P}_{k|k-1}\mathbf{H}_k^{\mathrm{T}} + \mathbf{R}_k\right)}_{=\mathbf{I}_m}\mathbf{K}_k^{\mathrm{T}} \\
&= (\mathbf{I}_n - \mathbf{K}_k\mathbf{H}_k)\,\mathbf{P}_{k|k-1}.
\end{aligned}
\tag{8.69}
$$

8.2.3.4 Summary of the Kalman Filter

The Kalman filter algorithm is summarized as follows:

Step 1   Process model:

$$\mathbf{x}_k = \boldsymbol{\Phi}_{k-1}\mathbf{x}_{k-1} + \mathbf{u}_{k-1} + \boldsymbol{\Gamma}_{k-1}\mathbf{w}_{k-1}, \ \ \mathbf{w}_k \sim \mathcal{N}\left(\mathbf{0}_{n\times 1}, \mathbf{Q}_k\right) \tag{8.70}$$

Measurement model:

$$\mathbf{z}_k = \mathbf{H}_k\mathbf{x}_k + \mathbf{v}_k, \ \ \mathbf{v}_k \sim \mathcal{N}\left(\mathbf{0}_{m\times 1}, \mathbf{R}_k\right) \tag{8.71}$$

where $\mathbf{w}_{k-1}$ and $\mathbf{v}_k$ are independent, zero-mean, Gaussian noise processes with covariance matrices being $\mathbf{Q}_k$ and $\mathbf{R}_k$, respectively.

Step 2   Initial state:

$$
\begin{aligned}
\hat{\mathbf{x}}_0 &= \mathrm{E}\,(\mathbf{x}_0) \\
\mathbf{P}_0 &= \mathrm{E}\left[(\mathbf{x}_0 - \mathrm{E}\,(\mathbf{x}_0))\,(\mathbf{x}_0 - \mathrm{E}\,(\mathbf{x}_0))^{\mathrm{T}}\right].
\end{aligned}
\tag{8.72}
$$

Step 3   For $k = 0$, set $\mathbf{P}_{0|0} = \mathbf{P}_0$, $\hat{\mathbf{x}}_{0|0} = \hat{\mathbf{x}}_0$.

Step 4 $k = k + 1$.

Step 5   State estimate propagation:

$$\hat{\mathbf{x}}_{k|k-1} = \boldsymbol{\Phi}_{k-1}\hat{\mathbf{x}}_{k-1|k-1} + \mathbf{u}_{k-1}. \tag{8.73}$$

Step 6   Error covariance propagation:

$$\mathbf{P}_{k|k-1} = \boldsymbol{\Phi}_{k-1}\mathbf{P}_{k-1|k-1}\boldsymbol{\Phi}_{k-1}^{\mathrm{T}} + \boldsymbol{\Gamma}_{k-1}\mathbf{Q}_{k-1}\boldsymbol{\Gamma}_{k-1}^{\mathrm{T}}. \tag{8.74}$$

Step 7   Kalman gain matrix:

$$\mathbf{K}_k = \mathbf{P}_{k|k-1}\mathbf{H}_k^{\mathrm{T}}\left(\mathbf{H}_k\mathbf{P}_{k|k-1}\mathbf{H}_k^{\mathrm{T}} + \mathbf{R}_k\right)^{-1}. \tag{8.75}$$

Step 8   State estimate update:

$$\hat{\mathbf{x}}_{k|k} = \hat{\mathbf{x}}_{k|k-1} + \mathbf{K}_k \left(\mathbf{z}_k - \hat{\mathbf{z}}_{k|k-1} \right) \tag{8.76}$$

where $\hat{\mathbf{z}}_{k|k-1} = \mathbf{H}_k \hat{\mathbf{x}}_{k|k-1}$.

Step 9   Error covariance update:

$$\mathbf{P}_{k|k} = (\mathbf{I}_n - \mathbf{K}_k \mathbf{H}_k)\, \mathbf{P}_{k|k-1}. \tag{8.77}$$

Step 10   Go back to Step 4.

The Kalman filter is a kind of recursive algorithm in time domain. It can estimate the current state according to the previous estimate and the current measurement. Moreover, it does not need to store a large amount of prior data, and it is easy to realize with a computer. The essence of the estimation algorithm is to minimize the trace of state estimate error matrix $\mathbf{P}_{k|k}$. The structure of the Kalman filter algorithm is plotted in Fig. 8.1. In practice, how to determine the initial state $\hat{\mathbf{x}}_0$ and the initial variance matrix $\mathbf{P}_0$ has to be considered. The value of the initial state $\hat{\mathbf{x}}_0$ is often obtained by experience, but $\mathbf{P}_0$ cannot. Instead, it needs to be obtained through statistical methods with several initial measurements. Fortunately, the stable estimate result will not depend on the choice of the initial value of $\hat{\mathbf{x}}_0$ and $\mathbf{P}_0$ if the filtering algorithm is stable.

In order to better understand the Kalman filter, some remarks are given as follows:

(1) It is noticed that the error covariance matrix $\mathbf{P}_{k|k}$ can be obtained using the filter, which represents the estimation accuracy. Also, it can be used to evaluate the health of sensors.

(2) Generally speaking, if a reasonable sampling period is adopted and the continuous-time system is observable, then the corresponding discrete-time system is observable as well. On the other hand, the system may also lose controllability and observability when an improper sampling period is adopted. So, it is necessary to check the observability of the discrete system after sampling.

(3) The matrix $\mathbf{H}_k \mathbf{P}_{k|k-1} \mathbf{H}_k^{\mathrm{T}} + \mathbf{R}_k$ needs to be non-singular. Otherwise, the solution expressed by $\mathbf{K}_k = \mathbf{P}_{k|k-1} \mathbf{H}_k^{\mathrm{T}} \left(\mathbf{H}_k \mathbf{P}_{k|k-1} \mathbf{H}_k^{\mathrm{T}} + \mathbf{R}_k \right)^{-1}$ does not make sense. In general, $\mathbf{R}_k$ is required to be non-singular, i.e., the measurements must always contain noise, so that $\mathbf{H}_k \mathbf{P}_{k|k-1} \mathbf{H}_k^{\mathrm{T}} + \mathbf{R}_k$ is non-singular.

(4) If the system $(\boldsymbol{\Phi}_{k-1}, \mathbf{H}_k)$ is unobservable, then $\mathrm{rank}\,\mathscr{O}\,(\boldsymbol{\Phi}_{k-1}, \mathbf{H}_k) \neq n$. The filter will also work without causing numerical problems. Only, the unobservable mode will not be corrected. In an extreme case, the whole system is completely unobservable if $\mathbf{H}_k = \mathbf{0}_{m \times n}$. Then, the filter gain $\mathbf{K}_k = \mathbf{0}_{n \times m}$. As a result, the Kalman filter degenerates as

$$\begin{aligned} \hat{\mathbf{x}}_{k|k} &= \boldsymbol{\Phi}_{k-1} \hat{\mathbf{x}}_{k-1|k-1} + \mathbf{u}_{k-1} \\ \mathbf{P}_{k|k} &= \boldsymbol{\Phi}_{k-1} \mathbf{P}_{k-1|k-1} \boldsymbol{\Phi}_{k-1}^{\mathrm{T}} + \boldsymbol{\Gamma}_{k-1} \mathbf{Q}_{k-1} \boldsymbol{\Gamma}_{k-1}^{\mathrm{T}}. \end{aligned} \tag{8.78}$$

Fig. 8.1 Structure of the Kalman filter algorithm

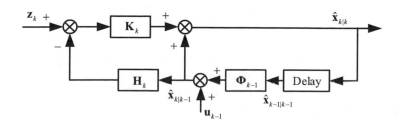

8.2.4   Multirate Kalman Filter

The above sections are about the classical Kalman filter, i.e., the so-called *single rate sampled system*. Sampled systems are the discretization of the actual systems in continuous time. The systems are called *multirate sampled systems* when each sampler or holder has a different sampling period. The multirate Kalman filter algorithm in a multirate sampling system is an important way to fuse multisource observations [9, 10]. According to the type of sampling interval applied, the multirate system can be divided into two categories, i.e., the one with uniform and non-uniform sampling. For a uniform sampling system, all the sampling periods is assumed to be an integer multiple of the basic sampling period, so there exists a least common multiple. On the other hand, as for a non-uniform sampling system, it is not cyclical for the uneven distribution of the data on the timeline. Compared with uniform sampling problem, the non-uniform sampling problem is more general and complex. For simplicity, in practice, the non-uniform sampling problem can be approximated into a uniform sampling problem.

A class of continuous-time linear systems under the uniform sampling considered here is the same as system (8.38) except that the measurement model is switched according to the sensors adopted. For simplicity, it is assumed that the basic sampling period is T_0, and two different sensors' sampling periods are T_i ($i = 1, 2$) which are integer multiples of T_0, namely $T_i = n_i T_0, n_i \in \mathbb{Z}_+, i = 1, 2$. Their measurement matrices are $\mathbf{H}_{ik} \in \mathbb{R}^{m_i \times n}$, and noise covariance matrices are $\mathbf{R}_{ik} \in \mathbb{R}^{m_i \times m_i}, i = 1, 2$, respectively. Moreover, αT_0 is the least common multiple of $T_i, i = 1, 2$, where $\alpha \in \mathbb{Z}_+$. When there is no observation data, let $\mathbf{H}_k = \mathbf{0}_{1 \times n}$ and $\mathbf{R}_k = 1$ in order to avoid the singularity without loss of generality. So the observation matrix $\mathbf{H}_k$ and noise covariance matrix $\mathbf{R}_k$ are changed according to

$$\mathbf{H}_k \triangleq \begin{cases} \mathbf{H}_{ik}, \text{ if } \mathrm{mod}\,(k, n_i) = 0\, \&\, \mathrm{mod}\,(k, \alpha) \neq 0 \\ \begin{bmatrix} \mathbf{H}_{1k} \\ \mathbf{H}_{2k} \end{bmatrix}, \text{ if } \mathrm{mod}\,(k, \alpha) = 0 \\ \mathbf{0}_{1 \times n}, \text{else} \end{cases} \tag{8.79}$$

and

$$\mathbf{R}_k \triangleq \begin{cases} \mathbf{R}_{ik}, \text{if } \mathrm{mod}\,(k, n_i) = 0\, \&\, \mathrm{mod}\,(k, \alpha) \neq 0 \\ \mathrm{diag}\,(\mathbf{R}_{1k}, \mathbf{R}_{2k})\,, \text{if } \mathrm{mod}\,(k, \alpha) = 0 \\ 1, \text{else} \end{cases} \tag{8.80}$$

where the expression $\mathrm{mod}\,(a, b)$ is the complementary operations, namely the remainder that a is divided by b. Then, the output is

$$\mathbf{z}_k \triangleq \begin{cases} \mathbf{z}_{ik}, \text{if } \mathrm{mod}\,(k, n_i) = 0\, \&\, \mathrm{mod}\,(k, \alpha) \neq 0 \\ \begin{bmatrix} \mathbf{z}_{1k} \\ \mathbf{z}_{2k} \end{bmatrix}, \text{ if } \mathrm{mod}\,(k, \alpha) = 0 \\ 0, \text{else.} \end{cases} \tag{8.81}$$

In the whole process, the dimensions of the observation matrix $\mathbf{H}_k$, observation noise covariance matrix $\mathbf{R}_k$, and observation $\mathbf{z}_k$ are changed, whereas the dimensions of $\hat{\mathbf{x}}_{k|k}$ and $\mathbf{P}_{k|k}$ are not changed and their values are updated at each step. And the prediction and update process is the same as the classical Kalman fliter.

8.3   Extended Kalman Filter

Kalman filter is an efficient recursive filter that estimates the internal state of a linear dynamical system from a series of noisy measurements. But the drawback of the Kalman filter is that the noise is assumed to be with Gaussian distribution and the algorithm is only applicable to linear systems. In fact, a larger class of estimation problems are involved in nonlinear systems. Thus, the state estimation for nonlinear systems is considerably more difficult and admits a wide variety of solutions than the linear problem. There are many possible ways to produce a linearized version of the Kalman filter. In this section, the most common approach, namely the EKF, is considered [11]. The EKF, though not precisely optimum, has been successfully applied to many nonlinear systems over past many years.

8.3.1   Basic Principle

The main idea of EKF is the linearization of nonlinear functions, which ignores the higher-order terms. The nonlinear problem is transformed into a linear problem through Taylor expansion and first-order linear truncation. Since the processing of linearization will cause additional error, the EKF is a suboptimal filter.

8.3.2   Theoretical Derivation

The following general nonlinear system is first considered, described by

$$
\begin{aligned}
\mathbf{x}_k &= \mathbf{f}\left(\mathbf{x}_{k-1}, \mathbf{u}_{k-1}, \mathbf{w}_{k-1}\right) \\
\mathbf{z}_k &= \mathbf{h}\left(\mathbf{x}_k, \mathbf{v}_k\right)
\end{aligned}
\tag{8.82}
$$

where the functions are defined as $\mathbf{f} : \mathbb{R}^n \times \mathbb{R}^m \times \mathbb{R}^n \to \mathbb{R}^n$ and $\mathbf{h} : \mathbb{R}^n \times \mathbb{R}^m \to \mathbb{R}^m$; the random vector $\mathbf{w}_{k-1}$ captures uncertainties in the system model and $\mathbf{v}_k$ denotes the measurement noise, both of which are temporally uncorrelated (white noise), zero-mean random sequences with covariance matrices $\mathbf{Q}_k$ and $\mathbf{R}_k$, respectively. Their properties can be found in Eq. (8.39) in Sect. 8.2.3.1. In the derivation of the EKF, $\mathbf{f}\left(\mathbf{x}_{k-1}, \mathbf{u}_{k-1}, \mathbf{w}_{k-1}\right)$ and $\mathbf{h}\left(\mathbf{x}_k, \mathbf{v}_k\right)$ are expanded by Taylor expansion. Assume that an optimal estimate $\hat{\mathbf{x}}_{k-1|k-1}$ with covariance matrix $\mathbf{P}_{k-1|k-1}$ at time $k-1$ is obtained. By ignoring the higher-order terms, $\mathbf{f}\left(\mathbf{x}_{k-1}, \mathbf{u}_{k-1}, \mathbf{w}_{k-1}\right)$ is expanded to Taylor series at the state $\hat{\mathbf{x}}_{k-1|k-1}$ that

$$
\begin{aligned}
\mathbf{f}\left(\mathbf{x}_{k-1}, \mathbf{u}_{k-1}, \mathbf{w}_{k-1}\right) = \ &\mathbf{f}(\hat{\mathbf{x}}_{k-1|k-1}, \mathbf{u}_{k-1}, \mathbf{0}_{n\times 1}) \\
&+ \left.\frac{\partial \mathbf{f}(\mathbf{x}, \mathbf{u}_{k-1}, \mathbf{w})}{\partial \mathbf{x}}\right|_{\mathbf{x}=\hat{\mathbf{x}}_{k-1|k-1}, \mathbf{w}=\mathbf{0}_{n\times 1}} \left(\mathbf{x}_{k-1} - \hat{\mathbf{x}}_{k-1|k-1}\right) \\
&+ \left.\frac{\partial \mathbf{f}(\mathbf{x}, \mathbf{u}_{k-1}, \mathbf{w})}{\partial \mathbf{w}}\right|_{\mathbf{x}=\hat{\mathbf{x}}_{k-1|k-1}, \mathbf{w}=\mathbf{0}_{n\times 1}} \mathbf{w}_{k-1}.
\end{aligned}
\tag{8.83}
$$

Similarly,

$$
\begin{aligned}
\mathbf{h}\left(\mathbf{x}_k, \mathbf{v}_k\right) = \ &\mathbf{h}\left(\hat{\mathbf{x}}_{k|k-1}, \mathbf{0}_{m\times 1}\right) \\
&+ \left.\frac{\partial \mathbf{h}(\mathbf{x}, \mathbf{v})}{\partial \mathbf{x}}\right|_{\mathbf{x}=\hat{\mathbf{x}}_{k|k-1}, \mathbf{v}=\mathbf{0}_{m\times 1}} \left(\mathbf{x}_k - \hat{\mathbf{x}}_{k|k-1}\right) \\
&+ \left.\frac{\partial \mathbf{h}(\mathbf{x}, \mathbf{v})}{\partial \mathbf{v}}\right|_{\mathbf{x}=\hat{\mathbf{x}}_{k|k-1}, \mathbf{v}=\mathbf{0}_{m\times 1}} \mathbf{v}_k.
\end{aligned}
\tag{8.84}
$$

It is noticed that $\mathbf{h}\left(\mathbf{x}_k, \mathbf{v}_k\right)$ is linearized at $\hat{\mathbf{x}}_{k|k-1}$ instead of $\hat{\mathbf{x}}_{k-1|k-1}$. In order to simplify the expression of the EKF, the following notation is defined

$$\begin{aligned}
\mathbf{\Phi}_{k-1} &\triangleq \left.\frac{\partial \mathbf{f}(\mathbf{x},\mathbf{u}_{k-1},\mathbf{w})}{\partial \mathbf{x}}\right|_{\mathbf{x}=\hat{\mathbf{x}}_{k-1|k-1},\mathbf{w}=\mathbf{0}_{n\times 1}} \\
\mathbf{H}_k &\triangleq \left.\frac{\partial \mathbf{h}(\mathbf{x},\mathbf{v})}{\partial \mathbf{x}}\right|_{\mathbf{x}=\hat{\mathbf{x}}_{k|k-1},\mathbf{v}=\mathbf{0}_{m\times 1}} \\
\mathbf{\Gamma}_{k-1} &\triangleq \left.\frac{\partial \mathbf{f}(\mathbf{x},\mathbf{u}_{k-1},\mathbf{w})}{\partial \mathbf{w}}\right|_{\mathbf{x}=\hat{\mathbf{x}}_{k-1|k-1},\mathbf{w}=\mathbf{0}_{n\times 1}} \\
\mathbf{u}'_{k-1} &\triangleq \mathbf{f}(\hat{\mathbf{x}}_{k-1|k-1},\mathbf{u}_{k-1},\mathbf{0}_{n\times 1}) - \mathbf{\Phi}_{k-1}\hat{\mathbf{x}}_{k-1|k-1} \\
\mathbf{z}'_k &\triangleq \mathbf{z}_k - \mathbf{h}\left(\hat{\mathbf{x}}_{k|k-1},\mathbf{0}_{m\times 1}\right) + \mathbf{H}_k\hat{\mathbf{x}}_{k|k-1} \\
\mathbf{v}'_k &\triangleq \left.\frac{\partial \mathbf{h}(\mathbf{x},\mathbf{v})}{\partial \mathbf{v}}\right|_{\mathbf{x}=\hat{\mathbf{x}}_{k|k-1},\mathbf{v}=\mathbf{0}_{m\times 1}} \mathbf{v}_k.
\end{aligned} \tag{8.85}$$

Based on the definitions above, the simplified system model becomes

$$\begin{aligned}
\mathbf{x}_k &= \mathbf{\Phi}_{k-1}\mathbf{x}_{k-1} + \mathbf{u}'_{k-1} + \mathbf{\Gamma}_{k-1}\mathbf{w}_{k-1} \\
\mathbf{z}'_k &= \mathbf{H}_k\mathbf{x}_k + \mathbf{v}'_k.
\end{aligned} \tag{8.86}$$

Here, the statistical property of $\mathbf{v}'_k$ is $\mathrm{E}\left(\mathbf{v}'_k\right) = \mathbf{0}_{m\times 1}$ and

$$\mathbf{R}_{\mathbf{v}'\mathbf{v}'}(k,j) \triangleq \mathrm{E}\left(\mathbf{v}'_k \mathbf{v}'^{\mathrm{T}}_j\right) = \left\{ \begin{array}{ll} \mathbf{R}'_k, & k=j \\ \mathbf{0}_{m\times m}, & k\neq j \end{array}\right.$$

where

$$\mathbf{R}'_k \triangleq \left.\frac{\partial \mathbf{h}(\mathbf{x},\mathbf{v})}{\partial \mathbf{v}}\right|_{\mathbf{x}=\hat{\mathbf{x}}_{k|k-1},\mathbf{v}=\mathbf{0}_{m\times 1}} \mathbf{R}_k \left(\left.\frac{\partial \mathbf{h}(\mathbf{x},\mathbf{v})}{\partial \mathbf{v}}\right|_{\mathbf{x}=\hat{\mathbf{x}}_{k|k-1},\mathbf{v}=\mathbf{0}_{m\times 1}}\right)^{\mathrm{T}}. \tag{8.87}$$

Then, the EKF algorithm is summarized as follows:

Step 1. Process model:

$$\mathbf{x}_k = \mathbf{f}\left(\mathbf{x}_{k-1},\mathbf{u}_{k-1},\mathbf{w}_{k-1}\right), \quad \mathbf{w}_k \sim \mathcal{N}\left(\mathbf{0}_{n\times 1},\mathbf{Q}_k\right) \tag{8.88}$$

Measurement model:

$$\mathbf{z}_k = \mathbf{h}\left(\mathbf{x}_k,\mathbf{v}_k\right), \quad \mathbf{v}_k \sim \mathcal{N}\left(\mathbf{0}_{m\times 1},\mathbf{R}_k\right) \tag{8.89}$$

where $\mathbf{w}_{k-1}$ and $\mathbf{v}_k$ are independent, zero-mean, Gaussian noise processes with covariance matrices being $\mathbf{Q}_k$ and $\mathbf{R}_k$, respectively. The definitions of $\mathbf{\Phi}_{k-1}, \mathbf{H}_k, \mathbf{\Gamma}_{k-1}, \mathbf{z}'_k$ are given as in Eq. (8.85).

Step 2. Initial state:

$$\begin{aligned}
\hat{\mathbf{x}}_0 &= \mathrm{E}\left(\mathbf{x}_0\right) \\
\mathbf{P}_0 &= \mathrm{E}\left[\left(\mathbf{x}_0 - \mathrm{E}\left(\mathbf{x}_0\right)\right)\left(\mathbf{x}_0 - \mathrm{E}\left(\mathbf{x}_0\right)\right)^{\mathrm{T}}\right].
\end{aligned} \tag{8.90}$$

Step 3. For $k=0$, set $\mathbf{P}_{0|0} = \mathbf{P}_0$, $\hat{\mathbf{x}}_{0|0} = \hat{\mathbf{x}}_0$.

Step 4. $k = k+1$.

Step 5. State estimate propagation:

$$\hat{\mathbf{x}}_{k|k-1} = \mathbf{f}\left(\hat{\mathbf{x}}_{k-1|k-1},\mathbf{u}_{k-1},\mathbf{0}_{n\times 1}\right). \tag{8.91}$$

Step 6. Error covariance propagation:

$$\mathbf{P}_{k|k-1} = \mathbf{\Phi}_{k-1}\mathbf{P}_{k-1|k-1}\mathbf{\Phi}^{\mathrm{T}}_{k-1} + \mathbf{\Gamma}_{k-1}\mathbf{Q}_{k-1}\mathbf{\Gamma}^{\mathrm{T}}_{k-1}. \tag{8.92}$$

Step 7. Kalman gain matrix:

$$\mathbf{K}_k = \mathbf{P}_{k|k-1}\mathbf{H}_k^{\mathrm{T}}\left(\mathbf{H}_k\mathbf{P}_{k|k-1}\mathbf{H}_k^{\mathrm{T}} + \mathbf{R}_k'\right)^{-1} \tag{8.93}$$

where $\mathbf{R}_k'$ is defined in (8.87).
Step 8. State estimate update:

$$\hat{\mathbf{x}}_{k|k} = \hat{\mathbf{x}}_{k|k-1} + \mathbf{K}_k\left(\mathbf{z}_k' - \hat{\mathbf{z}}_{k|k-1}\right). \tag{8.94}$$

where $\hat{\mathbf{z}}_{k|k-1} = \mathbf{H}_k\hat{\mathbf{x}}_{k|k-1}$.
Step 9. Error covariance update:

$$\mathbf{P}_{k|k} = \left(\mathbf{I}_n - \mathbf{K}_k\mathbf{H}_k\right)\mathbf{P}_{k|k-1}. \tag{8.95}$$

Step 10. Go back to Step 4.

8.3.3   Implicit Extended Kalman Filter

In some situations, the measurement model of the nonlinear system is the implicit function as

$$\mathbf{0}_{m\times1} = \mathbf{h}\left(\mathbf{x}_k, \mathbf{z}_k, \mathbf{v}_k\right), \quad \mathbf{v}_k \sim \mathcal{N}\left(\mathbf{0}_{m\times1}, \mathbf{R}_k\right). \tag{8.96}$$

In this case, the process of the IEKF can be summarized as follows:
Step 1. State estimation propagation:

$$\hat{\mathbf{x}}_{k|k-1} = \mathbf{f}\left(\hat{\mathbf{x}}_{k-1|k-1}, \mathbf{u}_{k-1}, \mathbf{0}_{n\times1}\right). \tag{8.97}$$

Step 2. Error covariance propagation:

$$\mathbf{P}_{k|k-1} = \mathbf{\Phi}_{k-1}\mathbf{P}_{k-1|k-1}\mathbf{\Phi}_{k-1}^{\mathrm{T}} + \mathbf{\Gamma}_{k-1}\mathbf{Q}_{k-1}\mathbf{\Gamma}_{k-1}^{\mathrm{T}}. \tag{8.98}$$

Step 3. Kalman gain matrix:

$$\mathbf{K}_k = \mathbf{P}_{k|k-1}\mathbf{H}_k'^{\mathrm{T}}\left(\mathbf{H}_k'\mathbf{P}_{k|k-1}\mathbf{H}_k'^{\mathrm{T}} + \mathbf{R}_k''\right)^{-1} \tag{8.99}$$

where

$$\mathbf{H}_k' \triangleq \left.\frac{\partial\mathbf{h}(\mathbf{x},\mathbf{z},\mathbf{v})}{\partial\mathbf{x}}\right|_{\mathbf{x}=\ddot{\mathbf{x}}_{k|k-1},\mathbf{z}=\mathbf{z}_k,\mathbf{v}=\mathbf{0}_{m\times1}}$$

$$\mathbf{R}_k'' \triangleq \left(\left.\frac{\partial\mathbf{h}(\mathbf{x},\mathbf{z},\mathbf{v})}{\partial\mathbf{z}}\right|_{\mathbf{x}=\hat{\mathbf{x}}_{k|k-1},\mathbf{z}=\mathbf{z}_k,\mathbf{v}=\mathbf{0}_{m\times1}}\right)\mathbf{R}_k\left(\left.\frac{\partial\mathbf{h}(\mathbf{x},\mathbf{z},\mathbf{v})}{\partial\mathbf{z}}\right|_{\mathbf{x}=\hat{\mathbf{x}}_{k|k-1},\mathbf{z}=\mathbf{z}_k,\mathbf{v}=\mathbf{0}_{m\times1}}\right)^{\mathrm{T}}.$$

Step 4. State estimate update:

$$\hat{\mathbf{x}}_{k|k} = \hat{\mathbf{x}}_{k|k-1} - \mathbf{K}_k\mathbf{h}\left(\hat{\mathbf{x}}_{k|k-1}, \mathbf{z}_k, \mathbf{0}_{m\times1}\right). \tag{8.100}$$

Step 5. Error covariance update:

$$\mathbf{P}_{k|k} = \left(\mathbf{I}_n - \mathbf{K}_k \mathbf{H}'_k\right) \mathbf{P}_{k|k-1} \left(\mathbf{I}_n - \mathbf{K}_k \mathbf{H}'_k\right)^{\mathrm{T}} + \mathbf{K}_k \mathbf{R}''_k \mathbf{K}_k^{\mathrm{T}}. \tag{8.101}$$

Step 6. Go back to Step 1.

The detailed derivation is stated in [12]. This is left as an exercise at the end of this chapter.

8.4 Summary

The concept of observability as well as how to evaluate the observability of linear and nonlinear systems is introduced in this chapter. Through these theories, it can be known whether the system state can be estimated based on the system model and the sensors. Without observability, the Kalman filter will not make sense. In practice, how to choose and place sensors often relies on experience. This problem could be solved by the *degree of observability* [13]. In the research of Kalman filters, more practical problems should be considered, such as how to reduce the calculation time for the model with a high dimension, how to reduce the dependence on models or noise characteristics, and how to work in the presence of unknown time delays or non-uniform sampling.

Exercises

8.1 Show the observability of the bilinear system in Example 8.7 based on Theorem 8.9.

8.2 Fill in the steps of the derivation of Eqs. (8.26)–(8.30).

8.3 Design a Kalman filter to estimate the value of a constant $x \in \mathbb{R}$, given discrete measurements of $x(k)$ corrupted by an uncorrelated gaussian noise sequence with zero mean and variance r_0.

8.4 Refer to [12] and give the detailed derivation of the IEKF.

8.5 Design a Kalman filter for a linear discrete-time-invariant system with time delay measurements:

$$\begin{aligned} \mathbf{x}_k &= \mathbf{\Phi}_{k-1}\mathbf{x}_{k-1} + \mathbf{u}_{k-1} + \mathbf{\Gamma}_{k-1}\mathbf{w}_{k-1} \\ \mathbf{z}_k &= \mathbf{H}_k\mathbf{x}_{k-d} + \mathbf{v}_k \end{aligned} \tag{8.102}$$

where $k \in \mathbb{Z}_+$ is the discrete time and $d \in \mathbb{Z}_+$ represents the measurement delay. The definition of $\mathbf{x}_k, \mathbf{z}_k, \mathbf{\Phi}_{k-1}, \mathbf{u}_{k-1}, \mathbf{\Gamma}_{k-1}, \mathbf{H}_k, \mathbf{w}_{k-1}, \mathbf{v}_k$ is the same as in system (8.38).

References

1. Chen CT (1999) Linear system theory and design, 3rd edn. Oxford University Press, New York
2. Banerjee S, Roy A (2014) Linear algebra and matrix analysis for statistics. CRC Press, London
3. Vidyasagar M (2002) Nonlinear systems analysis. Society for Industrial and Applied Mathematics, Philadelphia
4. Hermann R, Krener AJ (1977) Nonlinear controllability and observability. IEEE Trans Autom Control 22(5): 728–740
5. Leishman RC, Macdonald JC, Beard RW, McLain TW (2014) Quadrotors and accelerometers: state estimation with an improved dynamic model. IEEE Control Syst Mag 34(1):28–41

6. Kalman RE (1960) A new approach to linear filtering and prediction problems. J Basic Eng-T ASME 82(1):35–45
7. Gelb A. (Ed.) (1974) Applied optimal estimation. MIT Press, Cambridge, MA
8. Crassidis JL, Junkins JL (2011) Optimal estimation of dynamic systems. CRC Press, Boca Raton
9. Cristi R, Tummala M (2000) Multirate, multiresolution, recursive Kalman filter. Signal Process 80(9):1945–1958
10. Sun SL, Deng ZL (2004) Multi-sensor optimal information fusion Kalman filter. Automatica 40(6):1017–1023
11. Simon D (2006) Optimal state estimation: Kalman, H infinity, and nonlinear approaches. Wiley, New Jersey
12. Soatto S, Frezza R, Perona P (1996) Motion estimation via dynamic vision. IEEE Trans Autom Control 41(3): 393–413
13. Müller PC, Weber HI (1972) Analysis and optimization of certain qualities of controllability and observability for linear dynamical systems. Automatica 8(3):237–246

State Estimation

9

The state estimation is very important as it is the base for control and decision-making. Sensors of multicopters are like sensory organs of a human being, which can provide the necessary information. In order to reduce the cost of a multicopter, cheap sensors with poor precision are often used. With such kinds of sensors, some information such as acceleration, angular velocity and absolute position can be directly measured but subject to a lot of noise. Some information such as velocity, attitude angles, obstacle position may not be measured by sensors directly and requires to be estimated. Moreover, information measured by sensors is redundant. For instance, an accelerometer and a GPS receiver both contain information related to the position. Therefore, it is necessary to improve the accuracy and robustness of state estimation by fusing the redundant information of the sensors. This chapter aims to answer the question below:

How is the information from the multiple sensors fused?

The answer to this question involves the attitude estimation, position estimation, velocity estimation, and obstacle estimation.

© Springer Nature Singapore Pte Ltd. 2017

Q. Quan, *Introduction to Multicopter Design and Control*, DOI 10.1007/978-981-10-3382-7_9

Listen Makes You Enlightened

The ancient Chinese had already realized the importance of information integration. One proverb frequently cited illustrates this philosophy. "If you know the enemy and know yourself, you need not fear the result of a hundred battles (The Chinese is " 知己知彼，百战不殆", translated by Lionel Giles. (1910). Sun Tzu on the art of war: The oldest military treatise in the world. Champaign, IL: Project Gutenberg)". The book "*Comprehensive Mirror to Aid in Government*", recorded a conversation between Emperor Taizong of the Tang dynasty and his chancellor Zheng Wei. Zheng Wei responded to the emperor "Listen to both sides and you will be enlightened; heed only one side and you will be benighted (The Chinese is "兼听则明，偏信则暗")." This story emphasizes the necessity of information integration for a politician. Zhonglin Xu in the Ming dynasty wrote in his novel "*Creation of the Gods*" that "a general should constantly watch all directions and listen to everything around him (The Chinese is "为将之道：身临战场，务要眼观四处,耳听八方", translated by Zhizhong Gu (2000). Creation of the Gods. New world press)".

9.1 Attitude Estimation

The main purpose of attitude estimation is to estimate the attitude in the form of Euler angles, rotation matrix, or quaternions. Attitude information is often acquired by complementary filter or Kalman filter on data obtained by three-axis accelerometers, three-axis gyroscopes, and three-axis magnetometers. Dynamic response of a three-axis gyroscope is fast with high accuracy, but when the attitude is estimated through integration, the estimated error is prone to accumulate. Three-axis accelerometers and three-axis magnetometers can get stable attitude signals without cumulative error, but measurements are noisy and with poor dynamic response. In order to fuse these information, this section introduces three methods, i.e., the linear complementarity filter [1], nonlinear complementarity filter [2], and Kalman filter. Before describing these methods, the measuring principle of Euler angles is introduced first.

9.1.1 Measuring Principle

9.1.1.1 Pitch Angle and Roll Angle Measuring Principle

According to the aerodynamic drag model (6.33) in Chap. 6, by ignoring the cross term of velocity and angular velocity, the specific force $^b\mathbf{a}$ satisfies

$$\begin{bmatrix} a_{x_b} \\ a_{y_b} \end{bmatrix} = \begin{bmatrix} \dot{v}_{x_b} + g\sin\theta \\ \dot{v}_{y_b} - g\cos\theta\sin\phi \end{bmatrix} = \begin{bmatrix} -\frac{k_{drag}}{m}v_{x_b} \\ -\frac{k_{drag}}{m}v_{y_b} \end{bmatrix} \tag{9.1}$$

where $^b\mathbf{a} = [\,a_{x_b}\ a_{y_b}\ a_{z_b}\,]^T$. The Laplace transforms of $v_{x_b}, v_{y_b} \in \mathbb{R}$ are expressed as

$$v_{x_b}(s) = -\frac{g}{s+\frac{k_{drag}}{m}}(\sin\theta)(s)$$
$$v_{y_b}(s) = \frac{g}{s+\frac{k_{drag}}{m}}(\cos\theta\sin\phi)(s).$$

Furthermore, the Laplace transforms of $a_{x_b}, a_{y_b} \in \mathbb{R}$ are

$$\begin{bmatrix} a_{x_b}(s) \\ a_{y_b}(s) \end{bmatrix} = \begin{bmatrix} gH(s)(\sin\theta)(s) \\ -gH(s)(\cos\theta\sin\phi)(s) \end{bmatrix}$$

where $H(s) = \frac{k_{drag}}{m}\big/\left(s + \frac{k_{drag}}{m}\right)$ is a low-pass filter with $H(0) = 1$. Approximatively, the low-frequency specific forces along the o_bx_b and o_by_b axes are

$$a_{x_b} \approx g\sin\theta$$
$$a_{y_b} \approx -g\cos\theta\sin\phi.$$

The three-axis accelerometer is fixed to the multicopter, aligned with the Aircraft-Body Coordinate Frame (ABCF). Therefore, observation of low-frequency pitch and roll angle can be acquired by accelerometer measurement illustrated as

$$\theta_m = \arcsin\left(\frac{a_{x_bm}}{g}\right)$$
$$\phi_m = -\arcsin\left(\frac{a_{y_bm}}{g\cos\theta_m}\right) \tag{9.2}$$

where $^{b}\mathbf{a}_m = [\, a_{x_b m} \; a_{y_b m} \; a_{z_b m} \,]^{T}$ is the measurement from the accelerometer. Several further considerations are as follows:

(1) It is better to eliminate the slow time-varying drift of the accelerometer to get a more accurate angle.

(2) If the amplitude of the vibration is large, $a_{x_b m}$, $a_{y_b m}$ would be polluted by noise severely and further affect the estimation of θ_m, ϕ_m. Thus, the vibration damping is very important.

(3) The measuring method in Eq. (9.2) is applicable when the propeller disks are horizontal. In this case, the thrust f is parallel to the $o_b z_b$ axis without the component force along the $o_b x_b$, $o_b y_b$ axes. However, if the propellers tilt (Fig. 3.6 in Chap. 3), the thrust generated by propellers has component forces on all $o_b x_b$, $o_b y_b$, $o_b z_b$ axes. Then, the specific force is written as

$$\begin{bmatrix} a_{x_b} \\ a_{y_b} \end{bmatrix} = \begin{bmatrix} -\dfrac{k_{\mathrm{drag}}}{m} v_{x_b} - \dfrac{f_{x_b}}{m} \\ -\dfrac{k_{\mathrm{drag}}}{m} v_{y_b} - \dfrac{f_{y_b}}{m} \end{bmatrix} \tag{9.3}$$

where f_{x_b}, $f_{y_b} \in \mathbb{R}_+$ represent the component thrust along the $o_b x_b$, $o_b y_b$ axes, respectively. As a result, the low-frequency pitch angle and roll angle should be

$$\begin{aligned} \theta_m &= \arcsin\left(\frac{a_{x_b m} + \frac{f_{x_b}}{m}}{g} \right) \\ \phi_m &= -\arcsin\left(\frac{a_{y_b m} + \frac{f_{y_b}}{m}}{g \cos \theta_m} \right) \end{aligned} \tag{9.4}$$

where f_{x_b}, f_{y_b} are not easy to obtain. If θ_m, ϕ_m are still calculated by Eq. (9.2), then the measurements would have bias.

(4) Since $H(s)$ is considered as $H(s) \approx 1$ here, only low-frequency pitch and roll angle can be acquired. As indicated in [3], for a quadcopter with $m = 2.75$ kg and $k_{\mathrm{drag}} = 0.77$, the accelerometer would take more than 10 s to reach 95% of its steady state. Such a method cannot describe fast attitude changes. In Sect. 9.3.2, by taking blade flapping into consideration, a new measuring principle will be introduced to get more precise pitch angle and roll angle [3, 4].

9.1.1.2 Yaw Angle Measuring Principle

The Earth's magnetic field can be approximated with the dipole model as shown in Fig. 9.1. This figure illustrates that the Earth's field points down toward the north in the Northern Hemisphere and points up toward the north in the Southern Hemisphere. It is horizontal and points north at the equator. In all cases, the direction of the Earth's field is always pointing to the magnetic north, which is referred to as the Earth's magnetic pole position. But it differs from true, or geographic, north by about 11.5°. At different locations around the globe magnetic north, the true north can differ by 25°. This difference is called the *declination angle* and can be determined from a lookup table based on the geographic location. The key to accurately finding the true north is a two-step process [5]:

(1) First, determine the magnetic field direction in the horizontal plane of the vector and then obtain azimuth. Suppose that the magnetometer measurements are $^{b}\mathbf{m}_m = [\, m_{x_b} \; m_{y_b} \; m_{z_b} \,]^{T}$. Since a magnetometer is possibly not horizontally placed, the pitch/roll angles measured by a dual-axis tilt sensor are used to project the magnetometer measurement onto the horizontal plane as

$$\begin{aligned} \bar{m}_{x_e} &= m_{x_b} \cos \theta_m + m_{y_b} \sin \phi_m \sin \theta_m + m_{z_b} \cos \phi_m \sin \theta_m \\ \bar{m}_{y_e} &= m_{y_b} \cos \phi_m - m_{z_b} \sin \phi_m \end{aligned} \tag{9.5}$$

Fig. 9.1 Schematic diagram of Earth's magnetic field (photograph courtesy of Peter Reid from the University of Edinburgh)

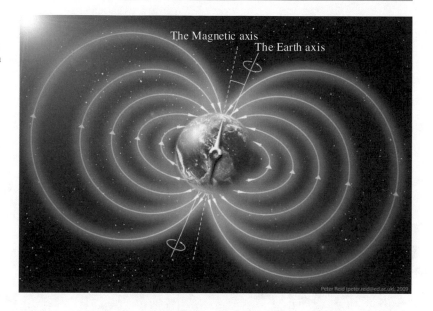

where $\bar{m}_{x_e}, \bar{m}_{y_e} \in \mathbb{R}$ are the horizontal projections of magnetometer readings. Let $\psi_{\text{mag}} \in [0, 2\pi]$. Then

$$\psi_{\text{mag}} = \begin{cases} \pi - \tan^{-1}\left(\frac{\bar{m}_{y_e}}{\bar{m}_{x_e}}\right), & \text{if } \bar{m}_{x_e} < 0 \\ 2\pi - \tan^{-1}\left(\frac{\bar{m}_{y_e}}{\bar{m}_{x_e}}\right), & \text{if } \bar{m}_{x_e} > 0, \bar{m}_{y_e} > 0 \\ -\tan^{-1}\left(\frac{\bar{m}_{y_e}}{\bar{m}_{x_e}}\right), & \text{if } \bar{m}_{x_e} > 0, \bar{m}_{y_e} < 0 \\ \pi/2, & \text{if } \bar{m}_{x_e} = 0, \bar{m}_{y_e} < 0 \\ 3\pi/2, & \text{if } \bar{m}_{x_e} = 0, \bar{m}_{y_e} > 0 \end{cases}.$$

If $\psi_{\text{mag}} \in [-\pi, \pi]$ is considered, then

$$\psi_{\text{mag}} = \arctan 2 \left(\bar{m}_{y_e}, \bar{m}_{x_e}\right). \tag{9.6}$$

If a multicopter turns clockwise, then the yaw angle is positive. The algorithm is realized by the electronic compass, which consists of a three-axis magnetometer, a dual-axis tilt sensor, and a micro-programmed control unit.

(2) Secondly, the yaw angle is corrected by adding or subtracting a *declination*. When multicopters are under the Semi-Autonomous Control (SAC) mode, the direction of the local magnetic field can be chosen as the $o_e x_e$ axis of the Earth-Fixed Coordinate Frame (EFCF), as shown in Fig. 9.2. Although the local magnetic direction is inclined to the west, it will have no effect on the control by remote pilots. During a mission under the Fully-Autonomous Control (FAC) mode, the $o_e x_e$ axis has to point to the north in order to be consistent with the latitude and longitude. Then, declination is required to correct the magnetic direction to the north direction. Figure 9.3 shows the contour map of 2015 world magnetic declination. For instance, the magnetic field orientation of Beijing is 6° west of north, and then, 6° are added to the magnetic field orientation to obtain the north direction.

Besides magnetometers, for a large size multicopter, a dual-antenna GPS receiver system can be used to estimate yaw angles measured by the two antennas on the multicopter head and rear, respectively. However, it is hard for a small-size multicopter to get a precise yaw angle limited by GPS measurement precision. Therefore, at present, yaw angles are mainly measured by magnetometers.

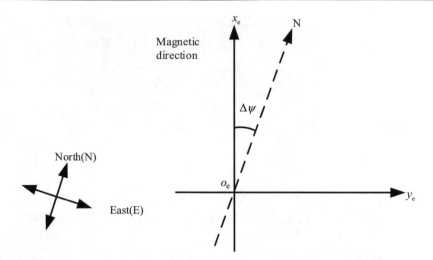

Fig. 9.2 Local magnetic direction and the north direction

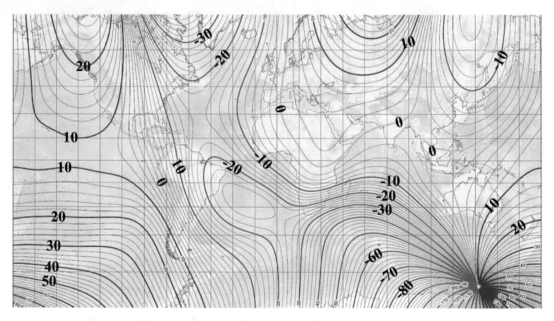

Fig. 9.3 2015 world magnetic field declination contour map (photo adapted from http://www.ngdc.noaa.gov/geomag/WMM/image.shtml)

9.1.2   Linear Complementary Filter

9.1.2.1 Model

According to Eq. (5.5) in Chap. 5, the attitude rates $\dot{\theta}$, $\dot{\phi}$, $\dot{\psi}$ and angular velocity $^b\boldsymbol{\omega}$ have the following relationship:

$$
\begin{bmatrix} \dot{\phi} \\ \dot{\theta} \\ \dot{\psi} \end{bmatrix} = \begin{bmatrix} 1 & \tan\theta\sin\phi & \tan\theta\cos\phi \\ 0 & \cos\phi & -\sin\phi \\ 0 & \sin\phi/\cos\theta & \cos\phi/\cos\theta \end{bmatrix} \begin{bmatrix} \omega_{x_b} \\ \omega_{y_b} \\ \omega_{z_b} \end{bmatrix}. \tag{9.7}
$$

Since multicopters often work under condition that θ, ϕ are small, the above equation is approximated as

$$\begin{bmatrix} \dot{\phi} \\ \dot{\theta} \\ \dot{\psi} \end{bmatrix} \approx \begin{bmatrix} \omega_{x_b} \\ \omega_{y_b} \\ \omega_{z_b} \end{bmatrix}. \tag{9.8}$$

In the previous section, it is shown that the attitude can be estimated by the accelerometers and magnetometers with large noise but small drifts. On the other hand, integrating the angular velocity will result in the attitude angle with small noise but large drifts. The basic idea of complementary filtering is to use their complementary characteristics to obtain more accurate attitude estimation. Here, the estimation of the pitch angle is taken as an example to describe the derivation of a linear complementarity filter in detail.

9.1.2.2 Pitch Angle

The Laplace transform of pitch angle $\theta \in \mathbb{R}$ is expressed as

$$\theta(s) = \frac{1}{\tau s + 1}\theta(s) + \frac{\tau s}{\tau s + 1}\theta(s) \tag{9.9}$$

where $1/(1 + \tau s)$ is the transfer function of a low-pass filter, $\tau \in \mathbb{R}_+$ is a time constant, and $\tau s/(1 + \tau s) = 1 - 1/(1 + \iota s)$ is the transfer function of a high-pass filter. Since the pitch angle obtained by an accelerometer has a large noise but a small drift, for simplicity, it is modeled as

$$\theta_m = \theta + n_\theta \tag{9.10}$$

where $n_\theta \in \mathbb{R}$ indicates high-frequency noise, and θ is the true value of the pitch angle. Considering that the pitch angle estimated by integrating angular velocity has a little noise but a large drift, the integration is modeled as

$$\frac{\omega_{y_b m}(s)}{s} = \theta(s) + c\frac{1}{s} \tag{9.11}$$

where $\omega_{y_b m}(s)/s$ is the Laplace transform of the integration of angular velocity $\omega_{y_b m}$, c/s represents the Laplace transform of the constant drift, and $\omega_{y_b m}$ is the gyroscope measurement. Therefore, for the pitch angle, the standard form of a linear complementary filter is expressed as

$$\hat{\theta}(s) = \frac{1}{\tau s + 1}\theta_m(s) + \frac{\tau s}{\tau s + 1}\left(\frac{1}{s}\omega_{y_b m}(s)\right). \tag{9.12}$$

Next, the explanation will be given why the linear complementary filter can obtain a more accurate attitude estimation. By using Eqs. (9.10) and (9.11), Eq. (9.12) becomes

$$\hat{\theta}(s) = \theta(s) + \left(\frac{1}{\tau s + 1}n_\theta(s) + \frac{\tau s}{\tau s + 1}c\frac{1}{s}\right) \tag{9.13}$$

where the high-frequency noise n_θ can be attenuated near zero after passing the low-pass filter $1/(1 + \tau s)$, and the low-frequency signal c/s can be eliminated after passing the high-pass filter $\tau s/(1 + \tau s)$. As a result, $\hat{\theta}(s) \approx \theta(s)$. During the process, the low-frequency filter keeps the advantage that θ_m has a small drift, while the high-frequency filter keeps the advantage that $\omega_{y_b m}(s)/s$ has a little noise. The process is shown as in Fig. 9.4.

Fig. 9.4 Structure of complementary filter

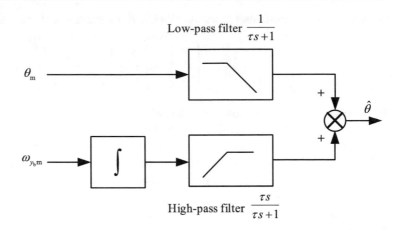

In order to realize the filter (9.12) with digital computers, the filter has to be transformed into a discrete-time differential form. For example, through the first-order backward difference [6, pp. 275–322], s is expressed as $s = \left(1 - z^{-1}\right)/T_s$, where $T_s \in \mathbb{R}_+$ is the filtering sampling period. Then, Eq. (9.12) becomes

$$\hat{\theta}(z) = \frac{1}{\tau \frac{1-z^{-1}}{T_s} + 1}\theta_m(z) + \frac{\tau}{\tau \frac{1-z^{-1}}{T_s} + 1}\omega_{y_b m}(z). \tag{9.14}$$

The above equation is further transformed into a discrete-time difference form as

$$\hat{\theta}(k) = \frac{\tau}{\tau + T_s}\left(\hat{\theta}(k-1) + T_s\omega_{y_b m}(k)\right) + \frac{T_s}{\tau + T_s}\theta_m(k). \tag{9.15}$$

If $\tau/(\tau + T_s) = 0.95$, then $T_s/(\tau + T_s) = 0.05$. In this way, the complementary filter for the pitch angle is

$$\hat{\theta}(k) = 0.95\left(\hat{\theta}(k-1) + T_s\omega_{y_b m}(k)\right) + 0.05\theta_m(k). \tag{9.16}$$

An experiment is performed, where a Pixhawk is used for measurement and the complementary filter (9.16) is used to estimate the pitch angle. Results are shown in Fig. 9.5. It is observed that the pitch angle estimated by the complementary filter is correct and smooth, while that derived by integrating the angular velocity is divergent. Except for the first-order backward difference method, there are also other difference methods, such as bilinear transformation (also known as Tustin's method).

9.1.2.3 Roll Angle
Similarly, the complementary filter of the roll angle is as follows:

$$\hat{\phi}(k) = \frac{\tau}{\tau + T_s}\left(\hat{\phi}(k-1) + T_s\omega_{x_b m}(k)\right) + \frac{T_s}{\tau + T_s}\phi_m(k). \tag{9.17}$$

Let $\tau/(\tau + T_s) = 0.95$, and $T_s/(\tau + T_s) = 0.05$. Then, Eq. (9.17) becomes

$$\hat{\phi}(k) = 0.95\left(\hat{\phi}(k-1) + T_s\omega_{x_b m}(k)\right) + 0.05\phi_m(k). \tag{9.18}$$

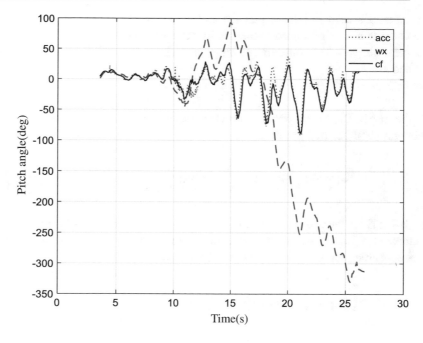

Fig. 9.5 Estimation of pitch angle by a linear complementary filter. In this figure, acc, wx, and cf. denote the pitch angle estimated by acceleration, angular velocity integration and complementary filter, respectively

9.1.2.4 Yaw Angle

Yaw angle can be measured by both GPS receiver and electronic compass, denoted by ψ_{GPS} and ψ_{mag}, respectively. A simple method of obtaining ψ_m is to sum the weighted measurement of the two sensors, written as

$$\psi_m = \left(1 - \alpha_\psi\right)\psi_{GPS} + \alpha_\psi\psi_{mag} \tag{9.19}$$

where $\alpha_\psi \in [0, 1]$ is the weighting factor. Since the sampling rates of an electronic compass and a gyroscope are often higher than that of a GPS receiver, the yaw angle can be obtained by

$$\psi_m(k) = \begin{cases} \left(1 - \alpha_\psi\right)\psi_{GPS}(k) + \alpha_\psi\psi_{mag}(k), & \text{when } \psi_{GPS} \text{ is updated} \\ \psi_{mag}(k), & \text{else.} \end{cases} \tag{9.20}$$

After getting ψ_m, the yaw angle is estimated by

$$\hat{\psi}(k) = \frac{\tau}{\tau + T_s}\left(\hat{\psi}(k-1) + T_s\omega_{z_bm}(k)\right) + \frac{T_s}{\tau + T_s}\psi_m(k). \tag{9.21}$$

Let $\tau/(\tau + T_s) = 0.95$, and $T_s/(\tau + T_s) = 0.05$. Then, the complementary filter of the yaw angle is

$$\hat{\psi}(k) = 0.95\left(\hat{\psi}(k-1) + T_s\omega_{z_bm}(k)\right) + 0.05\psi_m(k). \tag{9.22}$$

9.1.3   Nonlinear Complementary Filter

The idea of nonlinear complementary filters is similar to that of linear complementary filters. They both benefit from the advantage of complementary characteristics of accelerometers and gyroscopes. The difference is that nonlinear complementary filters are based on a nonlinear relationship between the angular velocity and the angle of rotation.

The notations used in this chapter are as follows. $\mathbf{R} \in SO(3)$ denotes a real rotation matrix, $\hat{\mathbf{R}} \in SO(3)$ is a rotation matrix estimated by the complementary filter, $\mathbf{R}_m \in SO(3)$ is a rotation matrix measured by accelerometers and magnetometers, and $\tilde{\mathbf{R}} \in SO(3)$ is the error between $\mathbf{R}_m$ and $\hat{\mathbf{R}}$, defined as

$$\tilde{\mathbf{R}} = \hat{\mathbf{R}}^T \mathbf{R}_m \tag{9.23}$$

where the matrix $\mathbf{R}_m$ can be obtained according to Eq. (5.9) in Chap. 5 and the $(\theta_m, \phi_m, \psi_m)$, or by the following optimization problem

$$\mathbf{R}_m = \arg \min_{\mathbf{R} \in SO(3)} \left(\lambda_1 \left\| \mathbf{e}_3 - \mathbf{R} \frac{^b\mathbf{a}_m}{\|^b\mathbf{a}_m\|} \right\| + \lambda_2 \left\| \mathbf{v}_m^* - \mathbf{R} \frac{^b\mathbf{m}_m}{\|^b\mathbf{m}_m\|} \right\| \right) \tag{9.24}$$

where $^b\mathbf{a}_m \in \mathbb{R}^3$ is the reading of a three-axis accelerometer, $^b\mathbf{m}_m \in \mathbb{R}^3$ is the reading of a three-axis magnetometer, $\mathbf{v}_m^* \in \mathbb{R}^3$ is the known direction of local magnetic field, and $\lambda_1, \lambda_2 \geq 0$ are adjustable weights depending on the reliability of devices. Define

$$\mathbf{x} \triangleq [\, x_1 \ x_2 \ x_3 \,]^T \in \mathbb{R}^3, [\mathbf{x}]_\times \triangleq \begin{bmatrix} 0 & -x_3 & x_2 \\ x_3 & 0 & -x_1 \\ -x_2 & x_1 & 0 \end{bmatrix}, \text{vex}\left([\mathbf{x}]_\times\right) \triangleq \mathbf{x}. \tag{9.25}$$

Then, according to the theory of nonlinear complementary filters, the rotation matrix is filtered out by the following equation

$$\dot{\hat{\mathbf{R}}} = \left[\mathbf{R}_m\,^b\boldsymbol{\omega}_m + k_p \hat{\mathbf{R}} \boldsymbol{\xi} \right]_\times \hat{\mathbf{R}} \tag{9.26}$$

where $\boldsymbol{\xi} \triangleq \text{vex}\left(\frac{1}{2}\left(\tilde{\mathbf{R}} - \tilde{\mathbf{R}}^T\right)\right)$, and $k_p \in \mathbb{R}_+$ is the feedback gain. Actually, the measured angular velocity has a drift, it can be estimated and then eliminated by the following filter

$$\begin{aligned} \dot{\hat{\mathbf{R}}} &= \left[\mathbf{R}_m \left({}^b\boldsymbol{\omega}_m - \hat{\mathbf{b}}_g \right) + k_p \hat{\mathbf{R}} \boldsymbol{\xi} \right]_\times \hat{\mathbf{R}} \\ \dot{\hat{\mathbf{b}}}_g &= -k_i \boldsymbol{\xi} \end{aligned} \tag{9.27}$$

where $k_i, k_p \in \mathbb{R}_+$. The detailed proof of this algorithm is described in [2]. After getting the rotation matrix, Euler angles can be calculated according to Eq. (5.11) in Chap. 5.

9.1.4 Kalman Filter

As shown in the previous section, a nonlinear complementary filter needs to use nine or twelve states, and it is also hard to choose optimal parameters. For such a purpose, a Kalman filter is adopted to estimate the attitude [7]. According to Eq. (5.9) in Chap. 5, the state vector $\mathbf{x} \in \mathbb{R}^3$ is chosen as the third column of $\left(\mathbf{R}_b^e\right)^T$, that is

$$\mathbf{x} = \begin{bmatrix} -\sin\theta \\ \sin\phi\cos\theta \\ \cos\phi\cos\theta \end{bmatrix}. \tag{9.28}$$

Let $\mathbf{R} = \mathbf{R}_b^e$ for simplicity in the following. Since $\mathbf{R}$ satisfies $\dot{\mathbf{R}} = \mathbf{R}\left[^b\boldsymbol{\omega}\right]_\times$, one has $\dot{\mathbf{R}}^T = -\left[^b\boldsymbol{\omega}\right]_\times \mathbf{R}^T$. Then, the process model is built as

$$\dot{\mathbf{x}} = -\left[^b\boldsymbol{\omega}\right]_\times \mathbf{x} \tag{9.29}$$

while the measurement model is built as

$$\mathbf{C}^{\mathrm{T}} \cdot {}^{b}\mathbf{a}_{\mathrm{m}} = -g\mathbf{C}^{\mathrm{T}}\mathbf{x} + \mathbf{n}_{\mathrm{a}} \tag{9.30}$$

where $\mathbf{C} = [\mathbf{I}_2 \ \mathbf{0}_{2\times1}]^{\mathrm{T}} \in \mathbb{R}^{3\times2}$ and $\mathbf{n}_{\mathrm{a}} \in \mathbb{R}^2$ is a noise. Furthermore, combining the drift model of gyroscope according to Eqs. (7.23) and (7.24) in Chap. 7, the process model of the Kalman filter is built as

$$\begin{aligned}
\dot{\mathbf{x}} &= -\left[{}^{b}\boldsymbol{\omega}_{\mathrm{m}} - \mathbf{b}_{\mathrm{g}} - \mathbf{n}_{\mathrm{g}}\right]_{\times} \mathbf{x} \\
\dot{\mathbf{b}}_{\mathrm{g}} &= \mathbf{n}_{\mathbf{b}_{\mathrm{g}}}
\end{aligned} \tag{9.31}$$

and the measurement model is

$$\mathbf{C}^{\mathrm{T}} \cdot {}^{b}\mathbf{a}_{\mathrm{m}} = -g\mathbf{C}^{\mathrm{T}}\mathbf{x} + \mathbf{n}_{\mathrm{a}}. \tag{9.32}$$

A Kalman filter can be adopted to obtain the estimation of $\mathbf{x}$. Furthermore, the estimated $\hat{\theta}, \hat{\phi}$ are obtained. The left yaw angle can be estimated by another method.

Recalling Eq. (5.51) in Chap. 5, it is the relationship between the derivative of the quaternions and the body's angular velocity, which is linear. This relationship can also be used to estimate attitude by Kalman filter. The model of the Kalman filter is built as

$$\dot{\mathbf{q}} = \frac{1}{2} \begin{bmatrix} 0 & -{}^{b}\boldsymbol{\omega}^{\mathrm{T}} \\ {}^{b}\boldsymbol{\omega} & -\left[{}^{b}\boldsymbol{\omega}\right]_{\times} \end{bmatrix} \mathbf{q} \tag{9.33}$$

while the measurement model can be built as

$$\mathbf{q}_{\mathrm{m}} = \mathbf{q} + \mathbf{n}_{\mathrm{q}} \tag{9.34}$$

where $\mathbf{q}_{\mathrm{m}}$ can be obtained according to Eq. (5.45) after getting $\theta_{\mathrm{m}}, \phi_{\mathrm{m}}, \psi_{\mathrm{m}}$, and $\mathbf{n}_{\mathrm{q}} \in \mathbb{R}^4$ is considered as a noise vector.

9.2   Position Estimation

Position information, including 2D location and altitude, is very important for the navigation of multicopters. It can be obtained from different sensors such as a GPS receiver, a camera, a barometer , and a laser range finder. Since a single sensor can only measure limited partial information or information with large noise for most of the time, information fusion is always required to acquire more precise position information. Two commonly used position estimations, namely GPS-based position estimation and SLAM-based position estimation, are introduced briefly in this section.

9.2.1   GPS-Based Position Estimation

GPS-based position estimation often utilizes an IMU, a GPS receiver, and a barometer. The kinematic model of a multicopter and different information obtained by these sensors are then fused by a Kalman filter.

Let ${}^{e}\mathbf{p} \triangleq [p_{x_{\mathrm{e}}} \ p_{y_{\mathrm{e}}} \ p_{z_{\mathrm{e}}}]^{\mathrm{T}} \in \mathbb{R}^3$ denote the absolute position of a multicopter and ${}^{e}\mathbf{v} \in \mathbb{R}^3$ denote the velocity in the EFCF. In addition to the dynamic state variables, the state vector also includes bias in the accelerometer and the barometer, which are modeled as simple random walks. By ignoring the cross term of velocity and angular velocity, the process model is formulated as

$$\begin{aligned}
{}^e\dot{\mathbf{p}} &= {}^e\mathbf{v} \\
{}^e\dot{\mathbf{v}} &= \mathbf{R}\left({}^b\mathbf{a}_m - \mathbf{b}_a - \mathbf{n}_a\right) + g\mathbf{e}_3 \\
\dot{\mathbf{b}}_a &= \mathbf{n}_{\mathbf{b}_a} \\
\dot{b}_{d_{\text{baro}}} &= n_{b_{d_{\text{baro}}}}
\end{aligned} \tag{9.35}$$

where $\mathbf{b}_a \in \mathbb{R}^3$ and $b_{d_{\text{baro}}} \in \mathbb{R}$ are the bias of acceleration and altitude, respectively.

The GPS receiver provides the measurement of 2D location, and the barometer measures the altitude. According to the measurement models of sensors described in Chap. 7, the measurement model is derived as

$$\begin{aligned}
p_{x\text{GPS}} &= p_{x_e} + n_{p_x\text{GPS}} \\
p_{y\text{GPS}} &= p_{y_e} + n_{p_y\text{GPS}} \\
d_{\text{baro}} &= -p_{z_e} + b_{d_{\text{baro}}} + n_{d_{\text{baro}}}
\end{aligned} \tag{9.36}$$

where $\left(p_{x\text{GPS}}, p_{y\text{GPS}}\right)$ is the measured 2D location by the GPS receiver, $d_{\text{baro}} \in \mathbb{R}$ is the measured altitude by the barometer, and $n_{p_x\text{GPS}}, n_{p_y\text{GPS}}, n_{d_{\text{baro}}} \in \mathbb{R}$ are the corresponding observation noise. Since the GPS receiver and the barometer have different sampling rates, a multirate Kalman filter described in Chap. 8 can be used. In addition, loss of GPS is quite common for multicopters because the environment around them is complex. If no more than four satellites are detected, then GPS localization would fail (refer to Sect. 7.7.1 in Chap. 7). To reduce the impact of the loss of GPS, a tightly coupled Kalman filter is recommended [8] which is directly based on GPS pseudorange rather than the 2D location provided by the GPS receiver.

9.2.2   SLAM-Based Position Estimation

Simultaneous Localization And Mapping (SLAM) is a computational problem of constructing or updating a map of an unknown environment while simultaneously keeping track of the agent's location with it. Related references could be found in [9, 10]. SLAM always uses several different types of sensors, including distance sensors such as ultrasonic range finders or laser range finders, and direction sensors such as cameras. Combination of distance sensors and direction sensors such as 3D cameras is also commonly used. Several SLAM systems and SLAM data sets are listed in Tables 9.1 and 9.2, respectively.

For multicopters, two kinds of SLAM including the laser-based SLAM and monocular vision-based SLAM will be introduced in the following.

9.2.2.1 Laser-Based SLAM

Most algorithms mentioned in Table 9.1 are designed for robots in flat, 2D environments, which are applicable to some special conditions for multicopters. For example, a multicopter flies at a fixed altitude, especially in an environment such as corridors where the horizontal cross section of the space at different altitudes can be regarded as the same. Then, 2D-SLAM algorithms are applicable. In [11], a 2D-Laser SLAM system is implemented by using EKF. When moving in a plane, the robot's position and map landmarks are acquired in real time. In practice, attitude change should be considered for non-planar scan. In [12], an algorithm is designed for a multicopter to compensate for the attitude change. As the flight space becomes larger, more landmarks should be stored, which will lead to a more time-consuming computation. This problem is one of the challenges in SLAM. In cases that a multicopter flies in a complex 3D environment, 3D-SLAM algorithms are required. A 3D-Laser or a LiDAR can be used to acquire data. Through extending 2D-SLAM algorithms, 3D-SLAM results can

Table 9.1 Open source SLAM algorithms

Author	Description	Web site
CyrillStachniss, Udo Frese, Giorgio Grisetti	OpenSLAM: A platform for SLAM researchers to publish their algorithms	http://openslam.org
Kai O. Arras	A GNU GPL licenced MATLAB toolbox for robot localization and mapping	http://www.cas.kth.se/toolbox
Tim Bailey	The code is written in MATLAB and performs EKF, UKF, FastSLAM 1, and FastSLAM 2	https://openslam.informatik.uni-freiburg.de/bailey-slam.html
Mark A. Paskin	Java and MATLAB hybrid programming SLAM system using thin junction tree filters	http://ai.stanford.edu/~paskin/slam/
Andrew Davison	An open source C++ library for SLAM designed and implemented	http://www.doc.ic.ac.uk/~ajd/Scene/index.html
Jose Neira	A simple SLAM simulation file	http://webdiis.unizar.es/~neira/software/slam/slamsim.htm
Dirk Hahnel	A grid-based FastSLAM system implemented in C	http://dblp.uni-trier.de/pers/hd/h/H=auml=hnel:Dirk.html
Various authors	MATLAB code of SLAM summer school 2002 in Sweden	http://www.cas.kth.se/SLAM/schedule.html

Table 9.2 Data sets for SLAM

Name	Description	Web site
Robotic 3D Scan Repository	3D point clouds from robotic experiments, log files of robot runs, standard 3D data sets for the robotics community	http://kos.informatik.uni-osnabrueck.de/3Dscans/
KITTI Vision Benchmark Suite	A large number of outdoor data sets generated for autonomous driving platforms	http://www.cvlibs.net/datasets/kitti/index.php
Radish (The robotics data set repository)	A collection of standard robotics data sets from real and simulated robots, including environment maps	http://radish.sourceforge.net
IJRR (The international journal of robotics research)	Web sites of research papers in IJRR which often contain data and video of results	http://www.ijrr.org

be obtained [13, 14]. Compared with fully-autonomous flight, it is easier for hovering because only a spot of landmarks should be stored.

By using the Vitoria Park data set,[1] the MATLAB code of 2D-Laser SLAM algorithm provided in [15] is tested. Results are shown in Fig. 9.6. With information of IMU and the 2D-Laser, landmarks are detected and updated and then location is estimated.

9.2.2.2 Monocular-Vision-Based SLAM

Vision-based SLAM consists of two core steps: tracking and mapping. During the tracking step, position and attitude are estimated from scene structure information, while during the mapping step, the 3D

---

[1]http://www-personal.acfr.usyd.edu.au/nebot/victoria_park.htm

Fig. 9.6 Results of
laser-based SLAM. Points
are landmarks and the line
is the trajectory of the
estimated motion

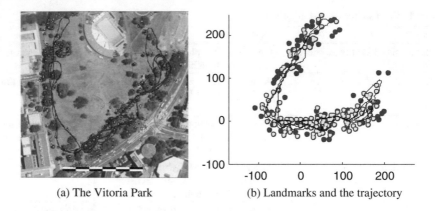

(a) The Vitoria Park   (b) Landmarks and the trajectory

map of the scene is built on the basis of position and attitude of the camera. Tracking and mapping
are alternately performed. Tracking relies on the scene structure information from the 3D map, and
mapping in turn needs the motion information provided by tracking. The SLAM method described
above is called Frame-by-Frame SLAM. In 2007, the monocular-vision-based SLAM algorithm was
proposed [16]. Monocular-vision-based SLAM uses only one camera to track and build the map
simultaneously. The paper and source code of this method can be downloaded in [17]. Since the
camera often moves in a known space with a built map, the mapping step has a relatively low real-
time requirement. Additionally, benefiting from the multi-CPU and parallel computing technology,
computing performance has been improved significantly. Furthermore, tracking and mapping can be
split into two separate tasks, processed in parallel threads on a dual-core computer at different rates.
Tracking works in a higher rate in order to ensure real-time performance, while mapping works in
a lower rate in order to get a high accuracy scene structure. High-accuracy mapping improves the
accuracy and robustness of the tracking step. This type of method is called Key-Frame SLAM, and a
well-performed system called Parallel Tracking And Mapping (PTAM) was firstly proposed in 2007
[18]. The associated code can be downloaded in [19]. Due to its better performance, the Key-Frame
SLAM is widely used in micro-UAVs. For instance, the authors in [20] realized PTAM using a 1.6 GHz
Intel Atom processor and a camera with a resolution of 752 pixels * 480 pixels. The mapping step works
at about 10 Hz. In this section, the PTAM source code provided online was tested using a camera with
a resolution of 640 pixels * 480 pixels and a dual-core processor. Figure 9.7 shows the initial plane
obtained from the linear motion at the beginning. Figure 9.8 shows the interface which is used to
observe the camera movement. Dots constitute the scene map, while each three-axis coordinate icon
presents a camera at one place. The camera is moved along an elliptical trajectory, and it has been
verified in the figure. Result shows that when moving smoothly, the PTAM system worked well.
However, sometimes, this system failed when occlusion happened or the camera moved too fast.

In addition, a monocular vision system cannot obtain the absolute scale information. To recover
the real motion of the camera, the information from IMU and altitude sensors is required. In [20], a
Kalman filter was adopted to recover the absolute scale. The process model is as follows[2]:

$$
\begin{aligned}
\dot{p}_{z_e} &= v_{z_e} \\
\dot{v}_{z_e} &= a_{z_e \mathrm{m}} + n_{a_{z_e}} + g \\
\dot{\lambda} &= n_\lambda \\
\dot{b}_{d_{\mathrm{baro}}} &= n_{b_{d_{\mathrm{baro}}}}
\end{aligned}
\tag{9.37}
$$

---

[2]In the book, the gravity is defined to point along the positive direction of the $o_e z_e$ axis. Therefore, the equation here is
slightly different from that in [20].

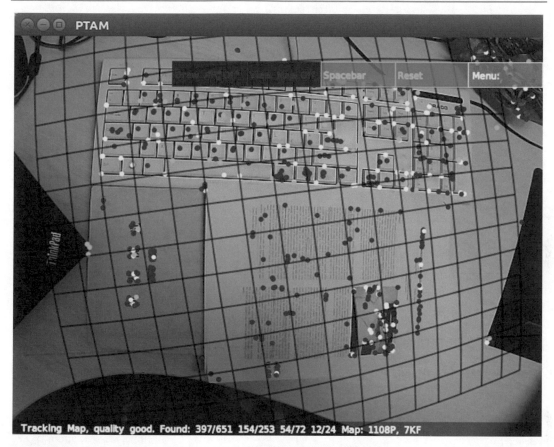

Fig. 9.7 Initial interface of PTAM algorithm. *Dots* in this figure are landmarks of the scene. These landmarks make up the scene map, and new *dots* are added to this map when the camera begins to move

where $p_{z_e}, v_{z_e}, \lambda, b_{d_{\text{baro}}} \in \mathbb{R}$ are the altitude, velocity in altitude direction, scale factor, and bias of the barometer, respectively, and $n_{a_{z_e}}, n_\lambda, n_{b_{d_{\text{baro}}}} \in \mathbb{R}$ are the corresponding noises. Generally, the accelerometer only provides acceleration in the ABCF. Then, the specific force in the EFCF is obtained by

$$a_{z_e\text{m}} = -\sin\theta \cdot a_{x_b\text{m}} + \sin\phi\cos\theta \cdot a_{y_b\text{m}} + \cos\phi\cos\theta \cdot a_{z_b\text{m}}. \tag{9.38}$$

Suppose that altitude information provided by SLAM is denoted as $p_{z\text{SLAM}}$ and that provided by the barometer is denoted as d_{baro}. The measurement model of the Kalman filter is

$$\begin{aligned} p_{z\text{SLAM}} &= \lambda \cdot p_{z_e} + n_{p_z\text{SLAM}} \\ d_{\text{baro}} &= -p_{z_e} + b_{d_{\text{baro}}} + n_{d_{\text{baro}}} \end{aligned} \tag{9.39}$$

For a low-altitude or an indoor flight, the altitude can be measured by an ultrasonic range finder. More accurate altitude information would provide a more accurate scale factor.

The scheme mentioned above is in fact a *loosely coupled* solution, which utilizes independent vision processing module for up-to-scale pose estimation. Another solution is called *tightly coupled* fusion [21]. It is expected to obtain more robust and better estimation.

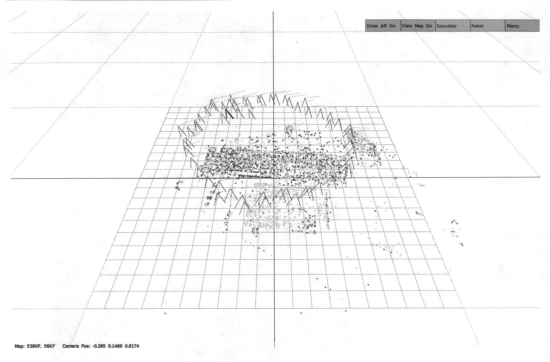

Map: 5380P, 56KF   Camera Pos: -0.285 0.1489 0.6174

Fig. 9.8   The updated map of PTAM algorithm

9.3   Velocity Estimation

For multicopters, the safety is very important, which requires an accurate and robust velocity estimation under all conditions, that is, because the velocity feedback will increase damping to improve the stability of multicopters for hovering. Moreover, it will make multicopters more tractable. Depending on different conditions, several kinds of velocity estimation methods are used on multicopters including the SLAM, GPS and optical flow. Taking the motion characteristics of multicopters into account, some researchers also proposed aerodynamic-drag-model-based algorithms. In this section, the optical-flow-based method and aerodynamic-drag-model-based method are mainly introduced.

9.3.1   Optical-Flow-Based Velocity Estimation Method

9.3.1.1 Optical Flow

As shown in Fig. 9.9, *optical flow* is the pattern of apparent motion of objects, surfaces, and edges in a visual scene caused by the relative motion between a camera and the scene. Sequences of ordered images allow the estimation of motion as either instantaneous image velocities or discrete-time image displacements.

Suppose that the intensity $I(x, y, t) \in \mathbb{R}$ denotes the intensity of image points (x, y) at time t. After an interval dt, an image point moves by dx, dy between the two successive image frames. The following *brightness constancy constraint* is given as

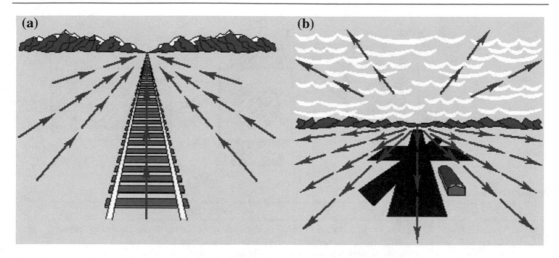

Fig. 9.9 Sketch map of optical flow (photograph from the book *The Ecological Approach to Visual Perception* by James Jerome Gibson). **a** Shows optical flow when the camera movement is away from mountains, while **b** shows optical flow when the camera movement is toward mountains

$$I (x + \mathrm{d}x, y + \mathrm{d}y, t + \mathrm{d}t) = I (x, y, t). \tag{9.40}$$

By assuming the movement to be sufficiently small, the image constraint at (x, y, t) with Taylor series is developed to get

$$I (x + \mathrm{d}x, y + \mathrm{d}y, t + \mathrm{d}t) = I (x, y, t) + \frac{\partial I}{\partial x}\mathrm{d}x + \frac{\partial I}{\partial y}\mathrm{d}y + \frac{\partial I}{\partial t}\mathrm{d}t + \varepsilon \tag{9.41}$$

where $\varepsilon \in \mathbb{R}$ is the second- or higher-order infinitely small term of $\mathrm{d}x, \mathrm{d}y, \mathrm{d}t$. According to Eqs. (9.40) and (9.41), one has

$$\frac{\partial I}{\partial x}\mathrm{d}x + \frac{\partial I}{\partial y}\mathrm{d}y + \frac{\partial I}{\partial t}\mathrm{d}t = 0 \tag{9.42}$$

which results in

$$I_x v_x + I_y v_y + I_t = 0 \tag{9.43}$$

where $v_x \triangleq \mathrm{d}x/\mathrm{d}t, v_y \triangleq \mathrm{d}y/\mathrm{d}t$ are the x and y components of the optical flow of image points (x, y) at time t, and $I_x \triangleq \partial I/\partial x, I_y \triangleq \partial I/\partial y, I_t \triangleq \partial I/\partial t$ are the partial derivatives at (x, y, t) in corresponding directions. Equation (9.43) is called the *optical flow constraint equation*. This equation links the temporal gradient and spatial gradient and finds the basis of almost all optical-flow-based methods. Equation (9.43) is an equation with two unknowns which cannot be solved. This is known as the *aperture problem* of the optical flow algorithms. As shown in Fig. 9.10, motion observed through little holes are the same even though the actual motion of A, B, C are completely different. In order to find the optical flow, another set of equations are needed, given by some additional constraints. Commonly used algorithms are, for instance, Horn-Schunck method [22], Lucas-Kanade method [23], and robust estimation methods. There exists many computer vision toolboxes and open source libraries which contain a large number of Application Programming Interfaces (APIs) to estimate image optical flow. For more details, please refer to Table 9.3.

Fig. 9.10 Aperture
problem

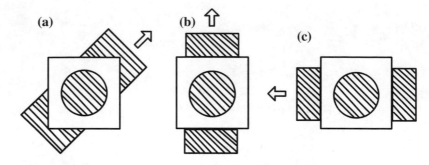

Table 9.3 Some toolboxes of optical flow

Toolbox	Description	Web site
Computer vision system toolbox	This toolbox is used in MATLAB R 2012a or higher version, optical flow is encapsulated to vision. Optical Flow class	http://cn.mathworks.com/help/vision/index.html
OpenCV	Open source libraries of computer vision, lots of APIs are provided	http://opencv.org
Machine vision toolbox	Vision toolbox emphasized computer vision and 3D vision	http://www.petercorke.com/Machine_Vision_Toolbox.html
VLFeat	Computer vision/image processing open source project, written in C and MATLAB with a large number of computer vision algorithms	http://www.vlfeat.org/download.html
Peter Kovesi's toolbox	Consists of lots of computer vision algorithms written in MATLAB, support Octave	http://www.peterkovesi.com/matlabfns

9.3.1.2 Optical-flow-based Velocity Estimation

Coordinate frames are built as shown in Fig. 9.11. The monocular camera is attached to the multicopter in a downward-looking direction. For simplicity, the camera coordinate frame is aligned with the ABCF, denoted as $o_b x_b y_b z_b$, and the ground is a plane, denoted as $p_{z_e} = 0$.

Let ${}^b\mathbf{p} \triangleq [\, p_{x_b} \; p_{y_b} \; p_{z_b} \,]^T$ denote the ground point $\mathbf{p}$ in the ABCF. Then, its corresponding normalized image point is given as

$$\bar{\mathbf{p}} \triangleq \begin{bmatrix} \bar{p}_x \\ \bar{p}_y \end{bmatrix} = \begin{bmatrix} \frac{p_{x_b}}{p_{z_b}} \\ \frac{p_{y_b}}{p_{z_b}} \end{bmatrix}. \tag{9.44}$$

Taking derivative of $\bar{\mathbf{p}}$ with respect to t yields

$$\begin{bmatrix} \dot{\bar{p}}_x \\ \dot{\bar{p}}_y \end{bmatrix} = \begin{bmatrix} \frac{\dot{p}_{x_b} - \bar{p}_x \dot{p}_{z_b}}{p_{z_b}} \\ \frac{\dot{p}_{y_b} - \bar{p}_y \dot{p}_{z_b}}{p_{z_b}} \end{bmatrix}. \tag{9.45}$$

Since the coordinate of the ground point $\mathbf{p}$ satisfies

$$^e\mathbf{p} = \mathbf{R} \cdot {}^b\mathbf{p} + {}^e\mathbf{T}_{o_b} \tag{9.46}$$

Fig. 9.11 A point **p** in ABCF $o_b x_b y_b z_b$ and EFCF $o_e x_e y_e z_e$, where d_{sonar} denotes the distance between camera center o_b and the image point **p**, and $x_e o_e y_e$ is on the plane where **p** is

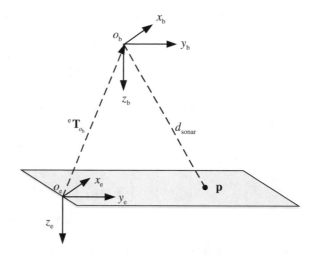

and $^e\mathbf{p}$ is constant, the derivative of Eq. (9.46) is

$$\mathbf{0}_{3\times 1} = \dot{\mathbf{R}} \cdot {}^b\mathbf{p} + \mathbf{R} \cdot {}^b\dot{\mathbf{p}} + {}^e\mathbf{v} \tag{9.47}$$

where $^e\dot{\mathbf{T}}_{o_b} = {}^e\mathbf{v}$. Moreover, since the rotation matrix $\mathbf{R}$ satisfies $\dot{\mathbf{R}} = \mathbf{R} \cdot \left[{}^b\boldsymbol{\omega}\right]_\times$, Eq. (9.47) becomes

$$^b\dot{\mathbf{p}} = -{}^b\mathbf{v} - \left[{}^b\boldsymbol{\omega}\right]_\times \cdot {}^b\mathbf{p} \tag{9.48}$$

where $^b\mathbf{v} = \mathbf{R}^T \cdot {}^e\mathbf{v}$. Equation (9.48) is further written as

$$\begin{aligned}
\dot{p}_{x_b} &= -v_{x_b} - \omega_{y_b} p_{z_b} + \omega_{z_b} p_{y_b} \\
\dot{p}_{y_b} &= -v_{y_b} - \omega_{z_b} p_{x_b} + \omega_{x_b} p_{z_b} \\
\dot{p}_{z_b} &= -v_{z_b} - \omega_{x_b} p_{y_b} + \omega_{y_b} p_{x_b}.
\end{aligned} \tag{9.49}$$

Substituting Eq. (9.49) into Eq. (9.45) results in

$$\underbrace{\begin{bmatrix} \dot{\bar{p}}_x \\ \dot{\bar{p}}_y \end{bmatrix}}_{\dot{\bar{\mathbf{p}}}} = \underbrace{\frac{1}{p_{z_b}} \begin{bmatrix} -1 & 0 & \bar{p}_x \\ 0 & -1 & \bar{p}_y \end{bmatrix}}_{\mathbf{A}(\bar{\mathbf{p}})} {}^b\mathbf{v} + \underbrace{\begin{bmatrix} \bar{p}_x \bar{p}_y & -\left(1 + \bar{p}_x^2\right) & \bar{p}_y \\ \left(1 + \bar{p}_y^2\right) & -\bar{p}_x \bar{p}_y & -\bar{p}_x \end{bmatrix}}_{\mathbf{B}(\bar{\mathbf{p}})} {}^b\boldsymbol{\omega}. \tag{9.50}$$

In Eq. (9.50), for an image point $\bar{\mathbf{p}}$, its optical flow vector is $\dot{\bar{\mathbf{p}}}$ which can be calculated by algorithms shown in Table 9.3. The angular velocity $^b\boldsymbol{\omega}$ can be obtained directly from a three-axis gyroscope. As shown in Fig. 9.11, the third component of $^e\mathbf{T}_{o_b}$ can be obtained from the reading of an ultrasonic range finder d_{sonar} and Euler angles that

$$\mathbf{e}_3^T \cdot {}^e\mathbf{T}_{o_b} = d_{sonar} \cos\theta \cos\phi \tag{9.51}$$

where $\mathbf{e}_3 = [0\ 0\ 1]^T$. Since the ground point **p** is on the plane $x_e o_e y_e$, this means

$$\mathbf{e}_3^T \cdot {}^e\mathbf{p} = 0. \tag{9.52}$$

By substituting Eqs. (9.46) and (9.51) into Eq. (9.52), the view depth p_{z_b} is obtained as

$$p_{z_b} = -\frac{d_{\text{sonar}} \cos \theta \cos \phi}{\mathbf{e}_3^{\mathsf{T}} \mathbf{R} \left[\, \bar{p}_x \;\; \bar{p}_y \;\; 1 \,\right]^{\mathsf{T}}}. \tag{9.53}$$

Suppose that $M \in \mathbb{Z}_+$ point pairs $(\bar{\mathbf{p}}_i, \dot{\bar{\mathbf{p}}}_i)$ are detected, $i = 1, 2, \ldots, M$. Then, according to Eq. (9.50), equations with respect to $^b\mathbf{v}$ are derived as

$$\underbrace{\begin{bmatrix} \dot{\bar{\mathbf{p}}}_1 \\ \vdots \\ \dot{\bar{\mathbf{p}}}_M \end{bmatrix}}_{\bar{\mathbf{p}}_a} = \underbrace{\begin{bmatrix} \mathbf{A}\,(\bar{\mathbf{p}}_1) \\ \vdots \\ \mathbf{A}\,(\bar{\mathbf{p}}_M) \end{bmatrix}}_{\mathbf{A}_a} \cdot {}^b\mathbf{v} + \underbrace{\begin{bmatrix} \mathbf{B}\,(\bar{\mathbf{p}}_1) \\ \vdots \\ \mathbf{B}\,(\bar{\mathbf{p}}_M) \end{bmatrix}}_{\mathbf{B}_a} \cdot {}^b\boldsymbol{\omega}. \tag{9.54}$$

Thus, the estimation of the velocity $^b\mathbf{v}$ can be obtained by

$$^b\hat{\mathbf{v}} = \left(\mathbf{A}_a^{\mathsf{T}}\mathbf{A}_a\right)^{-1} \mathbf{A}_a^{\mathsf{T}} \left[\bar{\mathbf{p}}_a - \mathbf{B}_a \left({}^b\boldsymbol{\omega}_{\mathrm{m}} - \hat{\mathbf{b}}_{\mathrm{g}}\right)\right] \tag{9.55}$$

where $^b\boldsymbol{\omega}_{\mathrm{m}} - \hat{\mathbf{b}}_{\mathrm{g}}$ is the de-biased angular velocity.

So far, the velocity estimation has been obtained based on optical flow. The following problems are still left to be solved: (1) Since the sampling periods of the camera, height sensor, and gyroscope are often different, time synchronization algorithms are required. (2) Lens distortion should be handled. (3) Rugged ground and moving background are required to be considered. (4) The mismatching of image pairs should be handled.

9.3.2 Aerodynamic-Drag-Model-Based Velocity Estimation Method

In this section, aerodynamic-drag-model-based velocity estimation methods are introduced. According to Eq. (6.32) in Sect. 6.2.2, the multicopter aerodynamic drag model is written as

$$\begin{aligned} \dot{v}_{x_b} &= -g \sin \theta - \frac{k_{\text{drag}}}{m} v_{x_b} + n_{a_x} \\ \dot{v}_{y_b} &= g \cos \theta \sin \phi - \frac{k_{\text{drag}}}{m} v_{y_b} + n_{a_y} \end{aligned} \tag{9.56}$$

where $n_{a_x}, n_{a_y} \in \mathbb{R}$ are the corresponding noises and $\dot{v}_{x_b}, \dot{v}_{y_b} \in \mathbb{R}$ are the accelerations of the multicopter in the ABCF. Here, the cross terms of velocity and angular velocity are ignored. By combining the measurement model of the gyroscope (7.23), (7.24) in Chap. 7 and Eq. (9.56), the process model of the Kalman filter is built as

$$\begin{aligned} \dot{\phi} &= \left(\omega_{x_b\mathrm{m}} - b_{\mathrm{g}_x} - n_{\mathrm{g}_x}\right) + \left(\omega_{y_b\mathrm{m}} - b_{\mathrm{g}_y} - n_{\mathrm{g}_y}\right) \tan \theta \sin \phi + \left(\omega_{z_b\mathrm{m}} - b_{\mathrm{g}_z} - n_{\mathrm{g}_z}\right) \tan \theta \cos \phi \\ \dot{\theta} &= \left(\omega_{y_b\mathrm{m}} - b_{\mathrm{g}_y} - n_{\mathrm{g}_y}\right) \cos \phi - \left(\omega_{z_b\mathrm{m}} - b_{\mathrm{g}_z} - n_{\mathrm{g}_z}\right) \sin \phi \\ \dot{b}_{\mathrm{g}_x} &= n_{b_{\mathrm{g}_x}} \\ \dot{b}_{\mathrm{g}_y} &= n_{b_{\mathrm{g}_y}} \\ \dot{b}_{\mathrm{g}_z} &= n_{b_{\mathrm{g}_z}} \\ \dot{v}_{x_b} &= -g \sin \theta - \frac{k_{\text{drag}}}{m} v_{x_b} + n_{a_x} \\ \dot{v}_{y_b} &= g \cos \theta \sin \phi - \frac{k_{\text{drag}}}{m} v_{y_b} + n_{a_y}. \end{aligned} \tag{9.57}$$

Here, the true angular velocities above have been replaced with

$$^b\boldsymbol{\omega} = {}^b\boldsymbol{\omega}_m - \mathbf{b_g} - \mathbf{n_g} \tag{9.58}$$

where $\mathbf{b_g} \triangleq [\, b_{g_x} \; b_{g_y} \; b_{g_z} \,]^T$, $\mathbf{n}_\omega \triangleq [\, n_{g_x} \; n_{g_y} \; n_{g_z} \,]^T$, and $\mathbf{n_{b_g}} \triangleq [\, n_{b_{g_x}} \; n_{b_{g_y}} \; n_{b_{g_z}} \,]^T$. The measurement model is built as

$$
\begin{aligned}
a_{x_b m} &= -\frac{k_{\mathrm{drag}}}{m} v_{x_b} + n_{a_x m} \\
a_{y_b m} &= -\frac{k_{\mathrm{drag}}}{m} v_{y_b} + n_{a_y m}
\end{aligned}
\tag{9.59}
$$

where $a_{x_b m}, a_{y_b m}$ are the measured specific force along the $o_b x_b$ and $o_b y_b$ axes, respectively; and $n_{a_x m}, n_{a_y m}$ are the corresponding noises. More details are shown in [4]. The filter model mentioned above is nonlinear and can be realized by an EKF. A simplified linear version is shown in [3] as follows:

$$
\begin{aligned}
\dot{\phi} &= \omega_{x_b m} - b_{g_x} - n_{g_x} \\
\dot{\theta} &= \omega_{y_b m} - b_{g_y} - n_{g_y} \\
\dot{b}_{g_x} &= n_{b_{g_x}} \\
\dot{b}_{g_y} &= n_{b_{g_y}} \\
\dot{v}_{x_b} &= -g\theta - \frac{k_{\mathrm{drag}}}{m} v_{x_b} + n_{a_x} \\
\dot{v}_{y_b} &= g\phi - \frac{k_{\mathrm{drag}}}{m} v_{y_b} + n_{a_y}
\end{aligned}
\tag{9.60}
$$

where higher-order terms of Eq. (9.57) are dropped. According to Eq. (9.59), the corresponding measurement model becomes

$$
\begin{aligned}
a_{x_b m} &= -\frac{k_{\mathrm{drag}}}{m} v_{x_b} + n_{a_x m} \\
a_{y_b m} &= -\frac{k_{\mathrm{drag}}}{m} v_{y_b} + n_{a_y m}.
\end{aligned}
\tag{9.61}
$$

The filter model mentioned above is linear and can be executed by Kalman filters. As shown in experiments of [3], the filter can obtain a more precise pitch angle and roll angle. Besides, the velocity is also obtained simultaneously.

There also exist some schemes which involve both the optical-flow-based velocity estimation method and the aerodynamic-drag-model-based velocity estimation method. AR. Drone is the first commercial product to adopt the optical flow method in velocity estimation of multicopters. Moreover, the aerodynamic drag model is also considered. The scheme was only described briefly in a passage of [24]: *"After inertial sensors calibration, sensors are used inside a complementary filter to estimate the attitude and de-bias the gyros; the de-biased gyros are used for vision velocity information combined with the velocity and altitude estimates from a vertical dynamics observer; the velocity from the computer vision algorithm is used to de-bias the accelerometers, the estimated bias is used to increase the accuracy of the attitude estimation algorithm; eventually the de-biased accelerometer gives a precise body velocity from an aerodynamics model."*

A question may arise why the velocity estimation by optical flow is only used to correct the acceleration but not as the final velocity estimate. This answer is probably that the time delay in optical-flow-based velocity estimate is quite small, but the estimation results are sensitive to the scene. As a result, the estimate is not always reliable. As shown in Fig. 2 in [24], there are some burrs in the optical-flow-based velocity estimation. In order to avoid polluting the final velocity, the optical-flow-based velocity estimate is only used to de-bias the acceleration. Then, de-biased acceleration and aerodynamic drag model are both used to estimate velocity of multicopters. In this case, even if the de-biased acceleration is not very accurate or corrected in real time sometimes, the velocity estimate can still be smooth only by using the IMU.

9.4   Obstacle Estimation

For moving objects, obstacle avoidance is very important. Numerous algorithms have been proposed for moving robots and aircraft. In this section, an obstacle avoidance method based on optical flow is introduced [25]. This method firstly recovers Time to Contact/Collision (TTC) from optical flow information and then guides multicopters to avoid obstacles. Generally, TTC is calculated from Focus of Expansion (FoE).

9.4.1   Focus of Expansion Calculation

In order to avoid obstacles, TTC is first calculated, and then, relative depths of obstacles are estimated. When a forward-looking camera moves horizontally, optical flow vectors of different image points intersect at one point, namely FoE. Optical flow vectors nearer to the FoE point have smaller length. In particular, the optical flow vector at the FoE point is zero. Moreover, the horizontal component of image points at the left-hand side of the column of FoE points to the left, while that at the right-hand side points to the right, as shown in Fig. 9.12.

By recalling Eq. (9.50), optical flows produced by a translation and a rotation are linear separable, where the scene depth only depends on the translation component. For multicopters, the angular velocity, measured by gyroscopes, is considered to be known. Then, Eq. (9.50) becomes

$$\Delta \bar{\mathbf{p}} = \mathbf{A} \left(\bar{\mathbf{p}} \right) \cdot {}^{b}\mathbf{v} \tag{9.62}$$

where

$$\Delta \bar{\mathbf{p}} \triangleq \dot{\bar{\mathbf{p}}} - \mathbf{B} \left(\bar{\mathbf{p}} \right) \cdot {}^{b}\boldsymbol{\omega}. \tag{9.63}$$

Fig. 9.12 FoE point

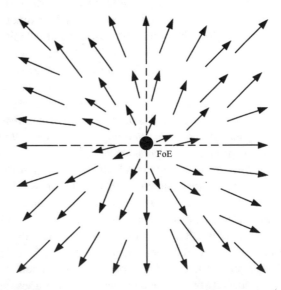

Further, let $\Delta\bar{\mathbf{p}} \triangleq [\,\Delta\bar{p}_x\ \Delta\bar{p}_y\,]^{\mathrm{T}}$, and rearranging Eq. (9.62) gives

$$\frac{\Delta\bar{p}_x}{\Delta\bar{p}_y} = \frac{\bar{p}_x - c_x}{\bar{p}_y - c_y} \tag{9.64}$$

where $c_x \triangleq v_{x_b}/v_{z_b}, c_y \triangleq v_{y_b}/v_{z_b}$ are independent of image point positions. Here, the image point (c_x, c_y) is the FoE point. Thus, given two arbitrary point pairs $(\Delta\bar{p}_{x,1}, \Delta\bar{p}_{y,1}, \bar{p}_{x,1}, \bar{p}_{y,1})$ and $(\Delta\bar{p}_{x,2}, \Delta\bar{p}_{y,2}, \bar{p}_{x,2}, \bar{p}_{y,2})$, the FoE point can be calculated according to Eq. (9.64). Since optical flow and angular velocity are often subject to noise, more point pairs are required to calculate the FoE. Let $p_i, i = 1, 2, \ldots, N$ be the chosen points. Then,

$$\underbrace{\begin{bmatrix} \Delta\bar{p}_{y,1} & -\Delta\bar{p}_{x,1} \\ \Delta\bar{p}_{y,2} & -\Delta\bar{p}_{x,2} \\ \vdots & \vdots \\ \Delta\bar{p}_{y,N} & -\Delta\bar{p}_{x,N} \end{bmatrix}}_{\mathbf{A}} \begin{bmatrix} c_x \\ c_y \end{bmatrix} = \underbrace{\begin{bmatrix} \bar{p}_{x,1}\Delta\bar{p}_{y,1} - \bar{p}_{y,1}\Delta\bar{p}_{x,1} \\ \bar{p}_{x,2}\Delta\bar{p}_{y,2} - \bar{p}_{y,2}\Delta\bar{p}_{x,2} \\ \vdots \\ \bar{p}_{x,N}\Delta\bar{p}_{y,N} - \bar{p}_{y,N}\Delta\bar{p}_{x,N} \end{bmatrix}}_{\mathbf{b}}. \tag{9.65}$$

The above equation can be solved by the least square method as

$$\begin{bmatrix} \hat{c}_x \\ \hat{c}_y \end{bmatrix} = \left(\mathbf{A}^{\mathrm{T}}\mathbf{A}\right)^{-1}\mathbf{A}^{\mathrm{T}}\mathbf{b}. \tag{9.66}$$

9.4.2 Time to Collision Calculation

Generally, the absolute depth cannot be recovered from the image sequence captured by a monocular camera. But the collision time TTC can be estimated from such an image sequence. TTC is the collision time when a multicopter flies at the current speed. The definition of TTC is

$$t_{\mathrm{TTC}} \triangleq \frac{p_{z_b}}{v_{z_b}}. \tag{9.67}$$

A long t_{TTC} means that it will cost longer time to collide with the obstacle. Thus, no special action is required. On the contrary, a short t_{TTC} means that the multicopter is about to collide with the object and avoidance should be made immediately. Therefore, the multicopter should turn to the direction with longer t_{TTC} for obstacle avoidance purpose. The collision time t_{TTC} can be estimated by the following algorithm.

According to Eqs. (9.50) and (9.63), one has

$$\begin{aligned} \Delta\bar{p}_x &= \frac{1}{p_{z_b}}\left(-v_{x_b} + \bar{p}_x v_{z_b}\right) \\ \Delta\bar{p}_y &= \frac{1}{p_{z_b}}\left(-v_{y_b} + \bar{p}_y v_{z_b}\right). \end{aligned} \tag{9.68}$$

Then, according to the definitions of $c_x \triangleq v_{x_b}/v_{z_b}, c_y \triangleq v_{y_b}/v_{z_b}$ and the definition of t_{TTC}, the following equations are easily obtained

$$\begin{aligned} \Delta\bar{p}_x t_{\mathrm{TTC}} &= \bar{p}_x - c_x \\ \Delta\bar{p}_y t_{\mathrm{TTC}} &= \bar{p}_y - c_y. \end{aligned} \tag{9.69}$$

Fig. 9.13 Optical flow and
collision time

(a) Optical flow   (b) Time to collision

The least square solution with respect to t_{TTC} is written as

$$\hat{t}_{TTC} = \sqrt{\frac{\left(\bar{p}_x - \hat{c}_x\right)^2 + \left(\bar{p}_y - \hat{c}_y\right)^2}{\Delta \bar{p}_x^2 + \Delta \bar{p}_y^2}} \tag{9.70}$$

where $\left(\hat{c}_x, \hat{c}_y\right)$ is the estimation of $\left(c_x, c_y\right)$ calculated by Eq. (9.66).

Experiments are undertaken to show the effect of the above obstacle avoidance algorithm. The capture rate of the camera is 30 Hz and the resolution is 640 * 480 pixels. Square grids with 16 pixels * 16 pixels uniformly distributed in image plane are tracked and then used to calculate optical flow and TTC. Optical flow is shown in Fig. 9.13a and corresponding TTC is shown in Fig. 9.13b. As shown, an optical flow vector with a large length corresponds to a small collision time. This is consistent with the observation. However, the calculation is not always correct. Also, from the shown optical flow, the mismatching of image pairs exists. Robustness problems related to optical-flow-based obstacle avoidance need to be further considered.

9.5   Summary

State estimation is the basis of control and decision-making. Actually, it is much more complicated and therefore deserves further research. The accurate and robust state estimation can make the controller design simpler and further improve the control performance. At present, the studies on multicopters are focused on the state estimation and environment sensing, especially based on vision, where SLAM and the optical flow methods mentioned are the main research areas. It is expected that the vision information can be used for target tracking, obstacle avoidance or visual odometry [26–28]. In order to improve its robustness, fusion with other sensors is also necessary. In a practical state estimation system, more problems need to be considered [29]:

(1) Computation performance. Since the computing ability of processors and the allowable calculation time are both limited, computing resources should be used efficiently.

(2) Abnormal data. Sensor failure or exception occurs frequently in multicopter flight. How to detect and deal with abnormal data remains an open problem and requires to be considered. In APM , about 90% of the code aims to detect and deal with abnormal data. This is very important for a reliable system.

(3) Measurement latency. In a real system, the phenomenon naturally exists that sensors are with different sampling rates and suffer measurement delay.

Exercises

9.1 In Sect. 9.1.2, the complementary filter of pitch angle is derived as Eq. (9.12). Then, it is transformed into a discrete-time differential form through first-order backward differencing as in Eq. (9.14). Besides the first-order backward difference method, there are also other difference methods, such as bilinear transformation (also known as Tustin's method). In bilinear transformation, s is expressed as $s = \frac{2}{T_s}\frac{1-z^{-1}}{1+z^{-1}}$. Please refer to the derivation of the discrete-time differential form of first-order backward difference method, and then, give the discrete-time differential form of the bilinear transform.

9.2 Download the IMU data from the Web site http://rfly.buaa.edu.cn/course and then implement the complementary filter algorithm of roll angle in C/C++/MATLAB.

9.3 The stability of an observer is very important. In Sect. 9.1.3, the nonlinear complementary filter of attitude angles is derived as in Eq. (9.27). Please refer to [2] and give the proof of the stability of nonlinear complementary filter.

9.4 In Sect. 9.2.1, the Kalman filter model is built for GPS-based position estimation. Please refer to Chap. 8 and check the observability of the Kalman filter model.

9.5 Please refer to Chap. 8 and check the observability of the Kalman filter model (9.37) (9.39) in Sect. 9.2.2.2.

9.6 Loss of GPS is quite common for multicopters because the environment around is complex. If no more than four satellites are detected, then GPS localization would fail. To reduce the impact of the loss of GPS signal, tightly coupled Kalman filter can be used. Please give a tightly coupled measurement model of GPS.

9.7 A GPS receiver and a barometer are adopted to estimate the position in Sect. 9.2.1. But in practice, there are the wind disturbance and occlusion. Please design a Kalman filter to estimate the height.

9.8 Download the MATLAB code and Vitoria Park data set provided in https://svn.openslam.org and test the 2D-Laser SLAM algorithm.

9.9 In Sect. 9.2.1, the principle of optical-flow-based velocity estimation method is described. In addition, another method called direct flow method can also estimate the velocity. Please refer to [30] and give the steps of velocity estimation.

References

1. Jung D, Tsiotras P (2007) Inertial attitude and position reference system development for a small UAV. In: AIAA Infotech @Aerospace 2007 Conference and Exhibit, Rohnert Park, California, 7-10 May, AIAA 2007-2763
2. Mahony R, Hamel T, Pflimlin JM (2008) Nonlinear complementary filters on the special orthogonal group. IEEE Trans Autom Control 53(5):1203–1218
3. Leishman RC, Macdonald JC, Beard RW et al (2014) Quadrotors and accelerometers: state estimation with an improved dynamic model. IEEE Control Syst Mag 34(1):28–41
4. Abeywardena D, Kodagoda S, Dissanayake G et al (2013) Improved state estimation in quadrotor MAVs: a novel drift-free velocity estimator. IEEE Robot Autom Mag 20(4):32–39
5. Caruso MJ (1997) Applications of magnetoresistive sensors in navigation systems. Prog in Technol 72:159–168
6. Perdikaris GA (1991) Computer controlled systems. Springer-Netherlands, Berlin

7. Kang CW, Park CG (2009) Attitude estimation with accelerometers and gyros using fuzzy tuned Kalman filter. In: Proceedings of IEEE European control conference, Hungary, Aug, pp 3713–3718
8. Wendel J, Trommer GF (2004) Tightly coupled GPS/INS integration for missile applications. Aerosp Sci and Technol 8(7):627–634
9. Durrant-Whyte H, Bailey T (2006) Simultaneous localization and mapping: part I. IEEE Robot Autom Mag 13(2):99–110
10. Bailey T, Durrant-Whyte H (2006) Simultaneous localization and mapping (SLAM): part II. IEEE Robot Autom Mag 13(3):108–117
11. Riisgard S, Blas MR (2005) SLAM for dummies: a tutorial approach to simultaneous localization and mapping. Technical report. ftp://revistafal.com/pub/alfredo/ROBOTICS/RatSLAM/1aslam_blas_repo.pdf. Accessed 25 Jan 2016
12. Wang F, Cui JQ, Chen BM et al (2013) A comprehensive UAV indoor navigation system based on vision optical flow and laser FastSLAM. Acta Autom Sin 39(11):1889–1899
13. Brenneke C, Wulf O, Wagner B (2003) Using 3D laser range data for SLAM in outdoor environments. In: Proceedings of IEEE international conference on intelligent robots and systems, Las Vegas, Oct, pp 188–193
14. Cole DM, Newman PM (2006) Using laser range data for 3D SLAM in outdoor environments. In: Proceedings of IEEE international conference on robotics and automation, Florida, May, pp 1556–1563
15. Kim C, Sakthivel R, Chung WK (2008) Unscented FastSLAM: a robust and efficient solution to the SLAM problem. IEEE Trans Robot 24(4):808–820
16. Davison AJ, Reid ID, Molton ND et al (2007) MonoSLAM: real-time single camera SLAM. IEEE Trans Pattern Anal 29(6):1052–1067
17. Andrew D (2003) SceneLib Homepage. http://www.doc.ic.ac.uk/~ajd/Scene/index.html. Accessed 21 Jan 2016
18. Klein G, Murray D (2007) Parallel tracking and mapping for small AR workspaces. In: 6th IEEE and ACM international symposium on mixed and augmented reality, Nara, Nov, pp 225–234
19. Klein G (2014) Parallel tracking and mapping for small AR workspaces-source code. http://www.robots.ox.ac.uk/~gk/PTAM. Accessed 21 Jan 2016
20. Achtelik M, Achtelik M, Weiss S et al (2011) Onboard IMU and monocular vision based control for MAVs in unknown in-and outdoor environments. In: Proceedings of IEEE international conference on robotics and automation, Shanghai, Aug, pp 3056–3063
21. Shen SJ, Michael N, Kumar V (2015) Tightly-coupled monocular visual-inertial fusion for autonomous flight of rotorcraft MAVs. In: IEEE international conference on robotics and automation, Washington, May, pp 5303–5310
22. Horn BK, Schunck BG (1981) Determining optical flow. Artif Intell 17(1–3):185–203
23. Lucas BD, Kanade T (1981) An iterative image registration technique with an application to stereo vision. In: Proceedings of 7th international joint conference on artificial intelligence, Vancouver, Aug, pp 674–679
24. Bristeau PJ, Callou F, Vissiere D, et al (2011) The navigation and control technology inside the AR drone micro UAV. In: 18th IFAC world congress 18(1): 1477–1484
25. Green WE, Oh PY (2008) Optic-flow-based collision avoidance. IEEE Robot Autom Mag 15(1):96–103
26. Nistér D, Naroditsky O, Bergen J (2004) Visual odometry. In: Proceedings of IEEE computer society conference on computer vision and pattern recognition, Washington, pp 652–659
27. Scaramuzza D, Fraundorfer F (2011) Visual odometry: part I - the first 30 years and fundamentals. IEEE Robot Autom Mag 18(4):80–92
28. Fraundorfer F, Scaramuzza D (2012) Visual odometry: part II - matching, robustness, and applications. IEEE Robot Autom Mag 19(2):78–90
29. Riseborough P (2015) Application of data fusion to aerial robotics. http://thirty5tech.com/vid/watch/Z3Qpi1Rx6HM. Accessed 21 Jan 2016
30. Horn BKP, Weldon EJ Jr (1988) Direct methods for recovering motion. Int J Comput Vision 2(1):51–76

Stability and Controllability

10

Stability and controllability are the basic properties of a dynamical system. Since a multicopter without control feedback is unstable, an autopilot is required to guarantee its stability and further make the multicopter hover automatically without any need for an external intervention. However, if a multicopter is uncontrollable at an equilibrium state, then no controller exists to stabilize it at the equilibrium state. Unlike general systems, the control inputs of multicopters are often constrained because of the fact that the thrust provided by each propeller is unidirectional. This will lead to a special controllability, called *positive controllability*. Without controllability, any controller, including the fault-tolerant controller, would not help. The controllability is to answer the question: Whether a multicopter can be controlled or not. However, multicopter design contains far more than merely the controllability issue. For such a purpose, the Degree of Controllability (DoC) is further introduced to quantify the degree to which a process can be controlled. For the same system subject to different control constraints, the DoC will be different. The DoC can help in choosing a fault-tolerant control strategy or evaluating a multicopter design in the sense of the controllability in presence of wind. This chapter will answer the questions below:

Why is a multicopter dynamical system unstable and how is the DoC of a multicopter?

In order to answer these questions, this chapter includes the definition of stability, stability criteria, basic concepts of controllability, and controllability of multicopters.

© Springer Nature Singapore Pte Ltd. 2017
Q. Quan, *Introduction to Multicopter Design and Control*, DOI 10.1007/978-981-10-3382-7_10

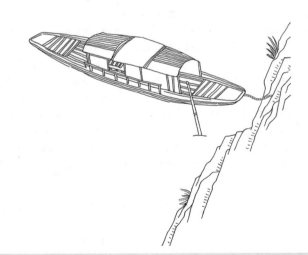

Boat Lying Low on the River

The ancient Chinese had observed the phenomenon of stability. Yingwu Wei, a poet in the Tang dynasty, wrote in one of his poems "*By the River to the West of Chuzhou*" that "Spring flood and a day's rain, by dusk the river runs swift, the country ferry deserted, the boat, by itself, lies low. (The Chinese is "春潮带雨晚来急，野渡无人舟自横" translated by Andrew W.F. Wong, from http://chinesepoemsinenglish.blogspot.fr/2011/12/wei-yingwu-by-river-towest-of-chuzhou.html)". Zhun Kou, a politician in the Northern Song Dynasty, described the same phenomenon in his poem "*Spring Went to Return*" that "Wild water no crossing, boat lying low on the river. (The Chinese is "野水无人渡，孤舟尽日横")". In these poems, they observed that the boat always lay at a particular position in the running water. So why does the boat always settle in such a position? In fact, this is a stability problem. Lots of experiences and theoretical analysis have shown that the position at which the major axis of the bottom of boat is perpendicular to the river flow, is more stable, which corresponds to the poets' observation.

10.1   Definition of Stability

In Fig. 10.1, an example is first given to introduce the stability concept. Intuitively, the ball in Fig. 10.1a is stable when it is at the bottom. If it is not at the bottom initially, it will roll and oscillate until it eventually settles at the bottom. The ball in Fig. 10.1b is unstable at the top as the ball will roll to the bottom and cannot get back to the top after it is disturbed. Similarly, there are other kinds of stability in real life, such as social stability, economic stability, and an athlete's performance stability. The following question arises: Is there a unified definition of stability?

This chapter considers the stability of a dynamical system about an *equilibrium state*. Stability is a concept associated with the equilibrium state. For example, the ball in Fig. 10.1 is stable at position 1, but it is unstable at position 3. Theoretically, the ball can stop at position 3, but it is not robust against any perturbation. This is like placing an egg upright on its end. Position 2 and position 4 are not equilibrium states as the ball cannot stay there. So, the stability is meaningless at these two positions. After an equilibrium state is defined, the stability about that equilibrium state can be discussed. By taking social stability for example, the state that people live and work in peace and comfort is considered as an equilibrium state. After someone destroys a stable society, the society will recover to the equilibrium state again. Similarly, the performance stability of an athlete is that, taking the normal training performance of the athlete as an equilibrium state, the athlete's performance keeps the same no matter the environment he/she is in. The stability of an aircraft is that the aircraft will recover to the equilibrium state after external disturbances disappear. If the aircraft cannot recover to the equilibrium state, then the aircraft is unstable. Based on these concepts mentioned above, a formal definition of stability will be given in the following.

Consider a nonlinear dynamical system described by

$$\dot{\mathbf{x}} = \mathbf{f}\,(t, \mathbf{x}, \mathbf{u}) \tag{10.1}$$

where $\mathbf{x} \in \mathbb{R}^n$ is the state, and $\mathbf{u} \in \mathbb{R}^m$ is the control input. In practice, the control input $\mathbf{u}$ is a function of the state $\mathbf{x}$ and time $t \in \mathbb{R}$, where

$$\mathbf{u} = \mathbf{g}\,(t, \mathbf{x})\,. \tag{10.2}$$

According to Eqs. (10.1) and (10.2), one has

$$\begin{aligned}\dot{\mathbf{x}} &= \mathbf{f}\,(t, \mathbf{x}, \mathbf{g}\,(t, \mathbf{x})) \\ &\triangleq \mathbf{f}_{\mathrm{c}}\,(t, \mathbf{x})\end{aligned} \tag{10.3}$$

The equilibrium state must satisfy the following equation

$$\dot{\mathbf{x}}^* = \mathbf{f}_{\mathrm{c}}\left(t, \mathbf{x}^*\right) = \mathbf{0}_{n \times 1}. \tag{10.4}$$

Fig. 10.1 Stability and instability

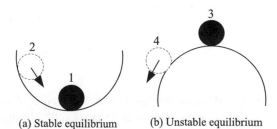

(a) Stable equilibrium   (b) Unstable equilibrium

Different systems have different equilibrium states. In order to normalize them, the error $\tilde{\mathbf{x}} \triangleq \mathbf{x} - \mathbf{x}^*$ is defined. Then

$$\dot{\tilde{\mathbf{x}}} = \tilde{\mathbf{f}}_c (t, \tilde{\mathbf{x}}) \tag{10.5}$$

where $\tilde{\mathbf{f}}_c (t, \tilde{\mathbf{x}}) \triangleq \mathbf{f}_c (t, \tilde{\mathbf{x}} + \mathbf{x}^*) - \mathbf{f}_c (t, \mathbf{x}^*)$ and $\tilde{\mathbf{f}}_c (t, \mathbf{0}_{n \times 1}) \equiv \mathbf{0}_{n \times 1}$. The new state variable $\tilde{\mathbf{x}} = \mathbf{0}_{n \times 1}$ is the equilibrium state of system (10.5). For simplicity, the dynamical system in the form of

$$\dot{\mathbf{x}} = \mathbf{f} (\mathbf{x}) \tag{10.6}$$

is discussed, where $\mathbf{f} (\mathbf{0}_{n \times 1}) = \mathbf{0}_{n \times 1}$, and $\mathbf{x} = \mathbf{0}_{n \times 1}$ is the equilibrium state. Moreover, it is a *time-invariant system*.

Definition 10.1 ([1, p. 48]) For system (10.6), the equilibrium state $\mathbf{x} = \mathbf{0}_{n \times 1}$ is said to be *stable* if, for any $R \in \mathbb{R}_+$, there exists $r \in \mathbb{R}_+$, such that if $\|\mathbf{x} (0)\| < r$, then $\|\mathbf{x} (t)\| < R$ for all $t \in \mathbb{R}_+$. Otherwise, the equilibrium state is *unstable*.

This definition of stability is also called stability in the sense of Lyapunov, or *Lyapunov stability*. The stability implies that the system trajectory can be kept arbitrarily close to the origin by starting sufficiently close to it.

Definition 10.2 ([1, p. 50]) For system (10.6), an equilibrium state $\mathbf{x} = \mathbf{0}_{n \times 1}$ is *asymptotically stable* if it is stable, and there exists some $r \in \mathbb{R}_+$ such that $\|\mathbf{x} (0)\| < r$, which implies that $\lim_{t \to \infty} \|\mathbf{x} (t)\| = 0$.

For Definition 10.1, the system trajectory may never reach the origin. But under Definition 10.2, a state starting close to the origin initially will converge to the origin eventually. Furthermore, Definition 10.3 will show how the system trajectory converges to the origin.

Definition 10.3 ([1, p. 51]) For system (10.6), an equilibrium state $\mathbf{x} = \mathbf{0}_{n \times 1}$ is *exponentially stable* if there exist $\alpha, \lambda \in \mathbb{R}_+$ such that $\forall t \in \mathbb{R}_+, \|\mathbf{x} (t)\| \leq \alpha \|\mathbf{x} (0)\| e^{-\lambda t}$ in some neighborhood around the origin.

The three definitions above have the following relationships: Definition 10.3 $\subset$ Definition 10.2 $\subset$ Definition 10.1. Local stability is related to the initial state $\|\mathbf{x} (0)\| < r$, determined by $r \in \mathbb{R}_+$, whereas global stability is independent of the initial state, or says $r = \infty$.

10.2 Stability Criteria

10.2.1 Stability of Multicopters

First, the attitude stability of multicopters is analyzed according to the attitude control model in Eq. (6.13) in Chap. 6. The effect of the term $-^b\boldsymbol{\omega} \times (\mathbf{J}^b\boldsymbol{\omega}) + \mathbf{G}_a$ is weak. Moreover, the pitch and roll angle are often small. Then, the attitude model becomes

$$\begin{aligned} \dot{\boldsymbol{\Theta}} &= {}^b\boldsymbol{\omega} \\ {}^b\dot{\boldsymbol{\omega}} &= \mathbf{J}^{-1}\boldsymbol{\tau}. \end{aligned} \tag{10.7}$$

Each propeller of the multicopter is fixed to the Aircraft-Body Coordinate Frame (ABCF). The control moment will not be changed if there is no active control. Then, if the control moment is zero, the attitude control model (10.7) becomes

$$\dot{\Theta} = {}^b\omega$$
$${}^b\dot{\omega} = \mathbf{0}_{3\times1} \tag{10.8}$$

where the closed-loop system is described by

$$\dot{\mathbf{z}} = \mathbf{A}\mathbf{z} \tag{10.9}$$

where

$$\mathbf{A} = \begin{bmatrix} \mathbf{0}_{3\times3} & \mathbf{I}_3 \\ \mathbf{0}_{3\times3} & \mathbf{0}_{3\times3} \end{bmatrix}, \mathbf{z} = \begin{bmatrix} \Theta \\ {}^b\omega \end{bmatrix}.$$

As the solution to Eq. (10.9) is $\mathbf{z}(t) = e^{\mathbf{A}t}\mathbf{z}(0)$ and the real parts of the eigenvalues of $\mathbf{A}$ are zeros. Then, the attitude dynamics are unstable at the equilibrium $\Theta = \mathbf{0}_{3\times1}, {}^b\omega = \mathbf{0}_{3\times1}$ according to [2, p. 130].

Physically, a multicopter is unstable at the hovering state no matter how the propellers are arranged. Because the forces on a multicopter are the total thrust and the gravity. Without controllers, the total thrust and the gravity will not be changed according to the attitude change, and then the external disturbance cannot be compensated for without feedback. However, the forces on a fixed-wing aircraft are thrust, gravity, drag and lift, where the lift is changed with the attitude. As a result, control forces will be changed with the attitude, and in turn, the attitude can be controlled by these feedback. That is why multicopters rely on autopilots to hover, whereas some fixed-wing aircraft do not need autopilots to keep forward flight.

10.2.2 Some Results of Stability

10.2.2.1 Invariant Set Theorems
In order to make this chapter self-contained, the invariant set theorems are introduced here. Most results are from [1, p. 69]. The invariant set theorems are often used to prove the stability of dynamical systems.

Definition 10.4 A set $\mathscr{S}$ is an *invariant set* for a dynamical system if every system trajectory which starts from the inside of $\mathscr{S}$ remains in $\mathscr{S}$ for all time thereafter.

The local version of the invariant set theorems is discussed first, and then the global version.

Theorem 10.5 (Local Invariant Set Theorem). *For system (10.6), let $V(\mathbf{x})$ be a scalar function with continuous first partial derivatives. Assume that*

(1) for some $l \in \mathbb{R}_+$, the region Ω_l defined by $V(\mathbf{x}) < l$ is bounded;
(2) $\dot{V}(\mathbf{x}) = \left(\frac{\partial V(\mathbf{x})}{\partial \mathbf{x}}\right)^{\mathrm{T}} \mathbf{f} \leq 0, \mathbf{x} \in \Omega_l$.

Let $\mathscr{R}$ be the set of all points within Ω_l where $\dot{V}(\mathbf{x}) = 0$, and $\mathscr{M}$ be the largest invariant set in $\mathscr{R}$. Then, every solution $\mathbf{x}(t)$ originating in Ω_l tends to $\mathscr{M}$ as $t \to \infty$.

In the theorem above, the word "largest" is understood in the sense of set theory, i.e., $\mathscr{M}$ is the union of all invariant sets (e.g., equilibrium states or limit cycles) within $\mathscr{R}$. In particular, if the set $\mathscr{R}$ is itself invariant (i.e., if once $\dot{V}(t) = 0$, then $\dot{V}(t) \equiv 0$ for all future time), then $\mathscr{M} = \mathscr{R}$. Also note that V, although often still referred to as a Lyapunov function, is not required to be positive definite. But the statement that "the region Ω_l defined by $V(\mathbf{x}) < l$ is bounded" is pretty much necessary. For example, consider the following system

$$\dot{x}_1 = -x_1^2$$
$$\dot{x}_2 = -x_2^2.$$

Let $V(\mathbf{x}) = x_1 + x_2$. Then, $\dot{V}(\mathbf{x}) \le -x_1^2 - x_2^2$. Since $V(\mathbf{x}) < l$ is not bounded, then the solution will not approach to zero as $t \to \infty$. In fact, the system trajectory will tend to be infinity.

Theorem 10.6 (Global Invariant Set Theorem) *For system (10.6), let $V(\mathbf{x})$ be a scalar function with continuous first partial derivatives. Assume that*

(1) $V(\mathbf{x}) \to \infty$ *as* $\|\mathbf{x}\| \to \infty$;
(2) $\dot{V}(\mathbf{x}) = \left(\frac{\partial V(\mathbf{x})}{\partial \mathbf{x}}\right)^{\mathrm{T}} \mathbf{f} \le 0$ *over the whole state space.*

Let $\mathscr{R}$ be the set of all points where $\dot{V}(\mathbf{x}) = 0$, and $\mathscr{M}$ be the largest invariant set in $\mathscr{R}$. Then, all states globally asymptotically converge to $\mathscr{M}$ as $t \to \infty$.

Because of the importance of this theorem, an additional example is given in the following.

Example 10.7 [1, p. 74] **Stability of second-order nonlinear systems**. Consider a second-order system of the form

$$\ddot{x} + b(\dot{x}) + c(x) = 0 \tag{10.10}$$

where $b(\dot{x})$ and $c(x)$ are continuous functions satisfying the sign conditions

$$\dot{x}b(\dot{x}) > 0, \dot{x} \ne 0$$
$$xc(x) > 0, x \ne 0. \tag{10.11}$$

Together with the continuity assumptions, the sign conditions in (10.11) imply that $b(0) = 0$ and $c(0) = 0$. A positive definite function for this system is defined as follows:

$$V = \frac{1}{2}\dot{x}^2 + \int_0^x c(y)\,\mathrm{d}y. \tag{10.12}$$

Taking the derivative of V along Eq. (10.10) results in

$$\dot{V} = \dot{x}\ddot{x} + c(x)\dot{x} = -\dot{x}b(\dot{x}) - \dot{x}c(x) + c(x)\dot{x} = -\dot{x}b(\dot{x}) \le 0.$$

Furthermore, by the condition (10.11), $\dot{x}b(\dot{x}) = 0$ only if $\dot{x} = 0$. Now $\dot{x} = 0$ implies that

$$\ddot{x} = -c(x)$$

which is nonzero as long as $x \ne 0$. Thus, the system cannot get "stuck" at an equilibrium value other than $x = 0$; in other words, with $\mathscr{R}$ being the set defined by $\dot{x} = 0$, the largest invariant set $\mathscr{M}$ in $\mathscr{R}$ contains only one point, namely $x = 0, \dot{x} = 0$. By using the local invariant set theorem, the origin is a locally asymptotically stable point.

10.2.2.2 Stability Criteria for Simple Systems

This section considers stability criteria for some simple systems. The results here will be used to analyze the stability of dynamical systems in later chapters. Consider a second-order dynamical system

$$
\begin{aligned}
\dot{\mathbf{x}} &= \mathbf{v} \\
\dot{\mathbf{v}} &= \mathbf{u}
\end{aligned}
\tag{10.13}
$$

where $\mathbf{x}, \mathbf{v}, \mathbf{u} \in \mathbb{R}^n$, and $\mathbf{u} \triangleq -k_1 (\mathbf{v} - \dot{\mathbf{x}}_d) - k_2 (\mathbf{x} - \mathbf{x}_d)$ is a Proportional-Derivative (PD) controller. Define

$$
\begin{aligned}
\tilde{\mathbf{x}} &\triangleq \mathbf{x} - \mathbf{x}_d \\
\tilde{\mathbf{v}} &\triangleq \mathbf{v} - \dot{\mathbf{x}}_d.
\end{aligned}
$$

Then, Eq. (10.13) is rearranged as

$$
\begin{aligned}
\dot{\tilde{\mathbf{x}}} &= \tilde{\mathbf{v}} \\
\dot{\tilde{\mathbf{v}}} &= -k_1 \tilde{\mathbf{v}} - k_2 \tilde{\mathbf{x}}.
\end{aligned}
\tag{10.14}
$$

Let $\mathbf{z} \triangleq [\tilde{\mathbf{x}}^T \tilde{\mathbf{v}}^T]^T \in \mathbb{R}^{2n}$. Then, system (10.14) is expressed by

$$
\dot{\mathbf{z}} = \mathbf{A}\mathbf{z}
\tag{10.15}
$$

where

$$
\mathbf{A} = \begin{bmatrix} \mathbf{0}_{n \times n} & \mathbf{I}_n \\ -k_2 \mathbf{I}_n & -k_1 \mathbf{I}_n \end{bmatrix}.
$$

Theorem 10.8 *For system (10.13), if $k_1, k_2 > 0$, then $\lim_{t \to \infty} \|\mathbf{z}(t)\| = 0$. Furthermore, $\mathbf{z} = \mathbf{0}_{2n \times 1}$ is globally exponentially stable.*

Proof The solution to system (10.15) is

$$
\mathbf{z}(t) = e^{\mathbf{A}t} \mathbf{z}(0).
\tag{10.16}
$$

If $k_1, k_2 > 0$, then the real parts of all the eigenvalues of $\mathbf{A}$ are negative, implying $\lim_{t \to \infty} \|\mathbf{z}(t)\| = 0$ according to [2, p. 130]. According to Eq. (10.16), $\mathbf{z} = \mathbf{0}_{2n \times 1}$ is globally exponentially stable. $\quad\square$

Theorem 10.8 is a well-known result [2, p. 130]. In practice, the control input is constrained. In the following, the stability of the closed-loop system (10.14) subject to control constraints is studied. Consider the following dynamical system

$$
\begin{aligned}
\dot{\mathbf{x}} &= \mathbf{v} \\
\dot{\mathbf{v}} &= \text{sat}_{\text{gd}}(\mathbf{u}, a)
\end{aligned}
\tag{10.17}
$$

where $\mathbf{u} = -k_1 \tilde{\mathbf{v}} - k_2 \tilde{\mathbf{x}}$. Here, the saturation function is defined as

$$
\text{sat}_{\text{gd}}(\mathbf{u}, a) \triangleq \begin{cases} \mathbf{u}, & \|\mathbf{u}\|_\infty \leq a \\ a \frac{\mathbf{u}}{\|\mathbf{u}\|_\infty}, & \|\mathbf{u}\|_\infty > a \end{cases}
\tag{10.18}
$$

where $a \in \mathbb{R}_+$, $\mathbf{u} \triangleq [u_1 \dots u_n]^{\mathrm{T}} \in \mathbb{R}^n$ and $\|\mathbf{u}\|_\infty \triangleq \max (|u_1|, \dots, |u_n|)$. The saturation function $\mathrm{sat}_{\mathrm{gd}}(\mathbf{u}, a)$ and the vector $\mathbf{u}$ are parallel all the time, which is called the *direction-guaranteed saturation function* (see Fig. 10.2) here. The function $\mathrm{sat}_{\mathrm{gd}}(\cdot)$ not only can prevent the controller from saturation caused by large $-k_1 \tilde{\mathbf{v}} - k_2 \tilde{\mathbf{x}}$, but also can keep the control direction the same as that of the original vector. It is rewritten as

$$\mathrm{sat}_{\mathrm{gd}}(\mathbf{u}, a) = \kappa_a(\mathbf{u})\,\mathbf{u}$$

where

$$\kappa_a(\mathbf{u}) \triangleq \begin{cases} 1, \|\mathbf{u}\|_\infty \le a \\ \frac{a}{\|\mathbf{u}\|_\infty}, \|\mathbf{u}\|_\infty > a \end{cases}.$$

It is obvious that $0 < \kappa_a(\mathbf{u}) \le 1$. Now, the stability of system (10.17) is considered.

Theorem 10.9 *For system (10.17), if $k_1, k_2 > 0$, then $\lim_{t \to \infty} \|\mathbf{z}(t)\| = 0$ for any $\mathbf{z}(0)$. Furthermore, $\mathbf{z} = \mathbf{0}_{2n \times 1}$ is locally exponentially stable.*

Proof First, system (10.17) is rearranged as

$$\dot{\tilde{\mathbf{x}}} = \tilde{\mathbf{v}}$$
$$\dot{\tilde{\mathbf{v}}} = \kappa_a(\mathbf{u})\,\mathbf{u} \tag{10.19}$$

where $\mathbf{u} = -k_1 \tilde{\mathbf{v}} - k_2 \tilde{\mathbf{x}}$. It is easy to see that $\kappa_a(\mathbf{u})\,\mathbf{u}$ is a vector field with continuous components defined along the smooth curve C_u parameterized by $\mathbf{u}(t)$. Then, the line integral of $\kappa_a(\mathbf{u})\,\mathbf{u}$ along C_u is $\int_{C_u} \kappa_a(\mathbf{u})\,\mathbf{u}^{\mathrm{T}} \mathrm{d}\mathbf{u}$. A Lyapunov function is designed for this system as follows:

$$V = \frac{1}{2}\tilde{\mathbf{v}}^{\mathrm{T}}\tilde{\mathbf{v}} + \frac{1}{k_2} \int_{C_u} \kappa_a(\mathbf{u})\,\mathbf{u}^{\mathrm{T}} \mathrm{d}\mathbf{u} \tag{10.20}$$

where $\mathbf{u} \triangleq [u_1 \dots u_n]^{\mathrm{T}}$, $\mathrm{d}\mathbf{u} \triangleq [\mathrm{d}u_1 \dots \mathrm{d}u_n]^{\mathrm{T}}$. Therefore, V being bounded implies that $\tilde{\mathbf{x}}, \tilde{\mathbf{v}}$ are bounded. According to Thomas' Calculus [3, p. 911], one has

$$V = \frac{1}{2}\tilde{\mathbf{v}}^{\mathrm{T}}\tilde{\mathbf{v}} + \frac{1}{k_2} \int_0^t \kappa_a(\mathbf{u})\,\mathbf{u}^{\mathrm{T}} \dot{\mathbf{u}} \mathrm{d}t.$$

Fig. 10.2 Direction-guaranteed saturation function

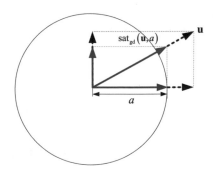

The derivative of V along system (10.19) is

$$\dot{V} = \tilde{\mathbf{v}}^T\dot{\tilde{\mathbf{v}}} + \frac{1}{k_2}\kappa_a(\mathbf{u})\mathbf{u}^T\dot{\mathbf{u}}$$

$$= \tilde{\mathbf{v}}^T\kappa_a(\mathbf{u})\mathbf{u} + \frac{1}{k_2}\kappa_a(\mathbf{u})\mathbf{u}^T(-k_1\kappa_a(\mathbf{u})\mathbf{u} - k_2\tilde{\mathbf{v}})$$

$$= -\frac{k_1}{k_2}\kappa_a^2(\mathbf{u})\mathbf{u}^T\mathbf{u} \le 0.$$

Furthermore, $\dot{V} = 0$ if and only if $\mathbf{u} = \mathbf{0}_{n\times1}$. Now $\mathbf{u} = \mathbf{0}_{n\times1}$ implies that $-k_1\tilde{\mathbf{v}} - k_2\tilde{\mathbf{x}} = \mathbf{0}_{n\times1}$. Then, one has

$$\dot{\tilde{\mathbf{x}}} = -\frac{k_2}{k_1}\tilde{\mathbf{x}} \tag{10.21}$$

according to (10.19). The system (10.21) cannot get "stuck" at an equilibrium value other than $\tilde{\mathbf{x}} = \mathbf{0}_{n\times1}$. As $-k_1\tilde{\mathbf{v}} - k_2\tilde{\mathbf{x}} = \mathbf{0}_{n\times1}$, one has $\tilde{\mathbf{v}} = \mathbf{0}_{n\times1}$. In other words, with $\mathscr{R}$ being the set defined by $\mathbf{u} = \mathbf{0}_{n\times1}$, the largest invariant set $\mathscr{M}$ in $\mathscr{R}$ contains only one point, namely $\tilde{\mathbf{x}} = \mathbf{0}_{n\times1}$ and $\tilde{\mathbf{v}} = \mathbf{0}_{n\times1}$. Use of the local invariant set theorem indicates that $\mathbf{z} = \mathbf{0}_{2n\times1}$ is locally asymptotically stable. Furthermore, $\|V(\mathbf{z})\| \to \infty$ as $\|\mathbf{z}\| \to \infty$. Then, $\mathbf{z} = \mathbf{0}_{2n\times1}$ is globally asymptotically stable by using the global invariant set theorem. Since the system (10.19) is equivalent to the system (10.14) in a neighborhood of the equilibrium $\mathbf{z} = \mathbf{0}_{2n\times1}$, $\mathbf{z} = \mathbf{0}_{2n\times1}$ is locally exponentially stable according to Theorem 10.8.   □

In the following, the reason why Proportional-Integral-Derivative (PID) controllers can compensate for constant disturbances and track constant desired signals is explained. In order to make this chapter self-contained, the *Barbara's lemma* is introduced as follows.

Barbara's Lemma [1, p. 123]. *If the differentiable function $f(t)$ has a finite limit as $t \to \infty$, and if $\dot{f}$ is uniformly continuous, then $\dot{f}(t) \to 0$ as $t \to \infty$.*

Consider the following system

$$\begin{aligned}\dot{\mathbf{x}} &= \mathbf{v} \\ \dot{\mathbf{v}} &= \mathbf{u} + \mathbf{d} \\ \mathbf{y} &= \mathbf{C}^T\mathbf{x}\end{aligned} \tag{10.22}$$

where $\mathbf{d} \in \mathbb{R}^n$ is a constant external disturbance, $\mathbf{y} \in \mathbb{R}^n$ is the output, $\mathbf{C} \in \mathbb{R}^{n\times n}$ is the output matrix. The control target is to make the output signal $\mathbf{y}$ track the desired signal $\mathbf{y}_d \in \mathbb{R}^n$. A PID controller is designed as follows:

$$\mathbf{u} = -k_1\mathbf{v} - k_2\mathbf{x} - k_3\int\tilde{\mathbf{y}} \tag{10.23}$$

where $k_1, k_2, k_3 \subset \mathbb{R}$, $\tilde{\mathbf{y}} \overset{\triangle}{=} \mathbf{y} - \mathbf{y}_d$. The following conclusion is obtained.

Theorem 10.10 *For system (10.22), a PID controller is designed as (10.23). Suppose that $\mathbf{d}$, $\mathbf{y}_d$ are constants, and the matrix*

$$\mathbf{A} = \begin{bmatrix} \mathbf{0}_{n\times n} & \mathbf{C}^T & \mathbf{0}_{n\times n} \\ \mathbf{0}_{n\times n} & \mathbf{0}_{n\times n} & \mathbf{I}_n \\ -k_3\mathbf{I}_n & -k_2\mathbf{I}_n & -k_1\mathbf{I}_n \end{bmatrix}$$

is stable (the real parts of all eigenvalues of $\mathbf{A}$ are negative). Then $\lim_{t\to\infty}\|\tilde{\mathbf{y}}(t)\| = 0$, and all the states are bounded.

Proof Let $\mathbf{e}_i \triangleq \int \tilde{\mathbf{y}}$. Then

$$\dot{\mathbf{e}}_i = \tilde{\mathbf{y}}. \tag{10.24}$$

According to system (10.22), one has

$$\dot{\mathbf{x}}_a = \mathbf{A}\mathbf{x}_a + \mathbf{d}_a \tag{10.25}$$

where $\mathbf{x}_a \triangleq [\mathbf{e}_i^T \ \mathbf{x}^T \ \mathbf{v}^T]^T$, $\mathbf{d}_a \triangleq [-\mathbf{y}_d^T \ \mathbf{0}_{1 \times n} \ \mathbf{d}^T]^T$. As the real parts of all the eigenvalues of $\mathbf{A}$ are negative and $\mathbf{d}_a$ is a constant signal, $\mathbf{x}_a$ will converge to a constant signal and all states are bounded. Then as one element of $\mathbf{x}_a$, the error $\mathbf{e}_i$ converges to a constant and $\ddot{\mathbf{e}}_i = \dot{\tilde{\mathbf{y}}} = \dot{y} - \dot{\mathbf{y}}_d$ is bounded. Consequently, $\dot{\mathbf{e}}_i$ is uniformly continuous. Then, $\dot{\mathbf{e}}_i$ converges to zero by *Barbara's Lemma*. According to Eq. (10.24), it has

$$\|\tilde{\mathbf{y}}\| = \|\dot{\mathbf{e}}_i\| .$$

Then $\lim_{t \to \infty} \|\tilde{\mathbf{y}}(t)\| = 0$. $\square$

10.3 Basic Concepts of Controllability

10.3.1 Classical Controllability

The concept of controllability is one of the fundamental outcomes of modern control theory and is an important tool in the design of control systems. As originally introduced by Kalman [4], the check for complete controllability via the well-known rank test on the controllability matrix provides an answer to the binary question of whether the system is completely controllable. Consider the following linear time-invariant system

$$\dot{\mathbf{x}} = \mathbf{A}\mathbf{x} + \mathbf{B}\mathbf{u} \tag{10.26}$$

where $\mathbf{x} \in \mathbb{R}^n$, $\mathbf{A} \in \mathbb{R}^{n \times n}$, $\mathbf{B} \in \mathbb{R}^{n \times m}$, $\mathbf{u} \in \mathbb{R}^m$.

Definition 10.11 ([2]) The system (10.26) is *controllable* if for any $\mathbf{x}(t_0)$, there exists a bounded admissible control $\mathbf{u}_{[t_0, t_1]}$ defined on the finite time interval $[t_0, t_1]$, which steers $\mathbf{x}(t_0)$ to zero. Or else, the system (10.26) is *uncontrollable*.

The controllability definition here is also called the *null-controllability*. It can be proved that the null-controllability is equivalent to the controllability which is defined by steering any $\mathbf{x}(t_0)$ to $\mathbf{x}(t_1)$ within a finite time interval. In this chapter, the null-controllability is considered only. The system (10.26) is controllable if and only if the controllability matrix

$$\mathscr{C}(\mathbf{A}, \mathbf{B}) \triangleq [\mathbf{B} \ \mathbf{AB} \ \dots \ \mathbf{A}^{n-1}\mathbf{B}] \tag{10.27}$$

is of full rank, namely rank $\mathscr{C}(\mathbf{A}, \mathbf{B}) = n$.

Example 10.12 Check the controllability of system (10.26), where

$$\mathbf{A} = \begin{bmatrix} -1 & 0 \\ 0 & -1 \end{bmatrix}, \mathbf{B} = \begin{bmatrix} 0 \\ 0 \end{bmatrix}. \tag{10.28}$$

According to definition (10.27), the controllability matrix becomes

$$\mathscr{C}(\mathbf{A}, \mathbf{B}) = \begin{bmatrix} 0 & 0 \\ 0 & 0 \end{bmatrix}. \tag{10.29}$$

Then, the pair $(\mathbf{A}, \mathbf{B})$ is uncontrollable. It is easy to understand this result as $\mathbf{B}u(t) \equiv \mathbf{0}_{2 \times 1}$ no matter what the value of $u(t)$ is. But the solution to Eq. (10.26) is

$$\mathbf{x}(t) = \begin{bmatrix} e^{-t} & 0 \\ 0 & e^{-t} \end{bmatrix} \mathbf{x}(0) \tag{10.30}$$

and then $\lim_{t \to \infty} \|\mathbf{x}(t)\| = 0$. It should be noticed that although any $\mathbf{x}(t_0)$ approaches to zero, the convergence time is not finite. So, system (10.26) is uncontrollable according to Definition 10.11. However, discrete-time linear systems can converge to zero in a finite time even if

$$\mathbf{B}u(k) \equiv \mathbf{0}_{2 \times 1}.$$

Because of this, both *controllability* and *reachability* are defined separately for discrete-time linear systems. The controllability of discrete-time linear systems is similar to the controllability of continuous-time linear systems, while the reachability of discrete-time systems requires the ability to steer any state to any other state rather than zero only.

10.3.2 Positive Controllability

Classical controllability theories on linear systems often require the origin $\mathbf{u} = \mathbf{0}_{m \times 1}$ to be an *interior point*[1] of the control constraint set, such as $\mathbf{u} \in [-1, 1]^m \subset \mathbb{R}^m$. However, the origin is not always inside control constraint set. A simple example is shown in Fig. 10.3b. This section considers the case that the control input is constrained by a positive constraint set. For example, each propeller of multicopters can only provide unidirectional thrust (upward or downward). As a result, the propulsor thrust is constrained by $\mathbf{u} \in [0, K]^{n_r}$, where $K \in \mathbb{R}_+$. In this case, the condition (10.27) is not enough

Fig. 10.3 Interior point

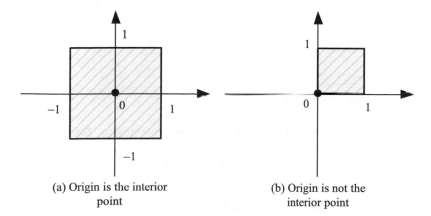

(a) Origin is the interior point

(b) Origin is not the interior point

---
[1] A point is an interior point of $\mathscr{S}$ if there exists an open ball centered at the point which is completely contained in $\mathscr{S}$.

to test the controllability of the dynamical system (10.26), because the origin $\mathbf{u} = \mathbf{0}_{n_r \times 1}$ is not an interior point of the set $[0, K]^{n_r}$. In this section, this kind of constraint set is called *positive constraint*. The controllability of the system with positive constraints is called *positive controllability*.

Definition 10.13 ([5]) The system (10.26) constrained by $\mathbf{u} \in \Omega$ is *completely controllable* if for each pair of points $\mathbf{x}_0, \mathbf{x}_1 \in \mathbb{R}^n$, there exists a bounded admissible control $\mathbf{u}_{[t_0,t_1]} \in \Omega$ defined on the finite time interval $[t_0, t_1]$, which steers $\mathbf{x}_0$ to $\mathbf{x}_1$. Or else, the system (10.26) is *uncontrollable*.

Definition 10.14 ([5]) The system (10.26) constrained by $\mathbf{u} \in \Omega$ is *null-controllable* if there exists an open set $\mathcal{V} \subset \mathbb{R}^n$ which contains the origin and any $\mathbf{x}_0 \in \mathcal{V}$ can be controlled to $\mathbf{x}_1 = \mathbf{0}_{n \times 1}$ in finite time by a bounded admissible controller.

The controllability theory of linear systems with control constraints is given as follows.

Theorem 10.15 ([5]) *If there exists $\mathbf{u} \in \Omega$ satisfying $\mathbf{Bu} = 0$, and the set $CH(\Omega)^2$ has nonempty interior in $\mathbb{R}^m$, then the following conditions are necessary and sufficient for the controllability of system (10.26):*

(1) rank $\mathscr{C}(\mathbf{A}, \mathbf{B}) = n$;
(2) There is no real eigenvector $\mathbf{v}$ of $\mathbf{A}^\mathrm{T}$ satisfying $\mathbf{v}^\mathrm{T}\mathbf{Bu} \leq 0$ for all $\mathbf{u} \in \Omega$.

The following example will explain why condition (2) in Theorem 10.15 is necessary. First, consider a simple system

$$\dot{x} = x + u, u \leq 0 \tag{10.31}$$

which is controllable based on the classical controllability test method. However, the state $x \to -\infty$ if the initial value $x_0 < 0$ no matter how close to zero it is. This is because $x + u \leq 0$ all the time. Similarly, suppose that there exists a real eigenvector $\mathbf{v}_0$ of $\mathbf{A}^\mathrm{T}$ satisfying $\mathbf{v}_0^\mathrm{T}\mathbf{Bu} \leq 0$ for all $\mathbf{u} \in \Omega$ and $\mathbf{A}^\mathrm{T}\mathbf{v}_0 = \lambda\mathbf{v}_0$. Then

$$\mathbf{v}_0^\mathrm{T}\dot{\mathbf{x}} = \mathbf{v}_0^\mathrm{T}\mathbf{Ax} + \mathbf{v}_0^\mathrm{T}\mathbf{Bu} \overset{y = \mathbf{v}_0^\mathrm{T}\mathbf{x}}{\Rightarrow} \dot{y} = \lambda y + \mathbf{v}_0^\mathrm{T}\mathbf{Bu}.$$

As $\mathbf{v}_0^\mathrm{T}\mathbf{Bu} \leq \mathbf{0}$ for all $\mathbf{u} \in \Omega$, the state $y(t) \to -\infty$ as $t \to \infty$ if $\lambda > 0$ and the initial value $y_0 < 0$.

10.4   Controllability of Multicopters

This section considers a class of multicopters shown in Fig. 10.4, which are often used in practice.

10.4.1 Multicopter System Modeling

According to [6], the linear dynamic model of a multicopter around the hovering state is given as

$$\dot{\mathbf{x}} = \mathbf{Ax} + \mathbf{B}\underbrace{(\mathbf{u}_f - \mathbf{g})}_{\mathbf{u}} \tag{10.32}$$

$^2 CH(\Omega)$ is the convex hull of Ω, which is defined as the intersection of all convex sets which includes Ω in $\mathbb{R}^m$.

Fig. 10.4 Different configurations of multicopters (the *white disc* denotes that the propeller rotates clockwise and the *black disc* denotes that the propeller rotates counterclockwise)

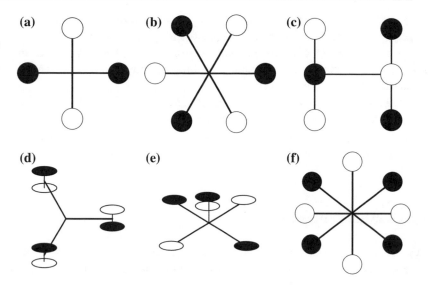

where

$$\mathbf{x} = \begin{bmatrix} p_{z_e} \ \phi \ \theta \ \psi \ v_{z_e} \ \omega_{x_b} \ \omega_{y_b} \ \omega_{z_b} \end{bmatrix}^{\mathrm{T}} \in \mathbb{R}^8$$

$$\mathbf{u}_f = \begin{bmatrix} f \ \tau_x \ \tau_y \ \tau_z \end{bmatrix}^{\mathrm{T}} \in \mathbb{R}^4$$

$$\mathbf{g} = [mg \ 0 \ 0 \ 0]^{\mathrm{T}} \in \mathbb{R}^4$$

$$\mathbf{A} = \begin{bmatrix} \mathbf{0}_{4\times4} & \mathbf{I}_4 \\ \mathbf{0}_{4\times4} & \mathbf{0}_{4\times4} \end{bmatrix} \in \mathbb{R}^{8\times8} \tag{10.33}$$

$$\mathbf{B} = \begin{bmatrix} \mathbf{0}_{4\times4} \\ \mathbf{J}_f^{-1} \end{bmatrix} \in \mathbb{R}^{8\times4}$$

$$\mathbf{J}_f = \mathrm{diag}\left(-m, J_{xx}, J_{yy}, J_{zz}\right) \in \mathbb{R}^{4\times4}.$$

For convenience of expression, the first element of $\mathbf{J}_f$ is $-m$ rather than m. In practice, the thrust of each propeller is $T_i \in [0, K_i]$, $i = 1, \ldots, n_r$ since the propellers are assumed to provide unidirectional thrust (upward or downward). As a result, the propeller thrust $\mathbf{f} = \begin{bmatrix} T_1 \ \ldots \ T_{n_r} \end{bmatrix}^{\mathrm{T}}$ is constrained by

$$\mathbf{f} \in \mathscr{U}_f = \Pi_{i=1}^{n_r} [0, K_i] \tag{10.34}$$

where $K_i \in \mathbb{R}_+, i = 1, \ldots, n_r$. Then, according to the geometry of multicopters, the mapping from the propeller thrust $T_i, i = 1, \ldots, n_r$ to the system total thrust/moment $\mathbf{u}_f$ is

$$\mathbf{u}_f = \mathbf{B}_f \mathbf{f}. \tag{10.35}$$

The matrix $\mathbf{B}_f \in \mathbb{R}^{4\times m}$ is the control effectiveness matrix. According to Eq. (6.25) in Chap. 6, the control effectiveness matrix $\mathbf{B}_f$ in parameterized form is

$$\mathbf{B}_f = \begin{bmatrix} \eta_1 & \cdots & \eta_{n_r} \\ -\eta_1 d_1 \sin \varphi_1 & \cdots & -\eta_{n_r} d_{n_r} \sin \varphi_{n_r} \\ \eta_1 d_1 \cos \varphi_1 & \cdots & \eta_{n_r} d_{n_r} \cos \varphi_{n_r} \\ \eta_1 k_\mu \sigma_1 & \cdots & \eta_{n_r} k_\mu \sigma_{n_r} \end{bmatrix} \tag{10.36}$$

where the parameters $\eta_i \in [0, 1]$, $i = 1, \ldots, n_r$ are used to represent wear/failure of the propulsion system, $k_\mu = c_M/c_T$ is the ratio between the reactive torque and thrust of propulsors. If the ith propulsor (including a propeller, an Electronic Speed Controller (ESC) and a motor) or the battery fails, then $\eta_i = 0$; if the ith propulsor and the battery are fully healthy, then $\eta_i = 1$. This section considers the controllability of the system (10.32).

Classical controllability theories of linear systems often require the origin to be an interior point of $\mathscr{U}$ so that $\mathscr{C}(\mathbf{A}, \mathbf{B})$ being of full row rank is a necessary and sufficient condition [5]. However, the origin is not always inside control constraint set $\mathscr{U}$ of system (10.32) under propulsor failures. Consequently, $\mathscr{C}(\mathbf{A}, \mathbf{B})$ being of full row rank is not sufficient to test the controllability of system (10.32).

10.4.2 Classical Controllability

The system (10.32) needs to be rewritten if the classical controllability analysis method is applied. First, the gravity $\mathbf{g}$ is allocated to each propulsor. If the pseudoinverse matrix method is used, then

$$\mathbf{g} = \mathbf{B}_f \mathbf{f}' \tag{10.37}$$

where $\mathbf{f}' = \mathbf{B}_f^T \left(\mathbf{B}_f \mathbf{B}_f^T\right)^{-1} \mathbf{g}$. According to Eq. (10.37), Eq. (10.32) is rewritten as

$$\begin{aligned}\dot{\mathbf{x}} &= \mathbf{A}\mathbf{x} + \mathbf{B}\left(\mathbf{B}_f \mathbf{f} - \mathbf{g}\right) \\ &= \mathbf{A}\mathbf{x} + \mathbf{B}\left(\mathbf{B}_f \mathbf{f} - \mathbf{B}_f \bar{\mathbf{f}}\right) \\ &= \mathbf{A}\mathbf{x} + \mathbf{B}'\mathbf{f}_u\end{aligned} \tag{10.38}$$

where $\mathbf{f}_u = \mathbf{f} - \mathbf{f}'$ and $\mathbf{B}' = \mathbf{B}\mathbf{B}_f \in \mathbb{R}^{8 \times n_r}$. By using the control allocation method, the control $\mathbf{f}'$ is obtained to compensate for the gravity. According to Eq. (10.34), the remaining control constraints are obtained as follows:

$$\mathbf{f}_u \in \mathscr{U}_{f_u} = \Pi_{i=1}^{n_r}\left[-f_i', K_i - f_i'\right] \tag{10.39}$$

where $f_i' \in \mathbb{R}$ is the ith element of $\mathbf{f}'$. If the control constraint set $\mathscr{U}_{f_u}$ has zero as its interior, and the controllability matrix

$$\mathscr{C}(\mathbf{A}, \mathbf{B}') = \left[\mathbf{B}' \ \mathbf{A}\mathbf{B}' \ \ldots \ \mathbf{A}^7\mathbf{B}'\right] \tag{10.40}$$

is of full rank, then system (10.32) is controllable.

As shown above, the classical controllability analysis method needs to know the constant disturbance $\mathbf{g}$ and the control effectiveness matrix $\mathbf{B}_f$. Moreover, it depends on the control allocation method. In practice, the disturbance is observed online, and the control effectiveness matrix is not fixed for the system with propulsor failures. It is hard to get the control constraint set expressed by (10.39). In the following, a new controllability analysis method is shown based on the positive controllability theory.

10.4.3 Positive Controllability

Designers may encounter a situation where a system is controllable but it becomes uncontrollable if one of the propulsors is moved by a distance $\varepsilon \in \mathbb{R}_+$ no matter how small ε is. In this case, although the system is controllable in theory, it is useless since the system will be extremely hard to tackle. In practice, the designers need to know DoC, based on which the number and locations of propulsors can be chosen in order to control the system easily. To some degree, it is easier to control a system with a higher DoC. The DoC was first addressed in an early study by Kalman [7]. Later, more research on the DoC was found in [8–10]. Also, the DoC concept is very important in flight control [11]. In this chapter, the DoC of multicopers with and without propulsor failures is focused on. Fault-tolerant control [12–16] may have the ability to maintain or gracefully degrade control objectives despite the occurrence of a fault. However, not all the faults can be addressed by fault-tolerant strategies if the faulty system has a low DoC [17, 18]. The DoC with different faults can be used to direct the design of a flight control system. By evaluating the DoC at different failure modes, a suitable fault-tolerant controller can be chosen [19].

In this chapter, a DoC is proposed based on the *available control authority* of system (10.32). By Eqs. (10.34) and (10.35), $\mathbf{u}_f$ is subject to

$$\Omega = \left\{\mathbf{u}_f | \mathbf{u}_f = \mathbf{B}_f \mathbf{f}, \mathbf{f} \in \mathscr{U}_f\right\}. \tag{10.41}$$

Then, $\mathbf{u}$ is constrained by

$$\mathscr{U} = \left\{\mathbf{u} | \mathbf{u} = \mathbf{u}_f - \mathbf{g}, \mathbf{u}_f \in \Omega\right\}. \tag{10.42}$$

From Eqs. (10.34), (10.41), and (10.42), $\mathscr{U}_f$, Ω, $\mathscr{U}$ are all convex and closed.

According to Definition 10.11, system (10.32) subject to constraint set $\mathscr{U} \subset \mathbb{R}^4$ is controllable if, for each pair of points $\mathbf{x}_0 \subset \mathbb{R}^8$ and $\mathbf{x}_1 \in \mathbb{R}^8$, there exists a bounded admissible control $\mathbf{u}(t) \in \mathscr{U}$, defined on some finite time interval $[0, t_1]$, which steers $\mathbf{x}_0$ to $\mathbf{x}_1$. Specifically, the solution to (10.32) described by $\mathbf{x}(t, \mathbf{u}(\cdot))$ satisfies the boundary conditions $\mathbf{x}(0, \mathbf{u}(\cdot)) = \mathbf{x}_0$ and $\mathbf{x}(t_1, \mathbf{u}(\cdot)) = \mathbf{x}_1$.

In this section, the controllability of system (10.32) is studied based on the positive controllability theory proposed in [5]. By applying Theorem 10.15 to system (10.32) directly, the following theorem is obtained.

Theorem 10.16 *The following conditions are necessary and sufficient for the controllability of system (10.32):*

(1) rank $\mathscr{C}(\mathbf{A}, \mathbf{B}) = 8$, where $\mathscr{C}(\mathbf{A}, \mathbf{B}) = \begin{bmatrix} \mathbf{B} \ \mathbf{AB} \ldots \mathbf{A}^7\mathbf{B} \end{bmatrix}$.
(2) There is no real eigenvector $\mathbf{v}$ of $\mathbf{A}^T$ satisfying $\mathbf{v}^T\mathbf{Bu} \leq 0$ for all $\mathbf{u} \in \mathscr{U}$.

In the following, an easy-to-use criterion is proposed to test the condition (2) in Theorem 10.16 intuitively. Before going further, a measure is defined as

$$\rho(\boldsymbol{\alpha}, \partial\Omega) \triangleq \begin{cases} \min\{\|\boldsymbol{\alpha} - \boldsymbol{\beta}\| : \boldsymbol{\alpha} \in \Omega, \boldsymbol{\beta} \in \partial\Omega\} \\ -\min\{\|\boldsymbol{\alpha} - \boldsymbol{\beta}\| : \boldsymbol{\alpha} \in \Omega^C, \boldsymbol{\beta} \in \partial\Omega\} \end{cases} \tag{10.43}$$

where $\partial\Omega$ is the boundary of Ω, and Ω^C is the complementary set of Ω. If $\rho(\boldsymbol{\alpha}, \partial\Omega) \leq 0$, then $\boldsymbol{\alpha} \in \Omega^C \cup \partial\Omega$, which means that $\boldsymbol{\alpha}$ is not an interior point of Ω. Otherwise, $\boldsymbol{\alpha}$ is an interior point of Ω. According to Eq. (10.43), if $\mathbf{g}$ is an interior point of Ω, one has

$$\rho(\mathbf{g}, \partial\Omega) = \min\{\|\mathbf{g} - \mathbf{u}_f\|, \mathbf{u}_f \in \partial\Omega\}$$

which is the radius of the biggest enclosed sphere centered at $\mathbf{g}$ in the attainable control set Ω. Suppose that Ω is two-dimensional and is shown in Fig. 10.5. The point $\mathbf{g}$ is in the interior of Ω. As shown in Fig. 10.5, $\rho(\mathbf{g}, \partial\Omega)$ is the biggest enclosed sphere centered at $\mathbf{g}$. In practice, it is the maximum control thrust/moment that can be produced in all directions. Therefore, it is an important quantity to

Fig. 10.5 The distance
from the **g** to the boundary
of Ω

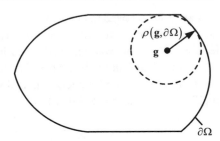

ensure controllability for wear/failure of the propulsion system. Then, $\rho\,(\mathbf{g}, \partial\Omega)$ is used to quantify the available control authority of system (10.32). Here, the available control authority is the control authority left after the system total thrust/moment $\mathbf{u}_f$ compensating for the disturbance $\mathbf{g}$ in system (10.32). The Available Control Authority Index (ACAI) for system (10.32) is expressed by $\rho\,(\mathbf{g}, \partial\Omega)$.

According to (10.42), all elements in $\mathscr{U}$ are given by translating all elements in Ω by a constant $\mathbf{g}$ (see Fig. 10.6). As a translation does not change the relative position of all the elements of Ω, the value of $\rho\,(\mathbf{0}_{8\times1}, \partial\mathscr{U})$ is equal to that of $\rho\,(\mathbf{g}, \partial\Omega)$. Here, the ACAI of system (10.32) is defined by $\rho\,(\mathbf{g}, \partial\Omega)$ because Ω is the attainable control set and more intuitive than $\mathscr{U}$ in practice.

With definition (10.43), the following lemma is obtained.

Lemma 10.17 ([20]) *The following three statements are equivalent to each other for system (10.32):*

(1) There is no real eigenvector $\mathbf{v}$ of $\mathbf{A}^{\mathrm{T}}$ satisfying $\mathbf{v}^{\mathrm{T}}\mathbf{B}\mathbf{u} \leq 0$ for all $\mathbf{u} \in \mathscr{U}$ or $\mathbf{v}^{\mathrm{T}}\mathbf{B}\,(\mathbf{u}_f - \mathbf{g}) \leq 0$ for all $\mathbf{u}_f \in \Omega$.
(2) $\mathbf{g}$ is an interior point of Ω.
(3) $\rho\,(\mathbf{g}, \partial\Omega) > 0$.

Proof See *Appendix* for details.   □

By Lemma 10.17, condition (2) in Theorem 10.16 can be tested by the value $\rho\,(\mathbf{g}, \partial\Omega)$. Now a new necessary and sufficient condition is derived to test the controllability of the system (10.32).

Theorem 10.18 *System (10.32) is controllable if and only if $\rho\,(\mathbf{g}, \partial\Omega) > 0$.*

Proof Because of the simple $(\mathbf{A}, \mathbf{B})$ structure, it is easy to see that rank $\mathscr{C}\,(\mathbf{A}, \mathbf{B}) = 8$ and then the condition (1) of Theorem 10.16 is satisfied. According to Lemma 10.17, the proof of Theorem 10.18 is straightforward by Theorem 10.16.   □

Fig. 10.6 Set translation

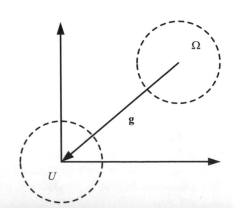

Actually, Theorem 10.18 is a corollary of Theorem 1.4 presented in [5]. To make this section more comprehensive and self-contained, condition (1.6) of Theorem 1.4 presented in [5] is extended, and Theorem 10.18 of this chapter is obtained based on the simplified structure of $(\mathbf{A}, \mathbf{B})$ pair and the convexity of $\mathscr{U}$. Then, the controllability of system (10.32) is determined by the value of $\rho\,(\mathbf{g}, \partial\Omega)$. If $\rho\,(\mathbf{g}, \partial\Omega) > 0$, system (10.32) is controllable, and if $\rho\,(\mathbf{g}, \partial\Omega) < 0$, system (10.32) is uncontrollable. A larger $\rho\,(\mathbf{g}, \partial\Omega)$ implies a more controllable system. This extension can enable the quantification of the controllability and also makes it possible to develop a step-by-step controllability test procedure for a multicopter system. A step-by-step controllability test procedure is given by [20].

10.4.4   Controllability of Multicopter Systems

10.4.4.1   Controllability Analysis For a Class of Hexacopters

In this section, the controllability test procedure developed in [20] is used to analyze the controllability of a class of hexacopters shown in Fig. 10.7, subject to propulsor wear/failures, to show its effectiveness. The parameters of the hexacopter are shown in Table 10.1. The propulsor arrangement of the considered hexacopter is the standard symmetrical plus-configuration shown in Fig. 10.7a. According to Eq. (10.36), the control effectiveness matrix $\mathbf{B}_f$ of the PNPNPN hexacopter configuration is

$$\mathbf{B}_f = \begin{bmatrix} \eta_1 & \eta_2 & \eta_3 & \eta_4 & \eta_5 & \eta_6 \\ 0 & -\frac{\sqrt{3}}{2}\eta_2 d & -\frac{\sqrt{3}}{2}\eta_3 d & 0 & \frac{\sqrt{3}}{2}\eta_5 d & \frac{\sqrt{3}}{2}\eta_6 d \\ \eta_1 d & \frac{1}{2}\eta_2 d & -\frac{1}{2}\eta_3 d & -\eta_4 d & -\frac{1}{2}\eta_5 d & \frac{1}{2}\eta_6 d \\ -\eta_1 k_\mu & \eta_2 k_\mu & -\eta_3 k_\mu & \eta_4 k_\mu & -\eta_5 k_\mu & \eta_6 k_\mu \end{bmatrix} \tag{10.44}$$

and the control effectiveness matrix $\mathbf{B}_f$ of the PPNNPN hexacopter configuration is

$$\mathbf{B}_f = \begin{bmatrix} \eta_1 & \eta_2 & \eta_3 & \eta_4 & \eta_5 & \eta_6 \\ 0 & -\frac{\sqrt{3}}{2}\eta_2 d & -\frac{\sqrt{3}}{2}\eta_3 d & 0 & \frac{\sqrt{3}}{2}\eta_5 d & \frac{\sqrt{3}}{2}\eta_6 d \\ \eta_1 d & \frac{1}{2}\eta_2 d & -\frac{1}{2}\eta_3 d & -\eta_4 d & -\frac{1}{2}\eta_5 d & \frac{1}{2}\eta_6 d \\ -\eta_1 k_\mu & -\eta_2 k_\mu & \eta_3 k_\mu & \eta_4 k_\mu & -\eta_5 k_\mu & \eta_6 k_\mu \end{bmatrix}. \tag{10.45}$$

By using the procedure proposed in [20], the controllability analysis results of the PNPNPN hexacopter subject to one propulsor failure are shown in Table 10.2. The PNPNPN hexacopter is uncontrollable when one propulsor fails, even though its controllability matrix is of full row rank.

Table 10.1 Hexacopter parameters

Parameter	Value	Units
m	1.535	kg
g	9.80	m/s^2
d	0.275	m
$K_i, i = 1, \ldots, 6$	6.125	N
J_{xx}	0.0411	kg · m^2
J_{yy}	0.0478	kg · m^2
J_{zz}	0.0599	kg · m^2
k_μ	0.1	–

Fig. 10.7 **a** Standard propulsor arrangement PNPNPN, **b** new propulsor arrangement PPNNPN, **c** the 1st propulsor of the PNPNPN hexacopter fails, **d** the 1st propulsor of the PPNNPN hexacopter fails. PNPNPN is used to denote the standard arrangement, where P denotes that a propulsor rotates clockwise and N denotes that a propulsor rotates counterclockwise

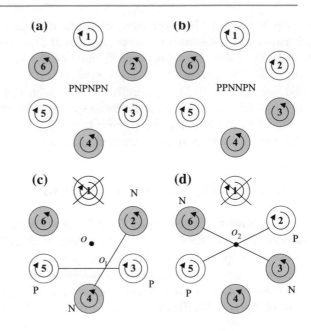

Table 10.2 Hexacopter (PNPNPN and PPNNPN) controllability with one propulsor failure

Propulsor failure	Rank of $\mathscr{C}(A, B)$	PNPNPN		PPNNPN	
		ACAI	Controllability	ACAI	Controllability
No wear/failure	8	1.4861	Controllable	1.1295	Controllable
$\eta_1 = 0$	8	0	Uncontrollable	0.7221	Controllable
$\eta_2 = 0$	8	0	Uncontrollable	0.4510	Controllable
$\eta_3 = 0$	8	0	Uncontrollable	0.4510	Controllable
$\eta_4 = 0$	8	0	Uncontrollable	0.7221	Controllable
$\eta_5 = 0$	8	0	Uncontrollable	0	Uncontrollable
$\eta_6 = 0$	8	0	Uncontrollable	0	Uncontrollable

A new propulsor arrangement (PPNNPN) of the hexacopter shown in Fig. 10.7b is proposed in [21], which is still controllable when one of some specific propulsors stops working. The controllability of the PPNNPN hexacopter subject to one propulsor failure is shown in Table 10.2.

From Table 10.2, ACAI is 1.4861 for the PNPNPN hexacopter without propulsor failures, whereas the ACAI is reduced to 1.1295 for the PPNNPN hexacopter. It is observed that the use of the PPNNPN configuration rather than the PNPNPN configuration improves the fault-tolerance capabilities, but in turn reduces the ACAI for the no failure condition. Similar to the results in [21], there is a trade-off between fault-tolerance and control authority. The PPNNPN system is not always controllable under a failure. From Table 10.2, it can be seen that the PPNNPN system is uncontrollable if the 5th propulsor or the 6th propulsor fails. Therefore, it is necessary to test the controllability of the multicopters before any fault-tolerant control strategy is employed. Moreover, the controllability test procedure can also be used to test the controllability of the hexacopter with different η_i, $i \in \{1, \ldots, 6\}$. Let η_1, η_2, η_5 vary in $[0, 1] \subset \mathbb{R}$, namely propulsor 1, propulsor 2, and propulsor 5 are worn. Then, the PNPNPN hexacopter retains controllability, while η_1, η_2, η_5 are in the grid region (where the grid spacing is 0.04) in Fig. 10.8. The corresponding ACAI near the boundaries of the projections shown in Fig. 10.8 is zero.

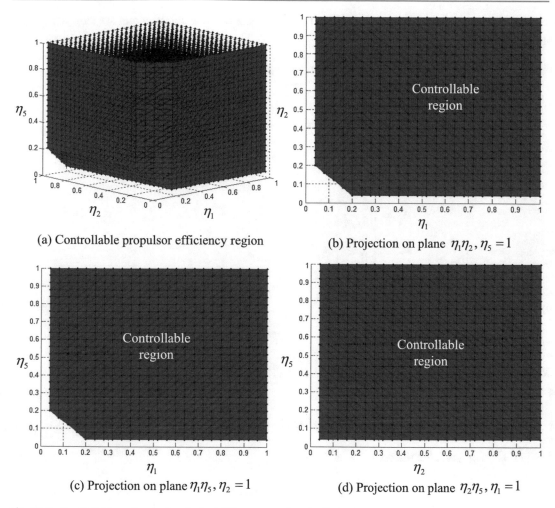

(a) Controllable propulsor efficiency region

(b) Projection on plane $\eta_1\eta_2, \eta_5 = 1$

(c) Projection on plane $\eta_1\eta_5, \eta_2 = 1$

(d) Projection on plane $\eta_2\eta_5, \eta_1 = 1$

Fig. 10.8 Controllable region (*red region*) of different propulsors' efficiency parameter for the PNPNPN hexacopter

The following provides some physical insight for the two configurations. For the PPNNPN configuration, if one of the propulsors (other than the 5th and 6th propulsor) of that system fails, the remaining propulsors are still composed of a basic quadcopter configuration that is symmetric about the mass center (see Fig. 10.7d). In contrast, if one propulsor of the PNPNPN system fails, although the remaining propulsors can make up a basic quadcopter configuration, the quadcopter configuration is asymmetric about the mass center (see Fig. 10.7c). Or, intuitively, if the 4th propulsor is stopped intentionally, the remaining propulsors are PPNN. Such a quadcopter will lose the controllability upon hovering. This is left as an exercise at the end of this chapter. The result is that the PPNNPN system under most single propulsor failures can provide the necessary thrust and moment control, while the PNPNPN system cannot.

10.4.4.2 Controllability for a Degraded System

This section considers a PNPNPN hexacopter. According to the analysis results in [19], the PNPNPN hexacopter system is uncontrollable if one propulsor fails. But the yaw states of the hexacopter can be left uncontrolled for safe landing [19]. The degraded system where the yaw and yaw rate are removed from Eq. (10.32) is given as

$$\dot{\bar{\mathbf{x}}} = \bar{\mathbf{A}}\bar{x} + \bar{B}\underbrace{(\bar{\mathbf{u}}_f - \bar{\mathbf{g}})}_{\mathbf{u}} \tag{10.46}$$

where

$$
\begin{aligned}
\bar{\mathbf{x}} &= \left[p_{z_e}\ \phi\ \theta\ v_{z_e}\ \omega_{x_b}\ \omega_{y_b} \right]^{\mathrm{T}} \in \mathbb{R}^6 \\
\bar{\mathbf{u}}_f &= \left[f\ \tau_x\ \tau_y \right]^{\mathrm{T}} \in \mathbb{R}^3 \\
\bar{\mathbf{g}} &= \left[mg\ 0\ 0 \right]^{\mathrm{T}} \in \mathbb{R}^3 \\
\bar{\mathbf{A}} &= \begin{bmatrix} \mathbf{0}_{3\times3} & \mathbf{I}_3 \\ \mathbf{0}_{3\times3} & \mathbf{0}_{3\times3} \end{bmatrix} \in \mathbb{R}^{6\times6} \\
\bar{\mathbf{B}} &= \begin{bmatrix} \mathbf{0}_{3\times3} \\ \bar{\mathbf{J}}_f^{-1} \end{bmatrix} \in \mathbb{R}^{8\times4} \\
\bar{\mathbf{J}}_f &= \mathrm{diag}\left(-m, J_{xx}, J_{yy} \right).
\end{aligned}
\tag{10.47}
$$

The propeller thrusts are constrained by

$$\mathbf{f} \in \mathscr{U}_f = [0, K_i]^6 \tag{10.48}$$

where K_i is the maximum thrust of each propulsor, $i = 1, \ldots, 6$. The mapping between $\mathbf{f}$ and $\bar{\mathbf{u}}_f$ is expressed by

$$\bar{\mathbf{u}}_f = \bar{\mathbf{B}}_f \mathbf{f} \in \bar{\Omega}. \tag{10.49}$$

Here, $\mathbf{f} = [T_1\ \ldots\ T_6]^{\mathrm{T}}$, and the matrix $\bar{\mathbf{B}}_f \in \mathbb{R}^{3\times6}$ is

$$\bar{\mathbf{B}}_f = \begin{bmatrix} \eta_1 c_{\mathrm{T}} & \eta_2 c_{\mathrm{T}} & \eta_3 c_{\mathrm{T}} & \eta_4 c_{\mathrm{T}} & \eta_5 c_{\mathrm{T}} & \eta_6 c_{\mathrm{T}} \\ 0 & -\frac{\sqrt{3}}{2}\eta_2 d c_{\mathrm{T}} & -\frac{\sqrt{3}}{2}\eta_3 d c_{\mathrm{T}} & 0 & \frac{\sqrt{3}}{2}\eta_5 d c_{\mathrm{T}} & \frac{\sqrt{3}}{2}\eta_6 d c_{\mathrm{T}} \\ \eta_1 d c_{\mathrm{T}} & \frac{1}{2}\eta_2 d c_{\mathrm{T}} & -\frac{1}{2}\eta_3 d c_{\mathrm{T}} & -\eta_4 d c_{\mathrm{T}} & -\frac{1}{2}\eta_5 d c_{\mathrm{T}} & \frac{1}{2}\eta_6 d c_{\mathrm{T}} \end{bmatrix} \tag{10.50}$$

where the parameters $\eta_i \in [0, 1]$, $i = 1, \ldots, 6$ are used to represent wear/failure of the propulsion system. If the ith propulsor or the battery fails, then $\eta_i = 0$. If system (10.46) is controllable, then a degraded control strategy can be designed to keep the multicopter level and undergo safe landing.

In the following, the controllability of system (10.46) is analyzed based on the ACAI in [20]. The parameters of the considered hexacopter are shown in Table 10.1. If the 2nd propulsor fails, then the ACAI $\rho\left(\bar{\mathbf{g}}, \partial\bar{\Omega}\right)$ for system (10.46) is 1.2882, namely system (10.46) is controllable. Thus, the hexacopter can land safely with a degraded control strategy. However, if the parameters in Table 10.1 vary, system (10.46) may be uncontrollable if the 2nd propulsor fails. It is assumed in this section that the maximum thrust of each propulsor is $K = \gamma mg, \gamma \in [0, 1]$. The ACAI of system (10.46) is shown in Table 10.3. From Table 10.3, it is observed that system (10.46) is uncontrollable if $K < 0.3$ mg. Although the thrust provided by the left five propulsors is still able to compensate for the weight of the hexacopter, almost no control strategy can land the multicopter safely. The ACAI of system (10.46) remains almost the same if $K \geq 0.60$ mg. The degraded control strategies can be found in [20, 22].

Table 10.3 ACAI with different γ

γ	$\rho\left(\bar{\mathbf{G}}, \partial\bar{\Omega}\right)$	γ	$\rho\left(\bar{\mathbf{G}}, \partial\bar{\Omega}\right)$
0.05	−0.1791	0.55	1.9704
0.10	−0.3583	0.60	2.0491
0.15	−0.1025	0.65	2.0491
0.20	0	0.70	2.0491
0.25	0	0.75	2.0491
0.30	0.4098	0.80	2.0491
0.35	0.8197	0.85	2.0491
0.40	1.2295	0.90	2.0491
0.45	1.6122	0.95	2.0491
0.50	1.7913	1.00	2.0491

10.4.5 Further Discussions

Further discussions are given as follows:

(1) Control constraints, as well as the configuration, determine the controllability of a multicopter system. In [19], the maximum thrust of the remaining five propulsors of the PNPNPN hexacopter needs to satisfy $K \geq 5\,mg/18\,(\approx 0.3\,mg)$ to achieve a safe landing theoretically. On the other hand, as mentioned in Table 10.3, $K \geq 0.3\,mg$ (which is computed numerically) is also a guidance to design a hexacopter for the degraded system (five propulsors work together). The two results are consistent with each other.

(2) Some researchers in [23] presented periodic solutions to control a quadcopter after losing one, two opposing or even three propellers. It seems a contradiction with the proposed results above. The quadcopter with propulsor failures is uncontrollable at hover based on the controllability analysis method, namely the damaged quadcopter cannot hover. It should be noticed that this does not mean that the quadcopter cannot stay in the air anymore. In [23], new equilibrium states (not constant positions) are established for the quadcopter, which is controllable around the new equilibrium states. This is also an example that controllability analysis requires to clarify the equilibrium states first, and the model afterward.

10.5 Summary

In practice, the designers often design the configuration, structure, and propulsion system first, then the flight control system. According to this procedure, the designed multicopter may be hard to control or even cannot be controlled. This chapter provides an example that small thrust provided by propellers may lead to uncontrollability, even if they can compensate for the weight of the multicopter. Furthermore, by revisiting Fig. 1.4, a compound multicopter was designed, where propellers with slow dynamic response were designed to provide the main thrust, while the propellers with fast dynamic response were designed for the attitude control. This is because the DoC can be improved compared with that of a counterpart with four propellers having slow dynamic response. Besides, the DoC can be used to evaluate the wind resistance of a multicopter, where wind is considered as a disturbance on the system. Based on the DoC, the configuration design can be evaluated before a prototype multicopter is produced. This will save lots of resources. Also, the DoC can be used to assess the safety of a multicopter in flight, namely to guarantee that the system is sufficiently controllable. If the DoC is lower than a threshold, then the multicopter is required to switch to a more controllable mode.

10.6   Appendix: Proof of Lemma 10.17

In order to make this chapter self-contained, the following lemma is introduced.

Lemma 10.19 ([24]) *If Ω is a nonempty convex set in $\mathbb{R}^4$ and $\mathbf{u}_{f0}$ is not an interior point of Ω, then there is a nonzero vector $\mathbf{k}$ such that $\mathbf{k}^{\mathrm{T}} \left(\mathbf{u}_f - \mathbf{u}_{f0} \right) \leq 0$ for each $\mathbf{u}_f \in cl\,(\Omega)$, where $cl\,(\Omega)$ is the closure of Ω.*

Proof of Lemma 10.17 (1) $\Rightarrow$ (2): Suppose that (1) holds. It is easy to see that all eigenvalues of $\mathbf{A}^{\mathrm{T}}$ are zero. By solving the linear equation $\mathbf{A}^{\mathrm{T}}\mathbf{v} = \mathbf{0}_{8\times 1}$, all eigenvectors of $\mathbf{A}^{\mathrm{T}}$ are expressed in the following form

$$\mathbf{v} = [0\,0\,0\,0\,k_1\,k_2\,k_3\,k_4]^{\mathrm{T}} \tag{10.51}$$

where $\mathbf{v} \neq \mathbf{0}_{8\times 1}$, $\mathbf{k} = [k_1\,k_2\,k_3\,k_4]^{\mathrm{T}} \in \mathbb{R}^4$, and $\mathbf{k} \neq \mathbf{0}_{4\times 1}$. With it, one has

$$\mathbf{v}^{\mathrm{T}}\mathbf{B}\mathbf{u} = -k_1 \frac{f - mg}{m} + k_2 \frac{\tau_x}{J_{xx}} + k_3 \frac{\tau_y}{J_{yy}} + k_4 \frac{\tau_z}{J_{zz}}. \tag{10.52}$$

By Lemma 10.19, if $\mathbf{g}$ is not an interior point of Ω, then $\mathbf{u} = \mathbf{0}_{4\times 1}$ is not an interior point of $\mathscr{U}$. Then, there is a nonzero $\mathbf{k}_u = [k_{u1}\,k_{u2}\,k_{u3}\,k_{u4}]^{\mathrm{T}}$ satisfying

$$\mathbf{k}_u^{\mathrm{T}}\mathbf{u} = k_{u1}\,(f - mg) + k_{u2}\tau_x + k_{u3}\tau_y + k_{u4}\tau_z \leq 0$$

for all $\mathbf{u} \in \mathscr{U}$. Let

$$\mathbf{k} = \left[-k_{u1}m\; k_{u2}J_{xx}\; k_{u3}J_{yy}\; k_{u4}J_{zz} \right]^{\mathrm{T}}. \tag{10.53}$$

Then, $\mathbf{v}^{\mathrm{T}}\mathbf{B}\mathbf{u} \leq 0$ for all $\mathbf{u} \in \mathscr{U}$ according to (10.52), which contradicts Theorem 10.16.

(2) $\Rightarrow$ (1): It will be proved by a counterexample. All the eigenvectors of $\mathbf{A}^{\mathrm{T}}$ are expressed in the form of Eq. (10.51). Then,

$$\mathbf{v}^{\mathrm{T}}\mathbf{B}\mathbf{u} = \mathbf{k}^{\mathrm{T}}\mathbf{J}_f^{-1}\mathbf{u}$$

according to Eqs. (10.33) and (10.51), where $\mathbf{k} \neq \mathbf{0}_{4\times 1}$. Then, the statement that there is no nonzero $\mathbf{v} \in \mathbb{R}^8$ expressed by (10.51) satisfying $\mathbf{v}^{\mathrm{T}}\mathbf{B}\mathbf{u} \leq 0$ for all $\mathbf{u} \in \mathscr{U}$ is equivalent to that there is no nonzero $\mathbf{k} \in \mathbb{R}^4$ satisfying $\mathbf{k}^{\mathrm{T}}\mathbf{J}_f^{-1}\mathbf{u} \leq 0$ for all $\mathbf{u} \in \mathscr{U}$. Suppose that (2) is true. Then, $\mathbf{u} = \mathbf{0}_{4\times 1}$ is an interior point of $\mathscr{U}$. There is a neighborhood $\mathscr{B}\,(\mathbf{0}_{4\times 1}, u_r)$ of $\mathbf{u} = \mathbf{0}_{4\times 1}$ belonging to $\mathscr{U}$, where $u_r \in \mathbb{R}_+$ is a small constant.

Suppose that condition (1) does not hold. Then, there is a $\mathbf{k} \neq \mathbf{0}_{4\times 1}$ satisfying $\mathbf{k}^{\mathrm{T}}\mathbf{J}_f^{-1}\mathbf{u} \leq 0$ for all $\mathbf{u} \in \mathscr{U}$. Without loss of generality, let $\mathbf{k} = [k_1 \; * \; * \; *]^{\mathrm{T}}$ where $k_1 \neq 0$ and $*$ indicates an arbitrary real number. Let $\mathbf{u}_1 = [\varepsilon\,0\,0\,0]^{\mathrm{T}}$ and $\mathbf{u}_2 = [-\varepsilon\,0\,0\,0]^{\mathrm{T}}$, where $\varepsilon \in \mathbb{R}_+$. Then $\mathbf{u}_1, \mathbf{u}_2 \in \mathscr{B}\,(\mathbf{0}_{4\times 1}, u_r)$ if ε is sufficiently small. As $\mathbf{k}^{\mathrm{T}}\mathbf{J}_f^{-1}\mathbf{u} \leq 0$ for all $\mathbf{u} \in \mathscr{B}\,(\mathbf{0}_{4\times 1}, u_r)$, then $\mathbf{k}^{\mathrm{T}}\mathbf{J}_f^{-1}\mathbf{u}_1 \leq 0$ and $\mathbf{k}^{\mathrm{T}}\mathbf{J}_f^{-1}\mathbf{u}_2 \leq 0$. According to Eq. (10.33), one has

$$-\frac{k_1\varepsilon}{m} \leq 0,\; \frac{k_1\varepsilon}{m} \leq 0.$$

This implies that $k_1 = 0$ which contradicts the fact that $k_1 \neq 0$. Then, condition (1) holds.

(2) $\Leftrightarrow$ (3): According to the definition of ρ (g, $\partial\Omega$), if ρ (g, $\partial\Omega$) ≤ 0, then g is not in the interior of Ω; and if ρ (g, $\partial\Omega$) > 0, then g is an interior point of Ω.

Exercises

10.1 Consider the following system

$$\dot{x}_1 = x_1 \left(x_1^2 + x_2^2 - 2 \right) - 4x_1 x_2^2$$
$$\dot{x}_2 = 4x_1^2 x_2 + x_2 \left(x_1^2 + x_2^2 - 2 \right).$$

Use the *Local Invariant Set Theorem* to show that the origin is asymptotically stable.

10.2 Consider the system

$$\dot{x} = \begin{bmatrix} 0 & 1 \\ 0 & 0 \end{bmatrix} x + \begin{bmatrix} 0 \\ 1 \end{bmatrix} u.$$

Perform the following simulation:

(1) the target state is $x = [0\ 0]^T$, the initial state is $x_0 = [1000\ 0]^T$, and the saturated controller is $u = [-2\ -1] x \in [-1, 1]$, simulate the closed-loop system and plot the results;
(2) the target state is $x = [0\ 0]^T$, the initial state is $x_0 = [1000\ 0]^T$, and the saturated controller is $u = \text{sat}_{gd} ([-2\ -1] x, 1)$, simulate the closed-loop system and plot the results.

Furthermore, compare the two results and clarify the advantages of direction-guaranteed saturation function.

10.3

(1) Check the controllability of the following system

$$\dot{x} = Ax + Bu, u \in \mathbb{R}$$

where

$$A = \begin{bmatrix} 0 & 1 \\ 0 & 0 \end{bmatrix}, B = \begin{bmatrix} 1 \\ 0 \end{bmatrix}.$$

(2) Check the controllability of the following system

$$\dot{x} = \begin{bmatrix} 0 & 1 \\ -\lambda & 0 \end{bmatrix} x + \begin{bmatrix} 1 \\ 0 \end{bmatrix} u$$

where $\lambda \in \mathbb{R}_+, u \in [0, \infty)$.

10.4 Use the classical controllability theory to check the controllability of the PNPNPN hexacopter expressed by Eqs. (10.32), (10.33), and (10.44).

10.5 Check the controllability of the PPNN quadcopter at hover shown in Fig. 1.21 by computing the ACAI where the parameters are shown in Table 10.1. The ACAI toolbox is available on the Web site http://rfly.buaa.edu.cn/course

References

1. Slotine JJE, Li W (1991) Applied nonlinear control. Prentice Hall, New Jersey
2. Chen CT (1999) Linear system theory and design, 3rd edn. Oxford University Press, New York
3. Thomas GB, Weir MD, Hass J (2009) Thomas' calculus, Twelfth edn. Pearson Addison Wesley, Boston
4. Kalman RE (1960) On the general theory of control systems. In: Proceedings of the 1st world congress of the international federation of automatic control, pp 481–493
5. Brammer RF (1972) Controllability in linear autonomous systems with positive controllers. SIAM J Control 10(2):779–805
6. Du GX (2016) Research on the controllability quantification of multirotor systems. Dissertation, Beihang University. (In Chinese)
7. Kalman RE, Ho YC, Narendra KS (1962) Controllability of linear dynamical systems. Control Differ Equ 1(2):189–213
8. Johnson CD (1969) Optimization of a certain quality of complete controllability and observability for linear dynamical systems. Trans ASME (J Basic Eng) 91:228–238
9. Hadman AMA, Nayfeh AH (1989) Measures of modal controllability and observability for first-and second-order linear system. J Guidance Control Dyn 12(3):421–428
10. Viswanathan CN, Longman RW, Likins PW (1984) A degree of controllability definitions: fundamental concepts and application to modal systems. J Guidance Control Dyn 7(2):222–230
11. Du GX, Quan Q (2014) Degree of controllability and its application in aircraft flight control. J Syst Sci Math Sci 34(12):1578–1594 (In Chinese)
12. Pachter M, Huang YS (2003) Fault tolerant flight control. J Guidance Control Dyn 26(1):151–160
13. Cieslak J, Henry D, Zolghadri A, Goupil P (2008) Development of an active fault-tolerant flight control strategy. J Guidance Control Dyn 31(1):135–147
14. Zhang YM, Jiang J (2008) Bibliographical review on reconfigurable fault-tolerant control systems. Annu Rev Control 32(2):229–252
15. Ducard G (2009) Fault-tolerant flight control and guidance systems: practical methods for small unmanned aerial vehicles. Springer, London
16. Du GX, Quan Q, and Cai KY (2013) Additive-state-decomposition-based dynamic inversion stabilized control of a hexacopter subject to unknown propeller damages. In: Proceedings of the 32nd Chinese Control Conference, Xi'an China, July, pp 6231–6236
17. Wu NE, Zhou K, Salomon G (2000) Control reconfigurability of linear time-invariant systems. Automatica 36(11):1767–1771
18. Yang Z (2006) Reconfigurability analysis for a class of linear hybrid systems. In: Proceedings of the 6th IFAC SAFEPRO-CESS'06, Beijing, China
19. Du GX, Quan Q, Cai KY (2015) Controllability analysis and degraded control for a class of hexacopters subject to rotor failures. J Intell Robot Syst 78(1):143–157
20. Du GX, Quan Q, Cai KY (2015) Controllability analysis for multirotor helicopter rotor degradation and failure. J Guidance Control Dyn 38(5):978–984
21. Schneider T, Ducard G, Rudin K, and Strupler P (2012) Fault-tolerant control allocation for multirotor helicopters using parametric programming. In: International micro air vehicle conference and flight competition, Braunschweig, Germany
22. Quan Q, Du GX, Yang B, and Cai KY (2012) A safe landing method for a class of hexacopters subject to one rotor failure. Chinese patent, ZL201210398628.2. (In Chinese)
23. Mueller MW, D'Andrea R (2014) Stability and control of a quadcopter despite the complete loss of one, two, or three propellers. In: IEEE international conference on robotics and automation, Hong Kong, China, pp 45–52
24. Goodwin G, Seron M, Doná J (2005) Constrained control and estimation: an optimisation approach. Springer, London

Low-Level Flight Control

11

Control is an enabling technology, the essence of which is the feedback. Unstable multicopters can become stable with the help of feedback control. However, it is not enough to be just stable. A good control performance is also required, such as fast response, no overshoot and robustness against uncertainties. Therefore, effective control strategy is very important. In practice, traditional Proportional-Integral-Derivative (PID) control laws can tackle most problems for multicopters. Thus, this chapter mainly introduces PID control schemes in the position and attitude control. Moreover, in terms of different attitude representations, attitude control methods based on Euler angles and rotation matrix are presented, respectively. Different control methods exhibit their own advantages. This chapter aims to answer the question below:

How is a multicopter controlled by motors to achieve a desired position?

This chapter consists of six parts, namely the framework of low-level flight control, model simplification, position control, attitude control, control allocation and motor control. The effectiveness of these control algorithms is illustrated by comprehensive simulations.

© Springer Nature Singapore Pte Ltd. 2017
Q. Quan, *Introduction to Multicopter Design and Control*, DOI 10.1007/978-981-10-3382-7_11

Dujiangyan

The ancient Chinese had known how to use the idea of automatic control. For example, *Dujiangyan*, located in the Min River in Si Chuan province of China, is an irrigation infrastructure built around 256–251 BC. It was designed and constructed by Bing Li, a Chinese administrator and engineer in the Qin State, during the Warring States period of China. It is the only system in the world that is still in use and It has been irrigating over 5,300 km^2 of land for 2000 years. This system consists of three main components, namely Fish Mouth Levee, Flying Sand Weir and Bottle-Neck Channel, that work in harmony with one another to protect the farmland in case of flooding and keep the fields well supplied with water. By the natural landform, this system can automatically split the water flow, remove river sand and control the water flux.

11.1   Framework of Low-Level Flight Control of Multicopters

As shown in Fig. 11.1, the low-level flight control of multicopters is divided into four parts, namely the position control, attitude control, control allocation and motor control.

(1) Position control. It aims to calculate the desired roll angle ϕ_d, the desired pitch angle θ_d and the desired total thrust f_d according to the desired position $\mathbf{p}_d$.

(2) Attitude control. It aims to calculate the desired moment $\boldsymbol{\tau}_d$ according to the desired attitude angles ϕ_d, θ_d, ψ_d.

(3) Control allocation. It aims to allocate the desired propeller angular speeds $\varpi_{d,k}, k = 1, 2, \ldots, n_r$ to the n_r motors with the intention of obtaining the desired inputs $f_d, \boldsymbol{\tau}_d$.

(4) Motor control. It aims to calculate the desired throttle command of each motor $\sigma_{d,k}$ according to $\varpi_{d,k}, k = 1, 2, \ldots, n_r$.

The closed-loop control diagram is presented in Fig. 11.2. The multicopter is underactuated because the system has six outputs (position $\mathbf{p} \in \mathbb{R}^3$ and attitude $\boldsymbol{\Theta} \in \mathbb{R}^3$), but only four independent inputs (total thrust $f \in \mathbb{R}$ and three-axis moment $\boldsymbol{\tau} \in \mathbb{R}^3$). Thus, a multicopter can only track four desired commands ($\mathbf{p}_d \in \mathbb{R}^3$, $\psi_d \in \mathbb{R}$), leaving the remaining variables ($\phi_d, \theta_d \in \mathbb{R}$) determined by the desired commands $\mathbf{p}_d, \psi_d$.

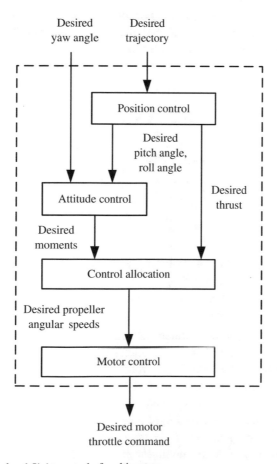

Fig. 11.1 Framework of low-level flight control of multicopters

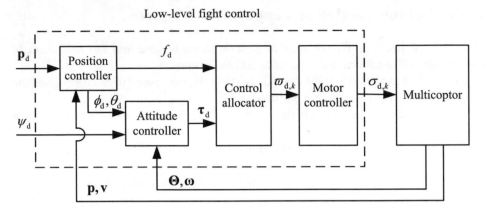

Fig. 11.2 Closed-loop structure of a low-level fight control system for multicopters

11.2   Linear Model Simplification

The nonlinear model has been established in Chap. 6. For the convenience of controller design, the nonlinear model needs to be simplified and linearized. Based on the nonlinear model (6.13) in Chap. 6, by ignoring the term $-^{\mathrm{b}}\boldsymbol{\omega} \times \left(\mathbf{J} \cdot {}^{\mathrm{b}}\boldsymbol{\omega}\right) + \mathbf{G}_{\mathrm{a}}$, the simplified model is

$$^{\mathrm{e}}\dot{\mathbf{p}} = {}^{\mathrm{e}}\mathbf{v} \tag{11.1}$$

$$^{\mathrm{e}}\dot{\mathbf{v}} = g\mathbf{e}_3 - \frac{f}{m}\mathbf{R}_{\mathrm{b}}^{\mathrm{e}}\mathbf{e}_3 \tag{11.2}$$

$$\dot{\boldsymbol{\Theta}} = \mathbf{W} \cdot {}^{\mathrm{b}}\boldsymbol{\omega} \tag{11.3}$$

$$\mathbf{J} \cdot {}^{\mathrm{b}}\dot{\boldsymbol{\omega}} = \boldsymbol{\tau}. \tag{11.4}$$

Here, position Eqs. (11.1) and (11.2) are rewritten as

$$\begin{aligned}
\ddot{p}_{x_{\mathrm{e}}} &= -\frac{f}{m}\left(\sin\psi\sin\phi + \cos\psi\sin\theta\cos\phi\right)\\
\ddot{p}_{y_{\mathrm{e}}} &= -\frac{f}{m}\left(-\cos\psi\sin\phi + \cos\phi\sin\theta\sin\psi\right)\\
\ddot{p}_{z_{\mathrm{e}}} &= g - \frac{f}{m}\cos\theta\cos\phi
\end{aligned} \tag{11.5}$$

where ${}^{\mathrm{e}}\mathbf{p} \triangleq \left[p_{x_{\mathrm{e}}}\, p_{y_{\mathrm{e}}}\, p_{z_{\mathrm{e}}}\right]^{\mathrm{T}} \in \mathbb{R}^3$. For attitude kinematics equation (11.3), besides the Euler angle representation, there is also the rotation matrix representation described in (6.14) in Chap. 6. The two representations are both considered in the attitude control. Here, for simplicity, the superscript and subscript are omitted, that is $\mathbf{p} = {}^{\mathrm{e}}\mathbf{p}$, $\mathbf{v} = {}^{\mathrm{e}}\mathbf{v}$, $\boldsymbol{\omega} = {}^{\mathrm{b}}\boldsymbol{\omega}$, $\mathbf{R} = \mathbf{R}_{\mathrm{b}}^{\mathrm{e}}$, where the specific definitions are referred to Chap. 6. Furthermore, let $\mathbf{p} \triangleq [p_x\, p_y\, p_z]^{\mathrm{T}}$ and $\mathbf{v} \triangleq \left[v_x\, v_y\, v_z\right]^{\mathrm{T}}$.

From Eqs. (11.1)–(11.4), the flight control system of a multicopter is a typical nonlinear system. This makes the analysis and design complex. Furthermore, the multicopter is underactuated, strong-coupling and highly dimensional. Therefore, the nonlinear model needs to be simplified according to the flight characteristic of multicopters. Then, based on the simplified model, controllers are designed. The pitch and roll angles being small is considered as the flight characteristic of multicopters. Moreover, the total thrust approximates to the weight of the multicopter. These assumptions are further written as

$$\sin\phi \approx \phi, \cos\phi \approx 1, \sin\theta \approx \theta, \cos\theta \approx 1 \text{ and } f \approx mg. \tag{11.6}$$

Then, the matrix $\mathbf{W}$ in Eq. (11.3) is approximated to the identity matrix $\mathbf{I}_3$, and the term $\mathbf{Re}_3$ in Eq. (11.2) becomes

$$\mathbf{Re}_3 \approx \begin{bmatrix} \theta\cos\psi + \phi\sin\psi \\ \theta\sin\psi - \phi\cos\psi \\ 1 \end{bmatrix}.$$

Finally, the original model (11.1)–(11.4) is decoupled into three linear models, namely the horizontal position channel model, altitude channel model and attitude model. They are introduced in the following.

11.2.1 Horizontal Position Channel Model

According to the small-angle assumption in (11.6), by ignoring the higher-order terms, the first two equations of Eq. (11.5) are simplified as

$$\begin{aligned} \dot{\mathbf{p}}_h &= \mathbf{v}_h \\ \dot{\mathbf{v}}_h &= -g\mathbf{A}_\psi \mathbf{\Theta}_h \end{aligned} \tag{11.7}$$

where

$$\mathbf{p}_h = \begin{bmatrix} p_x \\ p_y \end{bmatrix}, \mathbf{R}_\psi = \begin{bmatrix} \cos\psi & -\sin\psi \\ \sin\psi & \cos\psi \end{bmatrix}, \mathbf{A}_\psi = \mathbf{R}_\psi \begin{bmatrix} 0 & 1 \\ -1 & 0 \end{bmatrix}, \mathbf{\Theta}_h = \begin{bmatrix} \phi \\ \theta \end{bmatrix}.$$

In the horizontal position channel model (11.7), since $-g\mathbf{A}_\psi$ is known, the term $-g\mathbf{A}_\psi \mathbf{\Theta}_h$ can be viewed as the input and $\mathbf{p}_h$ is the output. Thus, the horizontal position channel model (11.7) is, in fact, a linear model.

11.2.2 Altitude Channel Model

According to the small-angle assumption in (11.6), the third equation of Eq. (11.5) is simplified as

$$\begin{aligned} \dot{p}_z &= v_z \\ \dot{v}_z &= g - \frac{f}{m}. \end{aligned} \tag{11.8}$$

Unlike the simplification of the horizontal position channel model, the term $g - \frac{f}{m}$ is not a higher-order infinite small term. As a result, it cannot be ignored. Otherwise, there is no input in the altitude channel. Obviously, the altitude channel model (11.8) is also a linear model.

11.2.3 Attitude Model

The combination of Eqs. (11.3) and (11.4) is the attitude model, rewritten as

$$\begin{aligned}\dot{\boldsymbol{\Theta}} &= \boldsymbol{\omega} \\ \mathbf{J}\dot{\boldsymbol{\omega}} &= \boldsymbol{\tau}\end{aligned} \tag{11.9}$$

which is also a linear model.

11.3   Position Control

The feedback values used in controllers are the estimated values like $\hat{\mathbf{p}}$, $\hat{\boldsymbol{\Theta}}$. However, for simplicity, the true values $\mathbf{p}$, $\boldsymbol{\Theta}$ are used instead according to the *separation principle*.[1]

11.3.1 Basic Concept

In order to make this section reader-friendly, several basic concepts are first introduced. As shown in Fig. 11.3, according to the given position $\mathbf{p}_d$, the position control is divided into three types, i.e., the set-point control, trajectory tracking and path following.

(1) Set-point control. The desired position $\mathbf{p}_d \in \mathbb{R}^3$ is a constant point. The aim of this part is to design a controller to make $\|\mathbf{p}(t) - \mathbf{p}_d\| \to 0$ or $\mathbf{p}(t) - \mathbf{p}_d \to \mathscr{B}(\mathbf{0}_{3\times 1}, \delta)^2$ as $t \to \infty$, where $\delta \in \mathbb{R}_+ \cup \{0\}$. Here, it is assumed that the multicopter can fly to the given position, regardless of the trajectory. In practice, most applications fall into this category, shown as in Fig. 11.3a.

(2) Trajectory tracking [2, 3]. The desired trajectory $\mathbf{p}_d : [0, \infty) \to \mathbb{R}^3$ is a time-dependent trajectory. The aim of this part is to design a controller to make $\|\mathbf{p}(t) - \mathbf{p}_d(t)\| \to 0$ or $\mathbf{p}(t) - \mathbf{p}_d(t) \to \mathscr{B}(\mathbf{0}_{3\times 1}, \delta)$ as $t \to \infty$, where $\delta \in \mathbb{R}_+\cup\{0\}$. This kind of application includes the case that a multicopter is required to track a given trajectory over a moving car, shown as in Fig. 11.3b.

(3) Path following.[3] The desired path $\mathbf{p}_d(\gamma(t)) \in \mathbb{R}^3$ is a path directly determined by the parameter γ rather than the time t. It aims to design a controller to make $\|\mathbf{p}(t) - \mathbf{p}_d(\gamma(t))\| \to 0$ or $\mathbf{p}(t) - \mathbf{p}_d(\gamma(t)) \to \mathscr{B}(\mathbf{0}_{3\times 1}, \delta)$ as $t \to \infty$, where $\delta \in \mathbb{R}_+ \cup \{0\}$. This kind of application includes that a multicopter is required to fly along a profile over a hill, shown as in Fig. 11.3c.

Intuitively, the difference between a *trajectory tracking* problem and a *path following* problem is whether the curves to describe the trajectories or the paths depend upon time. For a trajectory tracking problem, the desired trajectory is a time-dependent curve. However, the curve in a path following problem is independent of time. Path following is also called *three-dimensional (3D) tracking*, while trajectory tracking is called *four-dimensional (4D) tracking* with the time being added as a new dimension. When $\gamma(t) = t$, the path following problem is degenerated to a trajectory tracking problem. Furthermore, if $\mathbf{p}_d(t) \equiv$ constant, then the trajectory tracking problem is degenerated to a *set-point control problem*. In other words, path following is degenerated to trajectory tracking when imposing the time constraint on it, while trajectory tracking is degenerated to set-point control when the desired

---

[1]It is a principle of separating the controller design into the state estimation and the certain feedback control. When applying this principle, the system states are firstly estimated according to the observation data, then the estimated values are viewed as the true values, and the controller is designed based on the certain system [1, p. 255].

[2]$\mathscr{B}(\mathbf{o}, \delta) \triangleq \{\boldsymbol{\xi} \in \mathbb{R}^3 \,|\, \|\boldsymbol{\xi} - \mathbf{o}\| \le \delta\}$, and the notation $\mathbf{x}(t) \to \mathscr{B}(\mathbf{o}, \delta)$ means $\min_{\mathbf{y} \in \mathscr{B}(\mathbf{o}, \delta)} \|\mathbf{x}(t) - \mathbf{y}\| \to 0$.

[3]Only parameterized path following is considered here.

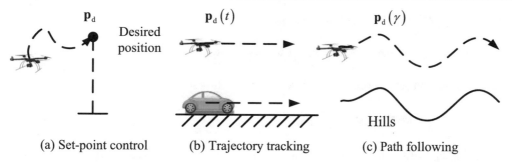

Fig. 11.3 Three types of position control

trajectory is further restricted. Thus, a set-point control problem is a special case of trajectory tracking problems. On the other hand, a trajectory tracking problem is a special case of path following problems. This chapter mainly focuses on the set-point control problem and trajectory tracking problem. In practice, path following problems can be transformed into trajectory tracking problems (see Chap. 13) or set-point control problems. So, it will not be included in this chapter. Readers who are interested in path following can refer to the Refs. [4–6].

Desired attitude is the output of position controllers. According to the attitude representation, the position control is divided into two types:

(1) Euler angles as output. The controller, which is designed according to the linear systems (11.7)–(11.8), generates the desired Euler angles ϕ_d, θ_d and thrust f_d as the output.

(2) Rotation matrix as output. The controller, which is designed according to the nonlinear coupled systems (11.1)–(11.2), generates the desired rotation matrix $\mathbf{R}_d$ and thrust f_d.

11.3.2 Euler Angles as Output

PID controllers are designed for the horizontal position channel and the altitude channel respectively according to the linear position model (11.7)–(11.8). In the following, traditional PID controllers are first designed. Moreover, PID controllers used in open source autopilots are introduced. Finally, PID controllers with saturation are further designed.

11.3.2.1 Traditional PID Controller

(1) Horizontal position channel
According to the linear position model (11.7), the desired attitude angle $\boldsymbol{\Theta}_{hd} \triangleq [\phi_d \ \theta_d]^T$ is expected to be designed such that

$$\lim_{t \to \infty} \left\| \mathbf{e}_{\mathbf{p}_h} (t) \right\| = 0$$

where $\mathbf{e}_{\mathbf{p}_h} \triangleq \mathbf{p}_h - \mathbf{p}_{hd}$. First, the transient process is expected as follows:

$$\ddot{\mathbf{e}}_{\mathbf{p}_h} = -\mathbf{K}_{\mathbf{p}_h d} \dot{\mathbf{e}}_{\mathbf{p}_h} - \mathbf{K}_{\mathbf{p}_h p} \mathbf{e}_{\mathbf{p}_h} \tag{11.10}$$

where $\mathbf{K}_{\mathbf{p}_h d}, \mathbf{K}_{\mathbf{p}_h p} \in \mathbb{R}^{2 \times 2} \cap \mathscr{D} \cap \mathscr{P}$. Then, according to the stability theory in Chap. 10, $\lim_{t \to \infty} \left\| \mathbf{e}_{\mathbf{p}_h} (t) \right\| = 0$. In order to realize the dynamics (11.10), the acceleration $\ddot{\mathbf{p}}_h$ is expected to satisfy

$$\ddot{\mathbf{p}}_h = \ddot{\mathbf{p}}_{hd} - \mathbf{K}_{\mathbf{p}_h d} \dot{\mathbf{e}}_{\mathbf{p}_h} - \mathbf{K}_{\mathbf{p}_h p} \mathbf{e}_{\mathbf{p}_h}. \tag{11.11}$$

Combining Eq. (11.7) with Eq. (11.11) results in

$$-g\mathbf{A}_\psi \mathbf{\Theta}_{hd} = \ddot{\mathbf{p}}_{hd} - \mathbf{K}_{\mathbf{p}_h d}\dot{\mathbf{e}}_{\mathbf{p}_h} - \mathbf{K}_{\mathbf{p}_h p}\mathbf{e}_{\mathbf{p}_h}. \tag{11.12}$$

According to the above equation, $\mathbf{\Theta}_{hd}$ is written explicitly as follows:

$$\mathbf{\Theta}_{hd} = -g^{-1}\mathbf{A}_\psi^{-1}\left(\ddot{\mathbf{p}}_{hd} - \mathbf{K}_{\mathbf{p}_h d}\dot{\mathbf{e}}_{\mathbf{p}_h} - \mathbf{K}_{\mathbf{p}_h p}\mathbf{e}_{\mathbf{p}_h}\right). \tag{11.13}$$

This implies that if $\mathbf{\Theta}_h = \mathbf{\Theta}_{hd}$ in (11.7), then $\lim_{t\to\infty}\|\mathbf{e}_{\mathbf{p}_h}(t)\| = 0$. In particular, when a set-point control problem is considered, namely $\dot{\mathbf{p}}_{hd} = \ddot{\mathbf{p}}_{hd} = \mathbf{0}_{2\times1}$, Eq. (11.13) becomes

$$\mathbf{\Theta}_{hd} = -g^{-1}\mathbf{A}_\psi^{-1}\left(-\mathbf{K}_{\mathbf{p}_h d}\dot{\mathbf{p}}_h - \mathbf{K}_{\mathbf{p}_h p}\left(\mathbf{p}_h - \mathbf{p}_{hd}\right)\right). \tag{11.14}$$

(2) Altitude channel
Similarly, according to the altitude channel model (11.18), the control objective is

$$\lim_{t\to\infty}\left|e_{p_z}(t)\right| = 0$$

where $e_{p_z} \triangleq p_z - p_{z_d}$. First, the transient process is expected as follows:

$$\ddot{e}_{p_z} = -k_{p_z d}\dot{e}_{p_z} - k_{p_z p}e_{p_z} \tag{11.15}$$

where $k_{p_z d}, k_{p_z p} \in \mathbb{R}_+$. From the stability theory in Chap. 10, it holds that $\lim_{t\to\infty}\left|e_{p_z}(t)\right| = 0$. In order to realize the dynamics (11.15), the acceleration $\ddot{p}_z$ is expected to satisfy

$$\ddot{p}_z = \ddot{p}_{z_d} - k_{p_z d}\dot{e}_{p_z} - k_{p_z p}e_{p_z}. \tag{11.16}$$

Combining Eq. (11.8) with Eq. (11.16) results in

$$g - \frac{f_d}{m} = \ddot{p}_{z_d} - k_{p_z d}\dot{e}_{p_z} - k_{p_z p}e_{p_z} \tag{11.17}$$

where f_d is the desired thrust. From the above equation, f_d is written explicitly as follows:

$$f_d = mg - m\left(\ddot{p}_{z_d} - k_{p_z d}\dot{e}_{p_z} - k_{p_z p}e_{p_z}\right). \tag{11.18}$$

This implies that if $f = f_d$ in Eq. (11.8), then $\lim_{t\to\infty}\left|e_{p_z}(t)\right| = 0$. In particular, when a set-point control problem is considered, namely, $\dot{p}_{z_d} = \ddot{p}_{z_d} = 0$, Eq. (11.18) becomes

$$f_d = mg - m\left(-k_{p_z d}\dot{p}_z - k_{p_z p}\left(p_z - p_{z_d}\right)\right). \tag{11.19}$$

11.3.2.2 PID Controllers in Open Source Autopilots
The design ideas of the position controller based on the present ArduPilot Mega (APM) open source autopilot is introduced in the following.

(1) Horizontal position channel
In order to satisfy $\lim_{t\to\infty}\left\|\mathbf{e}_{\mathbf{p}_\mathrm{h}}(t)\right\|=0$, according to

$$\dot{\mathbf{p}}_\mathrm{h}=\mathbf{v}_\mathrm{h} \tag{11.20}$$

the desired value of $\mathbf{v}_\mathrm{h}$, namely $\mathbf{v}_\mathrm{hd}$, is designed as

$$\mathbf{v}_\mathrm{hd}=\mathbf{K}_{\mathbf{p}_\mathrm{h}}\left(\mathbf{p}_\mathrm{hd}-\mathbf{p}_\mathrm{h}\right) \tag{11.21}$$

where $\mathbf{K}_{\mathbf{p}_\mathrm{h}}\in\mathbb{R}^{2\times2}\bigcap\mathscr{D}\bigcap\mathscr{P}$. Under the assumption that $\dot{\mathbf{p}}_\mathrm{hd}=\mathbf{0}_{2\times1}$, if $\lim_{t\to\infty}\left\|\mathbf{e}_{\mathbf{v}_\mathrm{h}}(t)\right\|=0$, then $\lim_{t\to\infty}\left\|\mathbf{e}_{\mathbf{p}_\mathrm{h}}(t)\right\|=0$, where $\mathbf{e}_{\mathbf{v}_\mathrm{h}}\triangleq\mathbf{v}_\mathrm{h}-\mathbf{v}_\mathrm{hd}$. In fact, Eqs. (11.20) and (11.21) build the horizontal position control loop. According to the following equation

$$\dot{\mathbf{v}}_\mathrm{h}=-g\mathbf{A}_\psi\mathbf{\Theta}_\mathrm{h} \tag{11.22}$$

the next task is to design the desired $\mathbf{\Theta}_\mathrm{h}$, namely $\mathbf{\Theta}_\mathrm{hd}$. Similar to Eq. (11.12), a PID controller is adopted as follows:

$$-g\mathbf{A}_\psi\mathbf{\Theta}_\mathrm{hd}=-\mathbf{K}_{\mathbf{v}_\mathrm{h}\mathrm{p}}\mathbf{e}_{\mathbf{v}_\mathrm{h}}-\mathbf{K}_{\mathbf{v}_\mathrm{h}\mathrm{i}}\int\mathbf{e}_{\mathbf{v}_\mathrm{h}}-\mathbf{K}_{\mathbf{v}_\mathrm{h}\mathrm{d}}\dot{\mathbf{e}}_{\mathbf{v}_\mathrm{h}} \tag{11.23}$$

where $\mathbf{K}_{\mathbf{v}_\mathrm{h}\mathrm{p}},\mathbf{K}_{\mathbf{v}_\mathrm{h}\mathrm{i}},\mathbf{K}_{\mathbf{v}_\mathrm{h}\mathrm{d}}\in\mathbb{R}^{2\times2}\bigcap\mathscr{D}$. Under the assumption $\dot{\mathbf{v}}_\mathrm{hd}=\mathbf{0}_{2\times1}$, if $\lim_{t\to\infty}\|\mathbf{\Theta}_\mathrm{h}(t)-\mathbf{\Theta}_\mathrm{hd}(t)\|=0$, then $\lim_{t\to\infty}\left\|\mathbf{e}_{\mathbf{v}_\mathrm{h}}(t)\right\|=0$. The desired attitude angle is derived from Eq. (11.23) as

$$\mathbf{\Theta}_\mathrm{hd}=g^{-1}\mathbf{A}_\psi^{-1}\left(\mathbf{K}_{\mathbf{v}_\mathrm{h}\mathrm{p}}\mathbf{e}_{\mathbf{v}_\mathrm{h}}+\mathbf{K}_{\mathbf{v}_\mathrm{h}\mathrm{i}}\int\mathbf{e}_{\mathbf{v}_\mathrm{h}}+\mathbf{K}_{\mathbf{v}_\mathrm{h}\mathrm{d}}\dot{\mathbf{e}}_{\mathbf{v}_\mathrm{h}}\right). \tag{11.24}$$

As a result, the designed attitude controller is required to make $\lim_{t\to\infty}\|\mathbf{\Theta}_\mathrm{h}(t)-\mathbf{\Theta}_\mathrm{hd}(t)\|=0$. Consequently, the horizontal position channel (11.7) can guarantee $\lim_{t\to\infty}\left\|\mathbf{e}_{\mathbf{p}_\mathrm{h}}(t)\right\|=0$. In fact, Eqs. (11.22) and (11.24) build the horizontal velocity control loop. If the set-point control problem is considered, then $\dot{\mathbf{p}}_\mathrm{hd}=\ddot{\mathbf{p}}_\mathrm{hd}=\mathbf{0}_{2\times1}$ in Eq. (11.24). To avoid the noise caused by the differential, the differential term $\mathbf{K}_{\mathbf{v}_\mathrm{h}\mathrm{d}}\dot{\mathbf{e}}_{\mathbf{v}_\mathrm{h}}$ can be omitted. So far, the controller design of the horizontal position channel is finished, which is composed of Eqs. (11.21) and (11.24).

A question is often asked that why Eq. (11.21) is written as a P controller, whereas Eq. (11.24) is in PID form. The reason is that the channel (11.20) is a kinematic model, which is uncertainty-free. So, the P controller (11.21) is satisfying. On the other hand, since the channel (11.22) is a dynamic model subject to various uncertainties, for which the PID controller (11.24) can compensate. The similar principle is applicable to the altitude channel and attitude channel in the following.

(2) Altitude channel
In order to satisfy $\lim_{t\to\infty}\left|e_{p_z}(t)\right|=0$, according to

$$\dot{p}_z=v_z \tag{11.25}$$

the desired value of v_z, namely $v_{z\mathrm{d}}$, is designed as

$$v_{z\mathrm{d}}=-k_{p_z}\left(p_z-p_{z\mathrm{d}}\right) \tag{11.26}$$

where $k_{p_z} \in \mathbb{R}_+$. Under the assumption $\dot{p}_{zd} = 0$, if $\lim_{t \to \infty} |e_{v_z}(t)| = 0$, then $\lim_{t \to \infty} |e_{p_z}(t)| = 0$, where $e_{v_z} \triangleq v_z - v_{zd}$. In fact, Eqs. (11.25) and (11.26) build the altitude control loop. According to the following equation

$$\dot{v}_z = g - \frac{f}{m} \qquad (11.27)$$

the next task is to design the desired thrust f, namely f_d. Similar to Eq. (11.17), a PID controller is adopted as follows:

$$g - \frac{f_d}{m} = -k_{v_z p} e_{v_z} - k_{v_z i} \int e_{v_z} - k_{v_z d} \dot{e}_{v_z} \qquad (11.28)$$

where $k_{v_z p}, k_{v_z i}, k_{v_z d} \in \mathbb{R}$. Under the assumption $\dot{v}_{zd} = 0$, if $\lim_{t \to \infty} |f(t) - f_d(t)| = 0$, then $\lim_{t \to \infty} |e_{v_z}(t)| = 0$. The desired thrust f_d is derived from Eq. (11.28) as

$$f_d = m \left(g + k_{v_z p} e_{v_z} + k_{v_z i} \int e_{v_z} + k_{v_z d} \dot{e}_{v_z} \right). \qquad (11.29)$$

In fact, Eqs. (11.27) and (11.29) build the altitude velocity control loop. If the set-point control problem is considered, then $\dot{p}_{zd} = \ddot{p}_{zd} = 0$ in Eq. (11.29). To avoid the noise caused by the differential, the differential term $k_{v_z d} \dot{e}_{v_z}$ can be omitted. So far, the controller design of altitude channel is completed, which is composed of Eqs. (11.26) and (11.29).

11.3.2.3 PID Controller with Saturation

So far, the position control problem seems to have been solved, i.e., the desired pitch angle and roll angle have been found. However, some other problems may arise in practical implementation, because $p_x - p_{x_d}$ or $p_y - p_{y_d}$ may be too large in practice. For example, $\|\mathbf{e}_{\mathbf{p}_h}(0)\| = 100$ km. This will lead to unacceptable θ_d, ϕ_d. As a result, the small-angle assumption is violated and then Eq. (11.7) does not hold anymore. If the attitude controller uses these unacceptable desired pitch angle and roll angle, then multicopters may crash. Therefore, PID controllers with saturation must be considered.

(1) Horizontal position channel
By considering the saturation, the traditional PID controller (11.14) becomes

$$\mathbf{\Theta}_{hd} = \text{sat}_{gd} \left(g^{-1} \mathbf{A}_{\psi}^{-1} \left(\mathbf{K}_{\mathbf{p}_h d} \dot{\mathbf{p}}_h + \mathbf{K}_{\mathbf{p}_h p} \left(\mathbf{p}_h - \mathbf{p}_{hd} \right) \right), a_0 \right) \qquad (11.30)$$

where $a_0 \in \mathbb{R}_+$ and the direction-guaranteed saturation function $\text{sat}_{gd}(\mathbf{x}, a)$ is defined in (10.18) in Sect. 10.2.2. The stability proof of the saturated system can also been found there. Similarly, the PID controller used in open source autopilots is rewritten as

$$\begin{aligned} \mathbf{e}_{\mathbf{v}_h} &= \text{sat}_{gd} \left(\mathbf{v}_h - \mathbf{v}_{hd}, a_1 \right) \\ \mathbf{\Theta}_{hd} &= \text{sat}_{gd} \left(g^{-1} \mathbf{A}_{\psi}^{-1} \left(\mathbf{K}_{\mathbf{v}_h p} \mathbf{e}_{\mathbf{v}_h} + \mathbf{K}_{\mathbf{v}_h i} \int \mathbf{e}_{\mathbf{v}_h} + \mathbf{K}_{\mathbf{v}_h d} \dot{\mathbf{e}}_{\mathbf{v}_h} \right), a_2 \right) \end{aligned} \qquad (11.31)$$

where $a_1, a_2 \in \mathbb{R}_+$. The parameters $a_0, a_1, a_2 \in \mathbb{R}_+$ related to saturation can be specified according to practical requirements. For example, if $\theta_d, \phi_d \in [-\theta_{max}, \theta_{max}]$, then the saturated horizontal position controller (11.31) becomes

$$\mathbf{\Theta}_{hd} = \text{sat}_{gd} \left(g^{-1} \mathbf{A}_{\psi}^{-1} \left(\mathbf{K}_{\mathbf{v}_h p} \mathbf{e}_{\mathbf{v}_h} + \mathbf{K}_{\mathbf{v}_h i} \int \mathbf{e}_{\mathbf{v}_h} + \mathbf{K}_{\mathbf{v}_h d} \dot{\mathbf{e}}_{\mathbf{v}_h} \right), \theta_{max} \right). \qquad (11.32)$$

Fig. 11.4 Comparison of the results of two saturation functions

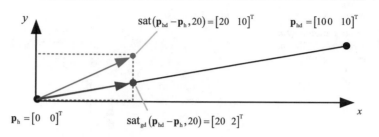

Next, the difference between the direction-guaranteed saturation function $\mathrm{sat_{gd}}\,(\mathbf{x}, a)$ and the traditional saturation function $\mathrm{sat}(\mathbf{x}, a)$ will be discussed. The traditional saturation function is defined as

$$\mathrm{sat}\,(\mathbf{x}, a) \triangleq \begin{bmatrix} \mathrm{sat}\,(x_1, a) \\ \vdots \\ \mathrm{sat}\,(x_n, a) \end{bmatrix}, \mathrm{sat}\,(x_k, a) \triangleq \begin{cases} x_k, & |x_k| \le a \\ a \cdot \mathrm{sign}\,(x_k), & |x_k| > a \end{cases}$$

whose direction may be different from that of $\mathbf{x}$. On the other hand, the direction-guaranteed saturation function $\mathrm{sat_{gd}}\,(\mathbf{x}, a)$ can not only confine that the absolute value of each element of the final vector is not greater than a, but also guarantee the direction to be the same as that of $\mathbf{x}$. For example, as shown in Fig. 11.4, if $\mathbf{p}_h = [0\,0]^T$ and $\mathbf{p}_{hd} = [100\,10]^T$, then

$$\mathrm{sat_{gd}}\,(\mathbf{p}_{hd} - \mathbf{p}_h, 20) = [20\,2]^T$$

but

$$\mathrm{sat}\,(\mathbf{p}_{hd} - \mathbf{p}_h, 20) = [20\,10]^T.$$

Here $\mathrm{sat_{gd}}\,(\mathbf{p}_{hd} - \mathbf{p}_h, 20)$ is still in the straight-line path from $\mathbf{p}_h$ to $\mathbf{p}_{hd}$, but $\mathrm{sat}(\mathbf{p}_{hd} - \mathbf{p}_h, 20)$ deviates from the straight-line path. The latter will not make multicopters fly along a straight line, and therefore more energy is consumed by multicopters.

The following simulation is performed to study flight trajectories based on the two saturation functions. In the simulation, the initial value is set as $\mathbf{p}_h\,(0) = [0\,0]^T$, and the desired position is set as $\mathbf{p}_{hd} = [4\,6]^T$. The parameters related to saturation are $a_1 = 3$, $a_2 = 12\pi/180$ in controller (11.31). Observed from the simulation results shown in Fig. 11.5, the direction-guaranteed saturation function can guarantee that the multicopter flies along a straight line, but the traditional saturation function cannot.

(2) Altitude channel

In order to avoid a throttle command out of range, the saturation needs to be considered as well. Thus, the traditional PID controller (11.19) becomes

$$f_d = \mathrm{sat_{gd}}\,(m\,(g + k_{p_z d}\dot{p}_z + k_{p_z p}\,(p_z - p_{zd}))\,, a_3) \tag{11.33}$$

where $a_3 \in \mathbb{R}_+$. Similarly, the PID controller design used in the open source autopilots (11.29) is rewritten as

$$\begin{aligned} e_{v_z} &= \mathrm{sat_{gd}}\,(v_z - v_{zd}, a_4) \\ f_d &= \mathrm{sat_{gd}}\,(m\,(g + k_{v_z p}e_{v_z} + k_{v_z i}\int e_{v_z} + k_{v_z d}\dot{e}_{v_z})\,, a_5) \end{aligned} \tag{11.34}$$

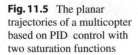
Fig. 11.5 The planar trajectories of a multicopter based on PID control with two saturation functions

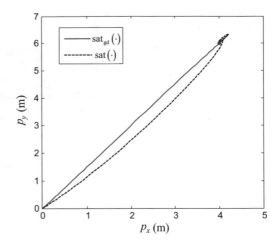

where $a_4, a_5 \in \mathbb{R}_+$. For a scale, the direction-guaranteed saturation function is the same as the traditional saturation function. Furthermore, if $f_d \in [f_{min}, f_{max}]$, then the saturated controller (11.34) becomes

$$f_d = \text{sat}_{gd}\left(m\left(g + k_{v_z p}e_{v_z} + k_{v_z i}\int e_{v_z} + k_{v_z d}\dot{e}_{v_z}\right) - \frac{f_{min} + f_{max}}{2}, \frac{f_{max} - f_{min}}{2}\right) + \frac{f_{min} + f_{max}}{2}$$
(11.35)

11.3.3 Rotation Matrix as Output

The rotation matrix based representation for the attitude kinematics equation (11.3) is as follows:

$$\dot{\mathbf{R}} = \mathbf{R}\left[{}^b\boldsymbol{\omega}\right]_\times$$
(11.36)

where $\mathbf{R} \in SO\,(3)$ is the rotation matrix. The desired attitude control command of Eq. (11.36) is $\mathbf{R}_d$. Thus, it is more beneficial if the corresponding position control provides $\mathbf{R}_d$ directly. The corresponding closed-loop control diagram is shown in Fig. 11.6. Next, how to obtain $\mathbf{R}_d$ will be focused on.

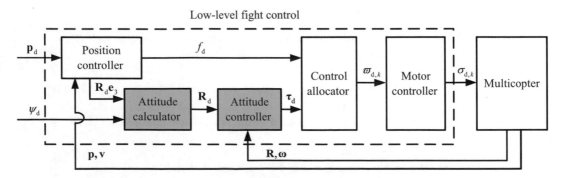

Fig. 11.6 The rotation matrix based closed-loop control diagram for multicopters

11.3.3.1 Desired Rotation Matrix $\mathbf{R_d}$

Define $\mathbf{R} \triangleq [\mathbf{r}_1 \ \mathbf{r}_2 \ \mathbf{r}_3]$ and $\mathbf{R}_d \triangleq [\mathbf{r}_{1,d} \ \mathbf{r}_{2,d} \ \mathbf{r}_{3,d}]$. According to the nonlinear model (11.1)–(11.2), the desired f_d and $\mathbf{R}_d$ are expected to satisfy

$$g\mathbf{e}_3 - \frac{f_d}{m}\mathbf{r}_{3,d} = \mathbf{a}_d \tag{11.37}$$

where $\mathbf{r}_{3,d} = \mathbf{R}_d \mathbf{e}_3$, and $\mathbf{a}_d$ is the desired acceleration. For example, it can be obtained as

$$\mathbf{a}_d = \begin{bmatrix} \mathrm{sat}_{gd}\left(\left(-\mathbf{K}_{v_h p}\mathbf{e}_{v_h} - \mathbf{K}_{v_h i}\int \mathbf{e}_{v_h} - \mathbf{K}_{v_h d}\dot{\mathbf{e}}_{v_h}\right), a_6\right) \\ \mathrm{sat}_{gd}\left(\left(-k_{v_z p}e_{v_z} - k_{v_z i}\int e_{v_z} - k_{v_z d}\dot{e}_{v_z}\right), a_7\right) \end{bmatrix}$$

where $\mathbf{e}_{v_h} = \mathrm{sat}_{gd}\left(\mathbf{v}_h - \mathbf{v}_{hd}, a_8\right)$ and $e_{v_z} = \mathrm{sat}_{gd}\left(v_z - v_{zd}, a_9\right)$, $a_6, a_7, a_8, a_9 \in \mathbb{R}_+$. The saturated acceleration makes the resulting rotation matrix feasible for control in practice. Equation (11.37) gives

$$\mathbf{r}_{3,d} = \frac{m\left(g\mathbf{e}_3 - \mathbf{a}_d\right)}{f_d}. \tag{11.38}$$

Since $\mathbf{r}_{3,d}$ is a column of an orthogonal matrix, it satisfies $\mathbf{r}_{3,d}^T \mathbf{r}_{3,d} = 1$. Thus, it is further corrected as

$$\mathbf{r}_{3,d} = \frac{g\mathbf{e}_3 - \mathbf{a}_d}{\|g\mathbf{e}_3 - \mathbf{a}_d\|}. \tag{11.39}$$

So far, the vector $\mathbf{r}_{3,d}$ is determined. In the following, two methods of obtaining $\mathbf{R}_d$ are introduced. One is based on the small-angle assumption, and the other is applicable to large-angle flight.

(1) Small-angle case [7]

In order to obtain $\mathbf{R}_d$, one of $\mathbf{r}_{1,d}$, $\mathbf{r}_{2,d}$ needs to be determined additionally. According to the definition of the rotation matrix, $\mathbf{r}_1 \in \mathbb{R}^3$ is

$$\mathbf{r}_1 = \begin{bmatrix} \cos\theta\cos\psi \\ \cos\theta\sin\psi \\ -\sin\theta \end{bmatrix}.$$

By using the small-angle assumption, the corresponding desired vector $\mathbf{r}_{1,d}$ is approximated by

$$\bar{\mathbf{r}}_{1,d} = \begin{bmatrix} \cos\psi_d \\ \sin\psi_d \\ 0 \end{bmatrix}$$

where $\theta \approx 0$ and $\cos\theta \approx 1$ have been used. With $\bar{\mathbf{r}}_{1,d}$, $\mathbf{r}_{3,d}$ in hand, $\mathbf{r}_{2,d}$ is determined as

$$\mathbf{r}_{2,d} = \frac{\mathbf{r}_{3,d} \times \bar{\mathbf{r}}_{1,d}}{\|\mathbf{r}_{3,d} \times \bar{\mathbf{r}}_{1,d}\|}.$$

According to the definition of cross-product, it has $\mathbf{r}_{2,d} \perp \mathbf{r}_{3,d}$. Finally, define $\mathbf{r}_{1,d} = \mathbf{r}_{2,d} \times \mathbf{r}_{3,d}$. Then, $\mathbf{r}_{1,d} \perp \mathbf{r}_{3,d}$, $\mathbf{r}_{1,d} \perp \mathbf{r}_{2,d}$. So far, $\mathbf{R}_d$ is obtained as

$$\mathbf{R}_d = \begin{bmatrix} \mathbf{r}_{2,d} \times \mathbf{r}_{3,d} & \mathbf{r}_{2,d} & \mathbf{r}_{3,d} \end{bmatrix} \tag{11.40}$$

which satisfies $\mathbf{R}_d^T \mathbf{R}_d = \mathbf{I}_3$. Differently, if define

$$\mathbf{R}_d = \left[\bar{\mathbf{r}}_{1,d} \ \mathbf{r}_{3,d} \times \bar{\mathbf{r}}_{1,d} \ \mathbf{r}_{3,d} \right]$$

directly, then the property of the orthogonal matrix is violated, i.e., $\mathbf{R}_d^T \mathbf{R}_d = \mathbf{I}_3$ does not hold anymore.

(2) Large-angle case

According to Eq. (5.9), the vector $\mathbf{r}_{3,d}$ is expressed as

$$\mathbf{r}_{3,d} = \begin{bmatrix} \cos\psi_d \sin\theta_d \cos\phi_d + \sin\psi_d \sin\phi_d \\ \sin\psi_d \sin\theta_d \cos\phi_d - \cos\psi_d \sin\phi_d \\ \cos\phi_d \cos\theta_d \end{bmatrix} = \begin{bmatrix} a_{11} \\ a_{12} \\ a_{13} \end{bmatrix}. \tag{11.41}$$

Because the vector $\mathbf{r}_{3,d}$ has been determined by the former controller design, i.e., a_{11}, a_{12}, a_{13} are known, Eq. (11.41) gives

$$\begin{aligned} \sin\phi_d &= \sin\psi_d \cdot a_{11} - \cos\psi_d \cdot a_{12} \\ \sin\theta_d \cos\phi_d &= \cos\psi_d \cdot a_{11} + \sin\psi_d \cdot a_{12} \\ \cos\phi_d \cos\theta_d &= a_{13}. \end{aligned} \tag{11.42}$$

The solutions to Eq. (11.42) are

$$\begin{aligned} \theta_d &= \theta_{d,0} \text{ or } \theta_{d,1} \\ \phi_d &= \phi_{d,0} \text{ or } \phi_{d,1} \end{aligned} \tag{11.43}$$

where

$$\begin{aligned} \theta_{d,0} &= \arctan2(\cos\psi_d \cdot a_{11} + \sin\psi_d \cdot a_{12}, a_{13}) \\ \theta_{d,1} &= \arctan2(-\cos\psi_d \cdot a_{11} - \sin\psi_d \cdot a_{12}, -a_{13}) \\ \phi_{d,0} &= \arcsin(\sin\psi_d \cdot a_{11} - \cos\psi_d \cdot a_{12}) \\ \phi_{d,1} &= \phi_{d,0} - \text{sign}\left(\phi_{d,0}\right)\pi. \end{aligned} \tag{11.44}$$

Although each Euler angle has two optional values in Eq. (11.44), the true value is uniquely determined by Eq. (11.41) in most cases. However, there are also some exceptions. For example, when $\psi_d = 0$, $\mathbf{r}_{3,d} = [0 \ 0 \ -1]^T$, the results consist of two cases: $\theta_d = 0$, $\phi_d = \pi$; $\theta_d = \pi$, $\phi_d = 0$. For the two cases, the corresponding rotation matrices are

$$\mathbf{R}_d = \begin{bmatrix} 1 & 0 & 0 \\ 0 & -1 & 0 \\ 0 & 0 & -1 \end{bmatrix} \text{ and } \mathbf{R}_d = \begin{bmatrix} -1 & 0 & 0 \\ 0 & 1 & 0 \\ 0 & 0 & -1 \end{bmatrix}$$

respectively. In practice, because of the continuity of rotation, the case which is close to the former value $(\theta_d (t - \Delta t), \phi_d (t - \Delta t), \Delta t \in \mathbb{R}_+$ is a small value) is selected as the true one. Then, $\mathbf{R}_d$ is determined by substituting the desired Euler angles ϕ_d, θ_d, ψ_d into Eq. (5.9).

11.3.3.2 Desired Thrust f_d

Since f_d and $\mathbf{r}_{3,d}$ are coupled, both sides of Eq. (11.37) are left multiplied by $\mathbf{r}_{3,d}^T$. Then

$$f_d = m\mathbf{r}_{3,d}^T (g\mathbf{e}_3 - \mathbf{a}_d) \tag{11.45}$$

where $\mathbf{r}_{3,d}$ is obtained from Eq. (11.39), and $\mathbf{r}_{3,d}^T \mathbf{r}_{3,d} = 1$. If f_d is required to confine as $f_d \in [f_{min}, f_{max}]$, then the altitude controller is written as

$$f_d = \mathrm{sat}_{gd}\left(m\mathbf{r}_{3,d}^T(g\mathbf{e}_3 - \mathbf{a}_d) - \frac{f_{min} + f_{max}}{2}, \frac{f_{max} - f_{min}}{2} \right) + \frac{f_{min} + f_{max}}{2}. \qquad (11.46)$$

11.4   Attitude Control

The successive loop closure is adopted in flight controller design for multicopters. The outer loop controller provides commands for the inner loop controller, i.e., the outputs of the horizontal position channel controller provide reference values for the attitude control system. Here, $\mathbf{\Theta}_{hd}$ or $\mathbf{R}_d$ is viewed as the reference. Subsequently, the attitude control objective is to accomplish $\lim_{t \to \infty} \|\mathbf{\Theta}_h(t) - \mathbf{\Theta}_{hd}(t)\| = 0$ or $\lim_{t \to \infty} \|\mathbf{R}^T \mathbf{R}_d - \mathbf{I}_3\| = 0$. It is required that the convergence rate of the attitude loop be 4–10 times higher than that of the horizontal position channel dynamics (11.7), or the bandwidth of the inner loop be 4–10 times higher than that of the outer loop. Then, it can be considered that the attitude control objective $\mathbf{\Theta}_h(t) = \mathbf{\Theta}_{hd}(t)$ or $\mathbf{R}(t) = \mathbf{R}_d(t)$ has already been achieved from the control of the horizontal position channel. Thus, the remaining control objective is handed on to the attitude control. As long as the attitude control is achieved, the horizontal position control problem can be solved. Next, the attitude control design will be discussed.

11.4.1   Basic Conception

For multicopters, the attitude control is the basis of the position control. At present, the commonly-used attitude representations for a rigid body include the Euler angles and rotation matrix. The advantages and disadvantages of the two commonly-used attitude representations are summarized in Table 11.1.

According to the two different attitude representations, two attitude tracking controllers are designed in this section. First, according to the Euler angle representation, under the small-angle assumption, a PID controller is designed. Next, according to the rotation matrix representation, an attitude controller based on the attitude error matrix is designed. In practice, the proper representation and the corresponding attitude controller should be selected according to the requirements.

Table 11.1 Comparison of two attitude representations

Attitude representations	Advantages	Disadvantages
Euler angles	No redundant parameter; Clear physical meaning	Singularity problem when the pitch angle is 90°; Plenty of transcendental function operation; Gimbal Lock
Rotation matrix	No singularity problem; No transcendental function operation; Applicability for the continuous rotation; Global and unique; Easy to interpolate	Six redundant parameters

11.4.2 Euler Angles Based Attitude Control

The following controller design is based on the attitude model (11.9). The attitude control objective is: given the desired attitude angle $\boldsymbol{\Theta}_d \triangleq [\boldsymbol{\Theta}_{hd}^T \, \psi_d]^T$, design a controller $\boldsymbol{\tau} \in \mathbb{R}^3$ such that $\lim_{t \to \infty} \|\mathbf{e}_{\boldsymbol{\Theta}}(t)\| = 0$, where $\mathbf{e}_{\boldsymbol{\Theta}} \triangleq \boldsymbol{\Theta} - \boldsymbol{\Theta}_d$. Here, $\boldsymbol{\Theta}_{hd}$ is generated by the position controller and ψ_d is given by the mission planner. To achieve that, according to

$$\dot{\boldsymbol{\Theta}} = \boldsymbol{\omega} \tag{11.47}$$

the desired angle velocity $\boldsymbol{\omega}_d$ is designed as

$$\boldsymbol{\omega}_d = -\mathbf{K}_{\boldsymbol{\Theta}} \mathbf{e}_{\boldsymbol{\Theta}} \tag{11.48}$$

where $\mathbf{K}_{\boldsymbol{\Theta}} \in \mathbb{R}^{3 \times 3} \cap \mathscr{D} \cap \mathscr{P}$. Then, Eqs. (11.47) and (11.48) build the angle control loop. Under the assumption $\dot{\boldsymbol{\Theta}}_d = \mathbf{0}_{3 \times 1}$, if $\lim_{t \to \infty} \|\mathbf{e}_{\boldsymbol{\omega}}(t)\| = 0$, then $\lim_{t \to \infty} \|\mathbf{e}_{\boldsymbol{\Theta}}(t)\| = 0$, where $\mathbf{e}_{\boldsymbol{\omega}} \triangleq \boldsymbol{\omega} - \boldsymbol{\omega}_d$. According to

$$\mathbf{J} \dot{\boldsymbol{\omega}} = \boldsymbol{\tau} \tag{11.49}$$

the remaining task is to design the desired moment $\boldsymbol{\tau}_d$ to make $\lim_{t \to \infty} \|\mathbf{e}_{\boldsymbol{\omega}}(t)\| = 0$. The PID controller is designed as

$$\boldsymbol{\tau}_d = -\mathbf{K}_{\boldsymbol{\omega} p} \mathbf{e}_{\boldsymbol{\omega}} - \mathbf{K}_{\boldsymbol{\omega} i} \int \mathbf{e}_{\boldsymbol{\omega}} - \mathbf{K}_{\boldsymbol{\omega} d} \dot{\mathbf{e}}_{\boldsymbol{\omega}} \tag{11.50}$$

where $\mathbf{K}_{\boldsymbol{\omega} p}, \mathbf{K}_{\boldsymbol{\omega} i}, \mathbf{K}_{\boldsymbol{\omega} d} \in \mathbb{R}^{3 \times 3} \cap \mathscr{D}$. Then, Eqs. (11.49) and (11.50) build the angular velocity control loop. Under the assumption $\dot{\boldsymbol{\omega}}_d = \mathbf{0}_{3 \times 1}$, if $\lim_{t \to \infty} \|\boldsymbol{\tau}(t) - \boldsymbol{\tau}_d(t)\| = 0$, then $\lim_{t \to \infty} \|\mathbf{e}_{\boldsymbol{\omega}}(t)\| = 0$. In order to avoid the noise caused by the differential, the differential term $\mathbf{K}_{\boldsymbol{\omega} d} \dot{\mathbf{e}}_{\boldsymbol{\omega}}$ can be omitted. So far, the Euler angles based attitude controller is completed, which is composed of Eqs. (11.48) and (11.50). In practice, saturation also needs to be considered in attitude control. Similar to position control, the PID controller is rewritten as

$$\begin{aligned} \mathbf{e}_{\boldsymbol{\omega}} &= \mathrm{sat}_{gd}\left(\boldsymbol{\omega} - \boldsymbol{\omega}_d, a_{10}\right) \\ \boldsymbol{\tau}_d &= \mathrm{sat}_{gd}\left(-\mathbf{K}_{\boldsymbol{\omega} p} \mathbf{e}_{\boldsymbol{\omega}} - \mathbf{K}_{\boldsymbol{\omega} i} \int \mathbf{e}_{\boldsymbol{\omega}} - \mathbf{K}_{\boldsymbol{\omega} d} \dot{\mathbf{e}}_{\boldsymbol{\omega}}, a_{11}\right) \end{aligned} \tag{11.51}$$

where $a_{10}, a_{11} \in \mathbb{R}_+$. The parameters $a_{10}, a_{11} \in \mathbb{R}_+$ related to saturation can be specified according to practical requirements. In particular, if $\boldsymbol{\tau}_d \in [-\tau_{\max}, \tau_{\max}]$, then the saturated attitude controller (11.51) becomes

$$\boldsymbol{\tau}_d = \mathrm{sat}_{gd}\left(-\mathbf{K}_{\boldsymbol{\omega} p} \mathbf{e}_{\boldsymbol{\omega}} - \mathbf{K}_{\boldsymbol{\omega} i} \int \mathbf{e}_{\boldsymbol{\omega}} - \mathbf{K}_{\boldsymbol{\omega} d} \dot{\mathbf{e}}_{\boldsymbol{\omega}}, \tau_{\max}\right). \tag{11.52}$$

11.4.3 Rotation Matrix Based Attitude Control

The rotation matrix based attitude control is introduced as follows. According to the rotation matrix $\mathbf{R}$ and the desired rotation matrix $\mathbf{R}_d$, the attitude error matrix is defined as

$$\tilde{\mathbf{R}} = \mathbf{R}^T \mathbf{R}_d.$$

From the definition above, if and only if $\mathbf{R} = \mathbf{R}_d$, then $\tilde{\mathbf{R}} = \mathbf{I}_3$. Based on this, the control objective of the rotation matrix based attitude control is $\lim_{t \to \infty} \left\| \tilde{\mathbf{R}}(t) - \mathbf{I}_3 \right\| = 0$.

According to the attitude kinematical equation (11.36) and the attitude dynamical equation (11.4), the attitude controllers are designed as follows. First, the attitude tracking error is defined as

$$\mathbf{e}_{\mathbf{R}} \triangleq \frac{1}{2} \mathrm{vex} \left(\mathbf{R}_d^\mathsf{T} \mathbf{R} - \mathbf{R}^\mathsf{T} \mathbf{R}_d \right). \tag{11.53}$$

Furthermore, the angular velocity tracking error is defined as

$$\mathbf{e}_{\omega} \triangleq \boldsymbol{\omega} - \mathbf{R}^\mathsf{T} \mathbf{R}_d \boldsymbol{\omega}_d. \tag{11.54}$$

Under the small-angle assumption, $\boldsymbol{\omega}_d = \dot{\boldsymbol{\Theta}}_d$ is the desired angular velocity, which can be ignored in general cases. Then, $\mathbf{e}_{\omega} = \boldsymbol{\omega}$. Based on the definitions above, the following PD controller is designed

$$\boldsymbol{\tau}_d = -\mathbf{K}_{\mathbf{R}} \mathbf{e}_{\mathbf{R}} - \mathbf{K}_{\omega} \mathbf{e}_{\omega} \tag{11.55}$$

where $\mathbf{K}_{\mathbf{R}}, \mathbf{K}_{\omega} \in \mathbb{R}^{3 \times 3} \bigcap \mathscr{P}$. The PD controller above can only guarantee the system stability within a small neighborhood of the hover position. In order to obtain a wider range of stability, through bringing in the correct error term, a nonlinear controller is designed as

$$\boldsymbol{\tau}_d = -\mathbf{K}_{\mathbf{R}} \mathbf{e}_{\mathbf{R}} - \mathbf{K}_{\omega} \mathbf{e}_{\omega} - \mathbf{J} \left([\omega]_\times \mathbf{R}^\mathsf{T} \mathbf{R}_d \boldsymbol{\omega}_d - \mathbf{R}^\mathsf{T} \mathbf{R}_d \dot{\boldsymbol{\omega}}_d \right). \tag{11.56}$$

By this controller, the resulting attitude control system is guaranteed to be exponentially stable for almost all rotation action. Readers can refer to the Ref. [7] for the detailed proof. In practice, the last term in the controller (11.56) is very small, so it can be ignored to achieve satisfactory performance. However, for an aggressive maneuver flight, the last term is dominant and must be considered. Readers who are interested in aggressive maneuver flight of multicopters can also refer to Refs. [8, 9].

11.4.4 Robust Attitude Control

11.4.4.1 Problem Formulation
In practice, the attitude control system is subject to external disturbances and parameter uncertainties, so the attitude model is rewritten as

$$\begin{aligned} \dot{\boldsymbol{\Theta}} &= \boldsymbol{\omega} \\ \mathbf{J}\dot{\boldsymbol{\omega}} &= \boldsymbol{\tau} + \mathbf{d} \end{aligned} \tag{11.57}$$

where $\mathbf{d} \in \mathbb{R}^3$ denotes an external disturbance, which may be related to the states and the inputs. In practice, there exists uncertainty in $\mathbf{J} \in \mathbb{R}^{3 \times 3} \bigcap \mathscr{P}$, or $\mathbf{J}$ may be changed during the flight. For example, the moment of inertia of an agriculture multicopter is decreased gradually as the payload pesticide is decreased. Thus, designing a controller to stabilize the attitude loop against uncertainties becomes a problem. There are a lot of anti-disturbance control methods [10–12]. A dynamic inverse method based on Additive State Decomposition (ASD) is introduced in the following [13].

11.4.4.2 Additive State Decomposition
Before proceeding further, ASD is introduced first. A commonly-used decomposition in the control field is to decompose a system into two or more lower-order subsystems, called as *lower-order subsystem*

decomposition here. In contrast, ASD is to decompose a system into two or more subsystems with the same dimension as that of the original system. Taking a system P for example, it is decomposed into two subsystems: P_p and P_s, where $\dim(P_p) = n_p$ and $\dim(P_s) = n_s$, respectively. The lower-order subsystem decomposition satisfies

$$n = n_p + n_s \quad \text{and} \quad P = P_p \oplus P_s.$$

By contrast, ASD satisfies

$$n = n_p = n_s \quad \text{and} \quad P = P_p + P_s.$$

(1) ASD on a dynamical control system
Consider an *original system* as follows:

$$\dot{\mathbf{x}} = \mathbf{f}(t, \mathbf{x}, \mathbf{u}), \mathbf{x}(0) = \mathbf{x}_0 \tag{11.58}$$

where $\mathbf{x} \in \mathbb{R}^n$. First, a *primary system* is brought in, having the same dimension as the original system

$$\dot{\mathbf{x}}_p = \mathbf{f}_p(t, \mathbf{x}_p, \mathbf{u}_p), \mathbf{x}_p(0) = \mathbf{x}_{p,0} \tag{11.59}$$

where $\mathbf{x}_p \in \mathbb{R}^n$. From the original system and the primary system, the following *secondary system* is derived

$$\dot{\mathbf{x}} - \dot{\mathbf{x}}_p = \mathbf{f}(t, \mathbf{x}, \mathbf{u}) - \mathbf{f}_p(t, \mathbf{x}_p, \mathbf{u}_p), \mathbf{x}(0) = \mathbf{x}_0.$$

New variables are defined as follows:

$$\mathbf{x}_s \triangleq \mathbf{x} - \mathbf{x}_p, \mathbf{u}_s \triangleq \mathbf{u} - \mathbf{u}_p. \tag{11.60}$$

Then the secondary system can be further written as follows:

$$\dot{\mathbf{x}}_s = \mathbf{f}(t, \mathbf{x}_p + \mathbf{x}_s, \mathbf{u}_p + \mathbf{u}_s) - \mathbf{f}_p(t, \mathbf{x}_p, \mathbf{u}_p)$$
$$\mathbf{x}_s(0) = \mathbf{x}_0 - \mathbf{x}_{p,0}. \tag{11.61}$$

From the definition (11.60), one has

$$\mathbf{x}(t) = \mathbf{x}_p(t) + \mathbf{x}_s(t), \quad t \geq 0.$$

The process is shown in Fig. 11.7.

Fig. 11.7 ASD of a dynamical control system results in: a primary system and secondary system

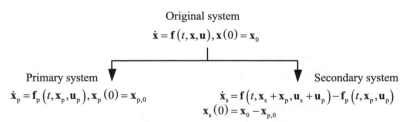

(2) Examples

Example 11.1   In fact, the idea of ASD has been implicitly mentioned in existing literature. An existing example is the tracking controller design, which often requires a reference system to derive error dynamics. For example, the reference system (primary system) is assumed to be given as follows:

$$\dot{\mathbf{x}}_r = \mathbf{f}\left(t, \mathbf{x}_r, \mathbf{u}_r\right), \mathbf{x}_r\left(0\right) = \mathbf{x}_{r,0}.$$

Based on the reference system, the error dynamics (secondary system) are derived as follows:

$$\dot{\mathbf{x}}_e = \mathbf{f}\left(t, \mathbf{x}_e + \mathbf{x}_r, \mathbf{u}\right) - \mathbf{f}\left(t, \mathbf{x}_r, \mathbf{u}_r\right), \mathbf{x}_e\left(0\right) = \mathbf{x}_0 - \mathbf{x}_{r,0}$$

where $\mathbf{x}_e \triangleq \mathbf{x} - \mathbf{x}_r$. This is a commonly-used step to transform a tracking problem into a stabilizing control problem when adaptive control is used.

Example 11.2   Consider a class of systems as follows:

$$\begin{aligned}
\dot{\mathbf{x}}\left(t\right) &= [\mathbf{A} + \Delta\mathbf{A}\left(t\right)]\mathbf{x}\left(t\right) + \mathbf{A}_d\mathbf{x}\left(t - T\right) + \mathbf{B}\mathbf{r}\left(t\right) \\
\mathbf{e}\left(t\right) &= -[\mathbf{C} + \Delta\mathbf{C}\left(t\right)]\mathbf{x}\left(t\right) + \mathbf{r}\left(t\right) \\
\mathbf{x}\left(\theta\right) &= \boldsymbol{\varphi}\left(\theta\right), \theta \in [-T, 0].
\end{aligned} \tag{11.62}$$

Choose (11.62) as the original system and design the primary system as follows:

$$\begin{aligned}
\dot{\mathbf{x}}_p\left(t\right) &= \mathbf{A}\mathbf{x}_p + \mathbf{A}_d\mathbf{x}_p\left(t - T\right) + \mathbf{B}\mathbf{r}\left(t\right) \\
\mathbf{e}_p\left(t\right) &= -\mathbf{C}\mathbf{x}_p + \mathbf{r}\left(t\right) \\
\mathbf{x}_p\left(\theta\right) &= \boldsymbol{\varphi}\left(\theta\right), \theta \in [-T, 0].
\end{aligned} \tag{11.63}$$

Then the secondary system is determined by the rule (11.61):

$$\begin{aligned}
\dot{\mathbf{x}}_s\left(t\right) &= [\mathbf{A} + \Delta\mathbf{A}\left(t\right)]\mathbf{x}_s\left(t\right) + \mathbf{A}_d\mathbf{x}_s\left(t - T\right) + \Delta\mathbf{A}\left(t\right)\mathbf{x}_p\left(t\right) \\
\mathbf{e}_s\left(t\right) &= -[\mathbf{C} + \Delta\mathbf{C}\left(t\right)]\mathbf{x}_s\left(t\right) - \Delta\mathbf{C}\left(t\right)\mathbf{x}_p\left(t\right) \\
\mathbf{x}_s\left(\theta\right) &= \mathbf{0}, \theta \in [-T, 0].
\end{aligned} \tag{11.64}$$

By ASD, $\mathbf{e}\left(t\right) = \mathbf{e}_p\left(t\right) + \mathbf{e}_s\left(t\right)$. Since $\|\mathbf{e}\left(t\right)\| \leq \|\mathbf{e}_p\left(t\right)\| + \|\mathbf{e}_s\left(t\right)\|$, the tracking error $\mathbf{e}\left(t\right)$ can be analyzed by $\mathbf{e}_p\left(t\right)$ and $\mathbf{e}_s\left(t\right)$, separately. If $\mathbf{e}_p\left(t\right)$ and $\mathbf{e}_s\left(t\right)$ are bounded and small, then so is $\mathbf{e}\left(t\right)$. Fortunately, note that (11.63) is a linear time-invariant system and is independent of the secondary system (11.64), for the analysis of which many tools such as the transfer function are available. By contrast, the transfer function tool cannot be directly applied to the original system (11.62) as it is time-varying. For more details, please refer to [14].

Example 11.3   Consider a class of nonlinear systems as follows:

$$\begin{aligned}
\dot{\mathbf{x}} &= \mathbf{A}\mathbf{x} + \mathbf{B}\mathbf{u} + \boldsymbol{\phi}\left(\mathbf{y}\right) + \mathbf{d}, \mathbf{x}\left(0\right) = \mathbf{x}_0 \\
\mathbf{y} &= \mathbf{C}^{\mathsf{T}}\mathbf{x}
\end{aligned} \tag{11.65}$$

where $\mathbf{x}, \mathbf{y}, \mathbf{u}$ represent the state, output and input, respectively; the function $\boldsymbol{\phi}\left(\cdot\right)$ is nonlinear. The objective is to design $\mathbf{u}$ such that $\lim_{t \to \infty} \|\mathbf{y}\left(t\right) - \mathbf{r}\left(t\right)\| = 0$. Choose system (11.65) as the original system and design the primary system as follows:

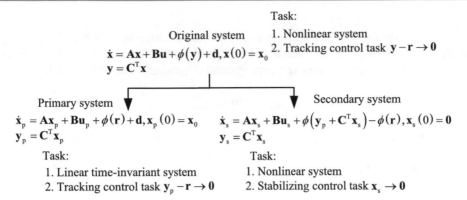

Fig. 11.8 Additive-state-decomposition-based tracking control

$$\dot{\mathbf{x}}_p = \mathbf{A}\mathbf{x}_p + \mathbf{B}\mathbf{u}_p + \boldsymbol{\phi}(\mathbf{r}) + \mathbf{d},\, \mathbf{x}_p(0) = \mathbf{x}_0$$
$$\mathbf{y}_p = \mathbf{C}^{\mathrm{T}}\mathbf{x}_p. \tag{11.66}$$

Then the secondary system is determined by the rule (11.61) that

$$\dot{\mathbf{x}}_s = \mathbf{A}\mathbf{x}_s + \mathbf{B}\mathbf{u}_s + \boldsymbol{\phi}\left(\mathbf{y}_p + \mathbf{C}^{\mathrm{T}}\mathbf{x}_s\right) - \boldsymbol{\phi}(\mathbf{r}),\, \mathbf{x}_s(0) = \mathbf{0}$$
$$\mathbf{y}_s = \mathbf{C}^{\mathrm{T}}\mathbf{x}_s \tag{11.67}$$

where $\mathbf{u}_s \triangleq \mathbf{u} - \mathbf{u}_p$. Then $\mathbf{x} = \mathbf{x}_p + \mathbf{x}_s$, $\mathbf{y} = \mathbf{y}_p + \mathbf{y}_s$ and $\mathbf{u} = \mathbf{u}_p + \mathbf{u}_s$. Here, the task $\mathbf{y}_p - \mathbf{r} \to \mathbf{0}$ is assigned to the linear time-invariant system (11.66) (A linear time-invariant system is simpler than a nonlinear one). On the other hand, the task $\mathbf{x}_s \to \mathbf{0}$ is assigned to the nonlinear system (11.67) (a stabilizing control problem is simpler than a tracking problem). If the two tasks are accomplished, then $\mathbf{y} = \mathbf{y}_p + \mathbf{y}_s \to \mathbf{r}$. The basic idea is to decompose an original system into two subsystems in charge of simpler subtasks. Then one designs controllers for two subtasks, and finally combines them to achieve the original control task. The process is shown in Fig. 11.8. For more details, please refer to [15–17].

(3) Comparison with superposition principle
Superposition Principle[4] is widely used in physics and engineering.

Superposition Principle: For all linear systems, the net response at a given place and time caused by two or more stimuli is the sum of the responses which would have been caused by each stimulus individually.

For a simple linear system

$$\dot{\mathbf{x}} = \mathbf{A}\mathbf{x} + \mathbf{B}(\mathbf{u}_1 + \mathbf{u}_2),\, \mathbf{x}(0) = \mathbf{0}$$

the statement of the superposition principle means $\mathbf{x} = \mathbf{x}_p + \mathbf{x}_s$, where

$$\dot{\mathbf{x}}_p = \mathbf{A}\mathbf{x}_p + \mathbf{B}\mathbf{u}_1,\, \mathbf{x}_p(0) = \mathbf{0}$$
$$\dot{\mathbf{x}}_s = \mathbf{A}\mathbf{x}_s + \mathbf{B}\mathbf{u}_2,\, \mathbf{x}_s(0) = \mathbf{0}.$$

[4]This principle was discussed first for dielectrics by Hopkinson [18]. The early papers [19–21] may serve for a proper orientation.

Table 11.2 Comparison of superposition principle and ASD

	Suitable systems	Emphasis
Superposition principle	Linear	Superposition
ASD	Linear\Nonlinear	Decomposition

Obviously, this result can also be derived from ASD. Moreover, the superposition principle and ASD have the relationship shown in Table 11.2. It is noted that ASD applies to not only linear systems but also nonlinear systems.

(4) Else
ASD is also used in stabilizing control [13]. Furthermore, ASD is extended to Additive Output Decomposition (AOD) [22], which will be used in Chap. 12.

11.4.4.3 Additive State Decomposition Based Dynamic Inverse Control [13]
The system (11.57) is further expressed as follows:

$$\dot{\mathbf{x}} = \mathbf{A}_0\mathbf{x} + \mathbf{B}\left(\mathbf{J}^{-1}\boldsymbol{\tau} + \mathbf{J}^{-1}\mathbf{d}\right)$$
$$\mathbf{y} = \mathbf{C}^{\mathrm{T}}\mathbf{x} \tag{11.68}$$

where

$$\mathbf{x} = \begin{bmatrix} \boldsymbol{\Theta} \\ \boldsymbol{\omega} \end{bmatrix}, \mathbf{A}_0 = \begin{bmatrix} \mathbf{0}_{3\times3} & \mathbf{I}_3 \\ \mathbf{0}_{3\times3} & \mathbf{0}_{3\times3} \end{bmatrix}, \mathbf{B} = \begin{bmatrix} \mathbf{0}_{3\times3} \\ \mathbf{I}_3 \end{bmatrix}.$$

The output matrix $\mathbf{C} \in \mathbb{R}^{6\times3}$ is specified later. The control objective is to design a controller $\boldsymbol{\tau} \in \mathbb{R}^3$ such that $\lim_{t\to\infty}\|\mathbf{y}(t) - \mathbf{y}_{\mathrm{d}}(t)\| = 0$, where $\mathbf{y}_{\mathrm{d}} \in \mathbb{R}^3$ is the desired output. Design a controller as follows:

$$\boldsymbol{\tau} = \mathbf{J}_0\left(\mathbf{K}^{\mathrm{T}}\mathbf{x} + \mathbf{u}\right)$$

where $\mathbf{J}_0 \in \mathbb{R}^{3\times3} \cap \mathscr{P}$ is the known nominal moment of inertial. Then, system (11.68) becomes

$$\dot{\mathbf{x}} = \mathbf{A}_0\mathbf{x} + \mathbf{B}\left(\mathbf{J}^{-1}\mathbf{J}_0\mathbf{K}^{\mathrm{T}}\mathbf{x} + \mathbf{J}^{-1}\mathbf{J}_0\mathbf{u} + \mathbf{J}^{-1}\mathbf{d}\right). \tag{11.69}$$

Since the pair $(\mathbf{A}_0, \mathbf{B})$ is controllable, a matrix $\mathbf{K} \in \mathbb{R}^{6\times3}$ can always be found such that $\mathbf{A} \triangleq \mathbf{A}_0 + \mathbf{B}\mathbf{K}^{\mathrm{T}}$ is stable. Then, Eq. (11.69) is written as

$$\dot{\mathbf{x}} = \mathbf{A}\mathbf{x} + \mathbf{B}\left(\mathbf{J}^{-1}\mathbf{J}_0\mathbf{u} + \left(\mathbf{J}^{-1}\mathbf{J}_0 - \mathbf{I}_3\right)\mathbf{K}^{\mathrm{T}}\mathbf{x} + \mathbf{J}^{-1}\mathbf{d}\right). \tag{11.70}$$

Considering system (11.70) as the original system, the primary system is designed as follows:

$$\dot{\mathbf{x}}_{\mathrm{p}} = \mathbf{A}\mathbf{x}_{\mathrm{p}} + \mathbf{B}\mathbf{u}$$
$$\mathbf{y}_{\mathrm{p}} = \mathbf{C}^{\mathrm{T}}\mathbf{x}_{\mathrm{p}}, \mathbf{x}_{\mathrm{p}}(0) = \mathbf{0}_{6\times1}. \tag{11.71}$$

Then, by subtracting Eq. (11.71) from Eq. (11.70), the secondary system is obtained as

$$\dot{\mathbf{x}}_s = \mathbf{A}\mathbf{x}_s + \mathbf{B}\left(\left(\mathbf{J}^{-1}\mathbf{J}_0 - \mathbf{I}_3\right)\mathbf{u} + \left(\mathbf{J}^{-1}\mathbf{J}_0 - \mathbf{I}_3\right)\mathbf{K}^T\mathbf{x} + \mathbf{J}^{-1}\mathbf{d}\right)$$
$$\mathbf{y}_s = \mathbf{C}^T\mathbf{x}_s, \mathbf{x}_s\left(0\right) = \mathbf{x}_0.$$

According to ASD, it follows

$$\mathbf{x} = \mathbf{x}_p + \mathbf{x}_s$$
$$\mathbf{y} = \mathbf{y}_p + \mathbf{y}_s.$$

Thus, the system (11.70) is rewritten as

$$\dot{\mathbf{x}}_p = \mathbf{A}\mathbf{x}_p + \mathbf{B}\mathbf{u} \tag{11.72}$$
$$\mathbf{y} = \mathbf{C}^T\mathbf{x}_p + \mathbf{d}_l$$

where $\mathbf{d}_l = \mathbf{y}_s$ is a lumped disturbance. Fortunately, since $\mathbf{y}_p = \mathbf{C}^T\mathbf{x}_p$ and the output $\mathbf{y}$ are all known, the lumped disturbance $\mathbf{d}_l$ can be obtained exactly by $\mathbf{d}_l = \mathbf{y} - \mathbf{y}_p$. Thus, the system (11.72) is completely known.

From the input-output relationship, the systems (11.70) and (11.72) are equivalent. Based on the system (11.72), a Dynamic Inverse Controller (DIC) is further designed to make $\lim_{t\to\infty}\|\mathbf{y}(t) - \mathbf{y}_d(t)\| = 0$. Furthermore, system (11.72) is written in the form of a transfer function as

$$\mathbf{y}(s) = \mathbf{G}(s)\mathbf{u}(s) + \mathbf{d}_l(s) \tag{11.73}$$

where $\mathbf{G}(s) = \mathbf{C}^T(s\mathbf{I}_6 - \mathbf{A})^{-1}\mathbf{B}$ denotes the transfer function. In order to make the output $\mathbf{y}(t)$ track $\mathbf{y}_d(t)$ rapidly, the Laplace transfer function of the DIC is designed as

$$\mathbf{u}(s) = \mathbf{G}^{-1}(s)\left(\mathbf{Q}_1(s)\mathbf{y}_d(s) - \mathbf{Q}_2(s)\mathbf{d}_l(s)\right) \tag{11.74}$$

where $\mathbf{Q}_1(s), \mathbf{Q}_2(s)$ are low-pass filter matrices. They guarantee that the order of the denominators of $\mathbf{G}^{-1}(s)\mathbf{Q}_1(s), \mathbf{G}^{-1}(s)\mathbf{Q}_2(s)$ are not lower than that of their numerators, respectively. Substituting the controller (11.74) into Eq. (11.73) results in

$$\mathbf{y}(s) = \mathbf{Q}_1(s)\mathbf{y}_d(s) + \left(\mathbf{I}_3 - \mathbf{Q}_2(s)\right)\mathbf{d}_l(s).$$

Since the low-frequency range is often dominant in a signal, it is expected that the output $\mathbf{y}_d(s)$ will be retained by $\mathbf{Q}_1(s)\mathbf{y}_d(s)$ very well, and $\mathbf{d}_l(s)$ will be largely attenuated by $(\mathbf{I}_3 - \mathbf{Q}_2(s))\mathbf{d}_l(s)$. Thus, the tracking objective will be achieved.

11.4.4.4   Design Example 1

In order to simplify $\mathbf{Q}(s)$, the output matrix is designed as $\mathbf{C} = [c_1\mathbf{I}_3 c_2\mathbf{I}_3]^T$. Then, the relative degree[5] of $\mathbf{G}(s)$ is one. If the characteristic roots of $\mathbf{A}$ are designed as $-a \pm aj$, then

$$\mathbf{G}(s) = \text{diag}\left(g(s), g(s), g(s)\right)$$

where $a \in \mathbb{R}_+$ and

$$g(s) = \frac{c_2 s + c_1}{s^2 + 2as + 2a^2}.$$

---

[5]The relative degree of a transfer function is the order of the denominator subtracting that of the numerator.

Furthermore, for simplicity, let $\mathbf{Q}(s) = \mathbf{Q}_1(s) = \mathbf{Q}_2(s)$. Because $g(s)$ is a transfer function whose relative degree is one, the first-order low-pass filter matrix is designed as

$$\mathbf{Q}(s) = \text{diag}(q(s), q(s), q(s))$$

where $q(s)$ can be simply designed as

$$q(s) = \frac{1}{\varepsilon s + 1}$$

and $\varepsilon \in \mathbb{R}_+$. Then

$$\mathbf{G}^{-1}(s)\mathbf{Q}(s) = \text{diag}\left(\frac{s^2 + 2as + 2a^2}{(c_2 s + c_1)(\varepsilon s + 1)}, \frac{s^2 + 2as + 2a^2}{(c_2 s + c_1)(\varepsilon s + 1)}, \frac{s^2 + 2as + 2a^2}{(c_2 s + c_1)(\varepsilon s + 1)}\right).$$

11.4.4.5   Design Example 2

Different $\mathbf{K}, \mathbf{C}$ and $\mathbf{Q}_1(s), \mathbf{Q}_2(s)$ will lead to different control performance. In [13], in order to solve the attitude stabilization problem for a quadcopter, namely designing a controller $\boldsymbol{\tau}$ such that $\lim_{t\to\infty} \|\mathbf{x}(t)\| = 0$, the following Proportional-Integral (PI) controller is generated through a special configuration

$$\mathbf{u}(s) = -\frac{1}{\varepsilon}\left(\mathbf{C}^{\mathrm{T}}\mathbf{B}\right)^{-1}\mathbf{C}^{\mathrm{T}}\mathbf{x}(s) - \frac{1}{\varepsilon s}\boldsymbol{\Lambda}\left(\mathbf{C}^{\mathrm{T}}\mathbf{B}\right)^{-1}\mathbf{C}^{\mathrm{T}}\mathbf{x}(s) \tag{11.75}$$

where $\boldsymbol{\Lambda}$ and $\mathbf{C}$ satisfy $\mathbf{C}^{\mathrm{T}}\mathbf{A} = -\boldsymbol{\Lambda}\mathbf{C}^{\mathrm{T}}, \boldsymbol{\Lambda} \in \mathbb{R}^{6\times6}$. In the design process, let

$$\mathbf{Q}(s) = \mathbf{Q}_1(s) = \mathbf{Q}_2(s) = \frac{1}{\varepsilon s + 1}\mathbf{I}_3.$$

The design procedure is as follows [13]:

Step 1. Select six characteristic roots with the negative real part $-\lambda_i < 0, i = 1, \ldots, 6$. Then, through the pole placement, a control gain matrix $\mathbf{K}$ is designed to obtain $\mathbf{A} \triangleq \mathbf{A}_0 + \mathbf{B}\mathbf{K}^{\mathrm{T}}$ with the selected six characteristic roots.

Step 2. Define $\mathbf{C} = [\mathbf{c}_1 \ \mathbf{c}_2 \ \mathbf{c}_3] \in \mathbb{R}^{6\times3}$, where $\mathbf{c}_i \in \mathbb{R}^6$ are the independent unit eigenvectors associated with eigenvalues $-\lambda_i < 0$ of $\mathbf{A}^{\mathrm{T}}, i = 1, \ldots, 6$, respectively. If $\det\left(\mathbf{C}^{\mathrm{T}}\mathbf{B}\right) = 0$, then go to Step 1 to reselect characteristic roots.

Step 3. Design the controller (11.75) with $\boldsymbol{\Lambda} = \text{diag}(\lambda_1, \lambda_2, \lambda_3, \lambda_4, \lambda_5, \lambda_6)$.

Step 4. Select a proper ε.

The stability of closed-loop dynamics is proved in [13]. It can be guaranteed that $\lim_{t\to\infty} \|\mathbf{x}(t)\| = 0$ as long as $\varepsilon \in \mathbb{R}_+$ is small enough. However, ε cannot be set too small, because too small ε will make the controller sensitive to time delay. Thus, the goal is to "select a proper ε" in the design procedure. The control method above is experimentally tested on a Quanser Qball-X4, a quadcopter developed by the Quanser. The controller introduced in this chapter is adopted only for attitude control, while an existing position controller offered by the Quanser is retained. The nominal inertial matrix of the Qball-X4 is $\mathbf{J}_0 = \text{diag}(0.03, 0.03, 0.04)$ kg $\cdot$ m^2. In order to test the robustness of this algorithm, a payload weighing about 10% of the aircraft weight is added to the aircraft, as shown in Fig. 11.9, which will change the weight and the moment of inertial of the aircraft. As a result, the designed controller can guarantee a satisfying attitude stability, as shown in Fig. 11.9. The video is available at https://www.youtube.com/watch?v=XE4plSkHYxc or http://rfly.buaa.edu.cn.

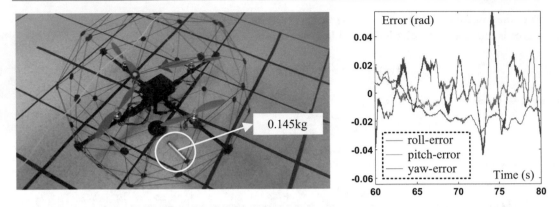

Fig. 11.9 A Quanser Qball-X4 with a 0.145 kg payload is presented in the *left*. The ASD based DICis applied to the Quanser Qball-X4, and the corresponding attitude controleffect is presented in the *right*

11.5   Control Allocation

11.5.1 Basic Concept

Control allocation was first proposed in the flight control system design, which has been generalized to some practical engineering at present. The key idea of control allocation is to allocate the control command to actuators according to some kinds of optimization objectives, and to guarantee that the constraint condition of actuators is satisfied. The study on the control allocation algorithm has gone through a developing process from the simple to the complex, from static optimization to dynamic optimization, from single object optimization to multi-objective optimization [23, 24]. From the mathematical point of view, control allocation is a method to solve equations with constraints. The input is the pseudo control command that needs to be produced, which is denoted by $\mathbf{u}_v (t) \in \mathbb{R}^n$, and the output is actual control input of each actuator $\boldsymbol{\delta} = [\delta_1 \cdots \delta_m]^\mathrm{T} \in \mathbb{R}^m$. Briefly, the control allocation problem can be described as: given $\mathbf{u}_v (t)$, seek $\boldsymbol{\delta} (t)$ to satisfy

$$\mathbf{u}_v (t) = \mathbf{g} (\boldsymbol{\delta} (t)) \tag{11.76}$$

where $\mathbf{g} : \mathbb{R}^m \rightarrow \mathbb{R}^n$ is the mapping from control inputs of actuators in controlled systems to the pseudo control command. In particular, the linear control allocation problem is based on

$$\mathbf{u}_v (t) = \mathbf{B}\boldsymbol{\delta} (t) \tag{11.77}$$

where $\mathbf{B} \in \mathbb{R}^{m \times n}$ is the known control efficiency matrix. In actual systems, actuators have structural constraint and load constraint, etc. So control inputs of actuators $\delta_k, k = 1, \ldots, m$ must satisfy

$$\begin{cases} \delta_{\min,k} \leq \delta_k \leq \delta_{\max,k} \\ \dot{\delta}_{\min,k} \leq \dot{\delta}_k \leq \dot{\delta}_{\max,k}, \quad k = 1, \ldots, m \end{cases} \tag{11.78}$$

where $\delta_{\min,k}, \delta_{\max,k}, \dot{\delta}_{\min,k}, \dot{\delta}_{\max,k}$ are the minimum, the maximum position constraint and the minimum, the maximum rate constraint of the actuators δ_k, respectively.

Multicopters have multiple propulsors. Thus, there are multiple combinations of propulsors, and the final reasonable and feasible combination must be given by control allocation. Control allocation module is located after position and attitude control modules, and before propulsors. The inputs of control allocator are the pseudo control variables generated by the position and attitude control modules, namely a total thrust and three-axis moments. Next, control allocator produces motion commands to propulsors. Then, propulsors track these commands to acquire the required force and moments. Benefits of using control allocator are as follows: (1) from a safety perspective, it makes multicopters adapt to different flight tasks and conditions, meanwhile, when certain propulsors fail, the rest propulsors can be recombined and cooperatively control multicopters, which improves the system robustness; (2) from a design perspective, the control allocation only considers how to optimally allocate the required control values to propulsors, which simplifies the controller design.

11.5.2 Implementation of Control Allocation in Autopilots

The control effectiveness models for a quadcopter with the plus-configuration and a multicopter are Eqs. (6.22) and (6.25) in Chap. 6, respectively. For a quadcopter, since $\mathbf{M}_4 \in \mathbb{R}^{4\times4}$ is nonsingular, the control allocation matrix can be directly obtained by matrix inversion, namely

$$\mathbf{P}_4 = \mathbf{M}_4^{-1}$$

where $\mathbf{P}_4 \in \mathbb{R}^{4\times4}$. The allocation is unique. Furthermore, if a multicopter has more than four propellers, the allocation may be non-unique. In open source autopilots, the control allocation matrix is often obtained by calculating the *pseudo-inverse*, that is

$$\mathbf{P}_{n_r} = \mathbf{M}_{n_r}^{\dagger} = \mathbf{M}_{n_r}^{\mathrm{T}} \left(\mathbf{M}_{n_r} \mathbf{M}_{n_r}^{\mathrm{T}} \right)^{-1} \tag{11.79}$$

where $\mathbf{P}_{n_r} \in \mathbb{R}^{n_r \times 4}$, $\mathbf{M}_{n_r} \in \mathbb{R}^{4 \times n_r}$. After the desired thrust f_d and the desired moment $\boldsymbol{\tau}_\mathrm{d}$ have been obtained, the desired propeller angular speeds are obtained as follows:

$$\begin{bmatrix} \varpi_{\mathrm{d},1}^2 \\ \varpi_{\mathrm{d},2}^2 \\ \vdots \\ \varpi_{\mathrm{d},n_r}^2 \end{bmatrix} = \mathbf{P}_{n_r} \begin{bmatrix} f_\mathrm{d} \\ \boldsymbol{\tau}_\mathrm{d} \end{bmatrix}. \tag{11.80}$$

In a particular allocation, some propeller angular speeds may exceed the saturation value. Thus, a good allocation algorithm is very important. Furthermore, some parameters in $\mathbf{M}_{n_r}$, namely $c_\mathrm{T}, c_\mathrm{M}, d_i, i = 1, \ldots, n_r$ (the concrete definitions can be referred to Sect. 6.1.3 in Chap. 6) are unknown. Under this condition, a question arises: how is the control allocation conducted? In order to answer this question, the control effectiveness matrix for a standard multicopter is defined as a function matrix that

$$\mathbf{M}_{n_r}\left(c_\mathrm{T}, c_\mathrm{M}, d\right) = \begin{bmatrix} c_\mathrm{T} & c_\mathrm{T} & \cdots & c_\mathrm{T} \\ -dc_\mathrm{T}\sin\varphi_1 & -dc_\mathrm{T}\sin\varphi_2 & \cdots & -dc_\mathrm{T}\sin\varphi_{n_r} \\ dc_\mathrm{T}\cos\varphi_1 & dc_\mathrm{T}\cos\varphi_2 & \cdots & dc_\mathrm{T}\cos\varphi_{n_r} \\ c_\mathrm{M}\sigma_1 & c_\mathrm{M}\sigma_2 & \cdots & c_\mathrm{M}\sigma_{n_r} \end{bmatrix}$$

Fig. 11.10 A hexacopter
with the plus-configuration

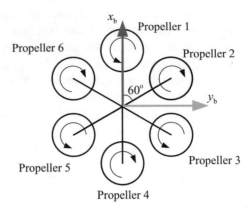

which satisfies

$$\mathbf{M}_{n_r}(c_T, c_M, d) = \mathbf{P}_a \mathbf{M}_{n_r}(1, 1, 1)$$

where $\mathbf{P}_a = \mathrm{diag}(c_T, \ dc_T, \ dc_T, \ c_M)$. Thus, the following relation exists

$$\mathbf{M}_{n_r}^{\dagger}(c_T, c_M, d) = \mathbf{M}_{n_r}^{\dagger}(1, 1, 1)\mathbf{P}_a^{-1}$$

Taking the hexacopter presented in Fig. 11.10 for example, $\mathbf{M}_6(c_T, c_M, d)$ is

$$\mathbf{M}_6(c_T, c_M, d) = \begin{bmatrix} c_T & c_T & c_T & c_T & c_T & c_T \\ 0 & -\frac{\sqrt{3}dc_T}{2} & -\frac{\sqrt{3}dc_T}{2} & 0 & \frac{\sqrt{3}dc_T}{2} & \frac{\sqrt{3}dc_T}{2} \\ dc_T & \frac{dc_T}{2} & -\frac{dc_T}{2} & -dc_T & -\frac{dc_T}{2} & \frac{dc_T}{2} \\ c_M & -c_M & c_M & -c_M & c_M & -c_M \end{bmatrix}.$$

Then

$$\mathbf{M}_6^{\dagger}(1, 1, 1) = \frac{1}{6}\begin{bmatrix} 1 & 0 & 2 & 1 \\ 1 & -\sqrt{3} & 1 & -1 \\ 1 & -\sqrt{3} & -1 & 1 \\ 1 & 0 & -2 & -1 \\ 1 & \sqrt{3} & -1 & 1 \\ 1 & \sqrt{3} & 1 & -1 \end{bmatrix}.$$

Thus

$$\begin{bmatrix} \varpi_{d,1}^2 \\ \varpi_{d,2}^2 \\ \vdots \\ \varpi_{d,6}^2 \end{bmatrix} = \mathbf{M}_6^{\dagger}(1, 1, 1)\mathbf{P}_a^{-1}\begin{bmatrix} f_d \\ \boldsymbol{\tau}_d \end{bmatrix} = \mathbf{M}_6^{\dagger}(1, 1, 1)\begin{bmatrix} f_d/c_T \\ \tau_{dx}/(dc_T) \\ \tau_{dy}/(dc_T) \\ \tau_{dz}/c_M \end{bmatrix}. \tag{11.81}$$

In most autopilots, controllers f_d, $\boldsymbol{\tau}_d$ are all PID controllers. So, the unknown variables c_T, c_M, d can be compensated for by adjusting PID parameters.

11.6   Motor Control

Motors drive propellers and further multicopters. Thus, a high performance motor speed control system is important. Rapid response of motors will produce high performance in attitude control and position control. Propellers are fixed to motor rotors, so the propeller angular speeds are the same as the rotor angular speeds. Suppose propeller angular speeds are $\varpi_k, k = 1, 2, \ldots, n_r$. Most multicopters are equipped with brushless Direct Current (DC) motors that use back ElectroMotive Force (EMF) sensing for rotor commutation and high-frequency PulseWidth Modulation (PWM) to control motor voltage. The controller design for the motor based on the motor throttle command is proposed in the following. The control objective is to design each motor throttle command $\sigma_{d,k}$ such that $\lim_{t \to \infty} |\varpi_k(t) - \varpi_{d,k}(t)| = 0, k = 1, \ldots, n_r$.

11.6.1 Closed-Loop Control

The propulsor model with throttle command as the input can be found in Eq. (6.28), that is

$$\varpi_k = \frac{1}{T_m s + 1} (C_R \sigma_k + \varpi_b) \tag{11.82}$$

where throttle command $\sigma_k \in [0, 1]$ is the input, propeller angular velocity $\varpi_k \in \mathbb{R}$ is the output, $k = 1, \ldots, n_r$. The motor control objective is to minimize the control error $\tilde{\varpi}_k \triangleq \varpi_k - \varpi_{d,k}$. Thus, a controller is designed as

$$\sigma_{d,k} = -k_{\varpi p}\tilde{\varpi}_k - k_{\varpi i} \int \tilde{\varpi}_k - k_{\varpi d}\dot{\tilde{\varpi}}_k \tag{11.83}$$

where $k_{\varpi p}, k_{\varpi i}, k_{\varpi d} \in \mathbb{R}, k = 1, \ldots, n_r$. Here, let $\sigma_k = \sigma_{d,k}, k = 1, \ldots, n_r$.

11.6.2 Open-Loop Control

Most autopilots usually adopt open-loop control. After f_d, τ_d being obtained, the desired propeller angular speeds $\varpi_{d,k}, k = 1, 2, \ldots, n_r$ can be obtained through control allocation, which is proportional to the motor throttle command. Thus, a controller is designed as

$$\sigma_{d,k} = a\varpi_{d,k} + b \tag{11.84}$$

where unknown parameters $a, b \in \mathbb{R}$ can be compensated for by PID parameters in the position and attitude controllers. Here, let $\sigma_k = \sigma_{d,k}, k = 1, \ldots, n_r$.

11.7   Comprehensive Simulation

11.7.1 Control Objective and System Parameter Setting

To demonstrate the effectiveness of the introduced controllers, simulations are carried out by using MATLAB. Two kinds of control schemes corresponding to Sects. 11.7.2 and 11.7.3 are adopted to realize one control task that: a quadcopter is required to fly a square trajectory starting from the central point in the horizontal xy plane, and then fly back to the central point to hover.

Table 11.3 System parameters for the quadcopter

g	9.8100 (m/s^2)	J_{yy}	0.0095 (kg · m^2)
d	0.2223 (m)	J_{zz}	0.0186 (kg · m^2)
m	1.0230 (kg)	T_m	0.0760 (s)
c_T	1.4865e-07 (N/RPM^2)	C_R	80.5840 (RPM)
c_M	2.9250e-09 (N · m/RPM^2)	ϖ_b	976.2000 (RPM)
J_{xx}	0.0095 (kg · m^2)		

Table 11.4 Time series of the desired trajectory commands

Time (s)	p_{x_d} (m)	p_{y_d} (m)	p_{z_d} (m)	ψ_d (°)
0	0	0	−3	0
5	10	0	−3	0
10	0	10	−3	0
15	−10	0	−3	0
20	0	−10	−3	0
25	10	0	−3	0
30	0	0	−3	0

The selected quadcopter is plus-configuration. In order to be consistent with practical scenario, the input saturation, propulsor dead-zone and motor dynamics are all considered in the simulation. The system parameters are selected as in Table 11.3.

The initial states of the system are selected as $\varpi(0) = [4000\ 4000\ 4000\ 4000]^T$ (RPM), $\mathbf{p}(0) = [0\ 0\ -3]^T$ (m), $\mathbf{v}(0) = [0\ 0\ 0]^T$ (m/s), $\Theta(0) = [0\ 0\ 0]^T$ (°), $\omega(0) = [0\ 0\ 0]^T$ (°/s). The desired trajectory commands are given by time series, as shown in Table 11.4.

11.7.2 Euler Angles Based Attitude Control Combined with Euler Angles Based Position Control

For simplicity, "Euler angles based attitude control combined with Euler angles based position control" is abbreviated as "EBAC+EBPC". A closed-loop system is built by combining the attitude controller (11.52), the horizontal position controller (11.32), the altitude controller (11.35) and the nonlinear model of the quadcopter. The controller parameters are selected as in Tables 11.5 and 11.6.

Simulation results are presented in Figs. 11.11, 11.12 and 11.13. Twelve state responses are presented in Fig. 11.11, which shows that position and attitude track the desired commands with an acceptable tracking error. The tracking lag is caused by the motor dynamics. Attitude tracking is fast, and position tracking lags behind attitude tracking for a little while. The variation ranges of velocity and angular

Table 11.5 Attitude controller parameters

	$\mathbf{K}_\Theta$	$\mathbf{K}_{\omega p}$	$\mathbf{K}_{\omega i}$	$\mathbf{K}_{\omega d}$
Roll channel	2	2	0	0.8
Pitch channel	2	2	0	0.8
Yaw channel	3	3	0	2.5

Table 11.6 Position
controller parameters

	K_p	K_{vp}	K_{vi}	K_{vd}
X channel	0.5	1	0.05	0.2
Y channel	0.5	1	0.05	0.2
Z channel	4	4	2	4

Fig. 11.11 State responses
of EBAC+EBPC

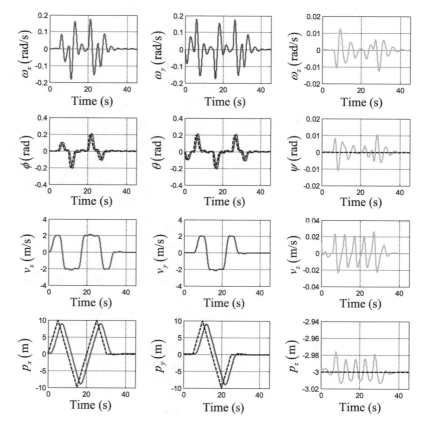

Fig. 11.12 Trajectory
tracking effect of
EBAC+EBPC

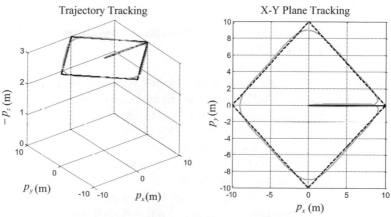

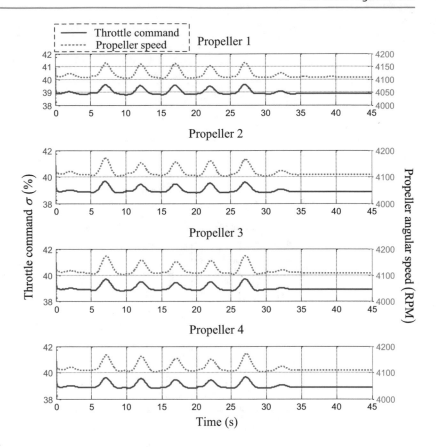

Fig. 11.13 Throttle commands and propeller angular speeds of EBAC+EBPC

velocity are all acceptable. In Fig. 11.12, trajectory tracking results are displayed in 3D space and in 2D plane separately. It is observed that the tracking error is small. throttle commands and propeller angular speeds are presented in Fig. 11.13, which shows that throttle commands change regularly during maneuvering and maintain at the value which is needed to balance the gravity during hovering. Propeller angular speeds have the similar changes with throttle commands.

11.7.3 Rotation Matrix Based Attitude Control Combined with Rotation Matrix Based Position Control

For simplicity, "Rotation matrix based attitude control combined with rotation matrix based position control" is abbreviated as "RBAC+RBPC". A closed-loop system is built by combining the attitude controller (11.55), the position controllers (11.40), (11.46) and the nonlinear model of the quadcopter. The attitude controller parameters are selected as $\mathbf{K_R} = \text{diag}(1, 1, 1)$, $\mathbf{K_\omega} = \text{diag}(1, 1, 2.5)$. The position controller parameters are selected as in Table 11.7.

Simulation results are presented Figs. 11.14, 11.15 and 11.16, which are similar to that in Sect. 11.7.2. The proposed controllers can achieve the target.

Compared with Euler angles based controller, the rotation matrix based controller has the advantages that the singularity problem is avoided, and full attitude flight can be realized. Thus, aggressive maneuver flights, such as air battle and vertical circle flight, can be realized by the rotation matrix based controller. The closed-loop system composed by the nonlinear model of the quadcopter, the attitude

Table 11.7 Position controller parameters

	K_p	K_{vp}	K_{vi}	K_{vd}
X channel	0.5	1	0	0
Y channel	0.5	1	0	0
Z channel	3.5	3	1	2

Fig. 11.14 State responses of RBAC+RBPC

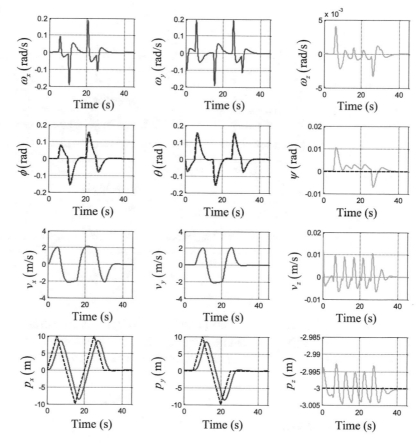

Fig. 11.15 Trajectory tracking effect of RBAC+RBPC

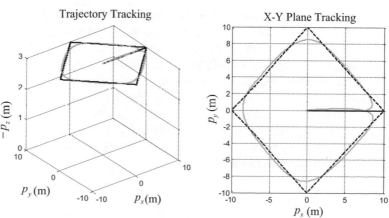

Fig. 11.16 Throttle
commands and propeller
angular speeds of
RBAC+RBPC

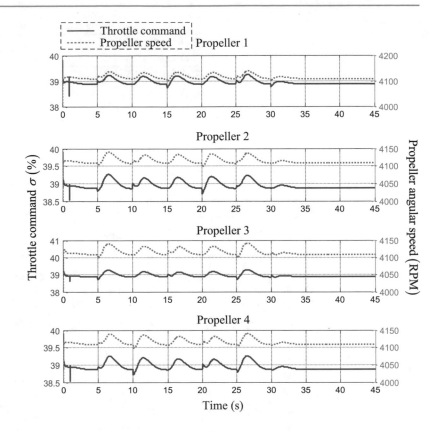

controller (11.56) and the position controllers (11.40), (11.46) can accomplish an aggressive maneuver flight. The corresponding simulation can be found in Ref. [7]. On the other hand, for general tasks, a rotation matrix based controller has a similar control performance as Euler angles based controller does.

11.7.4   Robust Attitude Control

The simulation for the design example in Sect. 11.4.4.4 is conducted in the following. Parameters are selected as $\mathbf{J}_0 = \mathrm{diag}(0.03,\ 0.03,\ 0.04)$ $(\mathrm{kg}\cdot\mathrm{m}^2)$, $\mathbf{d} = [0.5\ 0.5\ 0.5]^T$ $(\mathrm{N}\cdot\mathrm{m})$, $a = 5$, $c_1 = 10$, $c_2 = 1$, $\varepsilon = 0.01, 0.1, 1$. It can be calculated that $\mathbf{K} = [-50\mathbf{I}_3\ -10\mathbf{I}_3]^T$. The desired control command is assigned as $\mathbf{y}_d = [(10\pi/180)c_1\ (10\pi/180)c_1\ (10\pi/180)c_1]^T$, which is equivalent to $\boldsymbol{\Theta}_d = [10\pi/180\ 10\pi/180\ 10\pi/180]^T$ because $\mathbf{y}_d = c_1\boldsymbol{\Theta}_d + c_2\boldsymbol{\omega}_d$. It means that the three desired attitude angles are all $10°$. The simulation results are presented in Fig. 11.17, which show the system responses under the three conditions that $\varepsilon = 0.01, 0.1, 1$ respectively. The system states finally reach the reference, and the control input $\boldsymbol{\tau}$ finally settles at $[-0.5\ -0.5\ -0.5]^T$ $(\mathrm{N}\cdot\mathrm{m})$, which just compensates for the disturbance $\mathbf{d}$. It can be observed through the comparison that the smaller ε is, the faster the angles and angular velocities converge.

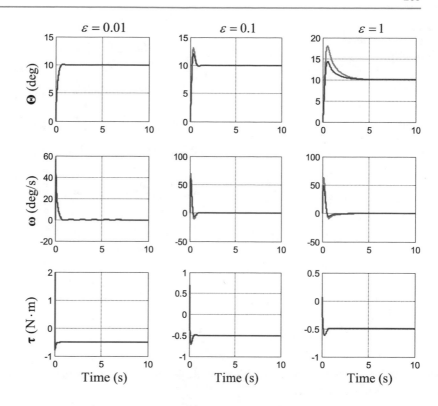

Fig. 11.17 Simulation results of robust attitude control based on ASD

11.8   Summary

The controller design for multicopters is based on specific control requirements. For different control requirements, different mathematical models and corresponding control methods are selected. From the discussion in this chapter, it is known that there are various mathematical representations of multicopters, and each of them has its own applications, advantages and disadvantages. The design using the successive loop closure is often adopted in flight control. Multiple closed-loop feedbacks work together to realize the flight control of multicopters. In order to simplify the controller design, the control objective for multicopters is decomposed into many subobjectives according to the controller design based on Euler angles, as shown in Fig. 11.18. As long as the subobjectives are realized one by one, the final objective is realized. The final objective obtained is the motor throttle command, which is expected to produce the desired angular speeds. This also has answered the question proposed at the beginning of this chapter.

Low-level flight control of multicopters mostly depends on the PID control, which is sufficient for most requirements. However, it is not enough for tough requirements. The stability region is hoped to be as large as possible. Thus, the flight can be as stable as possible in presence of disturbances. For ease-of-use, flight controllers should have adaptive features so that less or no parameters need to be tuned for most cases. For all multicopters, it is also expected to reduce vibration and save energy. Moreover, wind resistance is also necessary for most multicopters. In aerial photography, a high precision hover is needed especially for multicopters without stabilized platforms. If multicopters are used in aerobatic flight, the ability of aggressive maneuvers is wanted. Therefore, a question is being asked: how to design low-level flight controllers for multicopters to obtain a stability region as large as possible, reduce vibration, save energy, resist wind, achieve a high precision hover or achieve an aggressive

Fig. 11.18 Control
objective decomposition
for multicopters

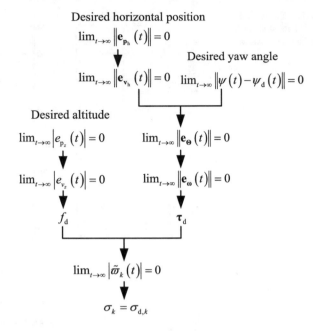

maneuver flight? In fact, a good controller should be the one that can achieve a trade-off between
stability and flight performance.

Exercises

11.1 Show the stability of the closed-loop system composed by the position model (11.7)–(11.8) and
controllers (11.30), (11.33).

11.2 There are many control allocation methods for multicopters. Except for the pseudo-inverse
method mentioned in this chapter, give several other control allocation methods, and explain their
advantages and disadvantages.

11.3 Except for Euler angles and rotation matrix, there is also another attitude representation, namely,
quaternion. According to the quaternion based attitude model, namely, Eqs. (6.4), (6.9) in Chap. 6,
design an attitude controller $\boldsymbol{\tau}_d \in \mathbb{R}^3$ to track the attitude command $\mathbf{q}_d = [q_{0d}\ \mathbf{q}_{vd}^T]$. The design
procedure can be as follows: (1) Define a unit quaternion based attitude tracking error $\tilde{\mathbf{q}} = \mathbf{q}_d^{-1} \otimes \mathbf{q}$,
where "$\otimes$" denotes the quaternion multiplication, $\mathbf{q}_d^{-1}$ is the reverse of the unit quaternion. Please refer
to Sect. 5.2.3 in Chap. 5. (2) Obtain the quaternion based attitude error system as follows:

$$\dot{\tilde{q}}_0 = -\tfrac{1}{2}\tilde{\mathbf{q}}_v^T \tilde{\omega}$$
$$\dot{\tilde{\mathbf{q}}}_v = \tfrac{1}{2}\left(\tilde{q}_0 \mathbf{I}_3 + [\tilde{\mathbf{q}}_v]_\times\right)\tilde{\omega}$$
$$\dot{\tilde{\omega}} = \mathbf{J}^{-1}\boldsymbol{\tau} - \left(\tilde{\mathbf{R}}\dot{\omega}_d - [\tilde{\omega}]_\times \tilde{\mathbf{R}}\omega_d\right).$$

(3) In the following, design a controller such that the quaternion based attitude tracking for multicopters
satisfies $\lim_{t\to\infty}\|\tilde{\mathbf{q}}_v(t)\| = 0$. (Hint: Lyapunov function can be defined as $V_q = (1 - \tilde{q}_0)^2 + \tilde{\mathbf{q}}_v^T \tilde{\mathbf{q}}_v$.)

11.4 The control effectiveness matrix for multicopters can be found in Eq. (6.25) in Chap. 6, and the implementation of control allocation in autopilots is presented in Sect. 11.5.2. Under the initial conditions $\mathbf{J} = \mathbf{J}_0, m = m_0, d = d_0, c_T = c_{T_0}, c_M = c_{M_0}$, suppose PD controllers $f_0, \boldsymbol{\tau}_0$ are designed as

$$
\begin{bmatrix} f_0 \\ \tau_{x,0} \\ \tau_{y,0} \\ \tau_{z,0} \end{bmatrix} = \begin{bmatrix} f_0^* + k_{z,\mathrm{p},0}\,(p_z - p_{z\mathrm{d}}) + k_{z,\mathrm{d},0}v_z \\ k_{\tau_x,\mathrm{p},0}\,(\phi - \phi_\mathrm{d}) + k_{\tau_x,\mathrm{d},0}\omega_x \\ k_{\tau_y,\mathrm{p},0}\,(\theta - \theta_\mathrm{d}) + k_{\tau_y,\mathrm{d},0}\omega_y \\ k_{\tau_z,\mathrm{p},0}\,(\psi - \psi_\mathrm{d}) + k_{\tau_z,\mathrm{d},0}\omega_z \end{bmatrix}.
$$

Then, if the parameters become $\mathbf{J} = \mathbf{J}_1, m = m_1, d = d_1, c_T = c_{T_1}, c_M = c_{M_1}$, how to change the PD controllers parameters, namely, $k_{z,\mathrm{p},0}, k_{z,\mathrm{d},0}, k_{\tau_x,\mathrm{p},0}, k_{\tau_x,\mathrm{d},0}, k_{\tau_y,\mathrm{p},0}, k_{\tau_y,\mathrm{d},0}, k_{\tau_z,\mathrm{p},0}$ and $k_{\tau_z,\mathrm{d},0}$, to achieve the same control performance.

References

1. Chen CT (1999) Linear system theory and design. Oxford University Press, New York
2. Hoffmann GM, Waslander SL, Tomlin CJ (2008) Quadrotor helicopter trajectory tracking control. In: Proceedings of AIAA guidance, navigation and control conference and exhibit, Honolulu, AIAA 2008-7410
3. Zuo Z (2010) Trajectory tracking control design with command-filtered compensation for a quadrotor. IET Control Theory A 4(11):2343–2355
4. Cabecinhas D, Cunha R, Silvestre C (2015) A globally stabilizing path following controller for rotorcraft with wind disturbance rejection. IEEE Trans Control Syst Technol 23(2):708–714
5. Aguiar AP, Hespanha JP (2007) Trajectory-tracking and path-following of underactuated autonomous vehicles with parametric modeling uncertainty. IEEE Trans Autom Control 52(8):1362–1379
6. Roza A, Maggiore M (2012) Path following controller for a quadrotor helicopter. In: Proceedings of American control conference, Montreal, pp 4655–4660
7. Lee T, Leoky M, McClamroch NH (2010) Geometric tracking control of a quadrotor UAV on SE (3). In: Proceedings of 49th IEEE conference on decision and control, Atlanta, pp 5420–5425
8. Mellinger D, Michael N, Kumar V (2014) Trajectory generation and control for precise aggressive maneuvers with quadrotors. In: Proceedings of international symposium on experimental robotics, Marrakech and Essaouira, pp 361–373
9. Yu Y, Yang S, Wang M et al (2015) High performance full attitude control of a quadrotor on SO(3). In: Proceedings of IEEE international conference on robotics and automation, Seattle, pp 1698–1703
10. Chen WH (2004) Disturbance observer based control for nonlinear systems. IEEE-ASME Trans Mech 9(4):706–710
11. Han JQ (2009) From PID to active disturbance rejection control. IEEE Trans Ind Electron 56(3):900–906
12. Zhang RF, Quan Q, Cai KY (2011) Attitude control of a quadrotor aircraft subject to a class of time-varying disturbances. IET Control Theory A 5(9):1140–1146
13. Quan Q, Du GX, Cai KY (2016) Proportional-integral stabilizing control of a class of MIMO systems subject to nonparametric uncertainties by additive-state-decomposition dynamic inversion design. IEEE-ASME Trans Mech 21(2):1092–1101
14. Quan Q, Cai KY (2009) Additive decomposition and its applications to internal-model-based tracking. In: Proceedings of joint 48th IEEE conference on decision and control and 28th Chinese control conference, Shanghai, pp 817–822
15. Quan Q, Lin H, Cai KY (2014) Output feedback tracking control by additive state decomposition for a class of uncertain systems. Int J Syst Sci 45(9):1799–1813
16. Quan Q, Cai KY, Lin H (2015) Additive-state-decomposition-based tracking control framework for a class of nonminimum phase systems with measurable nonlinearities and unknown disturbances. Int J Robust Nonlinear 25(2):163–178
17. Wei ZB, Quan Q, Cai KY (2017) Output feedback ILC for a class of nonminimum phase nonlinear systems with input saturation: an additive-state-decomposition-based method. IEEE Trans Autom Control 62(1):502–508
18. Murnaghan FD (1928) The Boltzmann-Hopkinson principle of superposition as applied to dielectrics. J AIEE 47(1):41–43
19. Bromwich TJ (1917) Normal coordinates in dynamical systems. Proc Lond Soc s2-15(1):401–448
20. Carson JR (1917) On a general expansion theorem for the transient oscillations of a connected system. Phys Rev 10(3):217–225

21. Carson JR (1919) Theory of the transient oscillations of electrical networks and transmission systems. Trans Am Inst Electr Eng 38(1):345–427
22. Quan Q, Cai KY (2012) Additive-output-decomposition-based dynamic inversion tracking control for a class of uncertain linear time-invariant systems. In: Proceedings of 51st IEEE conference on decision and control, Maui, pp 2866–2871
23. Harkegard O, Glad ST (2005) Resolving actuator redundancy—optimal control vs. control allocation. Automatica 41(1):137–144
24. Johansen TA, Fossen TI (2013) Control allocation—a survey. Automatica 49(5):1087–1103

Position Control Based on Semi-Autonomous Autopilots

12

Remote pilots often control multicopters based on Semi-Autonomous Autopilots (SAAs) to accomplish some specified tasks, such as crop-dusting and inspection of power lines. During the process, most remote pilots do not know the flight control law in the autopilots. But, it is not an obstacle for them to accomplish these tasks. In fact, it is difficult for most people to design an autopilot starting from the motor control. On the other hand, many groups or companies have designed some open source SAAs or offered SAAs with Software Development Kits (SDKs). Please refer to Table 1.3 in Chap. 1. Therefore, it is useful to develop special applications based on existing SAAs directly. Not only can it avoid the trouble of modifying the low-level source code of autopilots, but also it can utilize commercial reliable autopilots to achieve targets. This will simplify the complete design. This chapter aims to answer the question as below:

How is a multicopter controlled based on the SAA to track a given target position?

To answer this question, this chapter mainly includes problem formulation, system identification, and controller design.

© Springer Nature Singapore Pte Ltd. 2017
Q. Quan, *Introduction to Multicopter Design and Control*, DOI 10.1007/978-981-10-3382-7_12

Climbing to a High Place

The ancient Chinese believed that 'learning from others' and 'practice' were key ways to knowledge acquisition. Xunzi, a Chinese philosopher, pointed out in his book "On Learning" that "One who makes use of a chariot and horses has not thereby improved his feet, but he can now go a thousand li. One who makes use of a boat and oars has not thereby learn to swim, but he can now cross rivers and streams. The gentleman is not different from others by birth. Rather, he is good at making use of things. (The Chinese is "假舆马者，非利足也，而致千里；假舟楫者，非能水也，而绝江河．君子生非异也，善假于物也", translated by Eric Hutton, from http://www.asiamarketingmanagement.com/onlearningbyxunzi.html.)". There is a proverb in "Analects" that "One must have good tools in order to do a good job". (The Chinese is "工欲善其事必先利其器") We also have another similar common saying that "Sharpening your axe will not delay your job of cutting wood (The Chinese is "磨刀不误砍柴功")" These proverbs indicate the importance of fundamental tools.

12.1   Problem Formulation

In general, the controller is designed for a multicopter, which is employed to control the propeller angular speeds $\varpi_k, k = 1, 2, \ldots, n_r$ directly to generate a resultant thrust and three moments. Then, the position of the multicopter is controlled through these variables. However, in this chapter, the position is controlled by the Radio Controlled (RC) commands $u_\theta, u_\phi, u_{\omega_z}, u_T \in \mathbb{R}$ directly. There are often three common modes in an SAA: the "stabilize mode", the "altitude hold mode", and the "loiter mode". They will be introduced in Sect. 13.2.1 in Chap. 13 in detail. In this chapter, a position controller is designed based on the stabilize mode, which is the most basic mode among the three commonly used modes.

12.1.1   Structure of Multicopter with SAA

The structure of a multicopter with an SAA is illustrated by Fig. 12.1. RC commands control the following variables of the multicopter directly:

(1) the vertical velocity v_{z_e};
(2) the attitude angular velocity ω_{z_b};
(3) the attitude angles θ, ϕ (or the velocities v_{x_b}, v_{y_b} in the Aircraft-Body Coordinate Frame (ABCF) of the multicopter).

Through these variables, the position $(p_{x_e}, p_{y_e}, p_{z_e})$ and the yaw angle ψ of the multicopter are further controlled.

12.1.2   Models of Three Channels

According to Chap. 11, in order to design the controller easily, the nonlinear model of multicopters is simplified appropriately through linearization. It means that the system shown in Fig. 12.1 is divided into three channels: the altitude channel which is from u_T to p_{z_e}, the yaw channel which is from

Fig. 12.1 The closed-loop diagram based on SAA

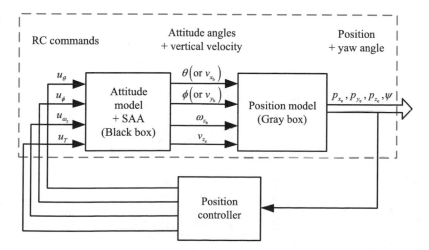

u_{ω_z} to ψ, and the horizontal position channel which is from $\mathbf{u_h}$ to $\mathbf{p_h}$, where $\mathbf{u_h} = [\, u_\phi \; u_\theta \,]^T$ and $\mathbf{p_h} = [\, p_{x_e} \; p_{y_e} \,]^T$. With SAAs, the models of these channels are assumed as follows.

Assumption 12.1 The model of the altitude channel is

$$
\begin{aligned}
\dot{p}_{z_e} &= v_{z_e} \\
\dot{v}_{z_e} &= -k_{v_z} v_{z_e} - k_{u_T} u_T
\end{aligned}
\tag{12.1}
$$

where $k_{v_z}, k_{u_T} \in \mathbb{R}_+$ are the parameters determined by the selected SAA which are considered to be unknown.

Assumption 12.2 The model of the yaw channel is

$$
\begin{aligned}
\dot{\psi} &= \omega_z \\
\dot{\omega}_z &= -k_{\omega_z} \omega_z + k_{u_{\omega_z}} u_{\omega_z}
\end{aligned}
\tag{12.2}
$$

where $k_{\omega_z}, k_{u_{\omega_z}} \in \mathbb{R}_+$ are the parameters determined by the selected SAA which are considered to be unknown.

Assumption 12.3 The model of the horizontal position channel is

$$
\begin{aligned}
\dot{\mathbf{p}}_h &= \mathbf{R}_\psi \mathbf{v}_{h_b} \\
\dot{\mathbf{v}}_{h_b} &= -\mathbf{K}_{\mathbf{v}_{h_b}} \mathbf{v}_{h_b} - g \begin{bmatrix} 0 & 1 \\ -1 & 0 \end{bmatrix} \mathbf{\Theta}_h \\
\dot{\mathbf{\Theta}}_h &= \boldsymbol{\omega}_{h_b} \\
\dot{\boldsymbol{\omega}}_{h_b} &= -\mathbf{K}_{\mathbf{\Theta}_h} \mathbf{\Theta}_h - \mathbf{K}_{\omega_{h_b}} \boldsymbol{\omega}_{h_b} + \mathbf{K}_{\mathbf{u}_h} \mathbf{u}_h
\end{aligned}
\tag{12.3}
$$

where $\mathbf{v}_{h_b} = [\, v_{x_b} \; v_{y_b} \,]^T$, $\mathbf{\Theta}_h = [\, \phi \; \theta \,]^T$, and $\boldsymbol{\omega}_{h_b} = [\, \omega_{x_b} \; \omega_{y_b} \,]^T$. Moreover, $\mathbf{K}_{\mathbf{v}_{h_b}}, \mathbf{K}_{\mathbf{\Theta}_h}, \mathbf{K}_{\omega_{h_b}}, \mathbf{K}_{\mathbf{u}_h} \in \mathbb{R}^{2 \times 2} \cap \mathscr{P} \cap \mathscr{D}$. These are the parameters determined by the used SAA, which are considered to be unknown. The definition of $\mathbf{R}_\psi$ is the same as in Eq. (11.7) in Chap. 11. It should be pointed out that if the horizontal velocity feedback is not considered by the SAA, then only the air damping is involved in this channel, which means $\mathbf{K}_{\mathbf{v}_{h_b}} \approx \mathbf{0}_{2 \times 2}$. Otherwise, $\mathbf{K}_{\mathbf{v}_{h_b}}$ would be a reasonable damping parameter.

Assumption 12.3 implies $\dot{\psi} \approx 0$. Since $\mathbf{v}_h = \mathbf{R}_\psi \mathbf{v}_{h_b}$, one has

$$
\dot{\mathbf{v}}_{h_b} = \mathbf{R}_\psi^T (\dot{\mathbf{v}}_h - \dot{\mathbf{R}}_\psi \mathbf{v}_{h_b}).
\tag{12.4}
$$

Without considering SAA, one further has

$$
\dot{\mathbf{v}}_{h_b} = -\mathbf{R}_\psi^T \dot{\mathbf{R}}_\psi \mathbf{v}_{h_b} - g \begin{bmatrix} 0 & 1 \\ -1 & 0 \end{bmatrix} \mathbf{\Theta}_h
\tag{12.5}
$$

according to Eq. (11.7). If $\dot{\psi} \approx 0$, then

$$
\dot{\mathbf{v}}_{h_b} = -g \begin{bmatrix} 0 & 1 \\ -1 & 0 \end{bmatrix} \mathbf{\Theta}_h.
\tag{12.6}
$$

Otherwise, the term $-\mathbf{R}_\psi^T \dot{\mathbf{R}}_\psi \mathbf{v}_{h_b}$ is involved.

12.1.3   Objective of Position Control

According to Eqs. (12.1)–(12.3), the channels from RC commands $u_\theta, u_\phi, u_T, u_{\omega_z}$ to the variables θ (or v_{x_b}), ϕ (or v_{y_b}), v_{z_e}, ω_z are stable respectively, because the controllers of the SAA provide damping for these channels. As a result, the multicopter is easily controlled. But, the SAA cannot accomplish the position control because the channels from θ (or v_{x_b}), ϕ (or v_{y_b}), ω_z, v_{z_e} to $p_{x_e}, p_{y_e}, p_{z_e}, \psi$ are still open-loop. That means the position control cannot be achieved only under the controllers of the SAA. Thus, under Assumptions 12.1–12.3, the objective of this chapter is to design extra position controllers to accomplish the following task. For a given desired trajectory $\mathbf{p}_d(t)$ and the desired yaw $\psi_d(t)$, it is expected to make $\|\mathbf{x}(t) - \mathbf{x}_d(t)\| \to 0$ or $\mathbf{x}(t) - \mathbf{x}_d(t) \to \mathscr{B}(\mathbf{0}_{4\times 1}, \delta)$ through the inputs $u_\theta, u_\phi, u_{\omega_z}, u_T$ as $t \to \infty$, where $\mathbf{x} = [\mathbf{p}^T \ \psi]^T$, $\mathbf{x}_d = [\mathbf{p}_d^T \ \psi_d]^T$ and $\delta \in \mathbb{R}_+$.

The internal structure of the SAA is a "black box" for designers. On the other hand, the control model of the multicopter is a "gray box", because the structure of the control model of the multicopter is known, but their parameters are unknown. Thus, the models in the "black box" and the "gray box" need to be analyzed through system identification, and then, the controllers are designed based on the results of the identification.

12.2   System Identification

As shown in Sect. 12.1.2, the linear models of different channels of multicopters are established, which are suitable to design the controllers. The linear models of different channels provide the prior knowledge for system identification, because the purpose of the identification is to obtain the parameters of the transfer functions. The transfer functions are from the inputs to the outputs of different channels, and their forms are derived from the linear models. Thus, the transfer functions of the altitude channel, the yaw channel, and the horizontal position channel of multicopters are identified, respectively, in this section. Before the concrete work, the system identification and the commonly used toolboxes are first introduced.

12.2.1   Procedure and Toolboxes of System Identification

System identification deals with the problem of building mathematical models of dynamical systems based on observed data from systems.

12.2.1.1   Procedure of System Identification [1, pp. 13–14]

The procedure of system identification is shown in Fig. 12.2. It contains the following main parts: prior knowledge, experiment design, data record, model selection, model calculation, and model validation.

(1) Prior knowledge. The prior knowledge includes the existing knowledge about the characteristics of the system, the data recording methods, and other aspects of the system. This knowledge is very helpful to choose the candidate models, design the experiment, and determine the calculation method and the criterion of validation. Because of the different goals of identifications, the required prior knowledge may be very different, even for the same system.

(2) Experiment design. The purpose of the experiment design is to obtain the input–output data of the system, which can reveal the system performance as much as possible under known conditions. The input–output data is sometimes recorded during a specifically designed identification experiment, where the user may determine which signals to measure and when to measure them. There are

Fig. 12.2 The procedure
of system identification [1]

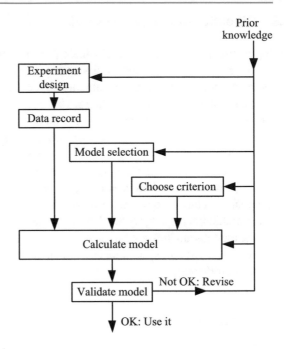

two kinds of experiments for system identification: open-loop experiments and closed-loop experiments, which are shown in Fig. 12.3a, b, respectively. As shown in Fig. 12.3b, some systems are unstable, and they must work with feedback controllers. As a result, the input signals are decided by the controllers. Thus, the excitation of open-loop experiments is richer than that of closed-loop experiments, because any input signal can be selected in open-loop experiments. Obviously, if systems can work without controllers, open-loop experiments are the better choices.

(3) Data record. Observe the input–output data through designed experiments.

(4) Model selection. Choose a set of candidate models and determine which one of them is the most suitable model through the validation. Through the mathematical modeling, a parametric model with unknown parameters is obtained. Then, the unknown parameters can be calculated by parameter identification methods.

(5) Model calculation. The model calculation means that the unknown parameters of the candidate model are determined by appropriate optimization methods.

(6) Model validation. A criterion is determined to test whether the candidate model with the calculated parameters is valid for the designed purpose. In general, this criterion relies on the observation data, the prior knowledge, and the application of the identified model. If the model with the parameters passes the validation, it will be chosen as the final model. If not, the previous steps need to be revised.

Fig. 12.3 Two kinds of
experiments for system
identification

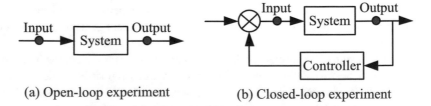

(a) Open-loop experiment   (b) Closed-loop experiment

12.2.1.2   Toolboxes of the System Identification

(1) System Identification Toolbox of MATLAB. This toolbox provides functions, Simulink blocks, and an application of constructing mathematical models of dynamical systems from measured input–output data [2]. It lets users obtain models of dynamical systems which are not easily modeled from physical principles. Time-domain and frequency-domain input–output data can be employed to identify continuous-time and discrete-time transfer functions, process models, and state-space models. The identification methods are also provided by the toolbox, such as maximum likelihood, Prediction-Error Minimization (PEM), and subspace system identification. If nonlinear systems are considered, Hammerstein–Weiner models, nonlinear ARX models with wavelet network, and other nonlinear models are available. The toolbox also provides the gray-box system identification. This means that a user-defined model can be employed in the estimation of the parameters. The identified model can be used for system response prediction and plant modeling. Please refer to Ref. [3] for further details.

(2) Comprehensive Identification from FrEquency Response (CIFER) toolbox. This toolbox is an integrated facility for system identification based on a comprehensive frequency-response approach. This toolbox is suited for difficult system identification problems, especially the problems related to the modeling and the controller design of aircraft. The foundation of the CIFER approach is the high-quality extraction of a complete Multi-Input Multi-Output (MIMO) set of nonparametric input-to-output frequency responses [4]. The coupled characteristics of the system can be extracted by these responses without priori assumptions. Advanced Chirp-Z transform and composite optimal window techniques, which are the theoretical basis of CIFER, provide significant improvement in frequency-response quality relative to standard Fast Fourier Transforms (FFTs). Various kinds of sophisticated nonlinear search algorithms are employed to extract the state-space model which matches the complete input/output frequency-response data set. Thus, this toolbox is widely used in the system identifications for different types of real aircraft. Please refer to the Web site [4] and the related book [5] for further details.

12.2.2   Model Used in System Identification

12.2.2.1   Outline

The identification of the altitude channel is introduced first and then the identification of the yaw channel. Only when these two channels are stable, the experiment of the horizontal position channel can obtain preferable input–output data. Thus, the horizontal position channel is introduced at last. It should be pointed out that if excellent system identification results are desired, then the channels which need to be identified are better to be stable. Therefore, it is important to investigate whether the channels are stable or not in the experiment design.

If the channel is stable, it can be identified directly. Otherwise, a stabilizing controller, such as a Proportional (P) or a Proportional-Derivative (PD) controller, needs to be designed to make the channel stable first. Then, the experiments for this channel are designed as shown in Fig. 12.4. For example, in the altitude channel (12.1), if the channel from u_T to v_{z_e} is stable, then the identification can be done without any controller design. However, if the channel from the thrust command u_T to the vertical position p_{z_e} is considered, then a controller is better to be designed before the identification. The reason is that this channel is unstable because of an integration term between v_{z_e} and p_{z_e}. Thus, a P controller is suitable for this channel. It should be noted that the purpose of controller design is only to make the channel stable rather than to get excellent performances. As for the question that which kind of controller is suitable for the system identification, it can be solved by the prior knowledge. Finally, the

Fig. 12.4 A modified open-loop experiment

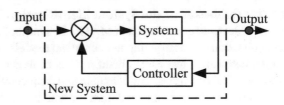

original channel with the stabilizing controller is regarded as a whole system, which becomes a new system for the system identification, as shown in Fig. 12.4.

12.2.2.2   Altitude Channel

In general, the SAA often controls the vertical velocity in the altitude channel. That means that the RC command u_T controls v_{z_e}, and when $u_T = 0$, the multicopter keeps the current height. The altitude channel with an SAA is described as shown in Eq. (12.1) and that is our prior knowledge. Then, according to Eq. (12.1), the transfer function model is obtained as follows:

$$p_{z_e}(s) = G_{p_z u_T}(s) u_T(s). \tag{12.7}$$

Here, $G_{p_z u_T}(s)$ is an unstable transfer function, because it contains a first-order integration. This is not very suitable for identification. Thus, a P controller is needed, which is designed as

$$u_T = k_{p_z} p_{z_e} + u_{p_z} \tag{12.8}$$

where $k_{p_z} \in \mathbb{R}_+$, and $u_{p_z} \in \mathbb{R}$ is defined as a new input. It should be noted that the parameter of the input of channel (12.1) is $-k_{u_T}$, which is negative. So, the controller is designed as Eq. (12.8). Under this controller, channel (12.1) is expressed by

$$\begin{aligned}
\dot{p}_{z_e} &= v_{z_e} \\
\dot{v}_{z_e} &= -k_{u_T} k_{p_z} p_{z_e} - k_{v_z} v_{z_e} - k_{u_T} u_{p_z}.
\end{aligned} \tag{12.9}$$

The corresponding transfer function is

$$p_{z_e}(s) = G_{p_z u_{p_z}}(s) u_{p_z}(s). \tag{12.10}$$

The channel from u_{p_z} to p_{z_e}, as shown in Eq. (12.10), is the parameterized model which needs to be identified. When the data is recorded for this model, the height data p_{z_e} (namely the output of the channel) is measured by altitude sensors, such as barometers on ultrasonic range finders. Simultaneously, the new defined input u_{p_z} (namely the input of the channel) is recorded. Then, the specific parameters of this model can be obtained by system identification tools. The concrete procedure is demonstrated in Sect. 12.4. It should be pointed out that if the vertical velocity v_{z_e} can be measured directly and the channel from u_T to v_{z_e} is considered, then this channel does not need a stabilizing controller and can be identified directly because it has been stable. In this situation, the corresponding transfer function of this channel is

$$v_{z_e}(s) = G_{v_{z_e} u_T}(s) u_T(s). \tag{12.11}$$

Consequently,

$$p_{z_e}(s) = \frac{1}{s} G_{v_{z_e} u_T}(s) u_T(s). \tag{12.12}$$

12.2.2.3   Yaw Channel

The yaw channel based on the feedback controller of an SAA is expressed in Eq. (12.2), and the corresponding transfer function model is expressed as follows:

$$\psi(s) = G_{\psi u_{\omega_z}}(s) u_{\omega_z}(s) \tag{12.13}$$

where $G_{\psi u_{\omega_z}}(s)$ is a transfer function containing a first-order integration. It can be stabilized by a P controller, such as

$$u_{\omega_z} = -k_\psi \psi + u_\psi \tag{12.14}$$

where $k_\psi \in \mathbb{R}_+$, and $u_\psi \in \mathbb{R}$ is a newly defined input. Then, system (12.2) becomes

$$\begin{aligned}
\dot{\psi} &= \omega_z \\
\dot{\omega}_z &= -k_{u_{\omega_z}} k_\psi \psi - k_{\omega_z} \omega_z + k_{u_{\omega_z}} u_\psi.
\end{aligned} \tag{12.15}$$

The corresponding transfer function is

$$\psi(s) = G_{\psi u_\psi}(s) u_\psi(s). \tag{12.16}$$

The yaw angle ψ and the newly defined input u_ψ are recorded at the same time. The compass and the motion capture system can be employed to measure the yaw angle ψ. Then, the specific parameters of this model can be obtained by system identification tools, and the concrete procedure can be found in Sect. 12.4. It should be pointed out that if ω_z is measured directly and the channel from u_{ω_z} to ω_z is considered, then this channel does not need a stabilizing controller and can be identified directly because it is stable. In this situation, the corresponding transfer function of this channel is

$$\omega_z(s) = G_{\omega_z u_{\omega_z}}(s) u_{\omega_z}(s). \tag{12.17}$$

Consequently,

$$\psi(s) = \frac{1}{s} G_{\omega_z u_{\omega_z}}(s) u_{\omega_z}(s). \tag{12.18}$$

12.2.2.4   Horizontal Position Channel

SAA controls the attitude angles θ, ϕ through its own sensors, which makes the channel from the RC commands u_θ, u_ϕ to θ, ϕ stable. This channel is expressed in Eq. (12.3), which is our prior knowledge.

The horizontal position channel is identified after the other two channels are well controlled. In this situation, under the controller of the yaw channel, $\psi \approx \psi_d$ is satisfied. In order to obtain preferable identification results, ψ_d is set as a constant ($\psi_d = 0$ is the best choice). This implies that ψ is also a constant approximately. Then, $\mathbf{R}_\psi$ is a constant matrix. According to Eq. (12.3), the channel is expressed as follows:

$$\begin{aligned}
\mathbf{p}_h(s) &= \operatorname{diag}\left(\tfrac{1}{s}, \tfrac{1}{s}\right) \mathbf{R}_\psi \mathbf{G}_{\mathbf{v}_{h_b} \mathbf{u}_h}(s) \mathbf{u}_h(s) \\
&= \mathbf{R}_\psi \operatorname{diag}\left(\tfrac{1}{s}, \tfrac{1}{s}\right) \mathbf{G}_{\mathbf{v}_{h_b} \mathbf{u}_h}(s) \mathbf{u}_h(s)
\end{aligned} \tag{12.19}$$

where $\mathbf{G}_{\mathbf{v}_{h_b}\mathbf{u}_h}(s)$ is the parametric model which needs to be identified. This channel is harder to control than the other two channels because of the coupling caused by $\mathbf{R}_\psi$. Thus, the horizontal velocity controller is designed in the following part.

For the SAA which contains a velocity feedback controller, $\mathbf{K}_{\mathbf{v}_{h_b}}$ is reasonable. The transfer function $\mathbf{G}_{\mathbf{v}_{h_b}\mathbf{u}_h}(s)$ is stable, and it can be identified directly. However, if the velocity feedback is not considered by the SAA, then $\mathbf{K}_{\mathbf{v}_{h_b}} \approx \mathbf{0}_{2\times2}$ according to Sect. 12.1. Thus, a controller needs to be designed to stabilize the system before identification, because $\mathbf{G}_{\mathbf{v}_{h_b}\mathbf{u}_h}(s)$ is unstable and contains first-order integrations. A controller is designed as follows:

$$\mathbf{u}_h = -\mathbf{K}'_{\mathbf{v}_{h_b}}\mathbf{v}_{h_b} + \mathbf{u}_{\mathbf{v}_h} \tag{12.20}$$

where $\mathbf{u}_{\mathbf{v}_h} = [\,u_{v_x}\; u_{v_y}\,]^T$ is a newly defined input and $\mathbf{K}'_{\mathbf{v}_{h_b}} \in \mathbb{R}^{2\times2}\cap\mathscr{D}\cap\mathscr{P}$. Substituting controller (12.20) into channel (12.3) results in

$$\begin{aligned}
\dot{\mathbf{v}}_{h_b} &= -\mathbf{K}_{\mathbf{v}_{h_b}}\mathbf{v}_{h_b} - g\begin{bmatrix}0 & 1\\ -1 & 0\end{bmatrix}\mathbf{\Theta}_h\\
\dot{\mathbf{\Theta}}_h &= \omega_{h_b}\\
\dot{\omega}_{h_b} &= -\mathbf{K}_{\mathbf{u}_h}\mathbf{K}'_{\mathbf{v}_{h_b}}\mathbf{v}_{h_b} - \mathbf{K}_{\mathbf{\Theta}_h}\mathbf{\Theta}_h - \mathbf{K}_{\omega_{h_b}}\omega_{h_b} + \mathbf{K}_{\mathbf{u}_h}\mathbf{u}_{\mathbf{v}_h}.
\end{aligned} \tag{12.21}$$

Then, the corresponding transfer function of the system is

$$\mathbf{v}_{h_b}(s) = \mathbf{G}_{\mathbf{v}_{h_b}\mathbf{u}_{\mathbf{v}_h}}(s)\,\mathbf{u}_{\mathbf{v}_h}(s). \tag{12.22}$$

The horizontal position is measured by a Global Positioning System (GPS) receiver or an indoor location system, such as an indoor motion capture system. The velocity can be obtained directly as some GPS receivers can output velocity or by estimating based on the GPS position data off-line. The newly defined input $\mathbf{u}_{\mathbf{v}_h}$ needs to be recorded simultaneously. Then, the specific parameters of this model can be obtained by system identification tools, and the concrete procedure can be found in Sect. 12.4. Finally, in practice, the desired horizontal velocity $\mathbf{v}_{h_b d}$ can be obtained by the desired horizontal position $\mathbf{p}_{h_e d}$, and it becomes an indirect objective of the controller design. This will be shown in Sect. 12.4.3.

12.3 Position Controller Design

Our purpose is to design a controller to make the output track the desired input. The PID controller is a commonly used choice, which can be used for each channel. However, if the identified system is accurate enough, some other choices can be adopted to achieve a better control performance. Here, the tracking controller is designed by employing a Dynamic Inversion Control (DIC) method which is based on Additive Output Decomposition (AOD) [6]. Through AOD, the original system with uncertainty is divided into a primary system without uncertainty which is close to the original system and a secondary system with uncertainty. Moreover, the output of the original system is equal to the sum of outputs of the two new systems. By regarding the output of the secondary system as a lumped disturbance, the output of the original system is equal to the sum of the output of the certain primary system and the lumped disturbance. Then, the DIC method is employed to suppress the lumped disturbance and make the output of the primary system track the input accurately [6].

12.3.1   PID Controller

It should be pointed out that the following PID controllers are used to make $\|\mathbf{x}(t) - \mathbf{x}_{\mathrm{d}}(t)\| \to 0$ or $\mathbf{x}(t) - \mathbf{x}_{\mathrm{d}}(t) \to \mathscr{B}(\mathbf{0}_{4\times 1}, \delta)$ as $t \to \infty$, not requiring system identification.

12.3.1.1   Altitude Channel
A PID controller is designed for the altitude channel (12.1) directly as

$$u_T = -k_{p_z\mathrm{p}}\left(p_{z_{\mathrm{e}}} - p_{z_{\mathrm{e}}\mathrm{d}}\right) - k_{p_z\mathrm{d}}\left(\dot{p}_{z_{\mathrm{e}}} - \dot{p}_{z_{\mathrm{e}}\mathrm{d}}\right) - k_{p_z\mathrm{i}}\int\left(p_{z_{\mathrm{e}}} - p_{z_{\mathrm{e}}\mathrm{d}}\right) \qquad (12.23)$$

where $p_{z_{\mathrm{e}}\mathrm{d}} \in \mathbb{R}$ is the desired height, and $k_{p_z\mathrm{p}}, k_{p_z\mathrm{d}}, k_{p_z\mathrm{i}} \in \mathbb{R}$ are the parameters which need to be tuned.

12.3.1.2   Yaw Channel
Likewise, a PID controller is designed for the yaw channel (12.2) as

$$u_{\omega_z} = -k_{\psi\mathrm{p}}\left(\psi - \psi_{\mathrm{d}}\right) - k_{\psi\mathrm{d}}\left(\omega_{z_{\mathrm{b}}} - \dot{\psi}_{\mathrm{d}}\right) - k_{\psi\mathrm{i}}\int\left(\psi - \psi_{\mathrm{d}}\right) \qquad (12.24)$$

where $\psi_{\mathrm{d}} \in \mathbb{R}$ is the desired yaw, and $k_{\psi\mathrm{p}}, k_{\psi\mathrm{d}}, k_{\psi\mathrm{i}} \in \mathbb{R}$ are the parameters which need to be tuned.

12.3.1.3   Horizontal Position Channel
Furthermore, a PID controller is designed for the horizontal position channel (12.3) as

$$\mathbf{u}_{\mathrm{h}} = -\mathbf{K}_{\mathrm{hp}}\mathbf{R}_\psi^{-1}\left(\mathbf{p}_{\mathrm{h}} - \mathbf{p}_{\mathrm{hd}}\right) - \mathbf{K}_{\mathrm{hd}}\mathbf{R}_\psi^{-1}\left(\dot{\mathbf{p}}_{\mathrm{h}} - \dot{\mathbf{p}}_{\mathrm{hd}}\right) - \mathbf{K}_{\mathrm{hi}}\int\mathbf{R}_\psi^{-1}\left(\mathbf{p}_{\mathrm{h}} - \mathbf{p}_{\mathrm{hd}}\right) \qquad (12.25)$$

where $\mathbf{p}_{\mathrm{hd}} \in \mathbb{R}^2$ is the desired horizontal position, and $\mathbf{K}_{\mathrm{hp}}, \mathbf{K}_{\mathrm{hd}}, \mathbf{K}_{\mathrm{hi}} \in \mathbb{R}^{2\times 2}$ are the parameters which need to be tuned.

The benefit of designing PID controllers is that they avoid modeling and identification of the system, and the structure of the controller is simple. However, when the controller is designed, their parameters need to be tuned repeatedly in experiments, and it is difficult to achieve an ideal transient process. In particular, for the multicopters whose channels are strongly coupled or their yaw angles are time-varying, the parameter tuning is much difficult.

12.3.2   Additive-Output-Decomposition-Based Dynamic Inversion Control

If the position control performance of the multicopter needs to be improved, the DIC method is adopted to design a controller based on the identified model of the multicopter.

12.3.2.1   Altitude Channel
After a P controller is employed, this channel is expressed by Eq. (12.10). Then, the model $\hat{G}_{p_z u_{p_z}}(s)$ is identified, which is close to $G_{p_z u_{p_z}}(s)$. Here, the controller u_{p_z} is designed by following the designing procedure. First, choose the primary system as follows:

$$p_{z_{\mathrm{e}}\mathrm{p}}(s) = \hat{G}_{p_z u_{p_z}}(s)\, u_{p_z\mathrm{p}}(s). \qquad (12.26)$$

Then, according to the AOD method [6], by subtracting Eq. (12.26) from Eq. (12.10), the secondary system is obtained as follows:

$$p_{z_eS}(s) = G_{p_zu_{p_z}}(s)\,u_{p_z}(s) - \hat{G}_{p_zu_{p_z}}(s)\,u_{p_zp}(s) \tag{12.27}$$

where $p_{z_eS} = p_{z_e} - p_{z_ep}$. In the following, choose $u_{p_zp} = u_{p_z}$ and denote $p_{z_eS} = d_{p_z l}$. Then,

$$\begin{aligned} p_{z_ep}(s) &= \hat{G}_{p_zu_{p_z}}(s)\,u_{p_z}(s) \\ p_{z_e}(s) &= p_{z_ep}(s) + d_{p_z l}(s) \end{aligned} \tag{12.28}$$

where $d_{p_z l}(s) = (G_{p_zu_{p_z}}(s) - \hat{G}_{p_zu_{p_z}}(s))u_{p_z}(s)$ is called the *lumped disturbance*. The lumped disturbance $d_{p_z l}$ includes uncertainties and input. Fortunately, since $\hat{G}_{p_zu_{p_z}}(s)\,u_{p_z}(s)$ and the output $p_{z_e}(s)$ are known, the lumped disturbance $d_{p_z l}$ is observed by

$$\hat{d}_{p_z l}(s) = p_{z_e}(s) - \hat{G}_{p_zu_{p_z}}(s)\,u_{p_z}(s). \tag{12.29}$$

It is easy to see that $\hat{d}_{p_z l} \equiv d_{p_z l}$. The input of systems (12.10) and (12.28) is the same. So, system (12.28) is an input–output equivalent system of system (12.10). So far, by AOD, system (12.10) has been transformed into a determinate system which is subject to a lumped disturbance. For system (12.28), the primary transfer function $\hat{G}_{p_zu_{p_z}}(s)$ will be designed to be minimum-phase to satisfy the requirement of the DIC method. Then, the tracking controller for the altitude channel is designed as follows:

$$u_{p_z}(s) = \hat{G}_{p_zu_{p_z}}^{-1}(s)\left(p_{z_ed}(s) - d_{p_z l}(s)\right). \tag{12.30}$$

However, the order of the numerator of $\hat{G}_{p_zu_{p_z}}^{-1}(s)$ may be not *proper*,[1] so it is physically unrealizable. Thus, a low-pass filter is adopted to make $Q_{p_zu_{p_z}}(s)\,\hat{G}_{p_zu_{p_z}}^{-1}(s)$ proper or strictly proper. Then, controller (12.30) becomes

$$u_{p_z}(s) = Q_{p_zu_{p_z}}(s)\,\hat{G}_{p_zu_{p_z}}^{-1}(s)\left(p_{z_ed}(s) - d_{p_z l}(s)\right) \tag{12.31}$$

where $Q_{p_zu_{p_z}}(0) = 1$. The closed-loop system is shown in Fig. 12.5.

Fig. 12.5 The closed-loop system based on AOD-based DIC

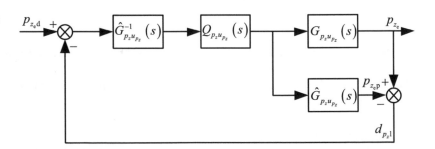

---

[1]In control theory, a proper transfer function is a transfer function in which the order of the numerator is not greater than the order of the denominator. A strictly proper transfer function is a transfer function where the order of the numerator is less than the order of the denominator.

Theorem 12.4 *For system (12.10), suppose*

(1) $\hat{G}_{p_z u_{p_z}}(s)$ *is a minimum-phase system*[2];
(2) $Q_{p_z u_{p_z}}(s)$ *and* $G_{p_z u_{p_z}}(s)$ *are stable and* $Q_{p_z u_{p_z}}(0) = 1$;
(3) $\sup_\omega \left| \left(1 - G_{p_z u_{p_z}}(j\omega) \, \hat{G}_{p_z u_{p_z}}^{-1}(j\omega) \right) Q_{p_z u_{p_z}}(j\omega) \right| < 1$;
(4) $p_{z_e d}$ *is a constant.*

Then, u_{p_z} *is bounded and* $\lim_{t\to\infty} \left| e_{p_z}(t) \right| = 0$, *where* $e_{p_z} \triangleq p_{z_e} - p_{z_e d}$.

Proof For convenience, the Laplace operator s is omitted in the proof, and this proof is organized as follows. *Step 1*, under conditions (1) (2) (3), one has

$$\bar{G} \triangleq \left[1 - Q_{p_z u_{p_z}} \left(1 - G_{p_z u_{p_z}} \hat{G}_{p_z u_{p_z}}^{-1} \right) \right]^{-1} \tag{12.32}$$

which is stable; *Step 2*, based on *Step 1* and condition (4), u_{p_z} is bounded; *Step 3*, based on *Step 1* and condition (4), $\lim_{t\to\infty} \left| e_{p_z}(t) \right| = 0$. Next, the three steps above are proven one by one in detail.

Step 1. Since $\hat{G}_{p_z u_{p_z}}(s)$ is minimum-phase and $Q_{p_z u_{p_z}}, G_{p_z u_{p_z}}$ are stable, the transfer functions $1 - G_{p_z u_{p_z}} \hat{G}_{p_z u_{p_z}}^{-1}$ and $Q_{p_z u_{p_z}}$ are stable. By the small gain theorem [7, pp. 96–98], $\bar{G}$ is stable under condition (3).

Step 2. Since $\hat{d}_{p_z 1} \equiv d_{p_z 1} = (G_{p_z u_{p_z}} - \hat{G}_{p_z u_{p_z}}) u_{p_z}$, the controller (12.31) is written as follows:

$$u_{p_z} = Q_{p_z u_{p_z}} \hat{G}_{p_z u_{p_z}}^{-1} \left[p_{z_e d} - \left(G_{p_z u_{p_z}} - \hat{G}_{p_z u_{p_z}} \right) u_{p_z} \right].$$

Rearranging the controller above results in

$$\begin{aligned} u_{p_z} &= \left[\hat{G}_{p_z u_{p_z}} - Q_{p_z u_{p_z}} \left(G_{p_z u_{p_z}} - \hat{G}_{p_z u_{p_z}} \right) \right]^{-1} Q_{p_z u_{p_z}} p_{z_e d} \\ &= \hat{G}_{p_z u_{p_z}}^{-1} \bar{G} Q_{p_z u_{p_z}} p_{z_e d}. \end{aligned} \tag{12.33}$$

Since $\hat{G}_{p_z u_{p_z}}$ is minimum-phase, $\hat{G}_{p_z u_{p_z}}^{-1}$ is exponentially stable. Moreover, according to the conclusion of *Step 1*, the transfer function $\bar{G}$ is stable, so $\hat{G}_{p_z u_{p_z}}^{-1} \bar{G} Q_{p_z u_{p_z}}$ is stable. In addition, since $p_{z_e d}$ is a constant which is bounded, the control input u_{p_z} is bounded according to (12.33).

Step 3. By employing the controller (12.31), the tracking error is represented as follows:

$$e_{p_z} = \left(G_{p_z u_{p_z}} \hat{G}_{p_z u_{p_z}}^{-1} \bar{G} Q_{p_z u_{p_z}} - 1 \right) p_{z_e d}. \tag{12.34}$$

The tracking error is further represented as follows:

$$e = \rho p_{z_e d} \tag{12.35}$$

where

$$\rho = \underbrace{\left[1 - \left(1 - G_{p_z u_{p_z}} \hat{G}_{p_z u_{p_z}}^{-1} \right) Q_{p_z u_{p_z}} \right]^{-1}}_{\rho_1} \underbrace{\left(1 - Q_{p_z u_{p_z}} \right)}_{\rho_2}. \tag{12.36}$$

---

[2]A system is called a minimum-phase system if all the zeros of its transfer function are in the left-half plane.

Since the signal $p_{z_e d}$ is constant and $Q_{p_z u_{p_z}}(0) = 1$, then $\mathscr{L}^{-1}\left[\rho_2 p_{z_e d}(s)\right]$ is only an impulse, where $\mathscr{L}^{-1}$ denotes the inverse Laplace transformation. Since $e = \rho_1 \rho_2 p_{z_e d}$ and according to the conclusion of *Step 1*, ρ_1 is stable, the tracking error $\lim_{t \to \infty} |e(t)| = 0$.   □

12.3.2.2   Yaw Channel

For system (12.15), by using the similar controller design for the altitude channel, the controller for the yaw channel is obtained as follows:

$$u_\psi(s) = Q_{\psi u_\psi}(s)\, \hat{G}^{-1}_{\psi u_\psi}(s)\left(\psi_d(s) - d_{\psi 1}(s)\right) \tag{12.37}$$

where $d_{\psi 1}(s) = \psi(s) - \hat{G}_{\psi u_\psi}(s)\, u_\psi(s)$.

12.3.2.3   Horizontal Position Channel

For system (12.21), by using the similar controller design for the altitude channel, the controller for the horizontal position channel is obtained as follows:

$$\mathbf{u}_{\mathbf{v}_h}(s) = \mathbf{Q}_{\mathbf{v}_{h_b} \mathbf{u}_{\mathbf{v}_h}}(s)\, \hat{\mathbf{G}}^{-1}_{\mathbf{v}_{h_b} \mathbf{u}_{\mathbf{v}_h}}(s)\left(\mathbf{v}_{h_b d}(s) - \mathbf{d}_{\mathbf{v}_h 1}(s)\right) \tag{12.38}$$

where $\mathbf{d}_{\mathbf{v}_h 1}(s) = \mathbf{v}_{h_b}(s) - \hat{\mathbf{G}}_{\mathbf{v}_{h_b} \mathbf{u}_{\mathbf{v}_h}}(s)\, \mathbf{u}_{\mathbf{v}_h}(s)$. Because the symbol $\mathbf{v}_{h_b}$ is redundant, $\mathbf{d}_{\mathbf{v}_h 1}$ is used to replace $\mathbf{d}_{\mathbf{v}_{h_b} 1}$ for convenience.

12.4   Simulation

In this section, the system identification (Sect. 12.2) and the controller design (Sect. 12.3) are implemented on a simulation platform which contains a multicopter with an SAA. This simulation platform is a modified form of an existing simulation platform [8]. A 6-DoF (Degree of Freedom) model is used in this platform based on AR. Drone. This 6-DoF model also considers the saturation of the actuator and the measurement noise and so on.

12.4.1   System Identification

The simulation for system identification is shown in Fig. 12.6. It should be noted that the outputs are measured by the sensors, so the controller design and the system identification use the measured outputs. The measurement errors of the sensors are approximately 0.01 m and 0.01 rad.

Fig. 12.6 The simulation for the system identification

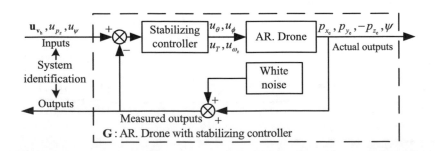

12.4.1.1 Prior Knowledge

The models of multicopters are established by Eqs. (12.1)–(12.3) in Sect. 12.1. They are the prior knowledge. Here, *fitness*[3] is chosen as the criterion of the model validation.

12.4.1.2 Experiment Design

Before the identification, stabilizing controllers will be employed if needed. The controllers of the altitude channel (also called as z channel) and the yaw channel (also called as ψ channel) are designed according to Eqs. (12.8) and (12.14). It should be pointed out that the output of the sensor in the altitude channel is height, namely $-p_{z_e}$. Moreover, the controller of the horizontal position channel is designed according to Eq. (12.20). Here, the desired yaw angle ψ_d is set to 0, so $\mathbf{R}_\psi \approx \mathbf{I}_2$ under a well-designed yaw channel controller and the horizontal position channel is decoupled into x channel and y channel. This means that u_{v_x} controls velocity v_{x_e} through θ, and u_{v_y} controls the velocity v_{y_e} through ϕ. Then, Eq. (12.22) is decoupled, and it is expressed as follows:

$$\mathbf{v}_{h_b}(s) = \mathbf{G}_{\mathbf{v}_{h_b}\mathbf{u}_{v_h}}(s)\,\mathbf{u}_{v_h}(s) = \begin{bmatrix} G_{v_x u_{v_x}}(s)\,u_{v_x}(s) \\ G_{v_y u_{v_y}}(s)\,u_{v_y}(s) \end{bmatrix}. \tag{12.39}$$

Then, choose a group of parameters simply as follows:

$$\mathbf{K}'_{\mathbf{v}_{h_b}} = \operatorname{diag}(0.1, 0.1). \tag{12.40}$$

Fig. 12.7 The actual step response of each channel under PID controller

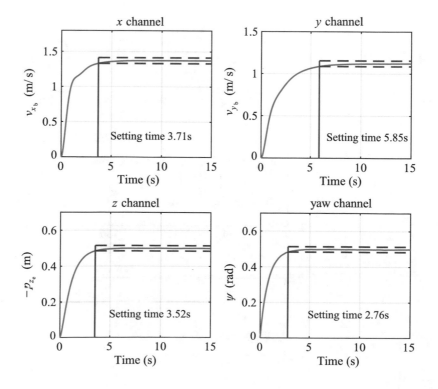

---

[3]Fitness is calculated by $1 - \lVert \mathbf{y} - \hat{\mathbf{y}} \rVert / \lVert \mathbf{y} - \bar{\mathbf{y}} \rVert$, where $\mathbf{y}, \hat{\mathbf{y}}, \bar{\mathbf{y}}$ are the real output, the estimation of the output, and the mean of the real output, respectively. The identification result is perfect, when fitness is close to 100%.

The step response of each channel with the corresponding stabilizing controller is shown in Fig. 12.7, which means that when the input of one channel is a step signal, the other inputs are set to be zero. It should be noted that the ordinate of the z channel in Fig. 12.7 is the height of the multicopter, which is equal to $-p_{z_e}$.

12.4.1.3 Data Record

Furthermore, Pseudo-Random Binary Signals (PRBSs) [1, pp. 418–422] are generated to stimulate these channels. The chosen amplitude of PRBS should prevent the saturation phenomenon from occurring in the system identification. Then, the newly defined inputs $u_{v_x}, u_{v_y}, u_{p_z}, u_\psi$, which are PRBSs, are recorded with the corresponding measured outputs $\hat{v}_{x_e}, \hat{v}_{y_e}, \hat{p}_{z_e}, \hat{\psi}$. As shown in Fig. 12.6, the measured outputs contain white noise caused by the sensors. The input–output data are the time-domain data set used in the system identification. Fig. 12.8 illustrates the input–output data of the four channels used in the identification.

Fig. 12.8 The input–output data of the four channels used in the identification (The output data is the measured outputs)

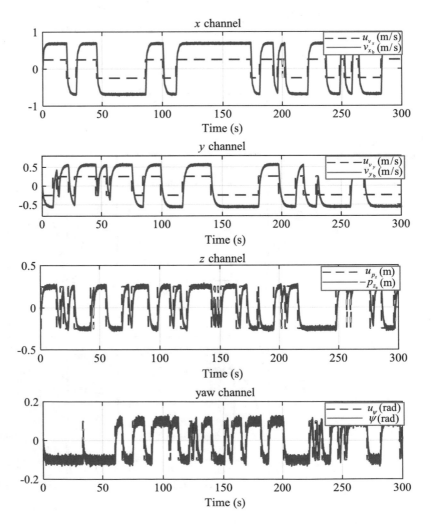

12.4.1.4 Model Selection

According to the prior knowledge, the candidate models are transfer functions as shown in Eqs. (12.10), (12.16), (12.22). The order of the denominator of the identified system can be chosen according to the prior knowledge which is introduced in Sect. 12.2.2, and it can be fine-tuned to obtain a suitable identification result. In addition, a small positive integer can be chosen for the order of the numerator at the beginning, and it can be increased gradually until a fine result is obtained. In this way, non-minimum-phase models should be avoided in the identification. The principles of choosing the order are listed as follows:

(1) The identification results can pass the model validation, as shown in Sect. 12.2.1. In this section, the fitness is used as the criterion.
(2) The identification results are hoped to be minimum-phase systems, because non-minimum-phase systems are hard to tackle for most controller designs.
(3) When the previous two principles are guaranteed, the order of the transfer function needs to be chosen as small as possible.

12.4.1.5 Model Calculation

Here, the simulation data is identified by using the time-domain tools for linear systems of System Identification Toolbox which is introduced in Sect. 12.2.1 and shown in Fig. 12.9. In the toolbox, choose "Time domain data" for "Import data," and choose "Remove means" for "Process." Then, start the identification by choosing "transfer function models" for "Estimate."

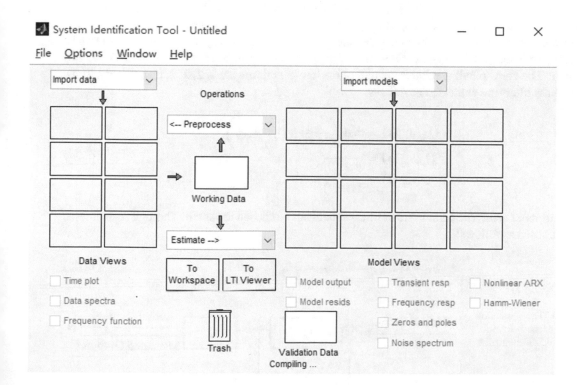

Fig. 12.9 System Identification Toolbox of MATLAB

12.4.1.6 Model Validation

The identification results are shown as follows:

$$\hat{G}_{v_x u_{v_x}}(s) = \frac{15.48s+29.9}{s^3+4.642s^2+16.09s+10.91}$$

$$\hat{G}_{v_y u_{v_y}}(s) = \frac{7.086s+17.02}{s^3+4.742s^2+15.49s+7.063}$$

$$\hat{G}_{p_z u_{p_z}}(s) = -\frac{6.25}{s^2+7.077s+6.249}$$

$$\hat{G}_{\psi u_\psi}(s) = \frac{1.277s+3.506}{s^2+4.045s+3.522}.$$

$$(12.41)$$

The fitnesses of the four channels are as follows: 98.56, 98.08, 95.84, and 90.40%. The identification results pass the validation, and transfer functions in Eq. (12.41) are chosen as the final models. It should be noted that because there is not any disturbance in the simulation, the high fitness can guarantee that the identified system captures the dynamics of the original system finely. But, in practice, the system contains some disturbances. Therefore, a 70% fitness is enough.

So far, the system identification is done. In the following part, according to the identification results in Eq. (12.41), the DIC will be designed.

12.4.2 Control Design

As shown in Fig. 12.10, the controllers are designed according to the AOD-based DIC method which is introduced in Sect. 12.3. The notation $\mathbf{G}$ is the AR. Drone control model with the stabilizing controllers as shown in Fig. 12.6, $\hat{\mathbf{G}}$ is the corresponding identified model, and $\hat{\mathbf{G}}^{-1}\mathbf{Q}$ means the DIC .

The relative degrees of the four channels are 2, 2, 2, 2, respectively, as shown in Eq. (12.41). According to Sect. 12.3.2, low-pass filters are adopted to make controllers (12.31), (12.37), (12.38) realizable. The corresponding relative degrees of the low-pass filters are 2, 2, 2, 2, respectively. Thus, the low-pass filters are expressed as follows:

$$\mathbf{Q}_{v_{h_b} u_{v_h}}(s) = \text{diag}\left(\frac{1}{(\eta_x s+1)^2}, \frac{1}{(\eta_y s+1)^2}\right)$$

$$Q_{p_z u_{p_z}}(s) = \frac{1}{(\eta_z s+1)^2}$$

$$Q_{\psi u_\psi}(s) = \frac{1}{(\eta_\psi s+1)^2}.$$

$$(12.42)$$

Because of input delay, the parameters of filters should not be chosen too small. Through several tests, they are chosen as follows:

$$\eta_x = 0.2, \eta_y = 0.2, \eta_z = 0.3, \eta_\psi = 0.3. \qquad (12.43)$$

Fig. 12.10 The simulation structure for the AOD-based DIC controller

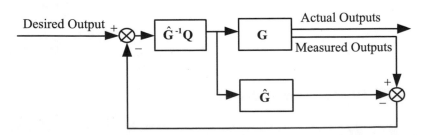

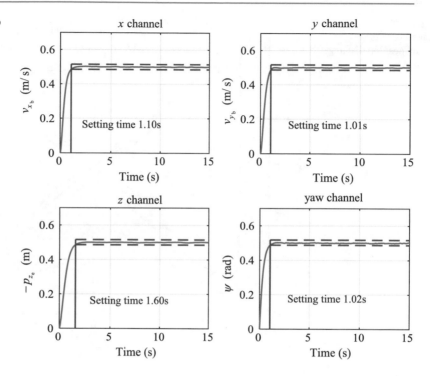

Fig. 12.11 The actual step responses of the channels with DIC

The controllers are designed by following Sect. 12.3, and under these controllers, the step responses of the four channels are shown in Fig. 12.11.

According to Fig. 12.11, the transient processes of the channels are improved remarkably. The settling time of the four channels is reduced and the steady-state errors of x, y channels disappear.

12.4.3   Comparison of Tracking Performance

A trajectory tracking is taken as an example to illustrate the performances of the controllers. Here, the height is held, and the horizontal position tracking is considered. As shown in Fig. 12.7, set $\psi_d = 0$, and let the multicopter track a trajectory as the dash line shows. The initial states of AR. Drone are $p_{x_e}(0) = p_{y_e}(0) = 0$, $\psi(0) = \pi/6$. One of the two considered controllers is the stabilizing controller as shown in Eqs. (12.23)–(12.25). The other one is the AOD-based DIC controller as shown in Eqs. (12.31), (12.37), (12.38) with the identified systems (12.41) and filters (12.42). It should be pointed out that in the horizontal position channel, the AOD-based DIC tracking controller is designed for controlling the horizontal velocity rather than the horizontal position. Thus, the desired horizontal velocity needs to be transformed from the desired horizontal position. First, the transient process is expected as follows:

$$\dot{\mathbf{p}}_h - \dot{\mathbf{p}}_{hd} = -\mathbf{K}_{\mathbf{p}_h}(\mathbf{p}_h - \mathbf{p}_{hd}) \tag{12.44}$$

where $\mathbf{K}_{\mathbf{p}_h} \in \mathbb{R}^{2 \times 2} \cap \mathscr{D} \cap \mathscr{P}$. If so, then $\lim_{t \to \infty} \|\mathbf{p}_h(t) - \mathbf{p}_{hd}(t)\| = 0$. Since $\dot{\mathbf{p}}_h = \mathbf{R}_\psi \mathbf{v}_{h_b}$ according to Eq. (12.3), the desired horizontal velocity should satisfy

$$\mathbf{R}_\psi \mathbf{v}_{h_bd} = \dot{\mathbf{p}}_{hd} - \mathbf{K}_{\mathbf{p}_h}(\mathbf{p}_h - \mathbf{p}_{hd}). \tag{12.45}$$

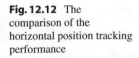

Fig. 12.12 The comparison of the horizontal position tracking performance

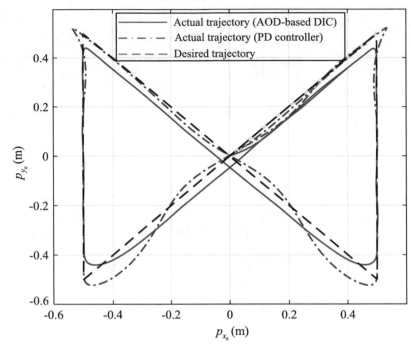

Because $\dot{\mathbf{p}}_{hd}$ is usually small, the desired horizontal velocity can be simplified as

$$\mathbf{v}_{h_bd} = -\mathbf{R}_\psi^{-1}\mathbf{K}_{\mathbf{p}_h}\left(\mathbf{p}_h - \mathbf{p}_{hd}\right). \tag{12.46}$$

In this simulation, $\mathbf{K}_{\mathbf{p}_h} = \mathrm{diag}\,(1,1)$. The obtained $\mathbf{v}_{h_bd}$ will be used in AOD-based DIC (12.38).

In order to compare the performances of different controllers, a PD tracking controller is also designed as follows:

$$\mathbf{u}_h = -\mathbf{K}_{hp}\mathbf{R}_\psi^{-1}\left(\mathbf{p}_h - \mathbf{p}_{hd}\right) - \mathbf{K}_{hd}\mathbf{R}_\psi^{-1}\left(\dot{\mathbf{p}}_h - \dot{\mathbf{p}}_{hd}\right) \tag{12.47}$$

where $\mathbf{K}_{hp} \in \mathbb{R}^{2\times2} \cap \mathscr{D} \cap \mathscr{P}$, and in this simulation, $\mathbf{K}_{hp} = \mathrm{diag}\,(1,1)$ and $\mathbf{K}_{hd} = \mathrm{diag}\,(0.1, 0.1)$. Since PD controller is enough for this example, the integral term in Eq. (12.25) is omitted here.

The tracking performance of the horizontal position is shown in Fig. 12.12, and the tracking error of the corresponding three channels is shown in Fig. 12.13. According to Fig. 12.13, when the desired trajectory changes, the transient process of the DIC is much better than that of the PD controller. Thus, Fig. 12.12 shows that the DIC has an excellent tracking performance. However, the tracking trajectories of the DIC lag behind those of the PD controller a little. The reason is that the gain $\mathbf{K}_{\mathbf{p}_h}$ in Eq. (12.46) is not large enough. If $\mathbf{K}_{\mathbf{p}_h}$ is increased, the time delay of the tracking will be decreased, but the transient process will get worse, vice versa. Thus, users need to adjust $\mathbf{K}_{\mathbf{p}_h}$ of the DIC according to actual requirements.

Fig. 12.13 The comparison of tracking performance of the three channels

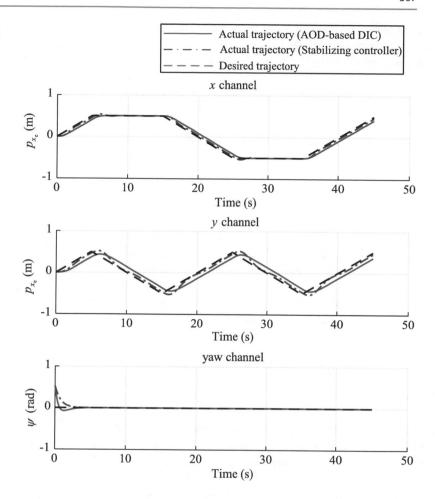

12.5 Summary

Recently, more and more multicopter companies open the SDKs of their autopilots (refer to Table 1.3 in Chap. 1). Based on these autopilots, tasks can be accomplished by secondary developments. If SDKs are not provided but the multicopter can be controlled by a general RC transmitter, then some interface devices, such as PCTX,[4] can be used to connect between RC transmitters and computers so that the computers can control the multicopter directly [9]. On the other hand, companies also provide some special types of multicopters for developers, such as "Matrice100" with "Guidance" provided by DJI[5] and "Hummingbird" provided by Ascending Technologies GmbH. These are very convenient for

---

[4]PCTX is an interface which connects the computer and the RC transmitter. By employing this device, the software applications can be used on the computer to control the RC multicopters. Readers can refer to http://www.endurance-rc.com/pctx.php for further details.

[5]Developers can customize and tailor this flight platform by using the DJI SDK and develop a system for any use by applying developers' knowledge and skills. Readers can refer to http://www.dji.com/product/matrice100 for further details.

research or prototypes development. In the future, one possibility of the development on multicopters is that some companies provide a stable and reliable multicopter platform or operating systems, based on which new applications can be developed easily.

Exercises

12.1 Design a PID tracking controller for the following system

$$\begin{cases} \dot{\mathbf{x}}(t) = \mathbf{A}\mathbf{x}(t) + \mathbf{b}u(t) + \boldsymbol{\phi}[y(t)] \\ y(t) = \mathbf{c}^{\mathrm{T}}\mathbf{x}(t), \mathbf{x}(0) = \mathbf{0}_{3\times1} \end{cases} \tag{12.48}$$

where

$$\mathbf{A} = \begin{bmatrix} 0 & 1 & 0 \\ 0 & 0 & 1 \\ -3 & -1 & -3 \end{bmatrix}, \mathbf{b} = \begin{bmatrix} 0 \\ 0 \\ 1 \end{bmatrix}, \mathbf{c} = \begin{bmatrix} 1 \\ 0 \\ 0 \end{bmatrix}, \boldsymbol{\phi}(y) = \begin{bmatrix} 1 - \cos y \\ 0 \\ \sin y \end{bmatrix}. \tag{12.49}$$

Only the output y of this system can be measured in the controller design, and the desired trajectory is $y_{\mathrm{d}}(t) \equiv 1$. The form of the PID tracking controller is

$$u = -k_{\mathrm{p}}(y - y_{\mathrm{r}}) - k_{\mathrm{d}}(\dot{y} - \dot{y}_{\mathrm{r}}) - k_{\mathrm{i}} \int (y - y_{\mathrm{r}}) \tag{12.50}$$

and $y_{\mathrm{r}} = y_{\mathrm{d}}$ in this exercise.

12.2 Use system identification to obtain the linear model of the system (12.48) with the controller designed in Exercise 12.1 at $(\mathbf{x}, u) = (\mathbf{0}_{3\times1}, 0)$. It means that the input and the output of the system identification are y_{r} in Eq. (12.50) and y in Eq. (12.49), respectively.

12.3 Based on the models identified in Exercise 12.2, design an AOD-based DIC tracking controller for system (12.48) to track the desired trajectory $y_{\mathrm{d}}(t) \equiv 1$.

12.4 Design an AOD-based DIC controller for system (12.48) with the parameters as follows:

$$\mathbf{A} = \begin{bmatrix} 0 & 1 & 0 \\ 0 & 0 & 1 \\ -3 & -1 & -3 \end{bmatrix}, \mathbf{b} = \begin{bmatrix} 0 \\ 0 \\ 1 \end{bmatrix}, \mathbf{c} = \begin{bmatrix} -1 \\ 0 \\ 1 \end{bmatrix}, \boldsymbol{\phi}(y) = \begin{bmatrix} 1 - \cos y \\ 0 \\ \sin y \end{bmatrix}. \tag{12.51}$$

Notice that this is a non-minimum-phase system. Readers need to identify an approximate linear system for the AOD-based DIC tracking controller.

12.5 Design a PID controller for horizontal position channel to replace Eq. (12.25) if the yaw angle ψ varies fast. (Hint: take the desired velocity as the control target.)

References

1. Ljung L (1998) System identification. Birkhauser, Boston
2. System identification toolbox. http://www.mathworks.com/help/ident/index.html. Accessed 20 Jan 2016

3. Ljung L (1995) System identification toolbox: user's guide. MathWorks, Natick
4. Flight control: CIFER. http://uarc.ucsc.edu/flight-control/cifer/index.shtml. Accessed 20 Jan 2016
5. Tischler MB, Remple RK (2006) Aircraft and rotorcraft system identification, 2nd edn. American Institute of Aeronautics and Astronautics, New York
6. Quan Q, Cai KY (2012) Additive-output-decomposition-based dynamic inversion tracking control for a class of uncertain linear time-invariant systems. In: Proceedings IEEE conference on decision and control, Hawaii, pp 2866–2871
7. Green M, Limebeer DJN (1995) Linear robust control. Prentice Hall, New Jersey
8. Zander J (2013) AR_Drone_Simulink. https://github.com/justynazander/AR_Drone_Simulink. Accessed 20 Jan 2016
9. How JP, Bethke B, Frank A, Dale D, Vian J (2008) Real-time indoor autonomous vehicle test environment. IEEE Control Syst Mag 28(2):51–64

Part V
Decision

Mission Decision-Making

13

A Flight Control System (FCS) or an autopilot includes not only a low-level flight control system as introduced in Chap. 11, but also a high-level decision-making module. The former just aims at solving the problem—"how to fly to a desired position," while the latter mainly aims at solving the problem—"how to determine the desired position." The decision-making process mainly includes the *mission decision-making* and *failsafe*. This chapter only considers the mission decision-making, aiming to provide a sequence of discrete-time desired waypoints or continuous-time desired trajectories in real time for a multicopter to follow. Currently, the majority of multicopter products and open source autopilots support two high-level control modes, namely Fully-Autonomous Control (FAC) and Semi-Autonomous Control (SAC). For each control mode, the mission decision-making has different functions. This chapter aims to answer the question below:

What are the mission decision-making mechanisms under the FAC and SAC modes, respectively?

In order to clearly explain the mission decision-making mechanism, this chapter describes the FAC in Sect. 13.1 and SAC in Sect. 13.2. In Sect. 13.1, the mission planning and path planning methods will be introduced, where the path following and obstacle avoidance are considered. In Sect. 13.2, the SAC, the Radio Control (RC), and Automatic Control (AC) are introduced. Furthermore, a switching logic between the RC and AC is summarized and analyzed.

© Springer Nature Singapore Pte Ltd. 2017
Q. Quan, *Introduction to Multicopter Design and Control*, DOI 10.1007/978-981-10-3382-7_13

Playing Go

The ancient Chinese had paid lots of attention to strategies during a Go or a war. There are many great strategists in China. Bang Liu, emperor Gaozu of the Han dynasty, praised his general Liang Zhang who could make better decisions in his tent a thousand miles away. This story became an idiom—"运筹帷幄", to describe people who are good at strategies. There is also an interesting idiom '老马识途', which means that old horses know the way home. This idiom originated from a story of the book "*Han Feizi*". Zhong Guan led the troops of the State of Qi to attack the State of Yan. They won the battle but lost their way while returning home. Zhong Guan selected several old horses and freed them to wander around. By following these old horses, they found the way back to the State of Qi at the end. This idiom emphasizes the value of experience—people who are experienced know a better way of doing things and therefore can guide other people.

13.1   Fully-Autonomous Control

13.1.1   Brief Introduction

In the FAC mode, the control algorithm in an autopilot can control a multicopter to take off, land, and accomplish waypoint-to-waypoint tasks automatically. As shown in Fig. 13.1, the FAC consists of the mission planning, path planning, and a low-level flight control system, among which the mission planning and path planning belong to the mission decision-making module. As the low-level flight control system has been discussed in detail in Chap. 11, this chapter focuses on the mission decision-making module.

(1) *Mission planning*
In a practical scenario, missions are often planned by remote pilots off-line. In general, a mission planning process consists of the mission stages-division and task-definition at each waypoint. By taking the ArduPilot Mega (APM) autopilot for example, as shown in Fig. 13.2, the mission stages-division specifies several waypoints $\mathbf{p}_{\mathrm{wp},k}$, $k = 1, \ldots, n_{\mathrm{wp}}$ before taking off. In some cases, the default heading is required to point to the next waypoint, i.e., the waypoints implicitly determine the desired yaw ψ_{d}. In some special cases, besides deciding the waypoints, remote pilots need to specify the desired yaw ψ_{d} to be a certain value. Here, two practical scenes are introduced where the yaw ψ_{d} needs to be specified. In the first case, the desired heading parallels to the flight path. For example, in trajectory tracking for complex terrain survey, the heading coinciding with the velocity in direction can provide a better observation [1]. In the other case, the desired heading has to point to an object all the time. For example, during a three-dimensional (3D) reconstruction for a tower, a multicopter needs to fly around the tower; meanwhile, its heading should point to the tower so that the forward-looking camera can capture the complete information about the tower.

(2) *Path planning*
The path planning aims to produce a flyable path between two waypoints. The flyable path is either a continuous-time trajectory $\mathbf{p}_{\mathrm{d}}(t)$ or a series of discrete-time goal waypoints ($\mathbf{p}_{\mathrm{d},k}$, $k = 1, 2, \ldots$). So far, current open source autopilots rarely take the obstacle avoidance and hazard avoidance into

Fig. 13.1 System architecture under FAC

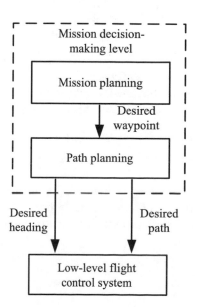

Fig. 13.2 Path planning interface in Mission Planner

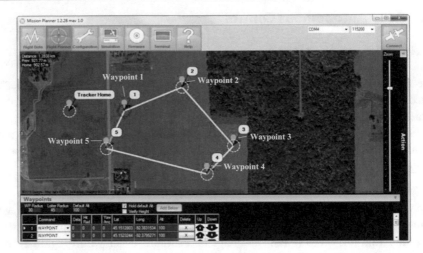

Fig. 13.3 Block diagram of the closed-loop control scheme under FAC for a multicopter

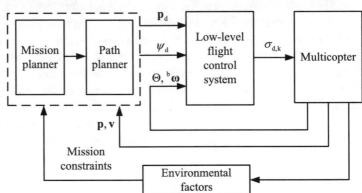

consideration during the path planning. As shown in Fig. 13.2, in Mission Planner (a Ground Control Station (GCS) software), the flight path is often defined by a set of waypoints joined by straight-line segments. In this case, the goal waypoint $\mathbf{p}_{d,k} = \mathbf{p}_{wp,k}$. Unlike this simple method, a practical path planning method is proposed in this chapter, where the path following and obstacle avoidance are both considered. The method is based on the idea of the Artificial Potential Field (APF) method [2].

(3) *Low-level flight control*
In Chap. 11, the low-level flight control system is introduced in detail. Here, a block diagram of the closed-loop control scheme under the FAC is shown in Fig. 13.3.

13.1.2   Mission Planning

Mission planning is often mission-oriented. Currently, the application of multicopters in agriculture industry has a promising development tendency [3, pp. 85–87]. So, for simplicity, a simple mission planner is designed for an agricultural multicopter in the following, which traverses a rectangle farm-land.

Fig. 13.4 Mission planning for farmland traversal

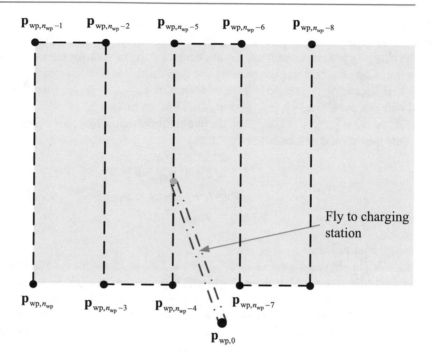

$\mathbf{p}_{\mathrm{wp},n_{\mathrm{wp}}-1}$ $\mathbf{p}_{\mathrm{wp},n_{\mathrm{wp}}-2}$ $\mathbf{p}_{\mathrm{wp},n_{\mathrm{wp}}-5}$ $\mathbf{p}_{\mathrm{wp},n_{\mathrm{wp}}-6}$ $\mathbf{p}_{\mathrm{wp},n_{\mathrm{wp}}-8}$

Fly to charging station

$\mathbf{p}_{\mathrm{wp},n_{\mathrm{wp}}}$ $\mathbf{p}_{\mathrm{wp},n_{\mathrm{wp}}-3}$ $\mathbf{p}_{\mathrm{wp},n_{\mathrm{wp}}-4}$ $\mathbf{p}_{\mathrm{wp},n_{\mathrm{wp}}-7}$

$\mathbf{p}_{\mathrm{wp},0}$

13.1.2.1 Mission Requirements

(1) *Given conditions*

1) As shown in Fig. 13.4, the desired waypoints $\mathbf{p}_{\mathrm{wp},0}, \mathbf{p}_{\mathrm{wp},1}, \ldots, \mathbf{p}_{\mathrm{wp},k}, \ldots, \mathbf{p}_{\mathrm{wp},n_{\mathrm{wp}}} \in \mathbb{R}^3, k = 0,$ $\ldots, n_{\mathrm{wp}}$ have been produced.
2) The real-time position $\mathbf{p} \in \mathbb{R}^3$ and the battery capacity are both known.

(2) *Mission requirements*

1) Starting from the base, the multicopter is required to traverse all the waypoints and follow an S-shaped path shown in Fig. 13.4.
2) The desired heading always points to the next waypoint.
3) When battery capacity falls below a threshold, the multicopter will record the current position and go back to the base to charge. After the charging completes, it will first fly back to the position where it just left and will continue to traverse the remaining waypoints.
4) After completing all waypoints, the multicopter will fly back to the base.

13.1.2.2 A Simple Protocol Design

Let $\mathbf{p}_{\mathrm{wp,cur}}$ be the current goal waypoint. A protocol is designed as the following:

Step 1. Determine the coordinate system of the farmland.

Step 2. Generate the waypoints $\mathbf{p}_{\mathrm{wp},0}, \mathbf{p}_{\mathrm{wp},1}, \ldots, \mathbf{p}_{\mathrm{wp},k}, \ldots, \mathbf{p}_{\mathrm{wp},n_{\mathrm{wp}}}, k = 0, \ldots, n_{\mathrm{wp}}$ based on the mission specifications and the coordinate system of farmland, where $\mathbf{p}_{\mathrm{wp},0}$ is the position of the base.

Step 3. Let $k = n_{\mathrm{wp}}, \mathbf{p}_{\mathrm{wp,cur}} = \mathbf{p}_{\mathrm{wp},k}$ and $\varepsilon > 0$.

Step 4. Update the position information $\mathbf{p}$ and battery capacity periodically. Proceed to Step 5 upon completion of the update and stay in Step 4 otherwise.

Step 5. Go to Step 6 if $\|\mathbf{p}_{wp,cur} - \mathbf{p}\| \leq \varepsilon$, and skip to Step 7 otherwise.

Step 6. End the mission if $\mathbf{p}_{wp,cur} = \mathbf{p}_{wp,0} \& k = 0$, and charge the battery if $\mathbf{p}_{wp,cur} = \mathbf{p}_{wp,0} \& k \neq 0$. When the charging process completes, the multicopter starts to wait for the takeoff command by the remote pilot. Once the multicopter receives the command, the mission planning protocol executes $k = k - 1$, $\mathbf{p}_{wp,cur} = \mathbf{p}_{wp,k}$ and goes back to Step 4. If $\mathbf{p}_{wp,cur} \neq \mathbf{p}_{wp,0}$, then the mission planning protocol will execute $k = k - 1$, $\mathbf{p}_{wp,cur} = \mathbf{p}_{wp,k}$. Then, go to Step 7.

Step 7. If the battery capacity falls below the threshold and the current goal waypoint $\mathbf{p}_{wp,cur} \neq \mathbf{p}_{wp,0}$, then execute the following instructions

$$k = k + 1, \mathbf{p}_{wp,k} = \mathbf{p}$$
$$k = k + 1, \mathbf{p}_{wp,k} = \mathbf{p}_{wp,0}$$
$$\mathbf{p}_{wp,cur} = \mathbf{p}_{wp,k}.$$

Otherwise go back to Step 4.

The idea of the above protocol is inspired by the Last-In First-Out (LIFO) data structure, which facilitates the programming. The desired flight path is $\mathbf{p}_{wp,n_{wp}} \to \mathbf{p}_{wp,n_{wp}-1} \to \cdots \to \mathbf{p}_{wp,1} \to \mathbf{p}_{wp,0}$. Step 6 is executed when the multicopter arrives at a waypoint. Then, it is to decide which mission stage the multicopter will enter: ending mission, charging the battery, or going to the next waypoint. Step 7 is to save the multicopter current position $\mathbf{p}$ and the base station position $\mathbf{p}_{wp,0}$ into the waypoint stack. By doing this, the multicopter can fly to position $\mathbf{p}$ to continue the mission after completing charging.

13.1.3 Path Planning

13.1.3.1 Basic Concepts

The primary aim of path planning is to provide flyable paths, i.e., to facilitate flying a multicopter from one position to another. In practice, there are various constraints involved in the path planning, most of them being multicopter-specific and the rest arising from obstacles in the environment. Therefore, the route defined in a mission planner may not be flyable because it may violate the constraints. In general, path planning is divided into two stages. The first stage is to produce a series of waypoints that are connected by straight lines. This stage is also called *global path planning* [4, pp. 10–12]. The second stage is to refine the path to be flyable according to local environmental conditions and kinematic or motion constraints of the multicopter. This stage is also called *local path planning*.

The two most important constraints for path planning of a multicopter are that the path must be *flyable* and *safe*. Flyable paths satisfy the maneuverability constraints of multicopters. The safety of the multicopters is related to avoiding obstacles, either fixed or moving, that intersect the path. Since multicopters often fly at a low altitude with a low speed, obstacles are common in the airspace. By taking this into consideration, path planning methods including straight-line path following and obstacle avoidance are designed, where the manoeuvrability is further considered by using saturation. The idea of the proposed method is similar to that of the APF method, which was first proposed in [5]. In this method, the airspace is formulated as an APF. The destination point and desired path are assigned *attractive potentials*, while the obstacles are assigned *repulsive potentials*. A multicopter in the field will be attracted toward the path and then fly to the destination, while being repelled by the obstacles. For compatibility with current autopilots, desired positions are produced for path planning. With these desired positions sent to the autopilot, multicopters can be guided to accomplish complex missions.

13.1.3.2 Straight-Line Path Following

When a multicopter sprays insecticide over a farmland or patrols power lines, it is often required to fly along straight-line paths. In the following, for simplicity, the multicopter is considered as a mass point, so that it satisfies the Newton's second law that

$$\dot{\mathbf{p}} = \mathbf{v}$$
$$\dot{\mathbf{v}} = \mathbf{u} \qquad (13.1)$$

where $\mathbf{u} \in \mathbb{R}^3$ is a pseudocontrol input (acceleration), $\mathbf{v} \in \mathbb{R}^3$ is the velocity, and $\mathbf{p}$ is the current position.

(1) *Problem formulation*

The problem is described as follows: Let $\mathbf{p} \in \mathbb{R}^3$ be the current position of a multicopter, $\mathbf{p}_{\mathrm{wp}} \in \mathbb{R}^3$ be the goal waypoint, and $\mathbf{p}_{\mathrm{wp,last}} \in \mathbb{R}^3$ be the previous waypoint. Design $\mathbf{u}$ to guide the multicopter to follow the straight line through $\mathbf{p}_{\mathrm{wp,last}} \in \mathbb{R}^3$ and $\mathbf{p}_{\mathrm{wp}} \in \mathbb{R}^3$ until it arrives at the goal waypoint $\mathbf{p}_{\mathrm{wp}}$.

(2) *Main results*

As shown in Fig. 13.5, the distance between the current position $\mathbf{p}$ and the path equals to $\left\| \mathbf{p} - \mathbf{p}_{\mathrm{wp,perp}} \right\|$, where $\mathbf{p}_{\mathrm{wp,perp}}$ is the foot point which is expressed as [6]:

$$\mathbf{p}_{\mathrm{wp,perp}} = \mathbf{p}_{\mathrm{wp}} + \left(\mathbf{p}_{\mathrm{wp,last}} - \mathbf{p}_{\mathrm{wp}}\right) \frac{\left(\mathbf{p} - \mathbf{p}_{\mathrm{wp}}\right)^{\mathrm{T}} \left(\mathbf{p}_{\mathrm{wp,last}} - \mathbf{p}_{\mathrm{wp}}\right)}{\left\| \mathbf{p}_{\mathrm{wp}} - \mathbf{p}_{\mathrm{wp,last}} \right\|^2}.$$

Then, the following equation

$$\mathbf{p} - \mathbf{p}_{\mathrm{wp,perp}} = \mathbf{A}\left(\mathbf{p} - \mathbf{p}_{\mathrm{wp}}\right) \qquad (13.2)$$

is derived with

$$\mathbf{A} = \mathbf{I}_3 - \frac{\left(\mathbf{p}_{\mathrm{wp,last}} - \mathbf{p}_{\mathrm{wp}}\right)\left(\mathbf{p}_{\mathrm{wp,last}} - \mathbf{p}_{\mathrm{wp}}\right)^{\mathrm{T}}}{\left\| \mathbf{p}_{\mathrm{wp}} - \mathbf{p}_{\mathrm{wp,last}} \right\|^2}$$

which is a positive semi-definite matrix. The equation $\mathbf{A}\left(\mathbf{p} - \mathbf{p}_{\mathrm{wp}}\right) = \mathbf{0}_{3 \times 1}$ implies that $\mathbf{p}$ is in the straight line through $\mathbf{p}_{\mathrm{wp,last}}$ and $\mathbf{p}_{\mathrm{wp}}$.

Fig. 13.5 Desired position generated for straight-line path following

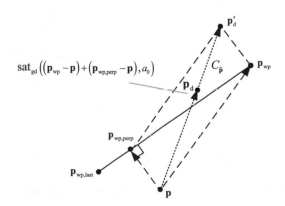

Theorem 13.1 *Suppose the pseudocontrol input is designed as*

$$\mathbf{u} = -\frac{1}{k_2} sat_{gd}\left(k_0\tilde{\mathbf{p}}_{wp} + k_1\mathbf{A}\tilde{\mathbf{p}}_{wp}, a_0\right) - \frac{1}{k_2}\mathbf{v} \tag{13.3}$$

where $\tilde{\mathbf{p}}_{wp} \triangleq \mathbf{p} - \mathbf{p}_{wp}$, $k_0, k_1 \in \{0\} \cup \mathbb{R}_+$, *and* $k_2 \in \mathbb{R}_+$. *Then* $\lim_{t\to\infty} \left\| (k_0\mathbf{I}_3 + k_1\mathbf{A})\,\tilde{\mathbf{p}}_{wp}(t) \right\| = 0$.

Proof If the Lyapunov function is defined as follows:

$$V_1 = \underbrace{k_0\tilde{\mathbf{p}}_{wp}^{\mathrm{T}}\tilde{\mathbf{p}}_{wp}}_{\text{Get close to the waypoint}} + \underbrace{k_1\tilde{\mathbf{p}}_{wp}^{\mathrm{T}}\mathbf{A}\tilde{\mathbf{p}}_{wp}}_{\text{Get close to the path}} + \frac{k_2}{2}\mathbf{v}^{\mathrm{T}}\mathbf{v} \tag{13.4}$$

where $k_0, k_1, k_2 \in \{0\} \cup \mathbb{R}_+$, then $V_1 \to 0$ implies that the multicopter gets close to the waypoint and the path simultaneously. However, the saturation on control needs to be considered. The Lyapunov function has to be modified. Define $sat_{gd}\left(k_0\tilde{\mathbf{p}}_{wp} + k_1\mathbf{A}\tilde{\mathbf{p}}_{wp}, a_0\right)$ as a vector field along a smooth curve $C_{\tilde{p}}$ parameterized by $\tilde{\mathbf{p}}_{wp}$. Then, the Lyapunov function (13.4) is modified as

$$V_1' = \int_{C_{\tilde{p}}} sat_{gd}\left(k_0\tilde{\mathbf{p}}_{wp} + k_1\mathbf{A}\tilde{\mathbf{p}}_{wp}, a_0\right)^{\mathrm{T}} \mathrm{d}\tilde{\mathbf{p}}_{wp} + \frac{k_2}{2}\mathbf{v}^{\mathrm{T}}\mathbf{v}$$

where $a_0 \in \mathbb{R}_+$. The definition of the saturation function $sat_{gd}\,(\mathbf{s}, a_0)$ has been given in Eq. (10.18) in Chap. 10. Similar to the proof of Theorem 10.9 in Chap. 10, the derivative of V_1' along the solution to system (13.1) is

$$\dot{V}_1' = sat_{gd}\left(k_0\tilde{\mathbf{p}}_{wp} + k_1\mathbf{A}\tilde{\mathbf{p}}_{wp}, a_0\right)^{\mathrm{T}}\mathbf{v} + k_2\mathbf{v}^{\mathrm{T}}\mathbf{u}.$$

If the pseudocontrol input satisfies

$$\mathbf{u} = -\frac{1}{k_2} sat_{gd}\left(k_0\tilde{\mathbf{p}}_{wp} + k_1\mathbf{A}\tilde{\mathbf{p}}_{wp}, a_0\right) - \frac{1}{k_2}\mathbf{v} \tag{13.5}$$

then $\dot{V}_1'$ becomes

$$\dot{V}_1' = -\mathbf{v}^{\mathrm{T}}\mathbf{v}.$$

Furthermore, $\dot{V}_1' = 0$ if and only if $\mathbf{v} = \mathbf{0}_{3\times 1}$. Now, $\mathbf{v} = \mathbf{0}_{3\times 1}$ implies that

$$-\frac{1}{k_2} sat_{gd}\left(k_0\tilde{\mathbf{p}}_{wp} + k_1\mathbf{A}\tilde{\mathbf{p}}_{wp}, a_0\right) = \mathbf{0}_{3\times 1}. \tag{13.6}$$

Then,

$$(k_0\mathbf{I}_3 + k_1\mathbf{A})\,\tilde{\mathbf{p}}_{wp} = \mathbf{0}_{3\times 1}. \tag{13.7}$$

Thus, the *invariant set theorem* (see Chap. 10) indicates that the system converges globally to $(\mathbf{p}, \mathbf{v})$, where $\mathbf{p}$ is the solution to equation (13.7) and $\mathbf{v} = \mathbf{0}_{3\times 1}$. $\qquad\square$

The parameters k_0, k_1, k_2 in Eq. (13.6) will achieve a trade-off convergence between to the path and to the goal waypoint.

(1) If $k_0 = 1, k_1 = 0, k_2 \in \mathbb{R}_+$, then the system converges globally to $(\mathbf{p}_{\mathrm{wp}}, \mathbf{0}_{3\times1})$. This implies that $\lim_{t\to\infty} \|\mathbf{v}(t)\| = 0$, $\lim_{t\to\infty} \|\tilde{\mathbf{p}}_{\mathrm{wp}}(t)\| = 0$. In this case, the multicopter will fly to $\mathbf{p}_{\mathrm{wp}}$ directly.

(2) If $k_0 = 0, k_1 = 1, k_2 \in \mathbb{R}_+$, then the system converges globally to $(\mathbf{p}_{\mathrm{p}}, \mathbf{0}_{3\times1})$, where $\mathbf{p}_{\mathrm{p}}$ is a value satisfying $\mathbf{A}(\mathbf{p}_{\mathrm{p}} - \mathbf{p}_{\mathrm{wp}}) = \mathbf{0}_{3\times1}$. By noticing Eq. (13.2), the relation $\mathbf{A}(\mathbf{p}_{\mathrm{p}} - \mathbf{p}_{\mathrm{wp}}) = \mathbf{0}_{3\times1}$ means that $\mathbf{p}_{\mathrm{p}}$ belongs to the path. This implies that $\lim_{t\to\infty} \|\mathbf{A}\tilde{\mathbf{p}}_{\mathrm{wp}}(t)\| = 0$, $\lim_{t\to\infty} \|\mathbf{v}(t)\| = 0$. In this case, the multicopter will fly to the path directly.

(3) If $k_0, k_1, k_2 \in \mathbb{R}_+$, then the multicopter will get close both to the $\mathbf{p}_{\mathrm{wp}}$ and to the path simultaneously. It is obvious that a higher k_0/k_1 means that the multicopter will get close to $\mathbf{p}_{\mathrm{wp}}$ faster. Otherwise, the multicopter will approach the path faster.

From the analysis above, the controller (13.5) can solve the path following problem and can be written as a PD controller as

$$\mathbf{u} = -\frac{1}{k_2}(\mathbf{p} - \mathbf{p}_{\mathrm{d}}) - \frac{1}{k_2}\mathbf{v}$$

where

$$\mathbf{p}_{\mathrm{d}} = \mathbf{p} + \mathrm{sat}_{\mathrm{gd}}\left(k_0(\mathbf{p}_{\mathrm{wp}} - \mathbf{p}) + k_1(\mathbf{p}_{\mathrm{wp,perp}} - \mathbf{p}), a_0\right).$$

Here, Eq. (13.2) is utilized. In order to figure out the physical meaning of $\mathbf{p}_{\mathrm{d}}$, let

$$k_0 = k_1 = 1.$$

Then, $\mathbf{p}_{\mathrm{d}}$ has a concise form

$$\mathbf{p}_{\mathrm{d}} = \mathbf{p} + \mathrm{sat}_{\mathrm{gd}}\left((\mathbf{p}_{\mathrm{wp}} - \mathbf{p}) + (\mathbf{p}_{\mathrm{wp,perp}} - \mathbf{p}), a_0\right)$$

where the physical meaning of $\mathbf{p}_{\mathrm{d}}$ is shown in Fig. 13.5. The local desired position $\mathbf{p}_{\mathrm{d}}$ can be sent to the low-level flight control system in real time to realize path following. The expression above is practical, as the path following function can be developed based on a Semi-Autonomous Autopilot (SAA) which is considered as a black box, namely commercial autopilots with an available SDK.

(3) *Simulation*

Here, a simulation example is given. Let $\mathbf{p}_{\mathrm{wp,last}} = [-8\ -8\ 0]^{\mathrm{T}}$, $\mathbf{p}_{\mathrm{wp}} = [8\ 8\ 8]^{\mathrm{T}}$, $a_0 = 3$, $a_1 = 10$, $k_0 = k_1 = 1$. The resulting APF is shown in Fig. 13.6a. As shown, each arrow stands for a potential field force. From the direction of the arrows, it can be predicted that the multicopter will get close to the path and, meanwhile, fly toward the goal waypoint $\mathbf{p}_{\mathrm{wp}}$. To investigate the relationship between parameters and the convergence, fix $k_0 = 0.1$, and further, let $k_1 = 0.5$, $k_1 = 1.0$, $k_1 = 2.0$, respectively. If the initial location is $\mathbf{p}_0 = [0\ -8\ 0]^{\mathrm{T}}$, then the local path planner will produce three flight paths which will guide the multicopter to fly. The flying trajectories in the simulation are shown in Fig. 13.6b. It is observed that a higher k_1 makes the multicopter approach the path faster. This is consistent with the conclusion of Theorem 13.1.

13.1.3.3 Obstacle Avoidance

In practice, a multicopter needs the ability to avoid both fixed and moving obstacles. Thus, in this section, a collision-free path planning is designed based on known waypoints and the position of obstacles.

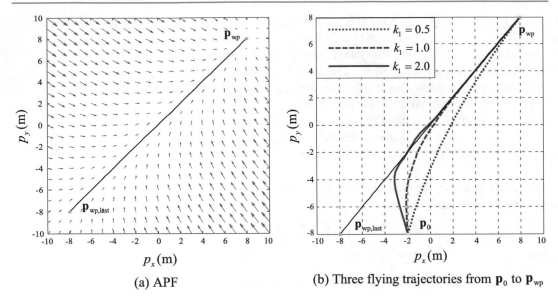

(a) APF

(b) Three flying trajectories from $\mathbf{p}_0$ to $\mathbf{p}_{wp}$

Fig. 13.6 Straight-line path following by the proposed method

(1) *Problem formulation*

The problem is described as follows: Let $\mathbf{p} \in \mathbb{R}^3$ be the current position of multicopter and $\mathbf{p}_{wp} \in \mathbb{R}^3$ be the goal waypoint. Based on the mass model (13.1), design $\mathbf{u}$ to guide the multicopter to fly toward the goal waypoint $\mathbf{p}_{wp} \in \mathbb{R}^3$, meanwhile avoiding a obstacle modeled as a sphere with radius $r_o \in \mathbb{R}_+$ and center $\mathbf{p}_o \in \mathbb{R}^3$.

Define $\boldsymbol{\xi} \triangleq \mathbf{p} + k_0\mathbf{v}$, where $k_0 \in \mathbb{R}_+$. According to Eq. (13.1), one has

$$\dot{\boldsymbol{\xi}} = \bar{\mathbf{u}}$$

where

$$\bar{\mathbf{u}} = \mathbf{v} + k_0\mathbf{u}. \tag{13.8}$$

Furthermore, define filtered errors as

$$\tilde{\boldsymbol{\xi}}_{wp} \triangleq \boldsymbol{\xi} - \boldsymbol{\xi}_{wp}$$
$$\tilde{\boldsymbol{\xi}}_o \triangleq \boldsymbol{\xi} - \boldsymbol{\xi}_o$$

where

$$\boldsymbol{\xi}_{wp} = \mathbf{p}_{wp} + k_0\mathbf{v}_{wp}$$
$$\boldsymbol{\xi}_o = \mathbf{p}_o + k_0\mathbf{v}_o$$

and $\mathbf{v}_{wp}, \mathbf{v}_o \in \mathbb{R}^3$. From the definition above, if $\left\| \tilde{\boldsymbol{\xi}}_{wp}(t) \right\| \to 0$, then $\left\| \tilde{\mathbf{p}}_{wp}(t) \right\| \to 0$ as $t \to \infty$. For simplicity, three assumptions are imposed as following.

Assumption 13.2 The velocities of the goal waypoint and obstacle are $\mathbf{v}_{wp} = \mathbf{v}_o = \mathbf{0}_{3 \times 1}$.

Assumption 13.3 The multicopter's initial position $\mathbf{p}\,(0) \in \mathbb{R}^3$ and velocity satisfy

$$\left\| \tilde{\boldsymbol{\xi}}_{\mathrm{o}}\,(0) \right\| - r_{\mathrm{o}} > 0$$
$$\left\| \mathbf{p}\,(0) - \mathbf{p}_{\mathrm{o}} \right\| - r_{\mathrm{o}} > 0.$$

Assumption 13.4 The goal waypoint $\mathbf{p}_{\mathrm{wp}}$ satisfies

$$\frac{1}{\left\| \left(\mathbf{p}_{\mathrm{wp}} - \mathbf{p}_{\mathrm{o}} \right) + k_0 \left(\mathbf{v}_{\mathrm{wp}} - \mathbf{v}_{\mathrm{o}} \right) \right\| - r_{\mathrm{o}}} \approx 0.$$

Assumption 13.2 implies that the goal waypoint and obstacle are both static. Assumption 13.3 implies that the multicopter is not located in the obstacle and its velocity is small or constrained initially. Based on Assumption 13.2, Assumption 13.4 implies that the goal waypoint $\mathbf{p}_{\mathrm{wp}}$ has a certain distance from the obstacle.

(2) *Main results*

Theorem 13.5 *Under Assumptions 13.2–13.4, suppose the pseudocontrol input is designed as*

$$\mathbf{u} = -\frac{1}{k_0} sat_{\mathrm{gd}} \left(a \tilde{\boldsymbol{\xi}}_{\mathrm{wp}} - b \tilde{\boldsymbol{\xi}}_{\mathrm{o}}, a_0 \right) - \frac{1}{k_0} \mathbf{v} \tag{13.9}$$

where

$$a = k_1$$
$$b = k_2 \frac{1}{\left(\left\| \tilde{\boldsymbol{\xi}}_{\mathrm{o}} \right\| - r_{\mathrm{o}} \right)^2} \frac{1}{\left\| \tilde{\boldsymbol{\xi}}_{\mathrm{o}} \right\|}$$

and $k_1, k_2 \in \mathbb{R}_+$. *Then* $\lim_{t \to \infty} \left\| \tilde{\mathbf{p}}_{\mathrm{wp}}\,(t) \right\| = 0$ *and* $\left\| \mathbf{p}_{\mathrm{o}} - \mathbf{p}\,(t) \right\| \geq r_{\mathrm{o}}, t \in [0, \infty)$ *for almost* $\tilde{\mathbf{p}}_{\mathrm{wp}}(0)$.[1]

Proof Based on Assumption 13.2, the derivative of filtered errors is

$$\dot{\tilde{\boldsymbol{\xi}}}_{\mathrm{wp}} = \bar{\mathbf{u}}$$
$$\dot{\tilde{\boldsymbol{\xi}}}_{\mathrm{o}} = \bar{\mathbf{u}}. \tag{13.10}$$

In order to investigate the convergence to the goal waypoint and the obstacle avoidance behavior, a function is defined as follows:

$$V_2 = \underbrace{\frac{k_1}{2} \tilde{\boldsymbol{\xi}}_{\mathrm{wp}}^{\mathrm{T}} \tilde{\boldsymbol{\xi}}_{\mathrm{wp}}}_{\text{Get close to the waypoint}} + \underbrace{k_2 \frac{1}{\left\| \tilde{\boldsymbol{\xi}}_{\mathrm{o}} \right\| - r_{\mathrm{o}}}}_{\text{Avoid the obstacle}} \tag{13.11}$$

where $k_1, k_2 \in \mathbb{R}_+$. By Assumption 13.3, the "avoid the obstacle" term in Eq. (13.11) is bounded at the initial moment of time. If the function V_2 is always bounded, then

---

[1] $\mathbf{p}_{\mathrm{wp}}$ is a stable equilibrium with probability 1, and other equilibriums are unstable with probability 1.

$$\left\|\tilde{\boldsymbol{\xi}}_\mathrm{o}\right\| - r_\mathrm{o} \neq 0$$

i.e.,

$$\left\|\tilde{\boldsymbol{\xi}}_\mathrm{o}\right\| > r_\mathrm{o}.$$

The derivative of V_2 along the solution to (13.10) is

$$\dot{V}_2 = k_1 \tilde{\boldsymbol{\xi}}_\mathrm{wp}^\mathrm{T} \bar{\mathbf{u}} - k_2 \frac{1}{\left(\left\|\tilde{\boldsymbol{\xi}}_\mathrm{o}\right\| - r_\mathrm{o}\right)^2} \frac{\tilde{\boldsymbol{\xi}}_\mathrm{o}^\mathrm{T}}{\left\|\tilde{\boldsymbol{\xi}}_\mathrm{o}\right\|} \bar{\mathbf{u}}$$
$$= \left(a\tilde{\boldsymbol{\xi}}_\mathrm{wp} - b\tilde{\boldsymbol{\xi}}_\mathrm{o}\right)^\mathrm{T} \bar{\mathbf{u}}.$$

Since the defined pseudoinput satisfies

$$\bar{\mathbf{u}} = -\mathrm{sat}_\mathrm{gd}\left(a\tilde{\boldsymbol{\xi}}_\mathrm{wp} - b\tilde{\boldsymbol{\xi}}_\mathrm{o}, a_0\right) \tag{13.12}$$

$\dot{V}_2$ becomes

$$\dot{V}_2 = -\left(a\tilde{\boldsymbol{\xi}}_\mathrm{wp} - b\tilde{\boldsymbol{\xi}}_\mathrm{o}\right)^\mathrm{T} \mathrm{sat}_\mathrm{gd}\left(a\tilde{\boldsymbol{\xi}}_\mathrm{wp} - b\tilde{\boldsymbol{\xi}}_\mathrm{o}, a_0\right) \leq 0$$

where

$$a = k_1$$
$$b = k_2 \frac{1}{\left(\left\|\tilde{\boldsymbol{\xi}}_\mathrm{o}\right\| - r_\mathrm{o}\right)^2} \frac{1}{\left\|\tilde{\boldsymbol{\xi}}_\mathrm{o}\right\|}.$$

First, the reason why the multicopter is able to avoid the obstacle is given. As $V_2(0) > 0$ and $\dot{V}_2 \leq 0$, the function V_2 satisfies $V_2(t) \leq V_2(0)$, $t \in [0, \infty)$. This implies that

$$\frac{1}{\left\|\tilde{\boldsymbol{\xi}}_\mathrm{o}\right\| - r_\mathrm{o}} \leq \frac{1}{k_2} V_2(0) < \infty.$$

Above equation can be rewritten as

$$0 < \frac{k_2}{V_2(0)} \leq \left\|\tilde{\boldsymbol{\xi}}_\mathrm{o}\right\| - r_\mathrm{o}.$$

Then, $\left\|\tilde{\boldsymbol{\xi}}_\mathrm{o}\right\| > r_\mathrm{o}$. Define $\tilde{\mathbf{p}}_\mathrm{o} \triangleq \mathbf{p} - \mathbf{p}_\mathrm{o}$. Next, the fact that $\|\tilde{\mathbf{p}}_\mathrm{o}\| > r_\mathrm{o}$ will be proved based on $\left\|\tilde{\boldsymbol{\xi}}_\mathrm{o}\right\| > r_\mathrm{o}$. According to the definition of $\tilde{\boldsymbol{\xi}}_\mathrm{o}$, one has

$$\tilde{\mathbf{p}}_\mathrm{o} + k_0 \dot{\tilde{\mathbf{p}}}_\mathrm{o} = \tilde{\boldsymbol{\xi}}_\mathrm{o}.$$

Based on it, one has

$$\frac{\mathrm{d}\tilde{\mathbf{p}}_\mathrm{o}^\mathrm{T}\tilde{\mathbf{p}}_\mathrm{o}}{\mathrm{d}t} = -\frac{2}{k_0}\tilde{\mathbf{p}}_\mathrm{o}^\mathrm{T}\tilde{\mathbf{p}}_\mathrm{o} + \frac{2}{k_0}\tilde{\mathbf{p}}_\mathrm{o}^\mathrm{T}\tilde{\boldsymbol{\xi}}_\mathrm{o}. \tag{13.13}$$

Since $\tilde{\mathbf{p}}_o^T \tilde{\mathbf{p}}_o = \|\tilde{\mathbf{p}}_o\|^2$, Eq. (13.13) further becomes

$$\frac{d \|\tilde{\mathbf{p}}_o\|}{dt} = -\frac{1}{k_0} \|\tilde{\mathbf{p}}_o\| + \frac{1}{k_0}\alpha \tag{13.14}$$

where $\alpha = \tilde{\mathbf{p}}_o^T \tilde{\boldsymbol{\xi}}_o / \|\tilde{\mathbf{p}}_o\|$. The solution to equation (13.14) is

$$\|\tilde{\mathbf{p}}_o(t)\| = e^{-\frac{1}{k_0}t} \|\tilde{\mathbf{p}}_o(0)\| + \frac{1}{k_0} \int_0^t e^{-\frac{1}{k_0}(t-s)} \alpha \, ds. \tag{13.15}$$

Since $\left\| \tilde{\boldsymbol{\xi}}_o \right\| > r_o$, the inequality $|\alpha| > r_o > 0$ holds. Furthermore, $\alpha > r_o > 0$. Otherwise, $\|\tilde{\mathbf{p}}_o(\infty)\| < 0$ according to (13.15). In this case, one has

$$\|\tilde{\mathbf{p}}_o(t)\| > e^{-\frac{1}{k_0}t} r_o + r_o \int_0^t e^{-\frac{1}{k_0}(t-s)} ds$$

$$= r_o.$$

This implies that the multicopter can avoid the obstacle with the controller given by Eq. (13.12). According to Eqs. (13.8) and (13.12), the pseudoinput is further written as

$$\mathbf{u} = -\frac{1}{k_0}\mathbf{v} - \frac{1}{k_0} \text{sat}_{\text{gd}} \left(a\tilde{\boldsymbol{\xi}}_{\text{wp}} - b\tilde{\boldsymbol{\xi}}_o, a_0 \right). \tag{13.16}$$

Next, the reason why the multicopter is able to arrive at the goal waypoint $\mathbf{p}_{\text{wp}}$ is given. Now, $\dot{V}_2 = 0$ if and only if $\bar{\mathbf{u}} = 0_{3\times 1}$. Then,

$$a \left(\mathbf{p} - \mathbf{p}_{\text{wp}} + k_0\mathbf{v} \right) + b \left(\mathbf{p} - \mathbf{p}_o + k_0\mathbf{v} \right) = 0_{3\times 1}. \tag{13.17}$$

Since

$$\dot{\mathbf{v}} = -\frac{1}{k_0}\mathbf{v} - \frac{1}{k_0}\bar{\mathbf{u}}$$

this implies that the system cannot get "stuck" at an equilibrium other than $\mathbf{v} = 0_{3\times 1}$. Thus, according to (13.17), the invariant set theorem indicates that the system converges globally to $(\mathbf{p}, \mathbf{v})$, where $\mathbf{p}$ is the solution to

$$a \left(\mathbf{p} - \mathbf{p}_{\text{wp}} \right) + b \left(\mathbf{p} - \mathbf{p}_o \right) = 0_{3\times 1} \tag{13.18}$$

and $\mathbf{v} = 0_{3\times 1}$. The parameters $k_0, k_1, k_2 \in \{0\} \cup \mathbb{R}_+$ will achieve a trade-off between convergence to the goal waypoint and obstacle avoidance.

According to Eq. (13.18), equilibrium points obviously sit on the straight line through $\mathbf{p}_{\text{wp}}$ and $\mathbf{p}_o$. As shown in Fig. 13.7a, the straight line is divided into "A half-line," "B segment," and "C half-line." The goal waypoint $\mathbf{p}_{\text{wp}}$ is assigned an attractive potential, while the obstacle $\mathbf{p}_o$ is assigned a repulsive potential.

In fact, there are two equilibrium points, one lying in "A half-line" and the other in "C half-line." By Assumption 13.4, the equilibrium point in "A half-line" is $\mathbf{p} = \mathbf{p}_{\text{wp}}$. The local stability of $\mathbf{p}_{\text{wp}}$ is discussed in the following. Besides, by Assumption 13.4, the error dynamics are

Fig. 13.7 Schematic illustration of equilibrium points

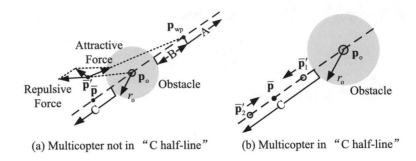

(a) Multicopter not in "C half-line"   (b) Multicopter in "C half-line"

$$\dot{\tilde{\mathbf{p}}} = \mathbf{v}$$

$$\dot{\mathbf{v}} = -\frac{1}{k_0}\mathbf{v} - \frac{k_1}{k_0}\tilde{\mathbf{p}}.$$

Therefore, equilibrium point $\mathbf{p}_{wp}$ is locally exponentially stable.

Without loss of generality, it is assumed that the other solution lying in "C half-line" is $\mathbf{p} = \bar{\mathbf{p}}_1'$, as shown in Fig. 13.7a. On the one hand, if the multicopter deviates from "C half-line," for instance it reaches $\mathbf{p} = \bar{\mathbf{p}}'$, then the sum of the attractive force and the repulsive force will make the multicopter further keep away from the "C half-line." On the other hand, as shown in Fig. 13.7b, when the multicopter gets close to the obstacle along "C half-line," for instance it reaches $\mathbf{p} = \bar{\mathbf{p}}_1'$, the repulsive force will dominate because of the term $1/\left(\left\|\tilde{\boldsymbol{\xi}}_o\right\| - r_o\right)^2$ in Eq. (13.16). As a result, a relatively large repulsive force will push it back to $\bar{\mathbf{p}}$. By contrary, if the multicopter reaches $\mathbf{p} = \bar{\mathbf{p}}_2'$, the attractive force will dominate. As a result, a relatively large attractive force will pull it back toward $\bar{\mathbf{p}}$. However, in practice, a multicopter will never strictly stay in "C half-line"; namely, the measure of "C half-line" in 3D space equals 0 or the probability is 0. Therefore, the equilibrium point $\bar{\mathbf{p}}$ is unstable with probability 1; i.e., any small deviation from the "C half-line" will drive the multicopter away from $\bar{\mathbf{p}}$. In conclusion, the solution $\mathbf{p} = \mathbf{p}_{wp}$ lying in "A half-line" is the only stable equilibrium point. It is also globally asymptotically stable with probability 1, namely $\lim_{t\to\infty}\left\|\tilde{\mathbf{p}}_{wp}(t)\right\| = 0$ for almost $\tilde{\mathbf{p}}_{wp}(0)$. $\qquad\square$

From the analysis above, the controller (13.16) can solve the obstacle avoidance problem and is rewritten as a PD controller

$$\mathbf{u} = -\frac{1}{k_0}(\mathbf{p} - \mathbf{p}_d) - \frac{1}{k_0}\mathbf{v}$$

where

$$\mathbf{p}_d = \mathbf{p} + \text{sat}_{gd}\left(-a\tilde{\boldsymbol{\xi}}_{wp} + b\tilde{\boldsymbol{\xi}}_o, a_0\right). \qquad (13.19)$$

In order to figure out the physical meaning of $\mathbf{p}_d$, let $k_0 = k_1 = k_2 = 1$ and $\mathbf{v} = \mathbf{0}_{3\times 1}$. Then, $\mathbf{p}_d$ has a concise form

$$\mathbf{p}_d = \mathbf{p} + \text{sat}_{gd}\left(\mathbf{p}_{wp} - \mathbf{p} - \frac{1}{(\|\tilde{\mathbf{p}}_o\| - r_o)^2}\frac{1}{\|\tilde{\mathbf{p}}_o\|}(\mathbf{p}_o - \mathbf{p}), a_0\right).$$

Here, the physical meaning of $\mathbf{p}_d$ is presented in Fig. 13.8b, where

$$\mathbf{p}_{od} - \mathbf{p} = -\frac{1}{(\|\tilde{\mathbf{p}}_o\| - r_o)^2}\frac{1}{\|\tilde{\mathbf{p}}_o\|}(\mathbf{p}_o - \mathbf{p}).$$

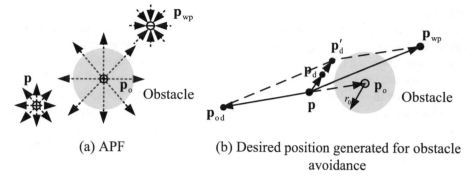

(a) APF | (b) Desired position generated for obstacle avoidance

Fig. 13.8 Schematic illustration of local path planning for obstacle avoidance. Here, $\mathbf{p}_{wp}$ and $\mathbf{p}_{od}$ are virtual goal waypoints produced by attractive potential field and repulsive potential field, respectively

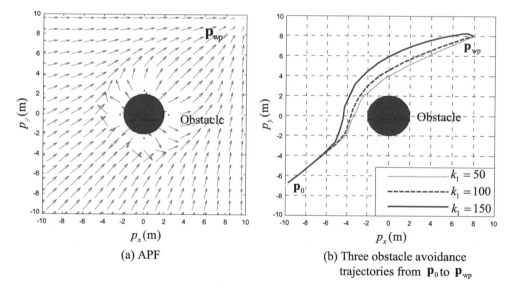

(a) APF | (b) Three obstacle avoidance trajectories from $\mathbf{P}_0$ to $\mathbf{P}_{wp}$

Fig. 13.9 Obstacle avoidance by the APF method

The local desired position $\mathbf{p}_d$ shown in (13.19) can be sent to the low-level flight control system in real time to realize obstacle avoidance.

(3) *Simulation*

Here, a simulation example is given. Let $\mathbf{p}_o = [0\ 0\ 0]^T$, $r_o = 2$, $\mathbf{p}_{wp} = [8\ 8\ 0]^T$, $a_0 = 1.5$, $k_0 = 2$, $k_1 = 0.5$, $k_2 = 10$. The resulting APF is shown in Fig. 13.9a. As shown, the obstacle is surrounded by repulsive potential field, while the goal waypoint is attractive to the multicopter. From the direction of arrows, it can be predicted that the multicopter will avoid the obstacle, meanwhile flying to the goal waypoint $\mathbf{p}_{wp}$. To investigate the relationship between parameters and the convergence performance, fix $k_0 = 2$ and $k_1 = 0.5$, and further, let $k_2 = 5$, $k_2 = 10$, $k_2 = 30$, respectively. If the initial location is $\mathbf{p}_0 = [-10\ -7\ 0]^T$, then the local path planner will produce three obstacle avoidance paths. The obstacle avoidance paths in simulation are shown in Fig. 13.9b. It is observed that a higher k_2 makes the multicopter circumvent the obstacle much earlier.

Besides the proposed methods, some references about collision avoidance are available, such as [7, 8].

13.1.3.4   Synthesis

In practice, when a multicopter sprays pesticides over a farmland, it needs not only to follow the desired path but also to avoid both fixed and moving obstacles. This problem is formulated as follows.

Let $\mathbf{p} \in \mathbb{R}^3$ be the current position of the multicopter, $\mathbf{p}_{\text{wp}} \in \mathbb{R}^3$ be the goal waypoint, and $\mathbf{p}_{\text{wp,last}} \in \mathbb{R}^3$ be the previous waypoint. Design a local path planner to produce the desired position $\mathbf{p}_{\text{d}} \in \mathbb{R}^3$ in real time, which can guide the multicopter to fly along the straight line connecting $\mathbf{p}_{\text{wp,last}} \in \mathbb{R}^3$ to $\mathbf{p}_{\text{wp}} \in \mathbb{R}^3$ until it arrives at the goal waypoint $\mathbf{p}_{\text{wp}}$, meanwhile avoiding the m_{o} obstacles modeled as spheres with radius $r_{\text{o},i} \in \mathbb{R}_+$ and center $\mathbf{p}_{\text{o},i} \in \mathbb{R}^3$, $i = 1, \ldots, m_{\text{o}}$.

The global potential field can be obtained by the superposition of potential fields of the path, the goal waypoint, and obstacles. Then, $\mathbf{p}_{\text{d}} \in \mathbb{R}^3$ is given by the following equation

$$\mathbf{p}_{\text{d}} = \mathbf{p} + \mathbf{p}_{\text{d,waypoint,path}} + \mathbf{p}_{\text{d,waypoint,obstacle}}$$

where

$$\mathbf{p}_{\text{d,waypoint,path}} = \text{sat}_{\text{gd}} \left(k_1 \left(\mathbf{p}_{\text{wp}} - \mathbf{p} \right) + k_2 (\mathbf{p}_{\text{wp,perp}} - \mathbf{p}), a_0 \right)$$

$$\mathbf{p}_{\text{wp,perp}} = \mathbf{p}_{\text{wp}} + \left(\mathbf{p}_{\text{wp,last}} - \mathbf{p}_{\text{wp}} \right) \frac{\left(\mathbf{p} - \mathbf{p}_{\text{wp}} \right)^{\text{T}} \left(\mathbf{p}_{\text{wp,last}} - \mathbf{p}_{\text{wp}} \right)}{\left\| \mathbf{p}_{\text{wp}} - \mathbf{p}_{\text{wp,last}} \right\|^2}$$

$$\mathbf{p}_{\text{d,waypoint,obstacle}} = \text{sat}_{\text{gd}} \left(k_3 \left(\mathbf{p}_{\text{wp}} - \mathbf{p} - k_0 \mathbf{v} \right) + \sum_{i=1}^{m_{\text{o}}} b_i \left(\mathbf{p} - \mathbf{p}_{\text{o},i} + k_0 \mathbf{v} \right), a_1 \right)$$

$$b_i = k_4 \frac{1}{\left(\left\| \mathbf{p} - \mathbf{p}_{\text{o},i} + k_0 \mathbf{v} \right\| - r_{\text{o},i} \right)^2} \frac{1}{\left\| \mathbf{p} - \mathbf{p}_{\text{o},i} + k_0 \mathbf{v} \right\|}, i = 1, 2, \ldots, m_{\text{o}}$$

and $k_0, k_1, k_2, k_3, k_4 \in \{0\} \cup \mathbb{R}_+$. As for the yaw, let it point to the next waypoint all the time. Define

$$\mathbf{e}_{\mathbf{p}} \triangleq \mathbf{p} - \mathbf{p}_{\text{d}}$$

Then

$$\psi_{\text{d}} = \text{atan2} \left(-\mathbf{e}_{\mathbf{p},1}, -\mathbf{e}_{\mathbf{p},2} \right)$$

where $\mathbf{e}_{\mathbf{p},i}$ is the ith element of $\mathbf{e}_{\mathbf{p}}$.

Although the proposed path following and collision avoidance algorithms are simple and can produce smooth desired path effectively, it has some problems for obstacle avoidance. When the goal waypoint is very close to obstacles, the multicopter may not be able to reach it. If there are more than one obstacle and they are close to each other, and the magnitudes of attractive and repulsive forces are equal, the multicopter may become stagnant or trapped. Some solutions to this problem have been proposed in [9–11]. To sum up, this proposed algorithm in this chapter is applicable under the condition where obstacles are sparse.

13.2   Semi-Autonomous Control

In the SAC, a multicopter's attitude stabilization or hover is realized by the autopilot, while the position control is often manually realized by a remote pilot with an RC transmitter. In the SAC manner, a Ground Control Station (GCS) is optional. Thus, the control command is a sum of two control inputs, one of which is generated by the RC transmitter and the other sent from the automatic controller of the autopilot. In fact, the SAC consists of AC and RC, and a multicopter in the SAC mode is controlled by either AC or RC. Generally, according to the degree of the AC, a multicopter in SAC can be in one of three main modes, i.e., "stabilize mode", "altitude hold mode", and "loiter mode".

13.2.1   Three Modes of SAC

13.2.1.1   Stabilize Mode

The *stabilize mode* allows a remote pilot to fly the multicopter manually, but self-levels the roll and pitch axis. Under RC, an RC transmitter is used to control its roll/pitch and then drive the multicopter to tilt toward the desired direction. When the remote pilot releases the roll and pitch control sticks, the multicopter automatically switches itself to AC. Then, its attitude will be stabilized, but the position drift will occur. During this process, the remote pilot will need to regularly give roll, pitch, and throttle commands to keep the multicopter in place as it is pushed around by wind. The throttle command controls the average motor speed to maintain the altitude. If the remote pilot puts the throttle control stick completely down, then the motors will operate at their minimum rate, and if the multicopter is in the air, it will lose altitude control and tumble. In addition, when the remote pilot releases the yaw control stick, the multicopter will maintain its current heading.

13.2.1.2   Altitude Hold Mode

In the *altitude hold mode*, a multicopter maintains a consistent altitude while allowing roll, pitch, and yaw to be controlled normally. When the throttle control stick is in the mid-throttle dead zone (40–60%), the multicopter automatically switches itself to AC (Fig. 13.10). Then, the throttle command is automatically given to maintain the current altitude, but the horizontal position drift will occur. AC can not only stabilize the attitude, but also keep the altitude. The remote pilot will need to regularly give roll and pitch commands to keep the multicopter in place. Going outside of the mid-throttle dead zone (i.e., below 40% or above 60% for example), the multicopter will enter RC, that is, the multicopter will descend or climb depending upon the deflection of the throttle control stick. When the throttle control stick is completely down, the multicopter will descend at a maximum permissible speed and if at the very top, it will climb at a maximum permissible speed. The altitude hold mode needs the support of height sensors, such as barometers or ultrasonic rang finders.

Fig. 13.10 Stick function and dead zone in RC transmitter

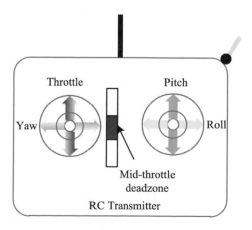

13.2.1.3 Loiter Mode

The *loiter mode* automatically attempts to maintain the current location, heading, and altitude. When the remote pilot releases the roll, pitch, and yaw control sticks and pushes the throttle control stick to the mid-throttle dead zone, the multicopter will automatically switch itself to AC and maintain the current location, heading and altitude. Precise GPS position, low magnetic interference on the compass, and low vibrations are all important in achieving a good hovering performance. The remote pilot can control the multicopter's position once by pushing the control sticks out of the midpoints. Horizontal location can be adjusted by the roll and pitch control sticks. When the remote pilot releases the control sticks, the multicopter will slow to a stop. Altitude can be controlled by the throttle control stick just as in the altitude hold mode. The heading can be set with the yaw control stick. The loiter mode needs the support from both the height sensors and position sensors such as GPS receiver and cameras.

From the analysis above, in the SAC, a multicopter is switched automatically between AC and RC. It is worth noting that RC has a higher priority than AC. This implies that if RC is active, AC will be disconnected to the multicopter. Only when a control stick is close to its midpoint, AC of the corresponding channel then takes over the control of the multicopter.

13.2.2 Radio Control

Define $(\cdot)_d$, $(\cdot)_{drc}$ to be desired values and RC values, respectively. According to symbols defined in Chap. 12, let $\theta_{drc} = u_\theta$, $\phi_{drc} = u_\phi$, $\dot{\psi}_{drc} = u_{\omega_z}$, and $f_{drc} = u_T$ for simplicity. In the SAC, the throttle/yaw control sticks and roll/pitch control sticks are manipulated to specify the desired total thrust $f_d = f_{drc}$, the desired yaw rate $\dot{\psi}_d = \dot{\psi}_{drc}$, the desired pitch angle $\theta_d = \theta_{drc}$, and the desired roll angle $\phi_d = \phi_{drc}$, respectively. For a direct-type RC transmitter, the throttle control stick can stop at any position when it is not pushed and the total thrust is proportional to the deflection of throttle control stick. For throttle control stick, the relationship between the deflection and the total thrust is shown in Fig. 13.11. Here, $\sigma_{drc} \in [0, 1]$ is the deflection of throttle control stick or throttle command, where $\sigma_{drc} = 0.5$ represents the midpoint of the throttle control stick. If $\sigma_{drc} \in [0.4, 0.6]$, then $f_{drc}(\sigma_{drc}) = 0$. Then, AC starts to introduce the feedback in altitude for altitude hold. Here, it is supposed that the feedforward mg has been given by AC, so f_{drc} starts from $-mg$ as shown in Fig. 13.11. The dead zone is used to reduce the effect by slight change of the throttle control stick. Similarly, other control sticks also have dead zones.

Fig. 13.11 The relationship between the stick deflection and the total thrust

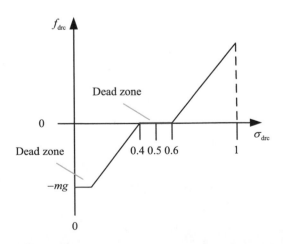

In addition, some multicopters, such as the Phantom quadcopter launched by DJI, use the increment-type throttle to control the total thrust. The increment-type throttle has two features. One is that the control stick can turn back to the midpoint automatically if it is released. The other is that the throttle control stick position is proportional to the desired vertical speed v_{zd} or thrust rate instead of the desired total thrust. For simplicity, this chapter just considers the direct-type throttle RC transmitter because this kind of RC transmitter is adopted by most open source autopilots.

13.2.3 Automatic Control

When all control sticks are close to their midpoints, as shown in Fig. 13.10, the AC takes over the control of the multicopter. In fact, no matter which mode of SAC is active, the decision-making module of an autopilot will produce the desired hover position $\mathbf{p}_{dac}$ and yaw ψ_{dac}. On the other hand, the state estimation module of the autopilot will produce the position estimate $\hat{\mathbf{p}} = [\,\hat{p}_x\ \hat{p}_y\ \hat{p}_z\,]^T$ and attitude estimate $\hat{\boldsymbol{\Theta}} = [\,\hat{\phi}\ \hat{\theta}\ \hat{\psi}\,]^T$. The AC aims to drive the multicopter so that $\lim_{t\to\infty}\left\|\hat{\mathbf{p}}(t) - \mathbf{p}_{dac}(t)\right\| = 0$ and $\lim_{t\to\infty}\left|\hat{\psi}(t) - \psi_{dac}(t)\right| = 0$ are satisfied. Thus, AC structures for the three modes of SAC could be the same. However, because different sensors are used, the estimation accuracies are different for the three modes, namely the stabilize mode, altitude hold mode, and loiter mode. In the following, the principle of the three modes is investigated. It should be noted that the proposed method is only one way to realize the three modes.

13.2.3.1 Stabilize Mode
The stabilize mode produces the desired thrust and moments according to the desired position $\mathbf{p}_d = \hat{\mathbf{p}}$ and desired yaw $\psi_d = \hat{\psi}$. As shown in Fig. 13.12, the autopilot can produce the desired position $\mathbf{p}_d$, and yaw angle ψ_d according to the RC commands.

By recalling the controllers proposed in (11.14) and (11.19) in Chap. 11, the horizontal position controller is

$$\boldsymbol{\Theta}_{hd} = -g^{-1}\mathbf{A}_\psi^{-1}\left(-\mathbf{K}_{phd}\dot{\hat{\mathbf{p}}}_h - \mathbf{K}_{php}\left(\hat{\mathbf{p}}_h - \mathbf{p}_{hd}\right)\right)$$

and the altitude controller is

$$f_d = mg - m\left(-k_{p_zd}\dot{\hat{p}}_z - k_{p_zp}\left(\hat{p}_z - p_{zd}\right)\right).$$

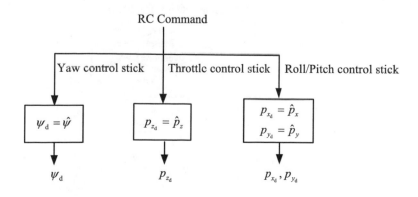

Fig. 13.12 The principle of producing the desired position and yaw angle in stabilize mode

Since $\mathbf{p}_d = \hat{\mathbf{p}}$, the horizontal position controller and altitude controller become

$$\Theta_{hd} = g^{-1}\mathbf{A}_\psi^{-1}\mathbf{K}_{phd}\dot{\hat{\mathbf{p}}}_h$$

$$f_d = mg + mk_{p_z d}\dot{\hat{p}}_z.$$

Generally, let $\Theta_{hd} = \begin{bmatrix} \phi_d & \theta_d \end{bmatrix}^T = \mathbf{0}_{2\times1}$, because $\dot{\hat{\mathbf{p}}}_h$ may be unavailable or inaccurate, or the magnetometer is unavailable. This implies that AC can make the multicopter self-level. Moreover, AC cannot hold the altitude anymore but just controls the vertical speed to zero. A simple yaw controller has the form as

$$\tau_z = -k_\psi \left(\hat{\psi} - \psi_d\right) - k_{\dot\psi}\dot{\psi}.$$

The yaw angle estimate $\hat{\psi}$ may be inaccurate and drifts slowly in a range because the magnetometer is sensitive to the environment. Therefore, if the estimate to the yaw angle is used as a feedback, the heading may oscillate. On the other hand, the yaw rate measured by gyroscopes is fairly accurate. So a simple way to keep the heading is to make the yaw rate be zero. This is realized by setting $\psi_d = \hat{\psi}$. As a result, the yaw controller becomes

$$\tau_z = -k_{\dot\psi}\dot{\psi}.$$

Stabilize mode cannot make the multicopter hover steadily due to lack of feedback of the horizontal position signal. This mode is often used when without GPS receiver and height sensors or they fail.

13.2.3.2 Altitude Hold Mode

The altitude hold mode produces the desired thrust and moment according to the desired altitude $p_{z_d} = p_{z_{dold}}$, horizontal position $\mathbf{p}_{hd} = \hat{\mathbf{p}}_h$ and yaw angle $\psi_d = \hat{\psi}$, where $\mathbf{p}_{hd} = \hat{\mathbf{p}}_h$ implies $\theta_d = \phi_d = 0$.

As shown in Fig. 13.13, the time, denoted by t_{z_d}, is recorded when the throttle control stick turns back to the midpoint, and then, the altitude estimate $\hat{p}_z\left(t_{z_d}\right)$ is saved as $p_{z_{dold}} = \hat{p}_z\left(t_{z_d}\right)$. At the same time, the altitude hold mode starts to hold the multicopter's altitude at $p_{z_d} = p_{z_{dold}}$. Just like the stabilize mode, the altitude hold mode cannot make the multicopter hover for lack of feedback in the horizontal position. The altitude hold mode is often used when the height sensors are available while position sensors or electronic compasses are unavailable.

Fig. 13.13 The principle of producing the desired position and yaw angle in altitude hold mode

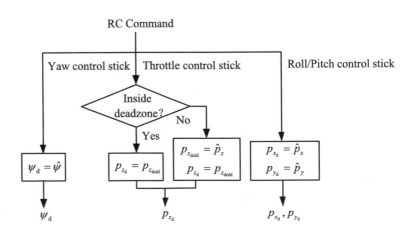

13.2.3.3 Loiter Mode

As shown in Fig. 13.14, the loiter mode produces the desired thrust and moment according to the desired position $\mathbf{p}_d = \mathbf{p}_{dold}$ and yaw angle $\psi_d = \psi_{dold}$. The time, denoted by $t_{\psi_d}, t_{z_d}, t_{x_d}, t_{y_d}$, is recorded when the four control sticks turn back to the midpoint, respectively.

The estimates are saved to be desired ones that

$$\psi_{dold} = \hat{\psi}\left(t_{\psi_d}\right)$$
$$p_{z_{dold}} = \hat{p}_z\left(t_{z_d}\right)$$
$$p_{x_{dold}} = \hat{p}_x\left(t_{x_d}\right)$$
$$p_{y_{dold}} = \hat{p}_y\left(t_{y_d}\right).$$

At the same time, the loiter mode starts to control the multicopter to hover at $\mathbf{p}_d = \mathbf{p}_{dold}$ and $\psi_d = \psi_{dold}$. The loiter mode is often used when the height sensors, position sensors, and electronic compasses are all available.

13.2.4 Switching Logic Between RC and AC

The closed-loop control block diagram under the SAC mode is shown in Fig. 13.15. Based on Fig. 13.15, the switching logic between RC and AC is discussed.

13.2.4.1 Yaw Command Switching Logic
In the SAC, as shown in Fig. 13.15, the total yaw command is

$$\psi_d = \psi_{dac} + \psi_{drc}$$

where ψ_{dac} is produced by AC in autopilot, and ψ_{drc} represents the command from the RC transmitter. As described in the previous section, $\psi_{dac} = \hat{\psi}$ or $\psi_{dac} = \psi_{dold}$ depends on which mode (stabilize mode, altitude hold mode, or loiter mode) the multicopter is in. RC command ψ_{drc} is presented as

$$\psi_{drc} = \dot{\psi}_{drc}\Delta t$$

where $\dot{\psi}_{drc}$ is the yaw control command, and Δt is a command period or can be considered as an adjustable gain.

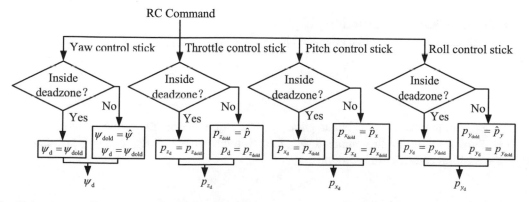

Fig. 13.14 The principle of producing the desired position and yaw angle in loiter mode

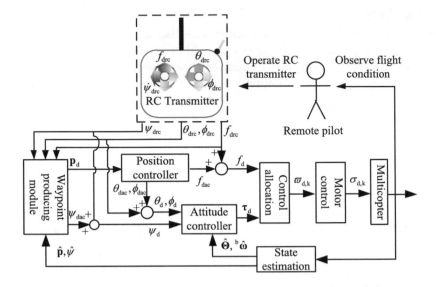

Fig. 13.15 The closed-loop control block diagram under the SAC mode

When the yaw control stick is at its midpoint, $\dot{\psi}_{drc} = 0$, then $\psi_{drc} = 0$, and the autopilot will control the yaw to ψ_{dac} depending on which mode the autopilot is in. If the yaw control stick leaves its midpoint, then $\psi_{dac} = \hat{\psi}$; namely, the RC will take over the control of the yaw channel completely.

13.2.4.2 Throttle Command Switching Logic

As shown in Fig. 13.15, the total throttle command is

$$f_d = f_{dac} + f_{drc}(\sigma_{drc})$$

where f_{dac} is produced by AC in autopilot, and $f_{drc}(\sigma_{drc})$ represents the command from the RC transmitter. More precisely, the waypoint producing module produces the desired position according to the current mode (stabilize mode, altitude hold mode, or loiter mode) the multicopter is in, and then, the position controller in autopilot produces the f_{dac}. RC command function $f_{drc}(\sigma_{drc})$ is presented in Fig. 13.11.

When the throttle control stick is at its midpoint, $f_{drc}(\sigma_{drc}) = 0$, the autopilot will control the position according to the total throttle command f_{dac}. Once the throttle control stick leaves its midpoint, $p_{z_d} = \hat{p}_z$, AC does not offer the feedback in altitude except for velocity feedback, and RC will take over the control of the altitude channel completely.

13.2.4.3 Roll/Pitch Command Switching Logic

As shown in Fig. 13.15, the total roll/pitch command is

$$\theta_d = \theta_{dac} + \theta_{drc}$$
$$\phi_d = \phi_{dac} + \phi_{drc}$$

where θ_{dac} and ϕ_{dac} are produced by AC in autopilot; θ_{drc} and ϕ_{drc} represent the command from the RC transmitter. Similar to the throttle command f_{dac}, the roll/pitch commands θ_{dac} and ϕ_{dac} are also produced by the position controller in autopilot.

Take the pitch control stick as an example. When the pitch control stick is at its midpoint, $\theta_{\text{drc}} = 0$, the autopilot will control the pitch to θ_{dac} depending on which mode the autopilot is in. Once the pitch control stick leaves its midpoint, $p_{x_{\text{d}}} = \hat{p}_x$. Then, the AC does not offer the feedback in horizontal location except for velocity feedback. This implies that RC will take over the control of the horizontal channel completely.

13.3   Summary

Mission planning or automatic flight control for single multicopter is relatively simple. In recent years, researchers have successfully presented the cooperative control among multiple multicopters in a controlled laboratory environment. Compared with a single multicopter, multiple multicopters can be cooperated more efficiently to accomplish a given task. The fully-autonomous mission planning will facilitate the manipulation of multiple multicopters in practice. In order to realize it, the following prerequisites have to be satisfied: (1) Each multicopter is adequately reliable; (2) the position and attitude of each multicopter are estimated accurately; (3) the endurance of flight is long enough, and each multicopter has a Battery Management System (BMS) to evaluate its health; and (4) fast-charging stations are built, and the charging is accomplished automatically.

Exercises

13.1 In Sect. 13.1.3.2, a straight-line path following algorithm was introduced. Similar to it, design a path following algorithm to guide the multicopter to follow a horizontal circular path which is modeled as a circle with radius $r_{\text{c}} \in \mathbb{R}_+$ and center $\mathbf{p}_{\text{c}} \in \mathbb{R}^2$.

13.2 Let the initial location $\mathbf{p}_{\text{wp,last}} = [-8 \ -8 \ 0]^{\text{T}}$, the target location $\mathbf{p}_{\text{wp}} = [8 \ 8 \ 10]^{\text{T}}$, and the obstacle be modeled as a cylinder of radius $r_{\text{o}} = 4$ with axis given by the line through $\mathbf{p}_{\text{o,1}} = [0 \ 0 \ -3]^{\text{T}}$ and $\mathbf{p}_{\text{o,2}} = [0 \ 0 \ 3]^{\text{T}}$. Design an algorithm to achieve the path following and obstacle avoidance.

13.3 As shown in Fig. 13.16, a multicopter is required to fly in a 2D rectangular area keeping away from the boundary. This is the so-called *geographic fence*. If the multicopter is considered as a mass point, then

$$\dot{\mathbf{p}} = \mathbf{v}$$
$$\dot{\mathbf{v}} = \mathbf{u}_{\text{ac}} + \mathbf{u}_{\text{rc}}$$

Fig. 13.16 2D geographic fence

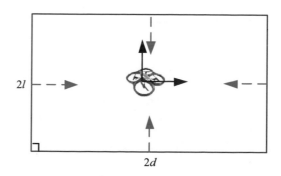

$2l$

$2d$

where $\mathbf{p}$, $\mathbf{v} \in \mathbb{R}^2$ stand for the position and velocity of the multicopter, respectively, $\mathbf{u}_{ac} \in \mathbb{R}^2$ is auxiliary force which can keep the multicopter within rectangular area, and $\mathbf{u}_{rc} \in \mathbb{R}^2$ is the control command by remote pilots. By using the APF method introduced in Sect. 13.1.3, design $\mathbf{u}_{ac}$ to realize the geographic fence.

13.4 If the magnetic interference is high, the yaw angle estimate will oscillate. What will happen to multicopters when they are in the stabilize mode, altitude hold mode, and loiter mode, respectively? Explain the reasons.

13.5 In Sect. 13.2.4.2, the throttle command switching logic with direct-type RC transmitter was introduced. Design the throttle command switching logic of altitude channel when an increment-type RC transmitter is used.

References

1. Zhang H, Wei ZB, Quan Q et al (2013) Ardrone quadrotor modeling and additive-output-decomposition based trajectory tracking. In: IEEE international conference of 32th Chinese control conference, Xi'an, pp 4946–4951. (In Chinese)
2. Gilbert EG, Johnson DW (1985) Distance functions and their application to robot path planning in the presence of obstacles. IEEE Robot Autom Mag 1(1):21–30
3. Barnhart RK, Hottman SB et al (2011) Introduction to unmanned aircraft systems. CRC Press, Boca Raton
4. Tsourdos A, White B, Shanmugavel M (2011) Cooperative path planning of unmanned aerial vehicles. Wiley, UK
5. Khatib O (1986) Real-time obstacle avoidance for manipulators and mobile robots. Int J Robot Res 5(1):90–98
6. Weisstein EW (2016) Point-line distance 3-dimensional. The Wolfram MathWorld. http://mathworld.wolfram.com/Point-LineDistance3-Dimensional.html. Accessed 14 May 2016
7. Hoffmann GM, Tomlin CJ (2008) Decentralized cooperative collision avoidance for acceleration constrained vehicles. In: IEEE conference on decision and control, Cancun, pp 4357–4363
8. Alonso-Mora J, Naegeli T, Siegwart R et al (2015) Collision avoidance for aerial vehicles in multi-agent scenarios. Auton Robot 39(1):101–121
9. Vadakkepat P, Tan KC, Wang ML (2000) Evolutionary artificial potential fields and their application in real time robot path planning. In: Proceedings of the IEEE congress on evolutionary computation, San Diego, pp 256–263
10. Warren CW (1990) Multiple robot path coordination using artificial potential fields. In: Proceedings of the IEEE international conference on robotics and automation, Cincinnati, pp 500–505
11. Luh GC, Liu WW (2007) Motion planning for mobile robots in dynamic environments using a potential field immune network. J Sys Control Eng 221(7):1033–1045

Health Evaluation and Failsafe

14

For multicopters, failures cannot be avoided, including communication breakdown, sensor failure and propulsion system anomaly, etc. These failures may abort missions, crash multicopters, and moreover, injure or even kill people. In order to guarantee safety, the multicopter's decision-making module should prevent or mitigate unsafe consequences of system's failures. For such a purpose, some flight modes are defined and switched to according to health evaluation results of each component. Concretely, if there are safety risks in key components of a multicopter before taking off, then previously arranged flight missions need to be aborted. Furthermore, if there is an anomaly or a fault in key components of a multicopter during the flight, then the multicopter should return home or land immediately. This chapter mainly considers the safety issue of multicopters and aims to answer the questions below:

What kind of events are involved in safety issue? How are these events dealt with?

The answer to these questions involves an introduction to potential safety issues, some health evaluation methods, failsafe suggestions, and a failsafe decision-making case.

© Springer Nature Singapore Pte Ltd. 2017

Q. Quan, *Introduction to Multicopter Design and Control*, DOI 10.1007/978-981-10-3382-7_14

Nest Before the Rain

There are many old idioms in China emphasizing the importance of safety, prevention and plan. "未雨绸缪", which means in fair weather preparing for foul, originated from "*Book of Poetry*" that "Before the sky was dark with rain, I gathered the roots of the mulberry tree, and bound round and round my window and door. Now ye people below, dare any of you despise my house?(The Chinese is "迨天之未阴雨，彻彼桑土，绸缪牖户", translated by James Legge, from http://ctext.org/book-of-poetry/chi-xiao)" This proverb refers to the value of precautions. Another idiom "居安思危" has the same meaning: it is better to prepare for danger in times of peace. This idiom originated from "*The Zuo's Commentary*" that "In a position of security, think of peril. If you think thus, you will make preparation against the danger, and with the preparation there will be no calamity.(The Chines is "居安思危，思则有备，有备无患，敢以此规". The English translation is available in http://www2.iath.virginia.edu)". "*Book of Rites—Doctrine of the Mean*" has the same opinion that "In all things success depends on previous preparation, and without such previous preparation there is sure to be failure.(The Chinese is"凡事预则立，不预则废", translated by James Legge, http://ctext.org/liji/zhong-yong/ens.)" Except prevention, we should also pay attention to some tiny problems. "防微杜渐" illustrates this opinion that we should solve the tiny problems before they develop to be big ones.

14.1   Purpose and Significance of Decision-Making Mechanism

The purpose and significance for multicopters to have decision-making modules (in the form of flight modes) are as follows:

(1) Bringing flight process under remote pilot's control. Since a multicopter needs to interact with the remote pilot, it is convenient to categorize the behavior of the multicopter into finite flight modes. This allows remote pilots to conveniently understand the behavior of the multicopter during flight and make a decision for the next action.

(2) Adapting to different flight missions. 1) Due to the ground effect, multicopter models are different during taking off, landing, and normal flight. As a result, different controllers need to be allocated according to different flight modes. 2) In Semi-Autonomous Control (SAC) mode, the multicopter mode switches between Radio Control (RC, Radio Controlled also designated as RC) mode and Automatic Control (AC) mode.

(3) Adapting to different anomalies. When a multicopter is flying, failures may occur in sensors, communication channels, or the power supply system. When a failure occurs, the multicopter should be reassigned to an appropriate control target by a corresponding controller. Both of these are realized by the mode transition for the sake of safety. This is also consistent with remote pilots' experience.

(4) Better interpretation of user demands. The working status of a multicopter can be displayed more visually by mode switches. Moreover, the user demands can be better demonstrated by flight modes. The latter can also facilitate the interpretation of the user demands.

14.2   Safety Issues

In the first place, the major types of failures that may cause accidents will be introduced. Here, the following three types of failures are mainly considered, which are communication breakdown, sensor failure, and propulsion system anomaly.

14.2.1   Communication Breakdown

Communication breakdown mainly refers to a contact anomaly between the RC transmitter and the multicopter, or between the Ground Control Station (GCS) and the multicopter. Such failures can be categorized as follows:

(1) RC transmitter not calibrated. An RC transmitter without calibration implies that the remote pilot does not calibrate the RC transmitter before the first flight of the multicopter. As a result, the flight control system cannot recognize the user instructions given by the sticks of the RC transmitter. This will lead to flight accidents due to the misinterpretation of the user instructions.

(2) Loss of RC. Loss of RC implies that the RC transmitter is unable to communicate with the corresponding RC receiver onboard before the multicopter takes off or during flight. The loss of RC will result in the multicopter going out of control and lead to an accident. The reasons of the loss of RC include but are not limited to: 1) The remote pilot turns off the RC transmitter. 2) The multicopter flies outside of the control range of the RC transmitter. 3) The RC transmitter or the receiver onboard loses power or fails. 4) The wires connecting the onboard RC receiver to the autopilot are broken.

(3) Loss of GCS. Loss of GCS implies that the GCS is unable to communicate with the corresponding multicopter before the multicopter takes off or during flight. The loss of GCS will cause the multicopter to fail to reach the mission position, and then the task fails. The reasons of the loss of GCS include but are not limited to: 1) The remote pilot turns off the GCS. 2) The multicopter flies outside of the control range of the GCS. 3) The GCS or the corresponding signal receiver onboard loses power or fails. 4) The wires connecting the radio telemetry receiver for the GCS to the autopilot are broken.

14.2.2 Sensor Failure

Sensor failure mainly implies that a sensor on the multicopter cannot measure accurately or cannot work properly. Such failures can be categorized as follows:

(1) Barometer failure. Barometer failure will cause a multicopter to fail to measure the flight altitude accurately. The reasons for the barometer failure include but are not limited to: 1) Barometer hardware failure. 2) Height measurement results from barometers and other height measurement sensors (ultrasonic range finder, etc.) are inconsistent.

(2) Compass failure. Compass failure will result in a multicopter's orientation going out of control, i.e., the yaw channel cannot be controlled effectively. The reasons for the compass failure include but are not limited to: 1) Compass hardware failure. 2) Compass not calibrated. 3) Compass offset too high, an error often caused by metal objects being placed too close to the compass. 4) Regional magnetic field too high or too low (for example, it is 35% above or below expected value). 5) The internal and external compasses[1] are pointing to different directions (For example, the difference is greater than 45°. This is normally caused by the external compass orientation being set incorrectly).

(3) GPS failure. GPS failure implies that a GPS receiver cannot measure the location information accurately. In this case, the multicopter cannot hover or complete the pre-programmed route. After losing the location information from the GPS receiver, the position estimation within several seconds is only acceptable with dead reckoning.

(4) Inertial Navigation System (INS) failure. An INS is a navigation aid that uses a computer, motion sensors (accelerometers), and rotation sensors (gyroscopes) to continuously calculate via dead reckoning the position, orientation, and velocity (direction and speed of movement) of a moving object without the need for external references [1]. In this chapter, INS failure mainly indicates anomalies in accelerometers and gyroscopes, which implies that the system cannot correctly measure attitude angle and attitude angular velocity. The reasons for the INS failure include but are not limited to:

(1) INS is not calibrated. If the accelerometers are not calibrated, then the measurements from accelerometers may be inconsistent.

(2) Accelerometer or gyroscope hardware failures. These may cause accelerometer or gyroscope measurements to be inconsistent.

(3) Gyroscopes calibrated unsuccessfully. It mainly implies that the gyroscope calibration fails to capture offsets. This is mostly caused by a multicopter being moved during the gyroscope calibration. Hardware failures of sensors can also cause such a failure. This may cause gyroscope measurement to be inconsistent.

---

[1]It may be a separate module, or a module embedded in the GPS.

14.2.3   Propulsion System Anomaly

Propulsion system anomaly mainly refers to either battery failure, or hardware failure of propulsors of the flight control system caused by Electronic Speed Controllers (ESCs), motors, or propellers.

(1) Battery failure. This usually refers to a lack of power caused by low battery capacity or a degradation in the battery life and is mainly reflected in the following three aspects. 1) The capacity is low and the output voltage is low, which implies that the motor cannot be driven properly. 2) The increase of battery internal resistance will shorten the battery life. 3) Over-charging, over-discharging, discharging under low battery voltage, and low temperature condition will decrease the battery's capacity.

(2) ESC failure. This is mainly reflected in the following two aspects. 1) An ESC cannot correctly recognize the Pulse Width Modulation (PWM) instructions given by the autopilot or RC transmitter. 2) An ESC cannot drive the motor as expected.

(3) Motor failure. This mainly means that the motor cannot work correctly under the given ESC control signal.

(4) Propeller failure. This is mainly caused by worn and broken blades, or a loose blade from the propeller shaft, etc.

Such a failure often occurs in the case that the motor and propeller are damaged due to a strong collision caused by the improper operation of remote pilots. These crashes will further cause the poor contact in the wires connecting the motor to ESC. Secondly, due to an aggressive maneuver or a motor rotation jam, the working current may be too high so that it damages these electronic components and related solder joints. Thirdly, these components have reached their life span. For motors, the phenomenon of *demagnetization* may occur under working condition with high temperature [2, pp. 43–49], [3, pp. 99–102]. This will also lead to motor failure during flight.

14.3   Health Evaluation

Health evaluation [4, 5] refers to the process of judging whether the system is working properly and whether there is an anomaly or a potential failure in the system during a certain period of time in the future. Such a process is important in order to guarantee the safety of a multicopter. This section contains two parts, i.e., the pre-flight health check (offline) and in-flight health evaluation (online).

14.3.1   Pre-Flight Health Check

Pre-flight health check means that a health check is undertaken for the key components and parameters of a multicopter to observe whether these components work properly. The scope of health check is listed in Table 14.1.

Before a user tries to *arm*[2] a multicopter, it is suggested that the autopilot automatically check the eleven items as shown in Table 14.1. If any of these items does not pass the health check, then the autopilot should give the corresponding warning using LED lights onboard. If the GCS and the

---

[2] *Arm* is the instruction that the propellers of the multicopter are unlocked. In this case, the multicopter can take off. Correspondingly, *disarm* is the instruction that the propellers of the multicopter are locked. In this case, the multicopter cannot take off.

Table 14.1   Pre-Flight health check issues

	Check item	Corresponding safety problem
1	Whether the RC has been calibrated	Communication breakdown
2	Whether the RC connection is normal	Communication breakdown
3	Whether the barometer hardware fails	Sensor failure
4	Whether the compass hardware fails	Sensor failure
5	Whether the compass has been calibrated	Sensor failure
6	Whether the GPS signal is normal	Sensor failure
7	Whether the INS has been calibrated	Sensor failure
8	Whether the accelerometer hardware fails	Sensor failure
9	Whether the gyroscope hardware fails	Sensor failure
10	Battery voltage check	Propulsion system anomaly
11	Whether the key parameter settings are correct	Parameter configuration mistake[a]

[a]Parameter configuration error is the incorrect user setting of autopilot parameters, such as PID controller parameters of attitude angles

multicopter are connected, then the occurrence and reasons of corresponding safety problems will be indicated by the GCS.

14.3.2   In-Flight Health Evaluation

In flight of a multicopter, the real-time health evaluation of the multicopter communication, sensors, propulsors, and batteries is also required.

14.3.2.1   Real-Time Health Evaluation for Communication Channels

Real-time health evaluation for communication channels is a process to evaluate whether the communication between the RC transmitter and the multicopter is normal. When a multicopter flies a pre-programmed route, it also needs to monitor whether the communication between the GCS and the multicopter is normal. Concretely, for example, if the multicopter has not received a signal from the RC transmitter for a certain period of time (e.g., 5 s), then it is inferred that the RC transmitter has lost contact with the multicopter. If the multicopter has not received the waypoint from the GCS for a certain period of time (e.g., 5 s), then it is inferred that the GCS has lost contact with the multicopter.

14.3.2.2   Real-Time Health Evaluation for sensors

Real-time health evaluation for sensors is to monitor whether the barometer, compass, GPS, and the INS work properly. The real-time health evaluation process of sensors often requires that the multicopter be preferably in a steady state such as the "Loiter mode" in ArduPilot Mega (APM) autopilot, because the output of each sensor is then stable. When the multicopter is in a steady state, if the height of the multicopter cannot be stabilized because of a severe fluctuation in the height values obtained from the barometer, then the possibility of an anomaly in the barometer is high. If the rotation phenomenon occurs in the multicopter, then the possibility of an unhealthy compass is high. If severe oscillations occur in the multicopter, then the possibility of an unhealthy INS is high. Methods for the health evaluation of the compass and GPS are given on the Web site http://copter.ardupilot.org, which are briefed as below.

(1) Compass health evaluation. 1) The magnetic interference from the propulsion system can be reflected by the "mag_field" value returned by the multicopter. When the propulsion system is working, the "mag_field" value will fluctuate, because the current will change the magnetic field. Concretely, when the throttle command to the multicopter is changed, it may be in one of the following three cases. First, if the fluctuation is between 10 and 20%, then it is considered that the compass works normally and there is no interference. Secondly, if the fluctuation is between 30 and 60%, then it is considered that there is some interference and the compass may work normally or abnormally. Finally, if the fluctuation is greater than 60%, then it is considered that there is some severe interference and the compass is out of operation. 2) The compensation dosage for each direction of the compass should be between −400 and 400 milligauss. If it is not in this range, then it is considered that there is a problem in the compass. The compensation dosage is reflected by parameters "COMPASS_OFS_X", "COMPASS_OFS_Y" and "COMPASS_OFS_Z" in APM autopilots.

(2) GPS health evaluation. The GPS health evaluation is based on the position estimation and position measurement from the GPS, where the estimation of the position is updated by using Extended Kalman Filter (EKF) combined with the data obtained by the Inertial Measurement Unit (IMU). If the error between the two values is less than the parameter "EKF_POS_GATE", then the GPS is considered healthy. Otherwise, it is considered unhealthy.

In addition, a method was proposed in [6] to use the state estimation residuals to determine whether a sensor is healthy. Here, it is assumed that the multicopter model is

$$\begin{aligned} \mathbf{x}_{k+1} &= \mathbf{f}(\mathbf{x}_k) + \mathbf{B}\mathbf{u}_k + \mathbf{\Gamma}\mathbf{w}_k \\ \mathbf{y}_k &= \mathbf{C}^{\mathrm{T}}\mathbf{x}_k + \mathbf{v}_k \end{aligned} \tag{14.1}$$

where $\mathbf{x}$ represents the state of the multicopter, $\mathbf{u}$ is the system input, $\mathbf{w}$ is the system noise, $\mathbf{\Gamma}$ is the noise driving matrix, and $\mathbf{v}$ is the measurement noise. To evaluate the sensor failures, an observer or filter is designed as shown in Fig. 14.1, where the observer can be a Kalman filter. The estimation residual is defined as follows:

$$\mathbf{r}_k \triangleq \mathbf{y}_k - \hat{\mathbf{y}}_k. \tag{14.2}$$

If the mean or variance of the ith estimation residual $\mathbf{r}_{i,k}$ is beyond the failure threshold defined, then the ith sensor is inferred to have failed. Readers can refer to [7, 8] for other related methods.

14.3.2.3   Real-Time Health Evaluation for the propulsion system

(1) *Health evaluation for the motor and propeller*

1) A method of propulsor health evaluation based on the EKF. It is assumed that the model of multicopters is shown in system (14.1). In order to measure the failure, an efficiency matrix is defined as

Fig. 14.1 Sensor fault diagnosis based on the observer

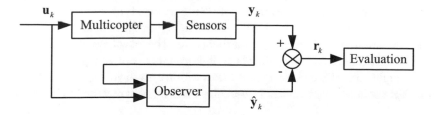

$$\mathbf{\Lambda} = \mathrm{diag}(\eta_1, \eta_2, \cdots, \eta_n) \tag{14.3}$$

where $\eta_i \in [0, 1]$ is the efficiency of the ith propulsor which reflects the total effect of the motor, propeller, ESC, and battery performance. In particular, $\eta_i = 1$ means that the ith propulsor is completely healthy, while $\eta_i = 0$ indicates that the ith propulsor has completely failed. The value $\eta_i \in (0, 1)$ represents that the ith propulsor is partly out of operation. Based on this, system (14.1) is rewritten as

$$\begin{aligned} \mathbf{x}_{k+1} &= \mathbf{f}(\mathbf{x}_k) + \mathbf{B}\mathbf{\Lambda}_k\mathbf{u}_k + \mathbf{\Gamma}\mathbf{w}_k \\ \mathbf{y}_k &= \mathbf{C}^{\mathrm{T}}\mathbf{x}_k + \mathbf{v}_k. \end{aligned} \tag{14.4}$$

Therefore, the real-time evaluation of the propulsor is based on the real-time estimation of η_i. Define $\boldsymbol{\eta} \triangleq [\eta_1\ \eta_2\ \cdots\ \eta_n]^{\mathrm{T}}$. It is assumed that $\boldsymbol{\eta}$ satisfies

$$\begin{aligned} \boldsymbol{\eta}_{k+1} &= \boldsymbol{\eta}_k + \boldsymbol{\xi}_k + \boldsymbol{\varepsilon}_{1,k} \\ \boldsymbol{\xi}_{k+1} &= \boldsymbol{\xi}_k + \boldsymbol{\varepsilon}_{2,k} \end{aligned} \tag{14.5}$$

where $\boldsymbol{\varepsilon}_{1,k}$ and $\boldsymbol{\varepsilon}_{2,k}$ are Gaussian white noise. Here, the real-time estimation can be realized by an EKF based on the process model extended from the system state equation (14.4). The extended system is

$$\begin{bmatrix} \mathbf{x}_{k+1} \\ \boldsymbol{\eta}_{k+1} \\ \boldsymbol{\xi}_{k+1} \end{bmatrix} = \begin{bmatrix} \mathbf{f}(\mathbf{x}_k) + \mathbf{B}\mathbf{\Lambda}_k\mathbf{u}_k + \mathbf{\Gamma}\mathbf{w}_k \\ \boldsymbol{\eta}_k + \boldsymbol{\xi}_k + \boldsymbol{\varepsilon}_{1,k} \\ \boldsymbol{\xi}_k + \boldsymbol{\varepsilon}_{2,k} \end{bmatrix}$$
$$\mathbf{y}_k = \begin{bmatrix} \mathbf{C}^{\mathrm{T}} & \mathbf{0} \end{bmatrix} \begin{bmatrix} \mathbf{x}_{k+1} \\ \boldsymbol{\eta}_{k+1} \\ \boldsymbol{\xi}_{k+1} \end{bmatrix} + \mathbf{v}_k. \tag{14.6}$$

Furthermore, the EKF can be used to estimate the value of η_i in real time. However, it should be pointed out that the observability condition has to be satisfied first.

2) A method of propulsor health evaluation based on the vibration signals. Features extracted from vibration signals of multicopter airframes are used to evaluate the propulsor health. By recalling the *dynamic balance* in Sect. 2.3.3, when a multicopter propulsor (such as a propeller, or a motor) is abnormal, the dynamic balance of the multicopter may be lost and the vibration signals of the multicopter airframe may be different from that in the normal state. Therefore, the propulsor anomaly can be detected by analyzing the features of the vibration data. Readers can refer to [9] for details.

(2) *Health evaluation of the flight control board voltage*

Before a multicopter takes off, the safety check of the flight control board voltage is necessary. If the flight control board voltage is within a given voltage range, then it is considered normal. Otherwise, if the voltage is out of a given voltage range, then it is considered problematic. Furthermore, if the flight control board is powered by a USB line, then it may be considered that there are some problems with the computer's USB port. If the flight control board is powered by a battery, then more attention and serious check to the flight control board voltage are needed. After the multicopter takes off, it is necessary to evaluate the flight control board voltage and battery state in real time. If the flight control board voltage fluctuates within a given voltage range, then it is considered that the flight control board voltage is normal. Otherwise, it is regarded to be abnormal.

(3) *Health evaluation for the battery*

Before a multicopter takes off, a safety check on the battery voltage is needed. If the battery voltage is lower than the threshold, then it is considered that the battery level is low and unhealthy. After the multicopter takes off, it is necessary to evaluate the battery state. Under normal circumstances, if the battery voltage is lower than a set value for longer than a certain time (several seconds), then it is considered that the battery level is low and unhealthy. Another way is to calculate the reserved capacity in real time. If the calculated value is lower than the specified Reserved Maximum Ampere-Hour (RMAH), then it is considered that the battery is low and unhealthy. However, there are some difficulties in using the battery voltage to estimate the state of the battery when the multicopter is in flight. First, the battery voltage cannot directly and accurately reflect the battery discharge status, because they are nonlinearly related. A small battery voltage measurement error may cause a large battery capacity estimation error [10]. Secondly, the calculation of the remaining capacity of the battery is an open-loop process. As a result, if there exist noises in the measurement of the current, then there will be accumulated errors in the calculation of the remaining power of the battery. For such a purpose, the State of Charge (SoC) is the third parameter that characterizes the charge and discharge states of the battery. In the process of battery discharge, the change of the SoC and battery impedance satisfies

$$
\begin{aligned}
S_{k+1} &= S_k - \frac{I_k \cdot T_s}{Q_{max}} + w_{1,k} \\
R_{k+1} &= R_k + w_{2,k}
\end{aligned}
\tag{14.7}
$$

where $S \in \mathbb{R}_+$ is the SoC of battery, $I \in \mathbb{R}$ is the discharge current (unit: A), $R \in \mathbb{R}_+$ is the battery impedance (unit: Ω), $Q_{max} \in \mathbb{R}_+$ is the nominal value of battery capacity (unit: Ah), $T_s \in \mathbb{R}_+$ is the sampling time (unit: h), and $w \in \mathbb{R}$ is the system noise. The measurement equation is designed as

$$
V_k = OCV(S_k) - I_k \cdot R_k + C + v_k
\tag{14.8}
$$

where $V \in \mathbb{R}$ is the battery terminal voltage (unit: V), $C \in \mathbb{R}_+$ is the constant error offset, $v \in \mathbb{R}$ is the measurement noise, and $OCV(S)$ is the curve of the Open Circuit Voltage and SoC (OCV-SoC). Generally, the values of OCV-SoC curve are derived from the battery charge and discharge tests. Based on Eqs. (14.7) and (14.8), the real-time value of SoC and R can be effectively estimated by using the EKF or Unscented Kalman Filter (UKF), which reflect the real-time battery capacity and health status. Here, only an abstract of this method is given. Readers can refer to [10, 11] for detailed implementation process of the method. Other similar methods can be found in [12–14]. Except for the battery resistance, other mathematical indicators are also used to evaluate battery health. For example, the *profust reliability* is used in [5] as a health indicator to evaluate the health status of a Li-ion battery and to predict its remaining charge cycles.

14.4   Failsafe Suggestions

Health check of key components of a multicopter will be carried out in the pre-flight process. If the components do not pass the check, then the remote pilot cannot arm the multicopter. During flight, failures will make the multicopter go out of control. Therefore, failsafe for key components will be considered in the autopilot design. This section provides several failsafe suggestions, most of which are summarized from [15].

14.4.1   Communication Failsafe

The communication failsafe includes: the RC system failsafe and the GCS failsafe. When a multicopter is in flight, it is recommended to perform the following protective measures if RC or GCS is lost:

(1) Do nothing if the multicopter is already disarmed.
(2) The multicopter will be immediately disarmed if it has landed or the remote pilot's throttle control stick is at the bottom.[3]
(3) Return-To-Launch (RTL) if the multicopter has a GPS lock and the straight-line distance from the home position is greater than the threshold[4] (2 m for APM autopilots).
(4) Immediately land if the multicopter has no GPS lock or the straight-line distance from the home position is less than the set threshold[5] (2 m for APM autopilots).
(5) In addition, if the contact between the RC transmitter and the onboard RC receiver is reestablished (or the contact between the GCS and the onboard signal receiver is reestablished), then it is suggested that the multicopter still stay in the current mode rather than being switched back to the resume task it left off before the failsafe was triggered.[6] However, if remote pilots manually manipulate the flight mode switch, they can retake the control of multicopters.

14.4.2   Sensor Failsafe

The sensor failsafe includes barometer failsafe, compass failsafe, GPS failsafe, and INS failsafe.

(1) The barometer failsafe. In the SAC mode, according to the user setting, it is suggested that the multicopter be switched from the loiter mode or the altitude hold mode to the stabilize mode,[7] or land directly. On the other hand, in the Fully-Autonomous Control (FAC) mode, it is suggested that the multicopter land immediately.

(2) The compass failsafe. In the SAC mode, according to the user setting, it is suggested that the multicopter be switched from the loiter mode to the altitude hold mode, or land directly. On the other hand, in the FAC mode, it is suggested that the multicopter land immediately.

(3) The GPS failsafe. In the SAC mode, according to the user setting, it is suggested that the multicopter be switched from the loiter mode to the altitude hold mode, or land directly. On the other hand, in the FAC mode, it is suggested that the multicopter land immediately.

(4) The INS failsafe. In both SAC and FAC modes, it is suggested that the multicopter land urgently by gradually reducing the throttle command.

14.4.3   Propulsion System Failsafe

If the propulsion system of a multicopter is evaluated to be abnormal, then

(1) Do nothing if the multicopter is already disarmed.

---

[3]This implies that the pilot has given up the control.

[4]This indicates that the multicopter does not arrive at the HOME point. Thus, RTL is executed.

[5]This indicates that the multicopter has already arrived at the HOME point. Thus, landing is executed.

[6]One of the reasons is that the remote pilot may put the throttle and other control sticks in improper positions after the RC is out of contact. Especially for beginners, they may be in panic facing the RC out of contact. Consequently, the improper throttle position will cause a sudden change to the multicopter speed. That may in turn cause an accident.

[7]The loiter mode, the altitude hold mode, and the stabilize mode have been explained in Sect. 13.2, Chap. 13.

(2) The multicopter will be immediately disarmed if it has landed or the remote pilot's throttle control stick is at the bottom for a given time.
(3) RTL if the multicopter has a GPS receiver and the straight-line distance from the home position is greater than a set threshold (2 m for APM autopilots). Moreover, the battery voltage is required to be higher than a set threshold (or other indices to show the battery with a sufficient capacity to support RTL).
(4) In other cases, it is suggested that the multicopter land directly.

If a multicopter has one propulsor (including a propeller, a motor, and an ESC) failed, it may lose the controllability at the hover state. Readers could recall the controllability of the multicopter in Chap. 10. In this case, it is suggested that the multicopter adopt a degraded control scheme immediately to land urgently by giving up the yaw [16] or follow methods given in [17]. If the multicopter is still controllable at the hover state, then the control reallocation [18, 19] is often adopted or robust stabilizing control is used by regarding the damage as a disturbance [16, 20].

14.5 A Safe Semi-Autonomous Autopilot Logic Design

In this section, for simplicity, a Semi-Autonomous Autopilot (SAA) logic is realized by using an Extended Finite State Machine (EFSM). The state automaton is a mathematical model to describe a discrete-event system or hybrid system. Generally, the following conditions are assumed to be true [21, 22]: 1) the system has a finite number of states; 2) system behavior in a specific state should remain the same; 3) the system always stays in a certain mode for certain period of time; 4) the number of conditions for states switch is finite; 5) a switch of the system state is the response to a set of events; 6) the time of state switch is negligible. In order to apply the EFSM, it is required to define the multicopter states, flight states, and the events that may occur. Based on these, the state transition conditions are constructed to guarantee the safety. It should be pointed out that the EFSM can not only realize the failsafe, but also make the user observe and understand the decision-making process clearly.

14.5.1 Requirement Description

(1) *System composition*
   Here, SAA design is considered. The main sensors on an autopilot include GPS receiver, compass, INS, and barometer. Here, the RC transmitter is only used without considering the GCS.

(2) *Functional requirements*

   1) The user can arm the multicopter by the RC transmitter and then allow it to take off. The user can manually control the multicopter to land and disarm it.
   2) The user can control the multicopter to fly, return, and land automatically through the RC transmitter.
   3) The multicopter can realize the spot hover (LOITER MODE in APM autopilots) if the GPS receiver, compass and barometer are all healthy.
   4) The multicopter can realize the altitude hold hover (ALTITUDE HOLD MODE in APM autopilots) if the GPS receiver or compass is not installed or is unhealthy.
   5) The multicopter can realize the attitude self-stabilization (STABILIZE MODE in APM autopilots), if the barometer is unhealthy.

(3) *Safety requirements*

This part mainly refers to the failsafe in Sect. 14.4 by using the SAC.

14.5.2   Multicopter State and Flight Mode Definition

The whole process from takeoff to landing of the multicopter is divided into three multicopter states and three flight modes. They form the basis of the logic design. First, three multicopter states are defined as follows:

(1) POWER OFF STATE. This state implies that a multicopter is out of power. In this state, the user can (possibly) disassemble, modify, and replace the hardware of a multicopter.

(2) STANDBY STATE. When a multicopter is connected to the power module, it enters the preflight state immediately. In this state, the multicopter is not armed, and users can arm the multicopter manually. Afterward, the multicopter will perform the safety check and then can transit into the next state according to results of the safety check.

(3) GROUND_ERROR STATE. This state indicates that the multicopter has a safety problem. In this state, the buzzer will turn on an alarm to alert the user that there are errors in the system.

Furthermore, three kinds of flight modes are defined.

(4) MANUAL FLIGHT MODE. This mode allows a remote pilot to manually control a multicopter. It further contains three submodes, namely LOITER MODE, ALTITUDE HOLD MODE, and STABILIZE MODE. Under normal circumstances, if the multicopter is in MANUAL FLIGHT MODE, then the default submode is in LOITER MODE. If the multicopter does not install the "GPS + compass", or "GPS + compass" are unhealthy, then the flight mode is degraded to ALTITUDE HOLD MODE. Furthermore, if the barometer is unhealthy, then the flight mode is further degraded to STABILIZE MODE.

(5) RTL MODE. Under this mode, the multicopter will return to the home location from the current position and hover there. In the process, if the current relative altitude comparing with that of the home location is higher than a set value (the default relative height is 15 m in the APM autopilot), the multicopter will maintain the current altitude and return home. Otherwise, it will first ascend to the preset height before returning home.

(6) AUTO-LANDING MODE. In this mode, the multicopter realizes the automatic landing by adjusting the throttle command according to the estimated height.[8]

14.5.3   Event Definition

Here, two kinds of events are defined: Manual Input Events (MIEs) and Automatic Trigger Events (ATEs). These events will cause the state or mode transition.

(1) MIEs

MIEs are instructions from remote pilots sent through the RC transmitter, including the following:

MIE1: Arm and Disarm instructions. These instructions are realized by manipulating the sticks of the RC transmitter. First, three deflections of the sticks in the RC transmitter are defined as shown in Fig. 14.2. The case that the two sticks are both at their middle position is defined as $MIE1 = -1$. The case is defined as $MIE1 = 1$ when the throttle/yaw control stick is at the lower left position, and the other at the lower right position. The case is defined as $MIE1 = 0$ when the throttle/yaw control stick

---

[8]Even if the barometer fails, the height estimation is acceptable within a short time. Similarly, the other estimates generated by filters will continue to be used for a short time, even if related sensors fail.

is at the lower right position, and the other at the lower left position. Based on the three positions, the arm and disarm instructions are defined. As shown in Fig. 14.2a, the two successive actions are $-1 \longrightarrow 1 \longrightarrow -1$, which produces an "arm" instruction. After given an arm instruction, the motors can be controlled by the RC transmitter. Similarly, as shown in Fig. 14.2b, a "disarm" instruction is $-1 \longrightarrow 0 \longrightarrow -1$, which will lock the multicopter, and make the motors uncontrolled by the RC transmitter. For simplicity, let $MIE1 = 1$ denote the arm instruction, and $MIE1 = 1$ denote the disarm instruction.

$MIE2$: Manual operation instruction (1: Switch to MANUAL FLIGHT MODE; 2: Switch to RTL MODE; 3: Switch to AUTO-LANDING MODE). This instruction is realized by the flight mode switch (a three-position switch) as shown in Fig. 14.3. When the multicopter is powered on, $MIE2$ is temporarily set as 0. Then, according to the position of the flight mode switch, the switch will correspondingly happen.

$MIE3$: Turn on or turn off the multicopter. (1: turn on; 0: turn off).

$MIE4$: Power cutoff for maintenance. (1: repaired; 0: repairing).

(2) ATEs

ATEs are independent of the remote pilot's operations, but mainly generated by the status of components on board and status of the multicopter.

$ATE1$: Health status of INS and status of multicopter (1: healthy; 0: unhealthy)

$ATE2$: Health status of GPS (1: healthy; 0: unhealthy)

$ATE3$: Health status of barometer (1: healthy; 0: unhealthy)

Fig. 14.2 Arm and disarm instruction

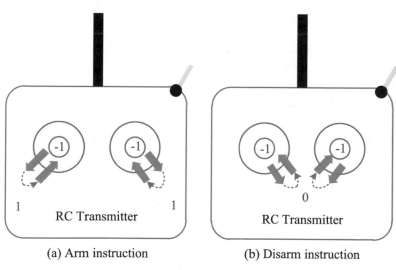

(a) Arm instruction   (b) Disarm instruction

Fig. 14.3 Manual operation instruction

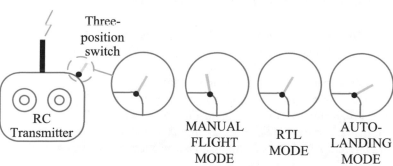

ATE4: Health status of compass (1: healthy; 0: unhealthy)

ATE5: Health status of propulsion system (1: healthy; 0: unhealthy)

ATE6: Status of connections of RC (1: normal; 0: abnormal)

ATE7: The status of battery's capacity (1: adequate; 0: inadequate, able to perform RTL; −1: inadequate, unable to perform RTL)

ATE8: Comparison of the multicopter's altitude and a specified threshold, (1: the multicopter's altitude is lower than the specified threshold, as $-p_{z_e} < -p_{z_T}$; 0: the multicopter's altitude is not lower than the specified threshold, as $-p_{z_e} \geqslant -p_{z_T}$.)

ATE9: Comparison of the multicopter's throttle command and a specified threshold over a time horizon, (1: the multicopter's throttle command is less than the specified threshold, as $\sigma_{drc} < \sigma_{drcT}$ for $t > t_T$; 0: otherwise)

ATE*10*: Comparison of the multicopter's distance from HOME point and a specified threshold, (1: the multicopter's distance from HOME point is greater than the specified threshold, as $d > d_T$; 0: the multicopter's distance from HOME point is not greater than the specified threshold, as $d \leqslant d_T$.)

The health status of each component above can be obtained by health evaluation methods mentioned in Sect. 14.3.

14.5.4   Autopilot Logic Design

Based on the multicopters' state, flight modes, and events, the autopilot logic is designed in the form of an EFSM as shown in Fig. 14.4, where Ci represents the transition condition, $i = 1, \cdots, 21$.

In Fig. 14.4, the transition conditions are described as follows:

C1: *MIE3* = 1

C2: *MIE3* = 0

C3:

$$(MIE1 = 1) \& (MIE2 = 1) \& (ATE1 = 1) \& (ATE5 = 1) \& (ATE6 = 1) \& (ATE7 = 1)$$

This condition implies a successful arm operation. This condition is true, when (1) the remote pilot tries to arm the multicopter (*MIE1* = 1); (2) the multicopter passes the check that the INS and propulsion system are both healthy (*ATE1* = 1&*ATE5* = 1); (3) the connection to the RC transmitter be normal (*ATE6* = 1); (4) the battery's capacity is adequate (*ATE7* = 1); and (5) the flight mode switch to MANUAL FLIGHT MODE happens (*MIE2* = 1). Then, the multicopter is armed and switched from STANDBY STATE to MANUAL FLIGHT MODE.

C4:

$$[(MIE1 = 0 \& ATE8 = 1)|(ATE9 = 1)] \& (ATE1 = 1) \& (ATE5 = 1) \& (ATE6 = 1)$$

This condition describes the disarm operation, including manual disarm operation and automatic disarm operation. The manual disarm operation requires that the remote pilot disarm the multicopter (*MIE1* = 0), and the multicopter be on the ground or lower than a given height (*ATE8* = 1). The automatic disarm operation is to disarm the multicopter automatically when the throttle command is less than a given value beyond a given time threshold (*ATE9* = 1). In this situation, it is required that the INS and propulsion system be both healthy (*ATE1* = 1&*ATE5* = 1), and the connection to the RC transmitter be normal (*ATE6* = 1). Then, the multicopter is switched from MANUAL FLIGHT MODE to STANDBY STATE. If there exists an unhealthy INS, or propulsion system, or the connection

Fig. 14.4 EFSM of
multicopters

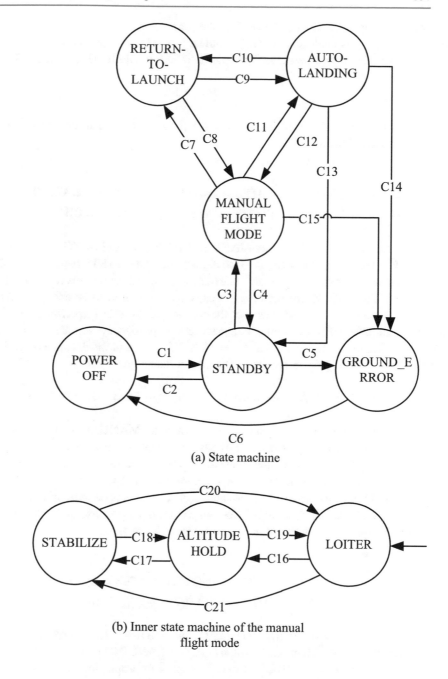

(a) State machine

(b) Inner state machine of the manual
flight mode

to the RC transmitter is abnormal, then the multicopter is switched from MANUAL FLIGHT MODE
to GROUND_ERROR STATE, for which condition C15 gives more details.
   C5:

$$(MIE1 = 1)\&[(ATE1 = 0)|(ATE5 = 0)|(ATE6 = 0)|(ATE7 = 0|ATE7 = -1)]$$

This condition implies an unsuccessful arm operation. When the remote pilot tries to arm the
multicopter ($MIE1 = 1$), the multicopter will check itself. If the INS is unhealthy ($ATE1 = 0$), or the

propulsion system is unhealthy ($ATE5 = 0$), or the connection to the RC is abnormal ($ATE6 = 0$), or the battery's capacity is inadequate ($ATE7 = 0|ATE7 = -1$), then the multicopter cannot be armed successfully and will be switched from STANDBY STATE to GROUND_ERROR STATE.

C6:

$$MIE4 = 1$$

This condition means that the power is cut off for maintenance in order to replace unhealthy component, or manual health check.

C7:

$$(ATE1 = 1\&ATE2 = 1\&ATE3 = 1\&ATE4 = 1\&ATE5 = 1\&ATE10 = 1)\&$$
$$[(MIE2 = 2\&ATE7 \geq 0)|(ATE6 = 0\&ATE7 \geq 0)|(ATE7 = 0)]$$

This condition implies a switch from MANUAL FLIGHT MODE to RTL. Such a switch can take place in one of the following three cases: 1) the flight mode switch to RTL happens ($MIE2 = 2$), where the battery's capacity is required to be adequate ($ATE7 \geq 0$); or 2) the connection to the RC transmitter is abnormal ($ATE6 = 0$), where the battery's capacity is required to be adequate ($ATE7 \geq 0$); or 3) the battery's capacity is inadequate, but the multicopter is able to perform RTL ($ATE7 = 0$). Furthermore, the INS, GPS, barometer, compass, and propulsion system are required to be healthy ($ATE1 = 1\&ATE2 = 1\&ATE3 = 1\&ATE4 = 1\&ATE5 = 1$), and the distance from the multicopter to the predefined HOME point is required to be greater than a given threshold ($ATE10 = 1$).

C8:

$$(MIE2 = 1)\&(ATE1 = 1\&ATE5 = 1)\&(ATE6 = 1)\&(ATE7 = 1)$$

This condition means that the flight mode switch to MANUAL FLIGHT MODE happens ($MIE2 = 1$) during the RTL process. In this case, the INS and propulsion system are required to be healthy($ATE1 = 1\&ATE5 = 1$), the connection to the RC transmitter is normal($ATE6 = 1$), and the battery's capacity is adequate ($ATE7 = 1$). Here, note that when the contact between the RC transmitter and onboard RC receiver is reestablished, the multicopter will stay in RTL. Only if the remote pilot manually switches the flight mode switch to another position, and then back to MANUAL FLIGHT MODE again, the control of the multicopter can be retaken.

C9:

$$(MIE2 = 3)|(ATE1 = 0|ATE2 = 0|ATE3 = 0|$$
$$ATE4 = 0|ATE5 = 0)|(ATE7 = -1)|(ATE10 = 0)$$

This condition implies a switch from RTL to AUTO-LANDING. This switch will occur in any of the following four cases: 1) the flight mode switch to AUTO-LANDING happens ($MIE2 = 3$); or 2) there exists a health problem in INS, GPS, barometer, compass, or propulsion system ($ATE1 = 0|ATE2 = 0|ATE3 = 0|ATE4 = 0|ATE5 = 0$); or 3) the battery's capacity is inadequate, and the multicopter is unable to perform RTL ($ATE7 = -1$); or 4) the distance from the multicopter to the HOME point is required to not greater than a given threshold beyond a given time threshold ($ATE10 = 0$).

C10:

$$(MIE2 = 2)\&(ATE1 = 1\&ATE2 = 1\&ATE3 = 1\&$$
$$ATE4 = 1\&ATE5 = 1)\&(ATE7 \geq 0)\&(ATE10 = 1)$$

This condition indicates that the flight mode switch to RTL happens ($MIE2 = 2$). It is required that 1) the INS, GPS, barometer, compass, and propulsion system be all healthy ($ATE1 = 1\&ATE2 = 1\&ATE3 = 1\&ATE4 = 1\&ATE5 = 1$); 2) the battery's capacity be adequate ($ATE7 \geq 0$); and (3) the distance from the multicopter to the HOME point be greater than a given threshold ($ATE10 = 1$).

C11:

$$(MIE2 = 3)|(ATE7 = -1)|(ATE1 = 0|ATE5 = 0)|$$
$$[(ATE6 = 0)\&(ATE7 \geq 0)\&(ATE2 = 0|ATE4 = 0|ATE10 = 0)]$$

This condition implies a switch from MANUAL FLIGHT MODE to AUTO-LANDING, including four situations: 1) the flight mode switch to AUTO-LANDING happens ($MIE2 = 3$); or 2) the battery's capacity is inadequate, and the multicopter is unable to perform RTL ($ATE7 = -1$); or 3) there exists a health problem in the INS, or propulsion system ($ATE1 = 0|ATE5 = 0$); or 4) the connection to the RC transmitter is abnormal ($ATE6 = 0$), where the battery's capacity is required to be adequate ($ATE7 \geq 0$), and the distance from the multicopter to the HOME point is not greater than a given threshold ($ATE10 = 0$), or the GPS or compass is unhealthy ($ATE2 = 0|ATE4 = 0$).

C12:

$$(MIE2 = 1)\&(ATE1 = 1\&ATE5 = 1)\&(ATE6 = 1)\&(ATE7 = 1)$$

This condition indicates that the flight mode switch to MANUAL FLIGHT MODE happens during the landing process ($MIE2 = 1$), where the INS and propulsion system are required to be healthy ($ATE1 = 1\&ATE5 = 1$), the connection to the RC transmitter is normal ($ATE6 = 1$), and the battery's capacity is adequate ($ATE7 = 1$). Here, note that when the contact between the RC transmitter and onboard RC receiver is reestablished, the multicopter will stay in AUTO-LANDING. Only if the remote pilot manually changes the flight mode switch to another position, and then back to MANUAL FLIGHT MODE again, the control of the multicopter can be retaken.

C13:

$$(ATE8 = 1|ATE9 = 1)\&[(ATE1 = 1\&ATE5 = 1)\&(ATE6 = 1)]$$

This condition indicates that during the landing process, the multicopter is considered to have landed successfully and be disarmed, when 1) the height is less than a predetermined threshold ($ATE8 = 1$); or 2) the throttle command is less than a given value beyond a given time threshold ($ATE9 = 1$). Furthermore, it is required that the INS and propulsion system be both healthy ($ATE1 = 1\&ATE5 = 1$), and the connection to the RC transmitter be normal ($ATE6 = 1$). In this case, the multicopter is switched from AUTO-LANDING to STANDBY STATE.

C14:

$$(ATE8 = 1|ATE9 = 1)\&(ATE1 = 0|ATE5 = 0|ATE6 = 0)$$

This condition indicates that during the landing process, the multicopter is considered to have landed and be disarmed, and its state is switched from AUTO-LANDING to GROUND_ERROR STATE, when 1) the height is lower than a predetermined threshold ($ATE8 = 1$); or 2) the throttle command is less than a given value beyond a given time threshold ($ATE9 = 1$). Furthermore, it is required that the INS or propulsion system be unhealthy, or the connection to the RC transmitter be abnormal ($ATE1 = 0|ATE5 = 0|ATE6 = 0$).

C15:

$$[(MIE1 = 0\&ATE8 = 1)|(ATE9 = 1)]\&(ATE1 = 0|ATE5 = 0|ATE6 = 0)$$

This condition describes the disarm operation, including manual disarm operation and automatic disarm operation. The manual disarm operation requires that the remote pilot disarm the multicopter ($MIE1 = 0$), and the multicopter be on the ground or lower than a given height ($ATE8 = 1$). The automatic disarm operation is to disarm the multicopter automatically when the throttle command is less than a given value beyond a given time threshold ($ATE9 = 1$). In this situation, if the INS or propulsion system is unhealthy ($ATE1 = 0|ATE5 = 0$), or the connection to the RC transmitter is abnormal ($ATE6 = 0$), then the multicopter is switched from MANUAL FLIGHT MODE to GROUND_ERROR STATE.

For the three submodes in MANUAL FLIGHT MODE, the switch conditions are given as follows:

C16: $ATE2 = 0|ATE4 = 0$

This condition indicates that if the GPS or compass is unhealthy ($ATE2 = 0|ATE4 = 0$), then the flight mode is switched from LOITER MODE to ALTITUDE HOLD MODE.

C17: $ATE3 = 0$

This condition indicates that if the barometer is unhealthy ($ATE3 = 0$), then the flight mode is switched from ALTITUDE HOLD MODE to STABILIZE MODE.

C18: $(ATE3 = 1)\&(ATE2 = 0|ATE4 = 0)$

This condition indicates that if the barometer is healthy ($ATE3 = 1$), and the GPS or compass is unhealthy ($ATE2 = 0|ATE4 = 0$), then the flight mode is switched from STABILIZE MODE to ALTITUDE HOLD MODE.

C19: $ATE2 = 1\&ATE4 = 1$

This condition indicates that if the GPS and compass are healthy ($ATE2 = 1\&ATE4 = 1$), then the flight mode is switched from ALTITUDE HOLD MODE to LOITER MODE.

C20: $ATE2 = 1\&ATE3 = 1\&ATE4 = 1$

This condition indicates that if the GPS, compass, and barometer are all healthy ($ATE2 = 1\&ATE3 = 1\&ATE4 = 1$), then the flight mode is switched from STABILIZE MODE to LOITER MODE.

C21: $ATE3 = 0$

This condition indicates that if the barometer is unhealthy ($ATE3 = 0$), then the flight mode is switched from LOITER MODE to STABILIZE MODE.

Finally, Table 14.2 presents the correspondence between functional requirements and decision-making procedure. Table 14.3 presents the correspondence between safety requirements and decision-making procedure, when the safety requirements are all achieved. However, despite the given charts and simulations of the EFSM, the logic of the EFSM is not guaranteed to be fully correct. Therefore, further scientific methods should be employed to design the safety decision-making logic. For example, based on the theory of discrete-event systems, the textual description of functional requirements can be transformed to multicopter plant modeled by a finite state automaton, while the textual description of safety requirements can be transformed to safety specifications modeled by some finite state automata. Then, a supervisor, generated by *supervisory control theory*, is the safety decision-making logic satisfying both the function and safety requirements. Readers can refer to [23, 24] for some useful information.

14.5.5   Demand and Decision-Making Table

Table 14.2 Functional demand and decision-making table

Functional term	Functional demand	Decision-making
Turn the power on	When the power is turned on, the various components of the multicopter are powered on	Refer to condition C1
Turn the power off	When the power is turned off, the various components of the multicopter are powered off	Refer to conditions C2 and C6
Arm and health check	When the multicopter is armed, the health check will be carried out	When the multicopter is in STANDBY STATE, if the user attempts to arm the multicopter manually, then the multicopter will perform self-check. If all the components are healthy (including the RC transmitter), then the multicopter enters MANUAL FLIGHT MODE, otherwise it will enter GROUND_ERROR STATE. Refer to conditions C3 and C5
Disarm	After landing, the multicopter needs to be disarmed	The multicopter disarm operation contains a manual disarm operation and an automatic disarm operation. The manual disarm operation should satisfy that the remote pilot tries to disarm the multicopter, and the multicopter should be on the ground or lower than a given height. The automatic disarm operation is to disarm automatically when the throttle command less than a given value is longer than a specified time period. In this situation, it is required that the INS and propulsion system be both healthy, the connection to the RC transmitter be normal, and then the multicopter be switched from MANUAL FLIGHT MODE to STANDBY STATE. Refer to condition C4
MANUAL FLIGHT MODE	MANUAL FLIGHT MODE contains the LOITER MODE, ALTITUDE HOLD MODE, and STABILIZE MODE, and these submodes can be switched among the three submodes according to its own health condition	MANUAL FLIGHT MODE contains three submodes, which are the LOITER MODE, ALTITUDE HOLD MODE, and STABILIZE MODE; and the three submodes can be switched from one to another. Refer to conditions C16-C21
	Under certain conditions, the multicopter can be switched from RTL to MANUAL FLIGHT MODE	The flight mode switch to MANUAL FLIGHT MODE happens during the RTL process. In this case, it is required that the INS and propulsion system be healthy, the connection to the RC transmitter be normal, and the battery's capacity be adequate. Refer to condition C8
	Under certain conditions, the multicopter can be switched from AUTO-LANDING to MANUAL FLIGHT MODE	The flight mode switch to MANUAL FLIGHT MODE happens during the landing process. It is required that the INS and propulsion system be healthy, the connection to the RC be normal, and the battery's capacity be adequate. Refer to condition C12

(continued)

Table 14.2 (continued)

Functional term	Functional demand	Decision-making
RTL	Under certain conditions, the multicopter can be switched from MANUAL FLIGHT MODE to perform RTL	When the INS, GPS, barometer, compass, and propulsion system are healthy and the distance from the multicopter to the HOME point is greater than a given threshold, the flight mode can be switched to RTL manually. Refer to condition C7
	Under certain conditions, the multicopter can be switched from AUTO-LANDING to RTL	The flight mode can be switched to RTL manually. It is required that the INS, GPS, barometer, compass, and propulsion system be all healthy, the battery's capacity be adequate, and the distance from the multicopter to the HOME point be greater than a threshold. Refer to condition C10
AUTO-LANDING	Under certain conditions, the multicopter can be switched from MANUAL FLIGHT MODE to AUTO-LANDING	The flight mode can be switched to AUTO-LANDING manually. Refer to condition C11
	Under certain conditions, the multicopter can be switched from RTL to AUTO-LANDING	The flight mode can be switched to AUTO-LANDING manually. Meanwhile, the switch happens automatically, when there exists a health problem in INS, GPS, barometer, compass, or propulsion system; the battery's capacity is inadequate; the multicopter is unable to perform RTL; or the distance from the multicopter to the HOME point is not greater than a given threshold beyond a given time threshold. Refer to condition C9
	The multicopter needs to be disarmed after AUTO-LANDING	During the landing process, if the height is lower than a predetermined threshold or the throttle command is less than a given value beyond a given time threshold, and the INS, barometer, propulsion system are all healthy, then the multicopter is considered to have landed successfully and be disarmed. In this case, the multicopter is switched from AUTO-LANDING to STANDBY STATE. If a health problem in components exists, then the multicopter is switched from AUTO-LANDING to GROUND_ERROR STATE and be disarmed. Refer to conditions C13 and C14

Table 14.3 Safety demand and decision-making table

Safety problem	Safety subproblem	Safety demand	Decision-making
Communication breakdown	The RC transmitter is not calibrated	Whether the RC transmitter has been calibrated should be checked in the pre-flight health check	When the multicopter is in STANDBY STATE, if the user tries to arm the multicopter manually, then the multicopter will perform self-check. If all the components are healthy (including the RC transmitter), then the multicopter enters MANUAL FLIGHT MODE. Otherwise, it will enter GROUND_ERROR STATE. Refer to conditions C3 and C5
	The RC transmitter loses contact	Do nothing if the multicopter is already disarmed	When the multicopter has been disarmed, it is in STANDBY STATE. In this case, if the remote pilot does not operate manually, then the multicopter will do nothing
		Multicopters will be immediately disarmed if the multicopter has landed or the remote pilot's throttle control stick is at the bottom	If the multicopter is landed or the remote pilot's throttle control stick is at the bottom, the multicopter will be switched from MANUAL FLIGHT MODE to GROUND_ERROR STATE when the RC transmitter loses contact. Refer to condition C15
		RTL if the multicopter has a GPS receiver and the straight-line distance from the home position is greater than the set threshold	When the multicopter is in MANUAL FLIGHT MODE, if the RC transmitter loses contact and the straight-line distance from the home position is greater than the set threshold, then the multicopter is switched to RTL. Refer to condition C7
		Multicopter will immediately land if the vehicle does not have a GPS receiver or the straight-line distance from the home position is less than the set threshold	When the multicopter is in MANUAL FLIGHT MODE, if the RC transmitter loses contact and the straight-line distance from the home position is not greater than the set threshold, then the multicopter is switched to AUTO-LANDING. Refer to condition C11
		If the contact between the RC transmitter and the onboard RC receiver is reestablished (or the contact between the GCS and the onboard signal receiver is reestablished), then it is suggested that the multicopter still stay in the current mode rather than being switched back to the resume task it left off before the failsafe was triggered. However, if remote pilots manually switch the flight mode switch, they can retake the control of the multicopter	When the contact between the RC transmitter and onboard RC receiver is reestablished, the multicopter will stay in RTL or AUTO-LANDING. Only if the remote pilot manually switches the flight mode switch to another position, and then back to MANUAL FLIGHT MODE, the control of the multicopter can be retaken. Refer to conditions C8 and C12

(continued)

Table 14.3 (continued)

Safety problem	Safety subproblem	Safety demand	Decision-making
Sensor failure	Barometer failure	If the multicopter is monitored that problems exist in the barometer, it is suggested that the throttle command remain constant, and MANUAL FLIGHT MODE degrade from LOITER MODE or ALTITUDE HOLD MODE to STABILIZE MODE	When the multicopter is in MANUAL FLIGHT MODE, if the barometer is unhealthy, then the multicopter will be switched from LOITER/ALTITUDE HOLD MODE to STABILIZE MODE, referring to conditions C17 and C21. If the problem is solved, then the multicopter can turn back to LOITER/ALTITUDE HOLD MODE, referring to conditions C18 and C20
	Compass failure	If the multicopter is monitored that problems exist in the compass, it is suggested that MANUAL FLIGHT MODE degrade from LOITER MODE to ALTITUDE HOLD MODE or immediately land according to user configuration	When the multicopter is in MANUAL FLIGHT MODE, if the compass is unhealthy, then the multicopter will be switched from LOITER MODE to ALTITUDE HOLD MODE, referring to condition C16. If the problem is solved, then the multicopter can turn back to LOITER MODE, referring to condition C18
	GPS failure	If the multicopter is observed that problems exist in the GPS, it is suggested that MANUAL FLIGHT MODE degrade from LOITER MODE to ALTITUDE HOLD MODE or immediately land according to user configuration	When the multicopter is in MANUAL FLIGHT MODE, if the GPS is unhealthy, then the multicopter will be switched from LOITER MODE to ALTITUDE HOLD MODE, referring to condition C16. If the problem is solved, then the multicopter can turn back to LOITER MODE, referring to condition C18
	INS failure	If the multicopter is observed that failures exist in the INS, it is suggested that an emergency landing be performed by gradually reducing the throttle command	When the multicopter is in MANUAL FLIGHT MODE, if the INS is unhealthy, then the multicopter will be switched to AUTO-LANDING. Refer to condition C11
Propulsion system anomaly		Do nothing if the multicopter is already disarmed	When the multicopter has been disarmed, it is in STANDBY STATE. In this case, if the remote pilot does not operate manually, then the multicopter will do nothing
		Multicopters will be immediately disarmed if it is landed or the remote pilot's throttle control stick is at the bottom for a given time	Suppose that the multicopter has landed and is disarmed, or the remote pilot's throttle control stick is at the bottom. In this case if, INS or propulsion system is unhealthy, or the RC transmitter loses contact, then the multicopter will be switched from MANUAL FLIGHT MODE to GROUND_ERROR STATE. Refer to condition C15

(continued)

Table 14.3 (continued)

Safety problem	Safety subproblem	Safety demand	Decision-making
		RTL if the multicopter has a GPS receiver and is greater than a threshold from the home position and battery voltage is higher than the set threshold	When the multicopter is in MANUAL FLIGHT MODE, if the battery power is low but can still support the multicopter to perform RTL, then the multicopter will be switched to RTL automatically. Refer to condition C7
		In other cases, the multicopter land directly	If the battery power falls suddenly and cannot support the multicopter to perform RTL, then the multicopter will be switched to AUTO-LANDING automatically. If the propulsion system is unhealthy, then the multicopter will also be switched to AUTO-LANDING. Refer to condition C11

14.6   Summary

The social negative impact of aircraft accidents is very significant. Therefore, more effort on design and testing should be devoted in order to guarantee flight safety. Events that affect safety are mainly from three aspects: communication, sensors, and the propulsion system. There are a lot of methods based on measurements and models, to identify unhealthy behaviors or components in the system during both pre-flight and in-flight phase. Once unhealthy events are detected, it is required that a failsafe be provided in order to guarantee the flight safety. For a commercial flight control software, most codes are about failsafe around the events and corner cases which will happen by a very small possibility.

The research on safety issues has a long way to go. For the multicopter control accuracy, an increase of the accuracy from 99.9 to 99.99% may be unnecessary, but a rise in the probability of safety from 99.9 to 99.99% is rather significant. Civil aviation requires the probability of occurrence of catastrophic accidents per flight hour to be 10^{-9} [25]. This probability is roughly equivalent to "even if someone takes a plane every day, on the average the plane has a crash until thousands of years later". After a rigorous airworthiness review, a manned aircraft can be allowed to fly in the public airspace. As far as the author knows, crashes happen frequently at present for small multicopters. So, a question often asked is how to guarantee a high reliability as that of civil aircraft. The distinctions for multicopters or small drones include the following [26]: 1) It is impractical to equip very expensive components and software on a commercial multicopter. 2) It is infeasible to equip a lot of redundant components on a small multicopter because of the limited payload capacity. 3) Small multicopters often fly at a low altitude, where the environment is more complex and challenging. 4) In civil aircraft, the onboard remote pilot is willing and able to minimize or eliminate hazards to other aircraft or personnels or property on the ground. That ability may be significantly different from multicopters. 5) The dynamic model of multicopters is much simpler than that of civil aircraft. It is possible to achieve functional hazard assessment by using system dynamic model rather than statical fault tree analysis.

Therefore, the improvement of the reliability and safety of multicopters cannot simply follow the manned aircraft airworthiness review methods. Thus, research on safety issues of multicopters, especially small multicopters has a long way to go. Safety will be the decisive factor in the popularity of multicopters worldwide. Returning to the main topic of this chapter, the author believes that there are some theoretical problems to be solved but are not limited to:

(1) Classification of safety states. The completeness of classification criteria of unsafe events is expected. In this chapter, the unsafe events are simply categorized and enumerated according to the type of component that fails, such as GPS failure, compass failure, and so on. What is the best classification criterion so that unsafe events can be covered as completely as possible?

(2) Design of transition conditions. The multicopter decision-making EFSM contains a lot of transition conditions, each of them being very complex and mainly designed based on experience. How should the transition condition be designed to improve the flight safety?

(3) Health evaluation and prediction. The decision-making EFSM for the multicopter is given, and it is assumed that the probability of occurrence of a single unsafe event is known. In this case, how could the probability of occurrence of a catastrophic accident be evaluated? Furthermore, how could the occurrence of a component failure and the result of evaluation to the multicopter's health condition be forecasted more accurately and timely?

Exercises

14.1 For a linear system

$$\dot{\mathbf{x}} = \begin{bmatrix} -1 & 0 & 0 \\ 0 & -2 & 0 \\ 0 & 0 & -2.5 \end{bmatrix} \mathbf{x} + \begin{bmatrix} 1 \\ \eta \\ 1 \end{bmatrix} u + \mathbf{w}$$

$$\mathbf{y} = \mathbf{x} + \mathbf{v}$$

where $\mathbf{w} \sim \mathcal{N} (\mathbf{0}_{3\times1}, 0.01\mathbf{I}_3)$ and $\mathbf{v} \sim \mathcal{N} (\mathbf{0}_{3\times1}, 0.01\mathbf{I}_3)$, $\eta \in \mathbb{R}$ is an unknown parameter, which satisfies $\dot{\eta} = \varepsilon, \varepsilon \sim \mathcal{N} (0, 10^{-5})$. Given a sequence of the control input u and the observation measurements $\mathbf{y}$, design a Kalman filter to estimate the true value of the unknown parameter η.

14.2 For a battery model in some references, the observation equation can be written as

$$U_k = K_0 - \frac{K_1}{S_k} - K_2 S_k + K_3 \ln (1 - S_k) - R I_k$$

where U_k is the battery output voltage, I_k is the current, S_k is the SoC value, R is the unknown internal resistance, and the parameters $\{K_0, K_1, K_2, K_3\}$ are unknown parameters. Given a sequence of $\{U_k, I_k, S_k\}$ for $k = 1, 2, \cdots , M$, use the least square method to estimate the unknown parameters $\{K_0, K_1, K_2, K_3\}$ and R.

14.3 Given a set of data collected from the accelerometer of the multicopter under the situation that one of the blades of the multicopter is faultless, fractured, and distorted. The data can be downloaded from http://rfly.buaa.edu.cn/course. Select the input and output for Artificial Neural Network (ANN) and design an ANN to identify different fault types.

14.4 For a multicopter, just consider the communication breakdown situation, and assume other modules such as sensors and propulsion system are all healthy. Design an autopilot logic such as the one shown in Fig. 14.4 to satisfy the communication failsafe in Sect. 14.4.1.

14.5 Except for the safety issues and failsafe suggestions given in this chapter, list other potential safety issues and failsafe suggestions.

References

1. Grewal MS, Weill LR, Andrews AP (2001) Global positioning systems, inertial navigation, and integration, 2nd edn. Wiley, New York
2. Gieras JF (2002) Permanent magnet motor technology: design and applications. CRC Press, Boca Raton
3. Tong W (2014) Mechanical design of electric motors. CRC Press, Boca Raton
4. Aaseng G, Patterson-Hine A, Garcia-Galan C (2005) A review of system health state determination methods. In: 1st space exploration conference, Orlando, AIAA 2005-2528
5. Zhao Z, Quan Q, Cai KY (2014) A profust reliability based approach to prognostics and health management. IEEE Trans Reliab 63:26–41
6. Heredia G, Ollero A, Bejar M, Mahtani R (2008) Sensor and actuator fault detection in small autonomous helicopters. Mechatronics 18:90–99
7. Isermann R (2005) Model-based fault-detection and diagnosis—status and applications. Annu Rev Control 29:71–85
8. Hwang I, Kim S, Kim Y, Seah CE (2010) A survey of fault detection, isolation, and reconfiguration methods. IEEE Trans Control Syst Technol 18:636–653
9. Yan J, Zhao Z, Liu H, Quan Q (2015) Fault detection and identification for quadrotor based on airframe vibration signals: a data-driven method. In: 34th CCC, Hangzhou, pp 6356–6361
10. He W, Williard N, Chen C, Pecht M (2013) State of charge estimation for electric vehicle batteries using unscented Kalman filtering. Microelectron Reliab 53:840–847
11. Sepasi S, Ghorbani R, Liaw BY (2014) A novel on-board state-of-charge estimation method for aged Li-ion batteries based on model adaptive extended Kalman filter. J Power Sources 245:337–344
12. Plett G (2004) Extended Kalman filtering for battery management systems of LiPB-based HEV battery packs, part1. Background. J Power Sources 134:252–261
13. Plett G (2004) Extended Kalman filtering for battery management systems of LiPB-based HEV battery packs, par2. Modeling and identification. J Power Sources 134:262–276
14. Plett G (2004) Extended Kalman filtering for battery management systems of LiPB-based HEV battery packs, part3. State and parameter estimation. J Power Sources 134:277–292
15. Failsafe. http://ardupilot.org/copter/docs/failsafe-landing-page.html. Accessed 11 Oct 2016
16. Du GX, Quan Q, Cai KY (2015) Controllability analysis and degraded control for a class of hexacopters subject to rotor failures. J Intell Robot Syst 78:143–157
17. DJI (2013) Failsafe of broken propeller. http://www.dji.com/cn/product/a2/feature. Accessed 11 Oct 2016
18. Zhou QL, Zhang YM, Rabbath CA, Apkarian J (2010) Two reconfigurable control allocation schemes for unmanned aerial vehicle under stuck actuator failures. In: AIAA guidance, navigation, and control conference, Toronto, AIAA 2010-8419
19. Zhang YM, Jiang J (2001) Integrated design of reconfigurable fault-tolerant control systems. J Guid Control Dyn 24:133–136
20. Michalska H, Mayne DQ (1993) Robust receding horizon control of constrained nonlinear systems. IEEE Trans Autom Control 38:1623–1633
21. Lin H, Antsaklis PJ (2014) Hybrid dynamical systems: an introduction to control and verification. Found Trends Syst Control 1:1–172
22. Bujorianu ML (2004) Extended stochastic hybrid systems and their reachability problem. Hybrid systems: computation and control. Springer, Berlin, pp 234–249
23. Wonham WM (2009) Supervisory control of discrete-event systems. Lecture notes, Department of Electrical and Computer Engineering, University of Toronto, Canada
24. Quan Q, Zhao Z, Lin LY, Wang P, Wonham WM, Cai KY (2017) Failsafe mechanism design of multicopters based on supervisory control theory. https://arxiv.org/abs/1704.08605. Accessed 17 May 2017
25. System Safety. Analysis and assessment for part 23, Technical Report 23.1309-1E, Small Airplane Directorate, Federal Aviation Administration, USA, 11/17/2011
26. Hayhurst KJ, Maddalon JM, Miner PS et al (2007) Preliminary considerations for classifying hazards of unmanned aircraft systems. Technical Report NASA TM-2007-214539, National Aeronautics and Space Administration, Langley Research Center, Hampton, Virginia

Outlook

15

Since multicopters entered the consumer market, the related companies have emerged like bamboo shoots after a spring rain, and scholars have also flocked. The largest market of multicopters remains the aerial photography area, which has been firmly occupied by a few well-known companies. The remaining markets seem relatively small or have a higher entering threshold. In recent years, there exists a dispute on the multicopter market whether it is a "Red Ocean" or a "Blue Ocean[1]". There is diminishing interest in academia because most of the fundamental problems, such as stabilizing or tracking, seem to have been solved. Therefore, everyone seems to be standing at a crossroads. This chapter aims to answer the question below:

Where will multicopters go?

The answer to this question involves the related technology development, innovation direction, risk analysis, opportunities, and challenges. Contents in this chapter are the revision and extension of a magazine paper in Chinese published by the author [1].

---

[1]A jargon for the uncontested market space for an unknown industry or innovation is coined in the book [2]. In an established industry, the competition is often so intense that some firms cannot sustain themselves and stop operating. This type of industry is described as a "red ocean" [3].

Q. Quan, *Introduction to Multicopter Design and Control*, DOI 10.1007/978-981-10-3382-7_15

Chinese Helicopter

According to Chinese history, it was Zhengming Xu who invented flying vehicles in the 17th century. Zhengming Xu was a talented craftsman. He began to study the mechanism of flying vehicles after being inspired by a book named "*Classic of Mountains and Seas*". After more than 10 years of effort, he finally succeeded. His flying vehicle was "made of wood, like an armchair with complex mechanical gears attached to the bottom; sitting on the chair and operating with the foot treadles, one could make the mechanical system turn around and then the chair could fly quickly in the air; the flying vehicle could lift off and pass through a small river (The Chinese is "其制如栲栳椅子式，下有机关，尺牙错合；人坐椅中，以两足击板，上下之机转，风旋疾驰而去，离地可尺余，飞渡港汉不由桥")". This story was recorded in "*A Gazetteer of Xiangshan* (香山小志)", a book recording the history of Xiangshan county in Jiangsu province. This flying vehicle might be one of the earliest prototypes of helicopter.

15.1   Related Technology Development

The performance of multicopters will be improved by following the advances of the related science and technology. As a result, the advantage over fixed-wing aircraft and helicopters will be further highlighted. The technology can be summarized as seven major components, as shown in Fig. 15.1.

15.1.1   Propulsion Technology

(1) New battery. On May 22, 2015, a company called EnergyOr Technologies Inc. demonstrated the world's longest multicopter flight, then creating a record of 3 h 43 min and 48 s [4]. Moreover, there are some potential new battery technologies although they are not applicable to multicopters for the time being, such as the Graphene battery [5], Aluminum-Air battery [6], and Nanodot battery [7]. There is an extremely urgent need for batteries with high capacity or fast-charging ability, especially in electric cars and smartphones. After new batteries being reliable enough to be applied to electric cars and smartphones, they can then be installed onto multicopters.

(2) Hybrid power. In 2015, a company called Top Flight Technologies created a world record that their hexacopter with 1 gallon of gasoline can fly for more than two and a half hours—or 160 km—carrying a payload weighing 9 kg [8].

(3) Power supply on the ground. A tethered hovering aerial system, such as HoverMast developed by a company called Sky Sapience in Israel [9], Tianshu-100 by Droneyee in China [10], and PARC by a company called CyPhy Works in the USA [11], could be driven by the power supply on the ground via a wire in the tether. As a result, these aerial systems can stay in the air for an unlimited period

Fig. 15.1 Related technologies on multicopters

of time in theory. Moreover, such a system could provide high quality, full frame rate, unbroken, and high definition video that no other small or micro UAV adopting wireless communication methods can match.

(4) Wireless charging. A company called Skysense in Berlin developed a landing pad with the capability as a wireless charging station [12]. A company called Reforges in Canada also developed a landing station for wireless charging of multicopters [13]. If the charging process is fast enough, then such wireless charging stations distributed over an area will ensure multicopters to fly a long distance autonomously by multiple takeoff and landing.

(5) Combustion engine. A German company called Airstier was building a quadcopter which was powered by a combustion engine. The launched quadcopter was expected to reach a top speed of 100 km/h, which could stay in the air for up to 60 min with a payload capacity of 5 kg [14].

15.1.2 Navigation Technology

15.1.2.1 Precise Positioning

(1) Real-Time Kinematics (RTK). A company called Swift Navigation released its first product Piksi, which was claimed to be a low-cost, high-performance GPS receiver with RTK functionality for centimeter-level relative positioning accuracy [15]. Since it is small in size and light in weight, it is convenient to be integrated into a small multicopter system. RTK is also supported by RTKLIB [16], an open source Global Navigation Satellite System (GNSS) toolkit for performing standard and precise positioning. On September 24, 2015, the fourth session of the Chinese satellite navigation and location-based services conference and exhibition was held in Beijing. In the exhibition, Chinese Beidou receivers with centimeter-level accuracy appeared for the first time.

(2) NAVigation via Signals of OPportunity (NAVSOP). UK defense firm BAE Systems has come up with a positioning system which "could make GPS obsolete". It is called NAVSOP and uses the different signals that populate the airwaves—including Wi-fi, TV, and mobile phone signals—to calculate the user's location, with the lower bound of detection a few meters [17].

(3) Ultra-WideBand (UWB). This technology is based on the Time Difference of Arrival (TDoA) of UWB wave pulses nanoseconds apart, achieving 2D and 3D position readings of within 10 cm accuracy [18]. Traditional location technology is based on signal strength and quality, often resulting in unsatisfactory results and high energy consumption. UWB location technology fills the gap by transmitting in a way that does not interfere with conventional signals, preventing jamming and the multipath effect, while delivering highly precise readings and requiring less power. The combination of UWB with IMU will further improve the accuracy and robustness against uncertainties [19, 20].

15.1.2.2 Velocity Measurement

For safety purpose, it is important to have an accurate and robust velocity estimate in all flight environments. First, velocity feedback will increase damping to improve the stability of multicopters during hovering. Moreover, it will make multicopters more tractable. A commonly used velocity estimation method is the optical flow-based scheme, i.e., estimation based on the combination of optical flow, ultrasonic range finders, and IMU. AR. Drone is the first multicopter product using optical flow to improve the performance, achieving a great success. PX4Flow, designed by ETH Zurich, is an optical flow smart camera. Unlike many mouse sensors, it can work indoors and in dim outdoor conditions. It can be freely reprogrammed to do any other basic, efficient low-level computer vision task [21, 22]. Such a sensor can help multicopters hover indoor and outdoor without GPS.

15.1.2.3 Obstacle Avoidance

(1) Depth camera. A depth camera often has a conventional camera, an infrared laser projector, and an infrared camera. The infrared projector projects an invisible infrared light grid onto the scene, which is recorded in order to compute the depth information. The first-generation Kinect was first introduced by Microsoft in November 2010 [23]. In the international Consumer Electronics Show (CES) 2015, a demonstration was given by Intel in which a number of multicopters were equipped with Intel RealSense cameras to avoid collision. Compared with Kinect, Intel's RealSense camera module is smaller and lighter, featured with as little as 8 grams in weight and 4 mm thick [24].

(2) Ultrasonic range finders. An obstacle avoidance system, eBumper4, was developed by a company called Panoptes. It had four sonar sensors providing the information about the physical environment to the front, right, left, and above. When eBumper4 senses an obstacle in its field of vision, it prevents the drone from getting too close to the obstacle [25].

(3) Camera & Memristor. A company called Bio Inspired Technologies was building a sense-and-avoid system using a memristor, which was a resistor with a memory. It was expected that a chip-sized neural system linked to the drone's existing camera can be trained to recognize aircraft and other hazards at a long distance [26].

(4) Binocular vision. A company called Skydio in its demonstration showed that a prototype quad-copter used two small cameras to sense objects and avoid collisions [27].

(5) Metamaterial Radar. A start-up called Echodyne was working on a solution for steering radar beams, and they claimed to be creating the world's lightest, smallest, and lowest cost high performance electronically scanning radar. This type of metamaterial radar can realize collision avoidance in most conditions. It was also expected by the company to equip onto quadcopters [28].

(6) LiDAR. Indoor mobile robots such as sweeping robots use 2D scanning laser sensors to navigate indoor environments including obstacle avoidance. The 2D scanning laser sensor can also be used for multicopters to do similar tasks. As a higher version, LiDARs are more and more widely used and are also made smaller and cheaper. For example, Velodyne's PUCK™ (VLP-16) is 0.83 kg in weight [29].

15.1.2.4 Tracking

A commonly used tracking method relies on a small GPS tracker which the user carries. However, for noncooperative targets, such as criminals, the GPS-based tracking method cannot work anymore. New tracking methods rely on vision and radar. In terms of visual tracking, a company called 3D Robotics released "Tower Drone Control App", which has the "Follow Me" mode. This mode can keep the camera centered on the user while the drone follows the user's movement and the camera's angle can also be adjusted as the drone follows you [30]. OpenCV, an open source library of programming functions, has many tracking algorithms [31]. Although visual tracking is cheap, it will be disturbed by complex weather or light. By contrast, radar-based tracking is a robust method. In 2015, a radar sensor developer called Oculii had launched the RFS-M. It was claimed to be the first real-time 4D tracking sensor designed for mobile platforms [32].

15.1.3 Interactive Technology

(1) Gesture Control. At CES 2014, a demonstration was shown that an AR. Drone's tilt direction was controlled by using just one arm with a Myo gesture-control armband [33]. Similarly, the wearable devices, like smartphones, smartbands, smartwatches, and smartrings, have IMU inside, which can be utilized to recognize gestures. They can then be used to control a multicopter as well.

(2) Brain–Computer Interface. The brain–computer interface is a direct communication pathway between the brain and an external device. Signals from an array of neurons are read by sensors and

then translated into actuation signals by using computer chips and programs. This is the basic principle of manipulating objects by mind. A company called Blackrock Microsystems LLC, announced that it had struck an agreement with Brown University to license and commercialize a wireless neural activity monitoring system to acquire high-fidelity neural data during animal behavior experiments [34]. The company called Emotiv Systems also offered a wireless brain–computer interface [35]. Based on these brain–computer interfaces, the potential applications to drones are further figured out. CCNT Lab in Zhejiang University released a demonstration that a quadcopter was controlled by a brain–computer interface. On February 25, 2015, it was demonstrated in Portugal that a drone was piloted from the ground using only a person's brain waves [36]. A research team in Fudan University also designed a hexacopter controlled by a combination of brain waves and the gestures. However, it should be pointed out such multicopters may be still miles away from commercial applications due to safety considerations.

15.1.4   Communication Technology

Communication technology will facilitate the information sharing and can be further used for traffic management or health monitoring of drones. With high-speed communication technologies, the data can be uploaded to cloud servers.

(1) 5G. In November, 2015, it was reported that the USA wanted a global deal to reallocate parts of the radio spectrum for the next "5G" generation of mobile devices, which will cause a boom in civilian drones and a worldwide flight tracking system. This was expected to become an US $80 billion business in the USA alone over the next decade [37].

(2) Wi-fi. A research team at the Karlsruhe Institute of Technology in Germany created a Wi-fi solution that used "better hardware" along with a high radio frequency of 240 GHz to produce a constant connection speed of 40 Gbit/s (5 GB/s) at a distance of about 0.6 miles [38]. This technology could offer a high-speed connection to devices on the ground through drones' video transmitters.

15.1.5   Chip Technology

On September 9, 2014, a company called 3D Robotics announced that it had partnered with Intel in the development of a new microcomputer, namely Edison. This provided PC power in postage stamp size but at an acceptable price [39]. At CES 2015, based on a Qualcomm Snapdragon processor, Qualcomm Research designed the Snapdragon Cargo—a flying and rolling robot with an integrated flight controller. The processor used could provide a low-power solution that integrates multicore processing, wireless communications, sensor integration, positioning, and real-time I/O for multiple robotics applications [40]. Brain-inspired chips or termed as neuromorphic chips are also attractive. DARPA's Systems of Neuromorphic Adaptive Plastic Scalable Electronics (SyNAPSE) program was launched in 2008 to develop neuromorphic or brain-like chips. It was expected that this chip could accomplish the own tasks such as drawing information from video feeds and other sensor data, and provide decision support. HRL, IBM, and Hewlett Packard received the initial grants [41]. HRL released a demonstration that, by a prototype chip with 576 silicon "neurons", a quadcopter was able to process data from its optical, ultrasound, and infrared sensors and recognize when it was in a new or a familiar room. IBM unveiled a neuromorphic chip with 5.4 billion transistors designed to perform tasks such as pattern recognition, audio processing, and motion control [41, 42]. It is expected that brain-like chips can push the autonomous systems to a new level.

15.1.6   Software Platform Technology

(1) On October 13, 2014, the Linux Foundation announced the founding of the Dronecode Project to advance development of drones. The ultimate goal is to maximize adoption of the project's code for the benefit of users by developing cheaper, better, and more reliable drone software. The platform has been adopted by many organizations on the forefront of drone technology [43, 44].

(2) Ubuntu was recently updated to version 15.04, which was smaller and more secure than any previous Ubuntu edition. This operating system has been adopted by Snapdragon Flight from Qualcomm and the Erle-Copter Ubuntu drone [45].

(3) A company called Airware launched its commercial drone operating system. It was claimed that this operating system could help businesses safely and reliably operate drones at scale, which moreover complied with the government and insurance requirements, and built industry-specific drone software solutions. Also, Airware hoped to enable customers to select vehicles and then mix and match hardware and software components to create drones for different jobs [46].

15.1.7   Air Traffic Control Technology

Airware had partnered with NASA to develop an Unmanned aerial system Traffic Management system (UTM) that will enable safe and efficient low-altitude Unmanned Aerial System (UAS) operations. The four-year program would create a series of prototype air traffic management systems and could shape how widely commercial drones can be used. NASA was also partnering with a company called Exelis to test a new air traffic control system called Symphony RangeVue for drone aircraft. It would transmit FAA data and drone-tracking information into a mobile APP, thus allowing Unmanned Aerial Vehicle (UAV) operators to see what aircraft are in the vicinity of their vehicles at any given time. While the system did not meet existing line-of-sight regulations, the company hoped that it can convince the FAA to relax those restrictions by demonstrating that it could solve traffic-related issues [47]. In 2015, a company called Transtrex released the beta version of the first dynamic geospatial restriction system for UAS, at the NASA UTM Convention at Moffett Airfield. The released system can provide dynamic 3D map data, including man-made objects and operational restrictions like permanent or temporal no-fly zones, based on Transtrex' dynamic 3DTMaps for UAS operations [48].

The NASA UAS Traffic Management Convention or UTM 2015 was held July 28–30, 2015 at NASA Ames Research Center at Moffett Field, California [49–51]. This convention offered an opportunity to engage in small group discussion and to shape the future of low-altitude UAS traffic management. White papers had been published in the convention by both Amazon [52, 53] and Google [54] to explore some of the strategies for managing the airspace and coordinating aerial vehicles through onboard system requirements such as ADS-B and V2V communication. In February 2016, researchers in [55] proposed a universal architecture and a vocabulary of concepts to describe the Internet of Drones (IoD). It is more concrete than the white papers. It was claimed that the proposed architecture could provide generic services for various drone applications such as package delivery, traffic surveillance, search and rescue, and more. In September 2015, at the 3rd AOPA International Flight Training Exhibition, AOPA (Aircraft Owners and Pilots Association) China announced and launched a monitoring system called "U Cloud" for small drones. It can help remote pilots know the drones nearby and will help to regulate the flight of drones in China.

15.1.8   Concluding Remark

The new technologies related to multicopters are mutually interdependent, as for chips, sensors, and algorithms, and so on. These technologies will create an ecological environment of drones (multicopters). In this case, the development of drones (multicopters) might be led by that of one technology or two. There exist various possibilities, many of which cannot be imagined now. Similarly, drones (multicopters) in turn will boost the development of the related industries and technology. During this process, many existing problems will be solved. Therefore, thanks to the development of the related technologies, there is a large space for the development of multicopters.

15.2   Demand and Technology Innovation

15.2.1   Innovation Level

Research and design of a multicopter start from the three levels: demand, scheme, and technology. Among the three levels, finding a new demand contributes the most. Under the same demand, a good scheme is the second best. While, under the same demand and scheme, improving the technology is the last resort.

The requirements of different innovation levels are not the same.

(1) Demand innovation requires that a person possess not only the capability grasping user demand, but also a comprehensive knowledge about the feasibility of possible schemes and technologies. Since demand innovation will bring new problems and features, it can immediately manifest differences from other products. Sequentially, new problems will stimulate a new design, crossing the threshold for a new product.

(2) From the point of view of solving a problem, the scheme is very important. Scheme innovation requires a wide knowledge of various technologies including software, hardware, and algorithms. Sometimes, an appropriate scheme will reduce the difficulty in technology greatly. For example, an IMU inside smart devices and a camera can both be used for gesture recognition, but the point is which scheme beats the other. The former will make it easier to develop a good and robust algorithm, whereas visual algorithms for gesture recognition are often restricted in its applications. In the end, it should be pointed out that the specific scenario has to be taken into account as to which choice of scheme is to be made.

(3) Technological innovation requires strong expertise in this field. It is the most difficult. The improvement in performance must be so significant that the users can see it immediately. For example, the optimal flow-based velocity estimation is an important technology, but the cost will be higher with an increase of one percentage point in accuracy.

In the following, innovation from applications and performances are introduced.

15.2.2   Application Innovation

Multicopters have many applications. There are different ways to classify multicopters, the majority of which is functionality-based. In the following, this chapter will classify multicopters from the payload they carry onboard, namely "Multicopter + X", where "X" is the payload. Based on this classification, here are some new demands and applications as examples.

(1) "Multicopter" itself. A multicopter can be used as a toy for education and fun purpose. The Kickstarter, the world's largest funding platform for creative projects, started a project about building

a multicopter by using LEGO Bricks [56]. Recently, FPV[2] Racing is becoming increasingly popular [57, 58]. Multicopters have become popular aircraft for such a sport.

(2) "Multicopter + Camera". In addition to the common areas like the aerial photography and aerial power line inspection, Renault, a car company, has unveiled the "KWID concept" at the Delhi Auto Show, i.e., a concept car featuring a built-in quadcopter. The flying companion can be used for a variety of purposes, including scouting traffic, taking landscape pictures, and detecting obstacles on the road ahead [59]. Hover camera passport, launched by Zero Robotics in China, is a quadcopter enclosed carbon fiber frame which is safe, portable, and foldable. It allows you to grab it and reposition it like a floating camera close to you [60].

(3) "Multicopter + Pesticide". On April 19, 2015, a Chinese Company called Xaircraft released an agricultural multicopter, namely XPLANET P20 system. It was reported that by using the agricultural multicopters, Xaircraft agricultural service teams accomplished effective chemical spraying for 1650 acres of rice field within just three days [61]. At the end of 2015, Chinese Companies DJI and Zero Tech also released their agricultural multicopters sequentially.

(4) "Multicopter + Parcel". As the first batch of enterprises to test drone delivery systems, Amazon has invested great effort and enthusiasm, and completed several rounds of drone test [62]. DHL Parcel has completed a comprehensive consultation and approval process led by Lower Saxony's Ministry for Economics, Labor and Transport, and it will soon launch a unique pilot project on the North Sea island of Juist for the first time [63]. The Red Line flying robot consortium aimed to set up the first cargo drone route in Africa starting in 2016. This route will include several affordable drone ports and drop facilities that will connect several towns and villages [64]. In terms of the cost of drone delivery, it is very cheap. Roughly, it is about 0.2 cent/km for a 2 kg payload [65]. In addition to the regular delivery, more importantly, drone can carry medical kits for emergency medical care to save lives [66].

(5) "Multicopter + Instrument of surveying and mapping". France-based company, Drones Imaging, integrated drone imagery with 3D printer technology. Based on photogrammetric approaches, the integration of gathered images results in 3D printed models for use in applications that include mining, archeology, construction, and various military functions [67].

(6) "Multicopter + Communication platform". A tethered hovering aerial system, such as HoverMast developed by Sky Sapience in Israel [9] and Tianshu-100 by Droneyee in China [10] can be used as a communication platform in the air. It is suitable for the emergency communication for the areas like earthquake areas.

(7) "Multicopter + Weapon". A video showed that a semiautomatic handgun mounted on a multicopter as a "flying gun" can shoot by remote control [68]. If the "flying gun" is equipped with an FPV and further improved, then this new weapon can be used by soldiers far from the battlefield. This can save soldiers' lives.

(8) "Multicopter + Light". The Ars Electronica Futurelab made up a drone swarm that could "draw" three-dimensional figures in midair [69]. In the Cirque du Soleil short film "SPARKED," a man interacted with a playful swarm of flying lampshades. The flying effect was achieved by ten quadcopters [70].

(9) "Multicopter + Voice". It was reported that 20-year-old Wellington engineering student herded about a thousand sheep with an RC quadcopter [71].

(10) "Multicopter + RFID reader". The Fraunhofer Institute's InventAIRy Project planned to develop flying inventory robots with RFID reader to keep tabs on stock. The autonomous flying robot was expected to navigate independently [72].

---

[2]First-person view (FPV), also known as remote-person view (RPV), or simply video piloting, is a method used to control a Radio Controlled (RC) vehicle from the driver or pilot's view point.

(11) "Multicopter + Rope". Researchers at ETH Zurich used multicopters in laboratory to build load-bearing structures that can support a person [73]. This idea could be applicable to rescue when without bridges.

(12) "Multicopter + Flamethrower". A team in Guangzhou Power Supply Bureau of China Southern Power Grid designed a quadcopter with a flamethrower. It can help to clear plastic films hanging on high-voltage wires.

Through the idea of "Multicopter Plus" or "Drone Plus", new applications or demands can be figured out by combining multicopters with other devices. Under new scenarios, new problems are on the way, and new technologies are in need as a result. For example, agricultural multicopters working on a farmland require a robust scheme or algorithm to obtain the accurate altitude, while a multicopter performance for entertainment requires a fast-and-accurate positioning technique and a precise trajectory tracking control algorithm. Besides the classification mentioned above, the market can be also classified with respect to customers, such as children, young men, seniors, or researchers, professionals, hobbyists. It is worthwhile to explore the potential demand.

At present, for start-ups, the market demand for a wide consumer group and high value-added applications looks more promising, such as mini toys, multicopters for education, wearable multicopters, the agricultural multicopters, and some commercial multicopters for special applications. The risks of these applications are under control, meanwhile the broad consumer market or high profit can help companies earn a reasonable profit, resulting in a positive feedback to facilitate the product to be further developed. Besides the business-to-customer fashion mentioned above, another way for start-ups is to focus on some special techniques, offering the service or products for the final producers, namely business-to-business.

15.2.3   Performance Innovation

Apart the demand, let us just consider the multicopter platform itself. In fact, designers can improve one of multicopter's performances to highlight the characteristics of products or research. As shown in Table 15.1, the eight key properties and five key technologies corresponding to main chapters of this book are listed.

Table 15.1 Relationship between performances and techniques (More "+" implies that the relationship is stronger, while "–" implies an irrelevant relationship)

Techniques performances	Configuration and structural design	Propulsion system	State estimation	Controller design	Health evaluation and failsafe
Minimum noise	+++	++	+	+	–
Minimum vibration	+++	++	+	+	–
Longest hover time	+	+++	+	++	–
Longest flight Distance	++	+++	+	+	–
Strongest capability of wind resistance	++	+++	+	++	–
Most precise trajectory tracking	–	–	+++	+++	–
Highest autonomy	–	–	+++	+	++
Highest safety factor	+	++	+++	++	+++

Each performance in Table 15.1 is a necessary topic, among which the performance "the highest safety factor" is the most important. However, it is very difficult to improve this performance. In order to improve the safety factor of multicopters, the health evaluation and failsafe mentioned in Chap. 14 are however far from enough. *Reliability* is often quantified by the mean time between successive failures, but the concept of *safety* is not easy to define clearly, which often depends on the concrete example. Roughly, safety related to multicopters can be considered as the protection of personnel ground, property, multicopter itself, and its accessory. Safety is a system property, not a component property, and must be controlled at the system level, not the component level [74, p.14]. At the system level, some interactions among components have to be considered. The reliable but unsafe example and the safe but unreliable example are given in [74, pp.8-11].

As a small complex system, a multicopter should be designed following the way of systems engineering to guarantee the flight safety. In 2007, the International Council on Systems Engineering (INCOSE) in "Systems Engineering Vision 2020" gave the definition of Model-Based Systems Engineering (MBSE): "MBSE is the formalized application of modeling to support system requirements, design, analysis, verification and validation activities beginning in the conceptual design phase and continuing throughout development and later life cycle phases [75]". The concept corresponds to that of Text-Based Systems Engineering (TSE). It is expected to replace the document-centric approach and further to eliminate uncertainty, ambiguity and incomputability and so on. The design process is changed from the "document-oriented supplemented by model" to "model-oriented supplemented by document" [76, 77]. This is also the aircraft design process adopted by some well-known international aircraft companies such as Boeing, Airbus, and Lockheed Martin. So far, this is an efficient way to guarantee a civil aircraft with a high safety factor and further to meet the international airworthiness standards. Some companies with sufficient resources can implement MBSE for drones or multicopters. With multicopters or drones being applied more and more widely, the safety factor standards or airworthiness standards are bound to become concerns to the public.

Compared with the concern about safety factor growing, autonomy is relatively easier to get the market's attention in a short term. Therefore, many new products feature the characteristic of autonomy, which is expected to become the selling point. Autonomy is about self-governing, also related to safety. "*When applied to UAS, autonomy can be defined as UAS's own ability of integrated sensing, perceiving, analyzing, communicating, planning, decision-making, and acting/executing, to achieve its goals as assigned by its human operator(s) through designed HRI or by another system that the UAS communicates with [77, p.59]*". The definition is also applicable to multicopter systems. The higher autonomous level implies that the flight is easier to realize. It was reported in "DoD Unmanned Aerial Vehicles Roadmap 2000–2025" that the US Air Force Research Laboratory (AFRL) defined ten levels of autonomous capability to serve as a standard for measuring progress [78]. In 2002, the US AFRL presented the results of a research study on how to measure the autonomy level of a UAV in detail [79]. This definition is for combat drones on the battlefield, hence not very suitable for micro and small multicopters mainly as personal use. For such a purpose, the level classification proposed in [77, p.65] is modified and improved, which is shown in Table 15.2.

From current products, most of the multicopter toys are in the autonomy Level 0. Most of the commercial products are in the autonomy Level 1. The multicopter equipped with the APM autopilot is a typical representative of this level. Both AR. Drone and DJI Phantom 3, with a capability to hover based on vision, have reached initial autonomy Level 2. Tasks in some competitions on UAVs, such as the International Aerial Robotics Competition (IARC),[3] have featured the characteristic of Level 4.

---

[3]http://www.aerialroboticscompetition.org/ or http://iarc.buaa.edu.cn/.

Table 15.2 Multicopter's autonomy levels

Level	Descriptor	Decision	Perception	Control	Typical scenario
4	Real-time obstacle/Event detection and path planning	Hazard avoidance, real-time path planning and re-planning, event driven decisions, robust response to mission changes	Perception capabilities for obstacle, risks, target and environment changes detection, real time mapping	Accurate and robust 3D trajectory tracking capability is desired	It can fly over a long distance using camera (no GPS) and return home automatically. Moreover, a quadcopter can land safely after one propulsor failure
3	Fault/Event adaptation	Health diagnosis; limited adaptation; onboard conservative and low-level decisions; execution of pre-programmed tasks	Most health and status sensing; detection of hardware and software faults	Robust flight controller; reconfigurable or adaptive control to compensate for most failures; mission and environment changes	It can evaluate its health and can analyze the reason that a failure arises. Moreover, a hexacopter can return home after one propulsor failure
2	ESI (External system independence) navigation (e.g., Non-GPS)	Same as in level 1	All sensing and state estimation by the multicopters (no external system such as GPS). A health assessment capabilities can perceive failure in advance to take failsafe strategies. All perception and situation awareness by the remote pilot	Same as in level 1	It can rely on a camera (no GPS) to accomplish hover, landing, and tracking a target. Moreover, it can report a typical failure in advance, including RC failure, sensor failure, and propulsion system failure and so on
1	Automatic flight control	Pre-programmed or uploaded flight plans (waypoints, reference trajectories, etc.); all analyzing, planning, and decision-making by ground stations or remote pilot; simple failsafe	Most sensing and state estimation by multicopters, and simple health sensing. All perception and situational awareness by the remote pilot	Control commands are computed by the flight control system	It can hover, land, and track a target by using GPS. It can also detect the RC failure, or GPS and electronic compass failure, or the battery power being low. Moreover, it can be switched to a mode for safe landing
0	Remote control	All guidance functions are mainly performed by remote pilots	Sensing may be performed by multicopters, all data is processed and analyzed by remote pilot	Control commands are given by a remote pilot	It can be controlled by a remote pilot to accomplish hover, landing, and tracking a target

15.3 Analysis

15.3.1 Risks

The risks will be different because the consequences caused by a failure are different. For example, an operating system crashes on a personal computer will probably not damage the user. However, if a multicopter falls from the air, the loss will be very serious, causing a series of negative impacts as follows:

(1) Personal safety. Although the multicopters are preferred to be light in weight, after carrying a variety of equipment, its weight may be considerable, causing it to injure or even kill people in case of a failure.

(2) Property. Unlike model aircraft, it may be equipped with some expensive equipment. A failure may destroy the expensive equipment.

(3) Ethics and public risk. Public safety or privacy issues about multicopters will inevitably attract media attention or be hyped. The resultant great public pressure will slow down the development of multicopters. The industry even faces legal restrictions. So far, multicopters do not seem much affected. This is because the number of multicopters in the market is not large enough. However, the signs of risk can be observed. For example, on January 29, 2015, a quadcopter crashed at the White House [80]. On April 24, 2015, a quadcopter landed on the roof of Japanese PM's office [81]. On June 26, 2015, a quadcopter crashed at the Milan Cathedral [82]. On September 5, 2015, a quadcopter slammed into the seating area at US Open [83].

15.3.2 Suggestions

In order to reduce the risk, if follow the standard of civil manned aircraft, multicopters have to obtain an airworthiness certificate issued by the national aviation authority, indicating that multicopters are qualified. However, this may strangle the industry of multicopters or civilian drones in the cradle, especially consumer-level multicopters. On the other hand, many countries in the world in fact still lack such standards. It will take some time to improve laws, regulations, and standards. So, in order to reduce the risk, multicopters have to be improved as safe as possible. The following are a few suggestions.

15.3.2.1 For Multicopter Producers

(1) Improve the reliability. With respect to hardware, qualified components are suggested to be used. In terms of software design, failsafe has to be considered in advance. MBSE is suggested to adopt, and software has to be tested intensively and comprehensively.

(2) Reduce the impact of multicopter falling. Reducing the multicopter's weight is the most effective method, which means the multicopter will be downsized with lighter equipment and materials. On the other hand, parachutes are also an option for multicopters.

(3) Label an ID number. Like the license plate number required of a vehicle, each multicopter also needs an ID number. This can effectively suppress misuse of multicopters.

(4) Set the no-fly zone. As long as the flight is not above the densely populated areas, the possibility to injure people will become very small. From population data of Beijing city in 2010, for example, in Xicheng District, there are 24517 people per square kilometer, while in Yanqing County, there are only 159 people per square kilometer. In other words, the probability of flight crashes hitting people in Xicheng District will be 154 times greater than that in Yanqing County. Thus, without special

permission, it cannot fly above densely populated areas. Even in sparsely populated areas, warning messages are suggested to notify to surrounding people to pay attention. This is relatively easy to do.

(5) Anti-spoofing and anti-intrusion. During the flight, multicopters or transmitted data may be stolen. Moreover, they may be invaded. In 2012, a demonstration was shown by the scientists with University of Texas at Austin that a flying drone was brought down with a device hacking its GPS system [84]. In August 2015, security experts at Def Con in Las Vegas demonstrated vulnerabilities in two consumer drones from Parrot company. The simplest of the attacks could make Parrot drones, including the company's Bebop model, fall from the sky with a keystroke [85].

15.3.2.2   For Multicopter Operators

(1) Training qualified multicopter remote pilots. In case of flying above sensitive areas or aircraft weight exceeds a certain limit, qualified multicopter remote pilots must be required.

(2) Insurance. Similar to cars, multicopters are required to buy compulsory insurance. Also, the operating companies need to buy another insurance for an expensive equipment to reduce their operating risks.

(3) Limit multicopter's flight range.

15.4   Opportunities and Challenges

15.4.1   Opportunities

Robots are solid requirement in the future. The opportunities lie on the following three facts.

(1) Hardware cost is decreasing. In the recent decade, the ascending market of civil or consumer drones is inseparable from the maturity of the hardware chain and cost of hardware reduced. With the rise of the mobile terminal, industry chains of chips, batteries, and inertial sensors are maturing rapidly, leading smart devices to be cheaper, more compact, and lower-power. This also lays a solid foundation for the development of multicopters.

(2) Human resource cost is increasing. Population aging is taking place in nearly all the countries of the world [86]. More and more work has to be done by robots. As a type of robot, the multicopter, such as agricultural multicopters or delivery multicopters, is expected to help humans in many places.

(3) UAS traffic management system will offer the safe and legal airspace for more drones. So far, such a system is not mature, the use of drones is inconvenient or illegal. As mentioned in Sect. 15.1.7, NASA has been preparing to build the UAS traffic management system to enable civilian low-altitude airspace and UAS operations. It is expected that four phase testing will be finished by March 2019 [51]. This UAS traffic management system will facilitate the development of drones, including multicopters.

In this environment, it can be foreseen that, in the next few years, laws and regulations will continue to be improved, multicopter-related demands will continue to be discovered, the platform will become more reliable, and multicopters will be accepted and used by more and more people. During this process, many opportunities will appear.

15.4.2   Challenges

The greatest challenge is to integrate multicopters or drones into the national airspace system. It is related to both policy and technology [87, 88].

(1) Challenges caused by policy. During the Chinese parade to commemorate the 70th anniversary of the end of World War II, major online shopping platforms have been told to stop selling drones

including multicopters. Moreover, all drones, unless having permission, were forbidden to fly. Similar bans also can be seen worldwide. With the number of multicopters increasing, government regulations of small drones will become increasingly stringent. It is pending to be solved how to make reasonable policies that can achieve a trade-off between the development of small drones and public pressure.

(2) Challenges from technical aspects. Although the development of multicopter seems fast, it is still not solid. So far, most multicopters' reliability cannot be guaranteed. How to design a highly reliable small multicopter is the greatest technological challenge. Most start-ups or small companies do not have such a capability. One possibility is that one or two companies provide a stable and reliable multicopter platform, based on which new applications are further developed. In this case, multicopter platforms have to open the corresponding interface for secondary development. If so, the multicopter safety standards are particularly important. Besides the problem of multicopter itself, technical challenges also include the safety problem related to collisions in airspace for various sizes and types of drones.

References

1. Quan Q (2015) Decryption of multirotor development. Robot Ind 2:72–83 (In Chinese)
2. Kim WC, Mauborgne R (2005) Blue ocean strategy: How to create uncontested market space and make competition irrelevant. Harvard Business School Press, Boston
3. Breaking down 'blue ocean'. http://www.investopedia.com/terms/b/blue_ocean.asp#ixzz3sqQ0pp1u/. Accessed 24 Jan 2016
4. EnergyOr demonstrates multirotor UAV flight of 3 hours, 43 MINUTES. https://www.youtube.com/watch?v=7rpxcCoycvA. Accessed 24 Jan 2016
5. Zhu YW et al (2011) Carbon-based supercapacitors produced by activation of graphene. Science 332(6037): 1537–1541
6. Edelstein S Aluminum-Air battery developer Phinergy partners With Alcoa. http://www.greencarreports.com/news/1090218_aluminum-air-battery-developer-phinergy-partners-with-alcoa. Accessed 5 Feb 2016
7. Hoopes HD (2015) StoreDot to scale up nanodot battery tech in pursuit of five minute-charging EVs. http://www.gizmag.com/storedot-nanodot-flashbattery-ev-five-minute-charge/39030/. Accessed 20 Jan 2016
8. Airborg^{TM} H6 1500 with top flight hybrid propulsion^{TM} for enhanced duration and extended payload applications. http://www.tflighttech.com/products.htm. Accessed 20 Jan 2016
9. Skysapience HoverMast. http://www.skysapience.com/. Accessed 20 Jan 2016
10. Tianshu-100. http://www.droneyee.com/. Accessed 7 Nov 2016
11. Cyphy Works PARC. http://cyphyworks.com/parc/. Accessed 5 Nov 2016
12. Skysense Deploying drones for the enterprise. http://skysense.co/. Accessed 20 Jan 2016
13. Reforges. Reforges Constellation. http://www.reforgesconstellation.com/. Accessed 5 Nov 2016
14. Airstier Yeair! Is the nest generation guadcopter solution. https://www.yeair.de/overview/. Accessed 20 Jan 2016
15. Swift Navigation Piksi. http://docs.swiftnav.com/wiki/Main_Page. Accessed 25 Jan 2016
16. RTKLIB: An open source program package for GNSS positioning. http://www.rtklib.com/. Accessed 25 Jan 2016
17. Brown M (2012) BAE systems' GPS rival Navsop uses radio signals to get your position. http://www.wired.co.uk/news/archive/2012-06/29/bae-gps. Accessed 25 Jan 2016
18. Pozyx Labs Pozyx. https://www.pozyx.io/. Accessed 6 June 2016
19. Hol JD, Dijkstra F, Luinge H, Schon T (2009) Tightly coupled UWB/IMU pose estimation. In: Proceedings IEEE International Conference on Ultra-Wideband, New York, USA, pp 688–692
20. Mueller MW, Hamer M, D'Andrea R (2015) Fusing ultra-wideband range measurements with accelerometers and rate gyroscopes for quadrocopter state estimation. In: Proceedings IEEE International Conference on Robotics and Automation, pp 1730–1736
21. Honegger D, Meier L, Tanskanen P, Pollefeys M (2013) An open source and open hardware embedded metric optical flow cmos camera for indoor and outdoor applications. In: Proceedings IEEE International Conference on Robotics and Automation, pp 1736–1741
22. PX4FLOW developer guide. https://pixhawk.org/dev/px4flow. Accessed 25 Jan 2016
23. Microsoft Meet Kinect for Windows. https://dev.windows.com/en-us/kinect. Accessed 25 Jan 2016
24. Developer zone. https://software.intel.com/en-us/realsense/home. Accessed 25 Jan 2016
25. Panoptes Meet eBumper4: explore with confidence. http://www.panoptesuav.com/ebumper/. Accessed 25 Jan 2016
26. New Scientist Smart drones that think and learn like us to launch this year. https://www.newscientist.com/article/mg22630172-000-smart-drones-that-think-and-learn-like-us-to-launch-this-year/. Accessed 25 Jan 2016

27. The Verge A tiny startup has made big strides in creating self-navigating drones. http://www.theverge.com/2015/1/15/7550669/skydio-drone-sense-and-avoid-camera-vision. Accessed 25 Jan 2016
28. Simonite T (2015) Metamaterial radar may improve car and drone vision. https://www.technologyreview.com/s/536341/metamaterial-radar-may-improve-car-and-drone-vision/. Accessed 5 Feb 2016
29. Velodyne PUCKTM (VLP-16). http://velodynelidar.com/vlp-16.html. Accessed 25 Jan 2016
30. 3DR releases tower drone control App, and 3DR services, "The App Store for Drones". https://3drobotics.com/3dr-releases-tower-drone-flight-control-app-3dr-services-app-store-drones/. Accessed 25 Jan 2016
31. OpenCV. http://opencv.org/. Accessed 25 Jan 2016
32. Oculii RFS-M. http://www.oculii.com/. Accessed 25 Jan 2016
33. When Parrot AR. Drone meets Myo armband, magic ensues (video). http://www.engadget.com/2014/01/10/parrot-ar-drone-thalmic-labs-myo/. Accessed 25 Jan 2016
34. Yin M, Borton DA, Komar J et al (2010) Wireless neurosensor for full-spectrum electrophysiology recordings during free behavior. Neuron 84(6):1170–1182
35. Emotiv Wearables for your brain. https://emotiv.com/. Accessed 25 Jan 2016
36. Tekever BBC News. Brain-controlled drone shown off by Tekever in Lisbon. http://www.bbc.com/news/technology-31584547. Accessed 25 Jan 2016
37. ETTelecom US pushes for spectrum for 5G, civil drones, flight tracking. http://telecom.economictimes.indiatimes.com/news/3g-4g/us-pushes-for-spectrum-for-5g-civil-drones-flight-tracking/49682589. Accessed 25 Jan 2016
38. Emspak J (2013) Wi-Fi network breaks speed record. http://mashable.com/2013/05/17/wi-fi-speed-record/#QGTFT0xxdEqw. Accessed 25 Jan 2016
39. 3D Robotics partners with Intel, develops new drone power. https://3drobotics.com/3d-robotics-partners-intel-develops-new-drone-power/. Accessed 25 Jan 2016
40. Jacobowitz PJ (2015) Introducing the Snapdragon Cargo. https://www.qualcomm.com/news/onq/2015/01/04/introducing-snapdragon-cargo-video. Accessed 25 Jan 2016
41. McCaney K (2014) Tiny drone with brain-like chip learns on the fly. https://defensesystems.com/articles/2014/11/05/darpa-hrl-brain-like-chip-tiny-drone.aspx. Accessed 25 Jan 2016
42. Merolla PA, Arthur JV, Alvarez-Icaza R et al (2014) A million spiking-neuron integrated circuit with a scalable communication network and interface. Science 345(6197):668–673
43. Dronecode. https://www.dronecode.org/. Accessed 25 Jan 2016
44. Linux Foundation Linux Foundation and leading technology companies launch open source Dronecode Project. http://www.linuxfoundation.org/news-media/announcements/2014/10/linux-foundation-and-leading-technology-companies-launch-open. Accessed 25 Jan 2016
45. Ubuntu Amazing autonomous things. http://www.ubuntu.com/internet-of-things. Accessed 25 Jan 2016
46. Airware. https://www.airware.com/. Accessed 25 Jan 2016
47. Bednar C (2015) NASA to create drone air traffic control system. http://www.redorbit.com/news/technology/1113350198/nasa-to-create-control-system-for-unmanned-drones-031115/. Accessed 25 Jan 2016
48. Transtrex tells drones where they can fly in real time. http://www.suasnews.com/2015/07/37440/transtrextellsdrones-where-they-can-fly-in-real-time. Accessed 25 Jan 2016
49. NASA UTM 2015: The next era of aviation. http://utm.arc.nasa.gov/utm2015.shtml. Accessed 17 April 2016
50. NASA UTM fact sheet. http://utm.arc.nasa.gov/docs/utm-factsheet-02-23-16.pdf. Accessed 17 April 2016
51. Kopardekar PH (2016) Safely enabling UAS operations in low altitude airspace. http://ntrs.nasa.gov/search.jsp?R=20160002414. Accessed 17 April 2016
52. Amazon Inc Determining safe access with a best-equipped, best-served model for small unmanned aircraft systems. http://utm.arc.nasa.gov/docs/Amazon_Determining%20Safe%20Access%20with%20a%20Best-Equipped,%20Best-Served%20Model%20for%20sUAS%5b2%5d.pdf. Accessed 17 April 2016
53. Amazon Inc Revising the airspace model for the safe integration of small unmanned aircraft systems. http://utm.arc.nasa.gov/docs/Amazon_Revising%20the%20Airspace%20Model%20for%20the%20Safe%20Integration%20of%20sUAS[6].pdf. Accessed 17 April 2016
54. Google Inc Google UAS airspace system overview. http://utm.arc.nasa.gov/docs/GoogleUASAirspaceSystemOverview5pager[1].pdf. Accessed 17 April 2016
55. Gharibi M, Boutaba R, Waslander SL (2016) Internet of drones. IEEE Access, 4:1148–1162
56. Brickdrones. http://www.brickdrones.com/. Accessed 25 Jan 2016
57. Multigp FPV multirotor racomg. http://www.multigp.com. Accessed 25 Jan 2016
58. Multirotorracing. http://multirotorracing.com/wordpress. Accessed 25 Jan 2016
59. Designboom Renault KWID concept: an off-road car with built-in drone quadcopter. http://www.designboom.com/technology/renault-kwid-concept-an-off-road-car-with-built-in-drone-quadcopter-02-11-2014. Accessed 25 Jan 2016
60. Hover camera passport. https://gethover.com/. Accessed 7 Nov 2016
61. Xaircraft 3 Days for 1650 acres, XAIRCRAFT reached a new milestone in agricultural industry. http://www.xaircraft.cn/en/news_detail/152. Accessed 25 Jan 2016

62. Amazon. http://www.amazon.com/b?node=8037720011. Accessed 25 Jan 2016
63. DHL parcelcopter launches initial operations for research purposes. http://www.dhl.com/en/press/releases/releases_2014/group/dhl_parcelcopter_launches_initial_operations_for_research_purposes.html. Accessed 25 Jan 2016
64. Afrotech EPFL Flying robots. http://afrotech.epfl.ch/page-99937.html. Accessed 25 Jan 2016
65. D'Andrea R (2014) Can drones deliver? IEEE Trans Autom Sci Eng 11(3):647–648
66. CNET Ambulance drone delivers help to heart attack victims. http://www.cnet.com/news/ambulance-drone-delivers-help-to-heart-attack-victims/. Accessed 25 Jan 2016
67. Hussenet L (2014) Integrated 3D visualization: drones, photogrammery tied to 3D printing. https://www.3dvisworld.com/features/feature-articles/9286-integrated-3d-visualization-drones-photogrammetry-tied-to-3d-printing.html. Accessed 25 Jan 2016
68. Flying gun. https://www.youtube.com/watch?v=xqHrTtvFFIs. Accessed 25 Jan 2016
69. Visnjic F (2014) Smart Atoms (Spaxels)—Flying building blocks. http://www.creativeapplications.net/environment/smart-atoms-spaxels-flying-building-blocks/. Accessed 25 Jan 2016
70. Coxworth B (2014) Cirque du Soleil and ETH Zurich collaborate on human/drone performance. http://www.gizmag.com/cirque-du-soleil-sparked-drone-video/33921/. Accessed 25 Jan 2016
71. Nicas J (2015) They're using drones to herd sheep. http://www.wsj.com/articles/theyre-using-drones-to-herd-sheep-1428441684. Accessed 25 Jan 2016
72. Szondy D (2014) Fraunhofer developing flying inventory robots to keep tabs on stock. http://www.gizmag.com/inventairy-fraunhofer/35006/. Accessed 25 Jan 2016
73. Lavars N (2015) Drones autonomously build a walkable rope bridge. http://www.gizmag.com/drones-build-bridge-that-can-bear-human-weight/39511/. Accessed 25 Jan 2016
74. Leveson N (2011) Engineering a safer world-systems thinking applied to safety. MIT Press, Cambridge, pp 14–15
75. Crisp HE (2007) Systems engineering vision 2020. Technical Report INCOSE-TP-2004-004-02, International Council on Systems Engineering
76. Haskins C, Forsberg K, Kruger M (2015) INCOSE systems engineering handbook: A guide for system life cycle processes and activities. Wiley, Hoboken
77. Nonami K, Kartidjo M, Yoon KJ, Budiyono A (2013) Autonomous control systems and vehicles: Intelligent unmanned systems. Springer-Verlag, Japan, pp 59–60
78. OSD Unmanned aerial vehicles roadmap 2000-2025. https://www.google.com.hk/url?sa=t&rct=j&q=&esrc=s&source=web&cd=1&ved=0ahUKEwjQrtSazsLKAhXGJI4KHRF4ADYQFggfMAA&url=http%3A%2F%2Fhandle.dtic.mil%2F100.2%2FADA391358&usg=AFQjCNHAkZfKQQ-fdu8PB0myFVtbkO0ctw&sig2=HIbgZoXIJoyfngOFSEBRkA. Accessed 25 Jan 2016
79. Clough BT (2002) Metrics, schmetrics! How the heck do you determine a UAV's autonomy anyway. In: Proceedings of the Performance Metrics for Intelligent Systems (PerMIS) conference, Gaithersburg, Maryland
80. Miller ZJ (2015) Drone that crashed at White House was quadcopter. http://time.com/3682307/white-house-drone-crash/. Accessed 25 Jan 2016
81. Russon MA (2015) Radioactive drone lands on roof of Japanese PM's office in 'silent protest' at nuclear plans. http://www.ibtimes.co.uk/radioactive-drone-lands-roof-japanese-pms-office-silent-protest-nuclear-plans-1497738. Accessed 25 Jan 2016
82. Dronefreaks (2015) Drone crashes into Milan Cathedral. http://dronefreaks.org/2015/06/26/drone-crashes-into-milan-cathedral/. Accessed 25 Jan 2016
83. Talanova J (2015) Drone slams into seating area at U.S. Open; teacher arrested. http://edition.cnn.com/2015/09/04/us/us-open-tennis-drone-arrest/. Accessed 25 Jan 2016
84. Drone hacked by university of texas at Austin research group. The world post. http://www.huffingtonpost.com/2012/06/29/drone-hacked-by-universit_n_1638100.html. Accessed 25 Jan 2016
85. Arstechnica Parrot drones easily taken down or hijacked, researchers demonstrate. http://arstechnica.com/security/2015/08/parrot-drones-easily-taken-down-or-hijacked-researchers-demonstrate/. Accessed 25 Jan 2016
86. Economic & Social Affairs World population ageing 2013. http://www.un.org/en/development/desa/population/publications/pdf/ageing/WorldPopulationAgeing2013.pdf. Accessed 25 Jan 2016
87. Cooney M (2008) Unmanned aircraft pose myriad problems to US airspace, GAO reports. http://www.networkworld.com/article/2344276/security/unmanned-aircraft-pose-myriad-problems-to-us-airspace--gao-reports.html. Accessed 22 July 2016
88. Dalamagkidis K, Valavanis KP, Piegl LA (2009) On integrating unmanned aircraft systems into the national airspace system: Issues, challenges, operational restrictions, certification, and recommendations. Springer-Verlag, Netherlands

Index

© Springer Nature Singapore Pte Ltd. 2017
Q. Quan, *Introduction to Multicopter Design and Control*, DOI 10.1007/978-981-10-3382-7

Printed in the United States
By Bookmasters